Carbon Dioxide Utilisation

University of Sheffield, UK

7–9 September 2015

FARADAY DISCUSSIONS

Volume 183, 2015

ROYAL SOCIETY OF CHEMISTRY

The Faraday Division of the Royal Society of Chemistry, previously the Faraday Society, founded in 1903 to promote the study of sciences lying between Chemistry, Physics and Biology.

Faraday Discussions are unique international discussion meetings that focus on rapidly developing areas of chemistry and its interfaces with other scientific disciplines.

Faraday Discussions (Print ISSN 1359-6640, Electronic ISSN 1364-5498) is published 9 times a year by the Royal Society of Chemistry, Thomas Graham House, Science Park, Milton Road, Cambridge, UK CB4 0WF.

Volume 183 ISBN-13: 978-1-78262-469-1

2015 annual subscription price: print+electronic £877, US $1635; electronic only £833, US $1553. Customers in Canada will be subject to a surcharge to cover GST. Customers in the EU subscribing to the electronic version only will be charged VAT. All orders, with cheques made payable to the Royal Society of Chemistry, should be sent to RSC Order Department, Royal Society of Chemistry, Thomas Graham House, Science Park, Milton Road, Cambridge, CB4 0WF, UK. Tel +44 (0) 1223 432398; E-mail **www.orders@rsc.org**

If you take an institutional subscription to any RSC journal you are entitled to free, site-wide web access to that journal. You can arrange access via Internet Protocol (IP) address at **www.rsc.org/ip**. Customers should make payments by cheque in sterling payable on a UK clearing bank or in US dollars payable on a US clearing bank.

Printed in the UK

Carbon Dioxide Utilisation

Faraday Discussions

www.rsc.org/faraday_d

A General Discussion on Carbon Dioxide Utilisation was held in Sheffield, UK on the 7th, 8th and 9th of September 2015.

CONTENTS

ISSN 1359-6640; ISBN: 978-1-78262-469-1

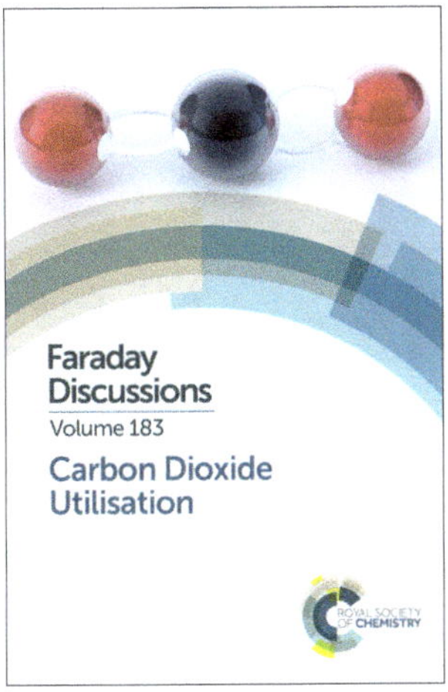

Cover
Front cover image courtesy of Shutterstock.

INTRODUCTORY LECTURE

PAPERS AND DISCUSSIONS

SPONSORS

cogent | Skills for Science Based Industries

Hiden Isochema
Advancing Sorption Analysis

SPONSORING ORGANISATION

IChemE ADVANCING CHEMICAL ENGINEERING WORLDWIDE

CONCLUDING REMARKS

ADDITIONAL INFORMATION

Faraday Discussions

PAPER

The Twelve Principles of CO_2 CHEMISTRY†

Martyn Poliakoff,*[a] Walter Leitner*[bc] and Emilia S. Streng[ab]

Received 9th October 2015, Accepted 13th October 2015
DOI: 10.1039/c5fd90078f

This paper introduces a set of 12 Principles, based on the acronym CO_2 CHEMISTRY, which are intended to form a set of criteria for assessing the viability of different processes or reactions for using CO_2 as a feedstock for making organic chemicals. The principles aim to highlight the synergy of Carbon Dioxide Utilisation (CDU) with the components of green and sustainable chemistry as well as briefly pointing out the connection to the energy sector.

Introduction

The world is changing rapidly. The global population is soaring, passing 7 billion in October 2011.[1] Human consumption is already greater than our planet can sustain; furthermore, 1.3 billion of the Earth's inhabitants are profoundly poor and, on every criterion, deserve to consume more.[2,3] Therefore, there is a need to increase chemical production to match the increased demand. However, it is generally accepted that petroleum reserves are finite and that it is necessary to find alternative feedstocks for making those chemicals. Although the planet's reserves of coal are much larger than those of oil, it is not clear whether these could be exploited to make chemicals without the release of large quantities of CO_2. Shale gas provides another large volume carbon source and, although its exploitation is associated with a large potential environmental burden, it is having a major impact on chemical production, often causing a shift from petroleum feedstocks. Nevertheless, the utilization of all fossil resources ultimately contributes to anthropogenic greenhouse gas emissions, either directly (methane) and/or indirectly by producing CO_2 as the end product of the value chain.

[a]*The School of Chemistry, The University of Nottingham, Nottingham, NG7 2RD, UK. E-mail: martyn.poliakoff@nottingham.ac.uk*

[b]*Institut für Technische und Makromolekulare Chemie, RWTH Aachen University, Worringerweg 2, 52074 Aachen, Germany. E-mail: leitner@itmc.rwth-aachen.de*

[c]*Max-Planck-Institut für Kohlenforschung, Mülheim an der Ruhr, Germany*

† Dedicated to the memory of our dear Nottingham colleague, Professor Daniel D Eley OBE FRS FRSC, the elder statesman of catalysis worldwide, who died peacefully on Sept 3rd, four days before the start of this Faraday Discussion and four weeks short of his 101st birthday.

Already, CO_2 levels are the highest in recorded history and there is a general scientific consensus that further increases in atmospheric CO_2 will have serious consequences for the Earth's climate.[4–6] The global demand for energy is rapidly increasing and CO_2 emissions are not yet under control. Therefore, all efforts have to be undertaken to reduce these emissions in every industrial sector including the production of chemicals.

A highly attractive scenario in this context is to make chemicals from CO_2, either captured before it is emitted to the atmosphere or by recovering it from the atmosphere *via* photosynthesis in biomass or *via* some equivalent artificial process. That is not to say that making chemicals from CO_2 could mitigate anthropogenic emissions of CO_2 because the scales of the two activities are totally mismatched: global CO_2 emissions are in the order of 10^{10} tons p.a. while the maximum potential consumption of CO_2 in chemical and industrial processes is estimated as 10^8 tons p.a.[7] So, even allowing for possible errors, these numbers are never likely to match.

However, these data can be seen as encouraging, rather than frustrating; they clearly indicate that CO_2 is and will be available from various industrial sources in more than sufficient quantities for its intensive use as chemical feedstock in Carbon Dioxide Utilisation (CDU). Capturing CO_2 from highly concentrated sources, *e.g. via* processes such as those developed for Carbon Capture and Storage (CCS) technologies,[8] may well provide an excellent source of CO_2 for making chemicals at least in the short to medium term. On the other hand, even if chemical processes require comparatively small amounts of CO_2 compared to the possible output of CCS, CDU does have the potential to add value to captured CO_2 which would otherwise be a largely valueless waste product. The idea of using CO_2 as a feedstock has fascinated chemists for decades, as indicated for example by the more than 20 years history of the International Conference on Carbon Dioxide Utilization (ICCDU). Potential synergies at the interface between chemistry and energy and the resulting opportunities for viable technical solutions associated with CDU have been a topic of lively discussion at this Faraday Discussion.

This Faraday Discussion includes 22 papers, apart from this one and the Concluding Remarks. The purpose of this introductory paper is to present a check list of the 12 Principles of CO_2 Chemistry to provide a rough set of metrics for gauging the value of a particular proposal in this area. The inspiration is the now

C	Catalysis is crucial
O	Origin of the CO_2?
2	Tomorrow's world may be different
C	Cleaner than existing process?
H	High volume or high value products?
E	E-factor must be low
M	Maximize integration
I	Innovative process technology
S	Sustainability is essential
T	Thermodynamics cannot be beaten
R	Renewable (& reasonable) energy input
Y	Your enthusiasm is not enough

Fig. 1 The proposed 12 Principles of CO_2 Chemistry. Their meaning is explained in the text.

widely known 12 Principles of Green Chemistry, introduced in 1998 by Anastas and Warner, which has now become an almost standard means of assessing green processes.[9] However, it has to be admitted that the full statement of the principles is quite lengthy. So, more recently, a condensed version of these principles was introduced, based on the acronym PRODUCTIVELY, to provide a list that could conveniently (and legibly) be shown on a single PowerPoint slide.[10]

Here we provide an equivalent condensed list for CO_2 CHEMISTRY in the hope that it will help to highlight the most pertinent issues and to structure future discussions of CDU. The list is shown in Fig. 1 as it was presented in the Introductory Lecture and below we expand these principles and explain how they might be applied.

Catalysis is crucial

CO_2 is notoriously unreactive in organic synthesis and in reductive processes. It can be considered a thermodynamic minimum with only highly energy rich compounds being able to react with it. Such reactions do not necessarily require a catalyst. For example, once heated, magnesium metal reacts with CO_2 in a highly exothermic reaction[11] ($2Mg + CO_2 \rightarrow 2MgO + C$) but the reaction yields elemental carbon which is not necessarily a useful starting point for making chemicals, especially given the financial and energy costs of obtaining metallic magnesium. Thus, nearly all reactions for making chemicals from CO_2 really need to use catalysts.[12,13] Heterogeneous catalysts are able to promote important transformations of carbon dioxide such as the famous Sabatier reaction $CO_2 + 4H_2 \rightarrow CH_4 + 2H_2O$.[14] Homogeneous catalysts are very selective for many synthetically useful C–C and C–X bond formation processes.[15,16] Electro- and photochemical transformations of CO_2 also require suitable catalyst materials or molecular catalysts due to the large over-potential for CO_2 reduction to its radical anion.[17] Nature's ingenious catalytic processes for construction of its entire material basis from CO_2 can serve as motivation to develop man-made catalysts for industrial implementation.

Origin of the CO_2?

CO_2 can be sequestered from many industrial processes and other sources.[18] The need for catalysis may well place quite severe restrictions on the purity of the CO_2. Trace impurities could easily poison catalysts or oxidize expensive ligands. Fortunately, the large disparity between global CO_2 production and the possible consumption by the chemical industry means that operators of chemical processes can afford to be quite choosy about the source of their CO_2. For example, some processes may require relatively dry CO_2 and could benefit from fears of pipeline corrosion in CCS that are leading to relatively stringent proposals for acceptable levels of H_2O currently for any captured CO_2 to be transported by pipeline.[19] At the same time, using atmospheric CO_2 will always carry a large energy cost because of the need to move very large masses of air to obtain a sufficient flux of CO_2 through one's chemical process. It may well be hard to beat photosynthesis by plants/trees/algae as a means of deriving carbon compounds from the atmosphere, even though such photosynthesis often requires very large amounts of fresh water, a commodity that may well be in short supply in the future. Furthermore if a process is based on sequestration of CO_2 by formation of biomass, one needs to consider questions of land usage, of "food *versus* chemicals" as well as water usage, all of which impact the overall sustainability.[20]

Tomorrow's world may be different

Unlike academic research, commercial chemical manufacture has to face economic reality. This means that any CO_2-based process will be fully exposed to changes in the financial environment. Government subsidies can change rapidly with changes of government or policy. In this context it is important to remember that, although CO_2-based chemistry cannot make a significant impact on overall anthropogenic CO_2 emissions, it could reduce the emissions from a particular plant or company; such a reduction could give a significant fiscal advantage to that company. Also, the global oil price, which is a key parameter in determining the commercial viability of any CO_2-based process, can fluctuate widely; for example the price of Brent Crude has dropped by nearly 50% over the last 12 months,[21] gravely affecting the economics of otherwise attractive processes using alternative carbon sources. On the other hand, the possibility of providing an outlet for fluctuating "excess" renewable energy could constitute a novel, quite unexpected business opportunity in the future.[22,23] Hence, the distinction between economically "useful" or "useless" processes is likely to depend on time and place, leading to different strategies when devising short-term and long-term goals in science.

Cleaner than existing process?

Just like in any chemical process, "greenness" should be an *a priori* criterion for CO_2 Chemistry: it seems only sensible that any new process should be cleaner (or at least not dirtier) than any existing process that it replaces. This is more than a mere gesture towards environmentalism. Disposal of waste has become a significant and increasing cost for chemical producers and the more toxic the waste, the more expensive is its disposal. Importantly, CO_2 holds great potential also in this context; it is a non-toxic, safe reagent as compared to many other C1-building blocks, often producing water as the only by-product. Ideally, the energetic and material aspects of the "green" approach work together to minimize both energy requirements and the waste produced, and it can be a very difficult task to balance the benefits and drawbacks if this is not the case.[24]

High volume or high value products?

This is an absolutely key question. If one is aiming to make high volume/low value products, any CO_2-based process must necessarily be low cost, thereby removing the option of using expensive catalysts or other costly processing options. Such processes will also be particularly vulnerable to fluctuations in the price of oil because feedstock price will be a large proportion of the cost of the final product. Conversely, if one is making low volume/high value products, the price of the feedstock will be only a small component in the overall cost and so any cost advantage of CO_2 as a feedstock may only be marginal. An attractive option, therefore, may be to aim for medium volume/medium value products where the cost of the feedstock is significant but low enough to dampen out effects of temporary swings in the oil price. A good example described below is the CO_2-based production of polyols for the manufacture of polyurethane polymers, which is currently being implemented on industrial scale in Germany.[25,26]

E-factor must be low

One of the key concepts of Green Chemistry is the E factor, originally proposed by Sheldon.[27] The factor is the weight of waste generated per unit weight of product. Its use has had real impact on chemical production, for example resulting in huge reductions in solvent usage in the production of Viagra.[28] However, it has even more importance in CO_2-based chemistry. This is because conversion of CO_2 into organic compounds is energy intensive and therefore waste organic compounds will also represent a waste of energy. In the introductory lecture, the Braskem process for making polyethene from cane sugar was mentioned, as an example of using CO_2 *via* photosynthesis.[29,30] The subsequent chemistry is simple: sucrose is fermented to ethanol, which is dehydrated to ethene that is then polymerized. The advantage is that the product slots into existing supply chains without the need to create a new commercial infrastructure. As regards the E-factor, 33% of the carbon fixed by photosynthesis is released as CO_2 during the fermentation (and, like all CO_2 utilization processes, it is vulnerable to falling oil prices which can render the cost of the product uncompetitive).

Maximize integration

Although research in CCS and research in CDU have often been carried out separately, it clearly makes sense to integrate carbon capture and its subsequent use. We have previously described how captured CO_2 might be exploited as a solvent for reactions in supercritical CO_2.[31] The rationale is that the energy costs of compressing CO_2 represent a substantial fraction of the cost of supercritical processing; however, compression of CO_2 is an integral part of the capture and transport technology. Consequently, power plants or industrial processes equipped with capture units could provide high pressure CO_2 at no or moderate additional cost. Although a supercritical CO_2 solvent is not strictly "utilised", because the CO_2 is still there at the end of the process, it could represent a substantial reduction in the usage of petroleum-based materials as solvents. A more chemically elegant example of integration is the process developed jointly by Bayer together with the CAT Catalytic Center at RWTH Aachen, which is now being commercialised by the company Covestro.[25] The catalytic transformation has been demonstrated to be compatible with captured CO_2 even from coal fired power plants and other processes; the process is now being industrialised with production using the CO_2 by-product from an ammonia synthesis plant located on the same industrial site. A Life Cycle Assessment is described in one of the papers in this Discussion (DOI: 10.1039/c5fd00067j).

Innovative process technology

Many proposed CO_2-based processes will require new chemical plants to implement them, immediately adding capital expenditure to other start-up costs.

Therefore, the introduction of innovative process technology, particularly continuous processing could have a significant impact on the economics of the process. The synthesis of formic acid from CO_2 and H_2 provides an illustrative example, where the thermodynamics of the reaction require additional stabilization of the product in the reaction stage that needs to be overcome during the downstream product separation.[32,33]

Sustainability is essential

Long term sustainability is a major reason, if not *the* major reason, for developing CO_2-based chemistry. Therefore, it has to be a criterion for judging the viability of any proposed process. However elegant the catalysis, one must always look at the whole life cycle; the energy and/or chemical resources needed to make the catalyst may well reveal that an otherwise attractive process is totally unsustainable. Furthermore, it is important to stress that sustainability involves more than demonstrating that a particular catalyst is based on abundant and easily obtained elements. Although the sustainability of a process can only be reasonably established by Life Cycle[34] or similar assessments (DOI: 10.1039/c5fd00038f), which are quite lengthy procedures, it may often be easier and quicker to recognize processes that are obviously unsustainable.

Thermodynamics cannot be beaten

As in other areas of chemistry, the thermodynamics of CO_2-based processes depend on where one draws the system boundary. For example, the hydrogenation of CO_2 mentioned above, $CO_2 + 4H_2 \rightarrow CH_4 + 2H_2O$, is exothermic, provided that one does not consider the energy involved in generating the hydrogen. Thermodynamics assume that systems are in equilibrium. Therefore, although thermodynamics cannot be beaten, it may be possible to side-step thermodynamic barriers by deliberately avoiding equilibrium situations, for example in plasma processing (DOI: 10.1039/c5fd00053j and DOI: 10.1039/c5fd00045a). Furthermore, even an endothermic process using CO_2 can be advantageous, if the alternative routes consume even more energy to arrive at the same product or function. This emphasises again that one must consider the whole system with clearly defined boundary conditions in making an evaluation, and most evaluations are only valid in comparison, rather than as absolute values.

Renewable (& reasonable) energy input

Since the aim of CO_2 Chemistry is to avoid use of fossil carbon in the manufacture of chemicals, it is clearly not sensible to incorporate less CO_2 into the chemicals than one creates by burning fossil carbon to generate the energy needed to make those chemicals. Otherwise, one might as well start with the fossil carbon as the feedstock. Therefore, renewable energy might seem to be the panacea for CO_2-based chemistry. If renewable energy sources are coupled with CO_2 as the carbon source, this could decouple the carbon emission from the energy sector and the feedstock.[22,23] Of course, this does not eliminate the challenge of devising energy efficient processes. Also renewable energy will not be "for free", either economically or environmentally. Furthermore, if renewable energy does become available on large scales, the chemical industry will not be the only potential outlet and its use may have an even larger impact on our carbon emissions in other sectors. Hence, the goal of minimizing the energy requirements of chemical production remains a core issue in CO_2 Chemistry, regardless of the primary energy source.

Your enthusiasm is not enough

In the context of this Discussion, the purpose of CDU is to devise processes that can lead to viable methods of using CO_2 as a chemical feedstock. Therefore, one needs to apply a certain degree of realism to any proposed CDU process whatever its intellectual and academic fascination. This is not to say that successful CDU processes will be mundane or boring. Merely, it means that processes will have the best chance of success if they arouse widespread interest, not only in academia but amongst those who make chemicals to sell. Even so, it must be stressed that society is wider than either academia or industry. Ethical issues and environmental consequences as well as the aesthetic impact on the landscape are also important, if not paramount issues. These are also issues that are strongly championed by non-governmental organizations (NGOs) and other lobby groups. Therefore, it is salutary to remember that no process, however elegant and potentially profitable, will be genuinely sustainable without the positive support from the wider community.

Conclusions

Humanity faces enormous challenges over how to maintain our quality of life as this century progresses. We face the threat of multiple shortages, increasing concentrations of greenhouse gases in the atmosphere, dwindling oil supplies, and increasingly expensive chemicals. CDU represents an excellent opportunity to reduce our dependency on fossil resources for making the chemicals that we will need for the foreseeable future. It also offers the possibility of harvesting renewable energy to benefit the chemical value chain. This possibility places CDU right at the interface between chemistry and energy.

The 12 Principles outlined here are intended to focus attention on the key issues that need to be addressed if CDU is to succeed in the way in which all of the participants in this Faraday Discussion hope. Those participants included both chemists and chemical engineers working on all aspects along the value chain. Success requires partnership across disciplinary boundaries. It is not guaranteed but it will not come if we do not all work together towards a common aim. In the longer term, it is the role of CO_2 Chemists and Engineers to get societal support for our joint endeavours so that our vision of sustainable CO_2-based chemical manufacture for the future can be fully realized.

Acknowledgements

We are grateful for the support of our funders in the UK and Germany, including UKCCSRC, EPSRC, National Grid, PSE, the Excellence Initiative of the German Federal and State Governments to promote science and research at German universities, the climate protection program of the German Ministry of Science and Education (BMBF), and the project "Sustainable Chemical Synthesis

(SusChemSys)" that is co-financed by the European Regional Development Fund (ERDF) and the state of North Rhine-Westphalia, Germany, under the Operational Programme "Regional Competitiveness and Employment 2007–2013". In particular, we acknowledge the EU SINCHEM project (FPA 2013-0037) for funding ESS to work jointly on her PhD thesis between our two universities.

References

1 *The State of World Population 2011: People and possibilities in a world of 7 billion*, United Nations Population Fund (UNFPA), New York, 2011, Available from http://www.unfpa.org/sites/default/files/pub-pdf/EN-SWOP2011-FINAL.pdf.
2 *People and the Planet*, The Royal Society, London, 2012. Available from https://royalsociety.org/~/media/Royal_Society_Content/policy/projects/people-planet/2012-04-25-PeoplePlanet.pdf.
3 *The Millennium Development Goals - Report 2015*, United Nations, New York, 2015, Available from http://www.un.org/millenniumgoals/2015_MDG_Report/pdf/MDG 2015 rev (July 1http://www.un.org/millenniumgoals/2015_MDG_Report/pdf/MDG%202015%20rev%20(July%201).pdf.
4 Climate Change: Evidence & Causes, An overview from the Royal Society and the US National Academy, 2014, Available from https://royalsociety.org/~/media/Royal_Society_Content/policy/projects/climate-evidence-causes/climate-change-evidence-causes.pdf.
5 IPCC, *Carbon Dioxide Capture and Storage - Special Report of the Intergovernmental Panel on Climate Change*, Cambridge University Press, Cambridge, United Kingdom, 2005, Available from https://www.ipcc.ch/pdf/special-reports/srccs/srccs_wholereport.pdf.
6 IPCC, *Climate Change 2014: Synthesis Report, Contribution of Working Groups I, II and III to the Fifth Assessment Report of the Intergovernmental Panel on Climate Change*, Geneva, Switzerland, 2015, Available from http://www.ipcc.ch/pdf/assessment-report/ar5/syr/SYR_AR5_FINAL_full.pdf.
7 M. Peters, B. Köhler, W. Kuckshinrichs, W. Leitner, P. Markewitz and T. E. Müller, *ChemSusChem*, 2011, **4**, 1216–1240.
8 Carbon capture and storage in Europe, European Academies Science Advisory Council (EASAC), Germany, 2013, Available from http://www.easac.eu/fileadmin/Reports/Easac_13_CCS_Web_Complete.pdf.
9 P. T. Anastas and J. C. Warner, *Green Chemisty: Theory and Practice*, Oxford University Press, New York, 1998.
10 S. L. Y. Tang, R. L. Smith and M. Poliakoff, *Green Chem.*, 2005, **7**, 761.
11 Carbon Dioxide (Part II) - Periodic Table of Videos, https://www.youtube.com/watch?v=0dSMzg0UPPo, accessed 06.10.15.
12 T. Sakakura, J.-C. Choi and H. Yasuda, *Chem. Rev.*, 2007, **107**, 2365–2387.
13 J. Klankermayer, K. Beydoun, S. Wesselbaum and W. Leitner, *Angew. Chem., Int. Ed.*, 2015, DOI: 10.1002/anie.201507458.
14 N. A. M. Razali, K. T. Lee, S. Bhatia and A. R. Mohamed, *Renewable Sustainable Energy Rev.*, 2012, **16**, 4951–4964.
15 M. Cokoja, C. Bruckmeier, B. Rieger, W. A. Herrmann and F. E. Kühn, *Angew. Chem., Int. Ed.*, 2011, **50**, 8510–8537.
16 W. Leitner, *Coord. Chem. Rev.*, 1996, **153**, 257–284.

17 W. H. Wang, Y. Himeda, J. T. Muckerman, G. F. Manbeck and E. Fujita, *Chem. Rev.*, 2015, DOI: 10.1021/acs.chemrev.5b00197.
18 M. Aresta, *Carbon Dioxide as Chemical Feedstock*, Wiley-VCH, Weinheim, 2010.
19 E. de Visser, C. Hendriks, M. Barrio, M. J. Mølnvik, G. de Koeijer, S. Liljemark and Y. Le Gallo, *Int. J. Greenhouse Gas Control*, 2008, **2**, 478–484.
20 S. Perathoner and G. Centi, *ChemSusChem*, 2014, **7**, 1274–1282.
21 http://www.nasdaq.com/markets/crude-oil-brent.aspx?timeframe=1y, accessed 22.09.15, 2015.
22 G. Centi, E. A. Quadrelli and S. Perathoner, *Energy Environ. Sci.*, 2013, **6**, 1711.
23 J. Klankermayer and W. Leitner, *Science*, 2015, DOI:10.1126/science.aac7997.
24 N. von der Assen, P. Voll, M. Peters and A. Bardow, *Chem. Soc. Rev.*, 2014, **43**, 7982–7994.
25 J. Langanke, A. Wolf, J. Hofmann, K. Böhm, M. A. Subhani, T. E. Müller, W. Leitner and C. Gürtler, *Green Chem.*, 2014, **16**, 1865–1870.
26 Bayer MaterialScience, *Weiterer Meilenstein für Kunststoff-Herstellung mit CO_2*, Press release, 29.05.15. Available from, http://presse.covestro.de/news.nsf/id/Weiterer-Meilenstein-fuer-Kunststoff-Herstellung-mit-CO2?Open&parent=Home_DE&ccm=000.
27 R. A. Sheldon, *Green Chem.*, 2007, **9**, 1273–1283.
28 P. J. Dunn, S. Galvin and K. Hettenbach, *Green Chem.*, 2004, **6**, 43.
29 http://www.braskem.com/site.aspx/How-it-is-produced, accessed 06.10.2015.
30 P. L. de Andrade Coutinho, A. T. Morita, L. F. Cassinelli, A. Morschbacker and R. Werneck Do Carmo, Braskem's Ethanol to Polyethylene Process Development, in *Catalytic Process Development for Renewable Materials*, Wiley-VCH, Weinheim, 2013.
31 J. G. Stevens, P. Gómez, R. A. Bourne, T. C. Drage, M. W. George and M. Poliakoff, *Green Chem.*, 2011, **13**, 2727.
32 T. Schaub and R. A. Paciello, *Angew. Chem., Int. Ed.*, 2011, **50**, 7278–7282.
33 S. Wesselbaum, U. Hintermair and W. Leitner, *Angew. Chem., Int. Ed.*, 2012, **51**, 8585–8588.
34 E4tech and LCAworks, *Environmental assessment of Braskem's biobased PE resin - Summary of the life cycle assessment, land-use change and water footprint reports*, 2013, Available from http://www.braskem.com.br/Portal/Principal/Arquivos/ModuloHTML/Documentos/1204/20131206-enviro-assessment-summary-report-final.pdf.

Faraday Discussions

ROYAL SOCIETY OF CHEMISTRY

PAPER

New catalysts for carboxylation of propylene glycol to propylene carbonate *via* high-throughput screening†

José A. Castro-Osma,[a] James W. Comerford,[a] Richard H. Heyn,*[b] Michael North[a] and Elisabeth Tangstad[b]

Received 4th May 2015, Accepted 28th May 2015

DOI: 10.1039/c5fd00061k

High throughput methodologies screened 81 different metal salts and metal salt combinations as catalysts for the carboxylation of propylene glycol to propylene carbonate, as compared to a 5 mol% $Zn(OAc)_2$/*p*-chlorobenzene sulfonic acid benchmark catalyst. The reactions were run with added acetonitrile (MeCN) as a chemical water trap. Two new catalysts were thereby discovered, zinc trifluoromethanesulfonate ($Zn(OTf)_2$) and zinc *p*-toluenesulfonate. The optimal reaction parameters for the former catalyst were screened. $Zn(OTf)_2$ gave an overall propylene carbonate yield of greater than 50% in 24 h, twice as large as the previous best literature yield with MeCN as a water trap, with 69% selectivity and 75% conversion of propylene glycol at 145 °C and 50 bar CO_2 pressure.

Introduction

Research on carbon dioxide (CO_2) utilization (CDU) has been gaining momentum over the past decade. While the motivation for this surge in interest may well be a response to the increasing levels of CO_2 in the atmosphere and the need for mitigating CO_2 emissions, the shear volumes of CO_2 emitted[1] and relatively modest contribution CDU can make toward mitigation of these emissions, at least in terms of chemical production volumes, strongly suggest that a greater motivation is the development of CO_2 as a sustainable C_1 source for the chemical industry. CO_2 is renewable, easily handled and stored, and essentially non-toxic. The challenge in its use arises from its thermodynamic and kinetic inertness. Advances in catalysis and process engineering are therefore necessary in order to overcome these barriers. Some of these issues can be and have been addressed by the use of high energy chemical reactants, as exemplified by the production of

[a]*University of York, Department of Chemistry, Green Chemistry Center of Excellence, York YO10 5DD, North Yorkshire, England, UK*

[b]*SINTEF Materials and Chemistry, P. O. Box 124 Blindern, 0314 Oslo, Norway. E-mail: rhh@sintef.no*

† Electronic supplementary information (ESI) available: ^{1}H NMR spectrum of a typical PG to PC product mixture; list of all HT screened catalyst formulations. See DOI: 10.1039/c5fd00061k

aliphatic polycarbonates from CO_2 and epoxides, which is on the cusp of industrial implementation.[2]

One class of compounds that have generated considerable interest as products for CDU is the organic carbonates, both open, linear carbonates such as dimethylcarbonate (DMC) and closed, cyclic carbonates such as propylene carbonate (PC). These molecules can be synthesized from CO_2 with two equivalents of an alcohol (methanol, for DMC) or one equivalent of a diol (1,2-propanediol, for PC). The latter can also be prepared from CO_2 and propylene oxide using the appropriate catalyst, and the cyclic carbonates can be undesired by-products in the production of aliphatic polycarbonates.[3]

The reaction between CO_2 and alcohols or diols is thermodynamically unfavorable,[4] with water as the other reaction product (see Scheme 1). Without any added water trap, the yield of PC from 1,2-propanediol (propylene glycol, PG) is 0.5% at 130 °C with $CeO_2 \cdot ZrO_2$ as a heterogeneous catalyst.[5] The addition of physical or chemical water traps will pull the equilibrium toward products, and this has most recently been successful with a CeO_2/2-cyanopyridine system (20 mol% CeO_2 and 10-fold excess 2-cyanopyridine based on PG), which is able to provide a nearly quantitative yield of PC from PG and CO_2 within 1 h at 130 °C and 50 bar CO_2, with 2-acetamidopyridine as the by-product from the chemical trapping of the co-produced water.[6]

The effect of water traps on the synthesis of organic carbonates from CO_2 and alcohols or diols has recently been reviewed.[7] The majority of the other investigations into this reaction have used acetonitrile (MeCN) as the chemical water trap. One drawback with MeCN is the formation of a number of by-products from the water-trapping reaction, including acetamide, acetic acid, and acetylated glycols (see Scheme 1). The most efficient catalytic system (as based on time-yield) reported thus far is 1.0 mol% (2.5 wt%) $Zn(OAc)_2$ with a 1.8 fold excess of MeCN at

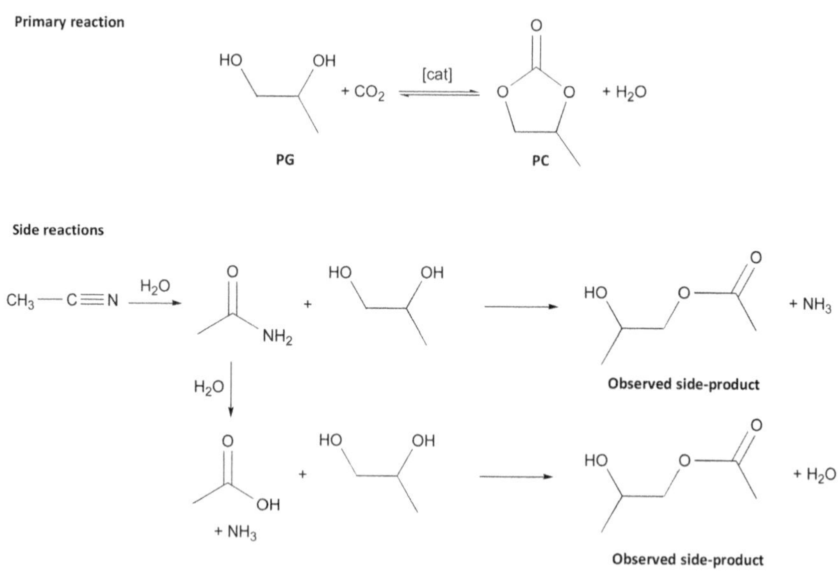

Scheme 1 Primary reaction and formation of the observed side products. Only the 2-hydroxypropyl acetate side-product is shown. 1-Hydroxypropyl acetate and 1,2-propyl diacetate side-products were also observed.

160 °C and an initial pressure of 30 bar CO_2, which provided a 12% yield of PC with a 64% selectivity after 2 h.[8] A second report also indicates $Zn(OAc)_2$ as the best $M(OAc)_2$ catalyst (M = Co, Ni, Cu, Mn, Mg, Ca), but the reported time-yield is poorer (24% yield and 62% selectivity within 12 h at 170 °C and 100 bar CO_2).[9] In addition to $Zn(OAc)_2$, inorganic carbonates (K_2CO_3 and $Cs_2CO_3/(NH_4)_2CO_3$),[10] organic bases (1,5,6-triazabicyclo[4.4.0]dec-5-ene)[11] and modified ZnO (KI/ZnO)[12] have been identified as the best catalysts in individual studies. Direct comparison between these catalysts is difficult since there is always variation in the reported reaction pressures, temperatures, times and catalyst loadings.

Considering the relatively limited number of compounds that have been tested for this reaction, we undertook a high throughput (HT) screening of 81 different metal salts and salt combinations. This contribution describes the selection of the benchmark catalyst for the HT studies, trends gleaned from the HT studies, and a parameter screening for an improved catalyst system found during the HT screening.

Results and discussion

Catalyst pre-screening and choice of benchmark

This study started with a small catalyst pre-screening for determination of a suitable benchmark catalyst for the carboxylation of PG into PC at 145 °C and

Table 1 Synthesis of PC from PG and CO_2 with different catalysts[a]

Entry	Catalyst	Solvent	Yield PC[b] (%)
1	$[SalenAl]_2O$[c]	MeCN	0
2[d]	$[SalenAl]_2O$ + TBAB	MeCN	0
3	Tetrabutylammonium bromide	MeCN	0
4	K_2CO_3	MeCN	0
5	$Bu_2Sn(OAc)_2$	MeCN	<1
6	$Mn(OAc)_2$	MeCN	7
7	$Mn(OAc)_2$	MeOH	0
8	$Fe(OAc)_2$	MeCN	<1
9	$Fe(OAc)_2$	MeOH	0
10	$Co(OAc)_2$	MeCN	8
11	$Ni(OAc)_2$	MeCN	0
12	$Cu(OAc)_2$	MeCN	5
13	$Cu(OAc)_2$	MeOH	0
14	$ZnBr_2$	MeCN	<1
15	ZnI_2	MeCN	<1
16	ZnI_2	MeOH	0
17	$Zn(acac)_2 \cdot H_2O$	MeCN	15
18	$Zn(OAc)_2$	MeOH	<1
19[e]	$Zn(OAc)_2$	MeOH	0
20[e]	$Zn(OAc)_2$	MeCN	1
21	$Zn(OAc)_2$	MeCN	27

[a] Reactions carried out at 145 °C and 60 bar CO_2 pressure for 16 hours using 2.5 mol% of catalyst and 10 ml MeCN or MeOH. [b] Yield of PC determined by ^1H NMR spectroscopy of the crude reaction mixture (see ESI). [c] See ref. 13. [d] 2.5 mol% tetra-butylammonium bromide (TBAB) used. [e] 13X molecular sieves added (200 mg).

60 bar CO_2. Reactions were carried out in the presence of either methanol or MeCN as the solvent for 16 hours using 2.5 mol% of catalyst (Table 1). These conditions were chosen as an average between the conditions provided in the two previous papers reporting the catalytic activity of $Zn(OAc)_2$.[8,9] As shown in Table 1, zinc salts displayed higher catalytic activity than other metal salts, and $Zn(OAc)_2$ was, perhaps unsurprisingly, the most active catalyst. Additionally, reactions carried out in MeCN showed higher conversions than those carried out in methanol, which supports previous results showing that MeCN not only acts as a solvent but also as a dehydrating agent, shifting the reaction to the formation of products. As a catalyst, $Zn(OAc)_2$ showed some problems with reproducibility; a series of 6 runs gave a 39 ± 7% conversion of PG and a 18 ± 5% yield of PC with a 41 ± 4% selectivity.

One of the main routes for the synthesis of cyclic carbonates is the trans-esterification reaction between DMC and diols.[14] Similarly, the synthesis of diphenyl carbonate (DPC) can be carried out by a transesterification reaction between DMC and phenol,[15] and it was found that using sulfonic acids as co-catalysts improved both the conversion of DMC and the isolated yield of DPC.[16] Therefore, the addition of a sulfonic acid as a co-catalyst for the synthesis of PC from PG was investigated. As shown in Table 2, the addition of 5 mol% *p*-chlorobenzene sulfonic acid (*p*-CBSA) gave higher PC yields and larger PG conversions. The yields of PC in Table 2 were the highest overall yields that have been reported in the literature for systems that employ MeCN as a water trap. The 5 mol % $Zn(OAc)_2$/*p*-CBSA system was therefore chosen as the benchmark catalyst for the high throughput screening.

High throughput screening

The HT screening experiments included a total of 81 different metal salts, combinations of metal salts and combinations of metal salts and strong acids. These screening experiments also included some metal salts tested in the pre-screening activity, as an additional check of the reproducibility of the HT system. The screened catalyst systems can be loosely grouped into four categories: simple Zn salts, other late transition metal salts, combinations of Zn salts with other

Table 2 Synthesis of PC from PG and CO_2 with $Zn(OAc)_2$ and sulfonic acids as catalysts[a]

Entry	Sulfonic acid	Yield PC (%)	Conversion PG (%)	Selectivity (%)	Isolated PC yield[b] (%)
1[c]	p-$CH_3C_6H_4SO_3H$	22	48	44	—
2[d]	p-$CH_3C_6H_4SO_3H$	28	69	41	18
3	$C_6H_5SO_3H$	29	72	40	17
4	p-$ClC_6H_4SO_3H$	33	73	45	25
5[e]	p-$ClC_6H_4SO_3H$	35	84	42	34
6[f]	p-$CH_3C_6H_4SO_3H$	26	63	41	27

[a] Reactions carried out with 5 mol% $Zn(OAc)_2$ and 5 mol% sulfonic acid, at 145 °C and 60 bar CO_2 for 16 h. Product distributions based on ^1H NMR spectra of the crude reaction mixtures. [b] Yield of PC isolated by flash chromatography. See Experimental section. [c] Used 2.5 mol% $Zn(OAc)_2$ and 2.5 mol% p-$CH_3C_6H_4SO_3$. Average values for 3 runs. [d] Average of three runs. [e] Dry MeCN. [f] 64 h reaction time.

reagents, and Lewis acidic triflate salts. The first three categories included combinations with strong acids. In addition to the *p*-CBSA and *p*-toluene sulfonic acid (*p*-TSA) used in the pre-screening, the HT studies also used 4-nitrobenzene sulfonic acid and dibenzenesulfonimide, as acids with different pK_a values (from 6.60 for 4-nitrobenzene sulfonic acid to 11.34 for dibenzenesulfonimide) in MeCN.[17] The Lewis acids and other reagents were chosen as they have been shown to be effective transesterification catalysts. Specifically, lanthanide triflates have been shown to catalyze the transesterification of DMC with ethanol.[18] The other reagents, KI, KOH, K_2CO_3 and NEt_4Br, have been shown to be catalysts or co-catalysts in the synthesis of DMC from CO_2, MeOH and epoxides.[19] The results from the catalysts tested in the pre-screening compared well with the results from the same catalysts tested in the HT experiments. Table 3 compares the PC yield of five catalysts studied both in the pre-screening and the HT experiments. The qualitative comparison as based on the yield of the benchmark catalyst is good, providing support of the validity of the results from the HT screening experiments. The larger differences for $Co(OAc)_2$ and $Zn(acac)_2 \cdot H_2O$ may be attributed to the different catalyst loadings in the different experiments.

All 81 screened catalyst formulations are provided in the ESI.† For HT screening experiments, an important decision point is the definition of a "hit". For this study, any screened catalyst formulation that gave a better PC yield than that of the benchmark was deemed a "hit" and worth potential follow-up in a bench-top reactor. It must be emphasized that the HT experiments are designed simply to identify potential lead catalyst candidates. The absolute PC yield for any one screened catalyst formulation was considered as merely suggestive. The list of catalyst formulations considered as lead candidates is shown in Table 4. These 16 catalyst formulations represent 20% of all screened catalysts. It is likely that the hits for zinc trifluoromethanesulfonate ($Zn(OTf)_2$) and zinc *p*-toluenesulfonate hydrate ($Zn(Tos)_2 \cdot H_2O$) with acids are simply due to the Zn salts, and therefore these successful catalyst combinations can essentially be reduced to four lead candidates: $Zn(OTf)_2$, $Zn(Tos)_2 \cdot H_2O$, zinc hexafluoroacetonylacetate dihydrate, and $Zn(OAc)_2$ with Lewis acidic triflate salts. On the basis of performance and the preference for a single-component catalyst, $Zn(OTf)_2$ and $Zn(Tos) \cdot H_2O$ were chosen for bench-top validation experiments.

Table 3 Comparison of selected catalysts in the pre-screening and HT experiments[a]

	PC yield relative to benchmark	
Catalyst	Pre-screeing	HT
$Co(OAc)_2$	24%	48%
$Cu(OAc)_2$	15%	10%
$Zn(acac)_2 \cdot H_2O$	45%	81%
$Zn(OAc)_2$	82%	97%
$Zn(OAc)_2$ + *p*-TSA[b]	85%	89%

[a] Pre-screening data with 2.5 mol% catalyst, HT data with 5 mol% catalyst. Entry 4 in Table 2 used as the benchmark catalyst yield. HT results based on HeadSpace GC analyses. [b] 5 mol% catalyst in the pre-screening experiment.

Table 4 Lead catalyst candidates from the HT screening[a]

Catalyst	Relative % PC yield[b]	Catalyst	Relative % PC yield
$Zn(OTf)_2$ + 4-NO_2-benzenesulfonic acid	1.01	$Zn(OAc)_2$ + $Sm(OTf)_3$	1.11
ZnI_2 + *p*-TSA	1.02	$Zn(OTf)_2$ + *p*-TSA	1.11
$Zn(F_6\text{-acac})_2$[c]	1.03	$Zn(OTf)_2$ + dibenzenesulfonimide	1.12
$Zn(F_6\text{-acac})_2$ + *p*-CBSA	1.07	$Zn(OAc)_2$ + LiOTf	1.18
$Zn(OTf)_2$ + *p*-CBSA	1.08	$Zn(OAc)_2$ + $Mg(OTf)_2$	1.25
$Zn(OTf)_2$	1.08	$Zn(OAc)_2$ + $Yb(OTf)_3$	1.30
$Zn(Tos)_2 \cdot H_2O$ + *p*-TSA	1.09	$Zn(Tos)_2 \cdot H_2O$	1.31
$Zn(OAc)_2$ + *p*-TSA[d]	1.09	$Zn(OAc)_2$ + $Ca(OTf)_2$	1.35

[a] HT reaction conditions: 145 °C, 50 bar CO_2, 16–18 h reaction time. [b] Data based on comparison to benchmark data, as determined from GC HeadSpace data. [c] F_6-acac = hexafluoroacetylacetonate. [d] 10 mol%.

Table 5 Lead candidate validation results[a]

Catalyst	Time (h)	$p(CO_2)$ bar	PC yield (%)	PG conversion (%)	Selectivity (%)
$Zn(OTf)_2$	24	53	52	75	69
$Zn(Tos)_2 \cdot H_2O$	25	47	48	77	62

[a] Conditions: 20.7 g (2.72 mmol) PG, 31.45 g (766 mmol) MeCN, 145 °C, time from start of heating, product ratios from integration of ^1H NMR spectra.

The data for the carboxylation of PG to PC using 5 mol% $Zn(OTf)_2$ and $Zn(Tos)_2 \cdot H_2O$ are shown in Table 5. In particular, $Zn(OTf)_2$ showed a significantly improved catalytic performance over previously studied catalysts. Over 50% yield with nearly a 70% selectivity was observed for the $Zn(OTf)_2$ catalyst. The observed PC yield for $Zn(OTf)_2$ is double the best reported overall yield in the literature for those systems that used MeCN as the chemical water trap.

Parameter screening for $Zn(OTf)_2$

Given the superior catalytic performance of $Zn(OTf)_2$ in the HT screening and validation studies, this catalyst was chosen for a parameter optimization. The effect of the reaction parameters of pressure, temperature and the amount of $Zn(OTf)_2$ on the conversion to PC and the overall selectivity of the reaction were screened in a bench-top pressure reactor, and the results are presented in Fig. 1–3, respectively.‡ There is very little effect of a change in pressure from 30 to 70 bar CO_2 on the product distribution. The yield of PC is essentially constant

‡ Note that the amount of MeCN and the reaction time in the parameter screening experiments are less than those for the validation experiments. These differences, in addition to differences in protocol and reactor set-up, are the likely reasons for the slightly poorer overall results in the parameter screening experiments, as compared to those in the validation experiments.

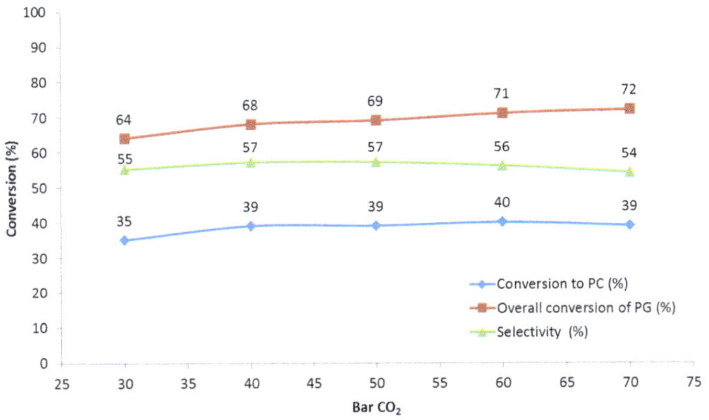

Fig. 1 Effect of different pressures on the yield of PC with 5 mol% $Zn(OTf)_2$ as the catalyst, 135 °C and 16 h reaction time.

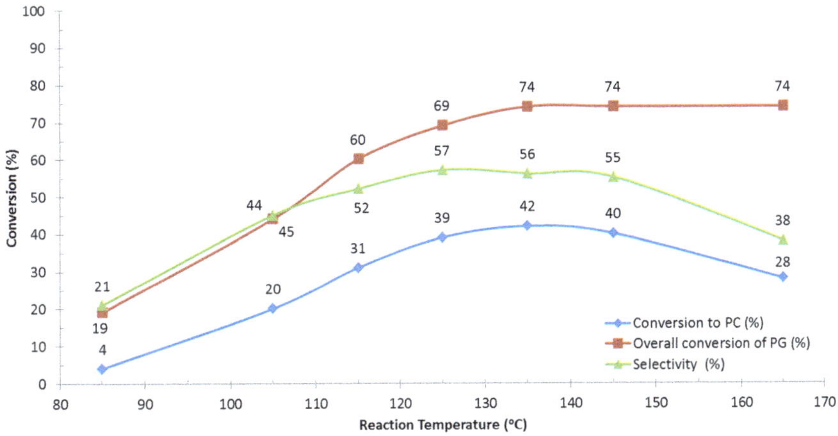

Fig. 2 Effect of different temperatures on the yield of PC with 5 mol% $Zn(OTf)_2$ as the catalyst, 50 bar CO_2 and 16 h reaction time.

above 40 bar, and there is a slight increase in the side products with increasing pressure, as indicated by the increase in the overall conversion of PG and the decrease in selectivity.

The effect of temperature is more profound. At 85 °C, the yield of PC is only 4%. The yield increases monotonically with an increase in temperature up to 135 °C; at this temperature the yield of PC, the overall selectivity and the conversion of PG are at their maximum values. Higher reaction temperatures show a decrease in PC yield and selectivity, while the conversion of PG remains constant, indicating that reaction temperatures higher than 135 °C only increase side product formation. Similarly, an increase in the mol% catalyst from 1 mol% to 5 mol% provides a monotonic increase in the PC yield, from 17% to 42%. There is a concomitant increase in PG conversion, while the selectivity remains essentially constant. Increasing the catalyst amount from 5 to 10 mol% does not appreciably increase the PC yield. The overall conversion of PG does slightly

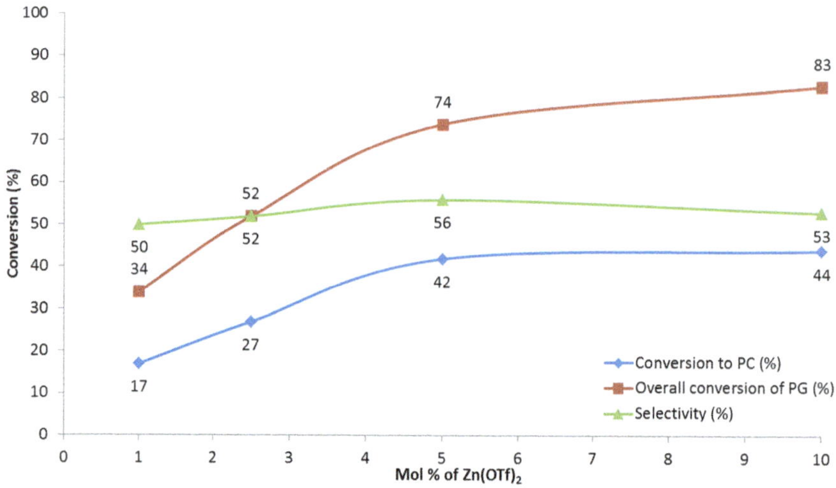

Fig. 3 Effect of different amounts of the catalyst $Zn(OTf)_2$ on the yield of PC, 135 °C, 50 bar CO_2 and 16 h reaction time.

increase, but this is mainly due to a greater increase in the formation of side products, as shown by the decrease in selectivity.

Discussion

On the basis of the many different catalyst formulations that have been screened, either in a bench-top reactor or in a HT system, certain trends can be identified in this work. One, transition metal salts, apart from the Zn salts, are not very good catalysts for this reaction. None of the Ni or Cu salts showed any significant catalytic activity, even in the presence of strong acids. Salts of Mn, Fe and Co showed better activities; while the screened salts did not fare as well as the Zn salts, examples with different coordination spheres may prove to be viable candidates. Two, Lewis acidic triflate salts, based on group 1, 2 and lanthanide metals, are also very poor catalysts, although some of these salts showed promise in combination with $Zn(OAc)_2$ and are under further study. Three, the pK_a of the added acid had in general no effect on catalytic activity. Four, Zn salts generally showed better activities in combination with a strong acid than they did alone. For example, both $ZnBr_2$ and ZnI_2 when combined with either *p*-CBSA or *p*-TSA showed much better catalytic activities in the HT screening experiments than the simple salts did in the pre-screening experiments.

One hypothesis for the improved catalytic activity of $Zn(OTf)_2$ and $Zn(Tos)_2 \cdot H_2O$ over the benchmark system is that they do not need other reagents to produce open metal coordination sites. The activity of the benchmark catalyst (and other Zn salts) in combination with sulfonic acids may well be attributed to the formation of HOAc and some Zn^{2+} species with a labile coordination sphere. That other Zn salts with ostensibly poorly coordinating anions such as $Zn(BF_4)_2$ fail to show appreciable catalytic activities, even in the presence of strong acids, suggests that $Zn(OTf)_2$ and $Zn(Tos)_2 \cdot H_2O$ may impart other, as-yet unknown, benefits towards the catalysis of this reaction.

Experimental

General considerations

All reagents were commercially available (Alfa Aeser, Sigma-Aldrich, Fluka, TCI or Acros) and were used as received. Metal salts for the high throughput studies were in general anhydrous. See ESI† for further information. The bimetallic aluminium salen complex was prepared as previously reported.[13] Carbon dioxide was purchased from BOC or Yara and used without further purification. ^{1}H and ^{13}C NMR spectra were recorded on a Jeol Oxford 400 spectrometer or on a Bruker UltraShield 400 spectrometer at resonance frequencies of 400 and 100 MHz, respectively.

Initial screening

Reactions were magnetically stirred in a 300 mL stainless steel pressure reactor heated in an oil bath. Product compositions were then analysed using ^{1}H NMR spectroscopy without any purification to determine the conversion and selectivity, as based on the integration of the Me signals of PC, side products and PG (see ESI.†)

Representative procedure for the synthesis of PC catalysed by $Zn(OAc)_2$ and *p*-chlorobenzene sulfonic acid

PG (7.35 mL, 100 mmol), $Zn(OAc)_2$ (0.917 g, 5 mmol), *p*-CBSA (1.07 g, 5 mmol) and MeCN (10 mL, 191 mmol) were placed in a 300 mL stainless steel pressure reactor heated in an oil bath. The reaction mixture was magnetically stirred and heated to 145 °C. Then, the reactor was pressurized to 60 bar of carbon dioxide and the reaction mixture was stirred for 16 h. The conversion of PG to PC was then determined by analysis of a sample using ^{1}H NMR spectroscopy. The remaining sample was evaporated *in vacuo* to give a mixture of PC, PG, propylene glycol acetate and propylene glycol diacetate. The mixture was purified by flash chromatography, using dichloromethane as the eluent, to give the pure PC.

High throughput screening

The high throughput experiments were conducted in an in-house constructed HT reactor featuring four rows of six reactor wells, with each row of reactors having a common gas feed. Each well was 9 cm high and had an inner diameter of 1.5 cm, with an effective reactor volume of approximately 11 mL. All wells were closed during the reaction to avoid well-to-well contamination. On a bench-top, the catalyst formulation (metal salts, acids, ligands) was added to each well of the reactor. The amount of catalyst was kept to *ca.* 5 mol% of the amount of PG. A 1 mL aliquot of PG (13.6 mmol) and a 2 mL aliquot of MeCN (38.3 mmol) were then added by autopipette. Three steel balls (5 mm diameter) were added to each well to ensure proper mixing. The set of 24 reactor wells was then attached to the gas inlet portion of the reactor under a CO_2 flow, placed in an oven and warmed up to 145 °C under 2–3 bar CO_2 pressure. After reaching the reaction temperature, the CO_2 pressure in each row of 6 reactors was sequentially increased to 50 bar. Mixing was accomplished with the help of a variable speed vortex (shaker). Reactions were run overnight, usually between 16–18 h from the time of pressurizing the reactors to 50 bar CO_2 until cooling was initiated. Approximately two hours were required to cool the reactor sufficiently before the CO_2 pressure could be released. Once the reactors reached a sufficiently low temperature, the pressure was carefully released overnight to a holding container. The initial HT experiment contained the benchmark catalyst system in each of the 24 reactor wells. No significant internal variation was observed, apart from a small, reproducible decrease in benchmark catalyst yields in the row of 6 reactors closest to the oven door. Comparison of PC yields for the screened catalysts were therefore made on a row-by-row basis, and in order to ensure reproducibility of each high-throughput experiment, the benchmark catalyst was placed in one randomly chosen well in each row of 6 reactors. The placement of the 20 other catalyst formulations in each high throughput experiment was also randomized.

Product analysis was conducted on a Teledyne Tekmar HT3 HeadSpace analyzer coupled to a Agilent 6890 gas chromatograph, using a DB-WAX column (30 m × 0.320 mm × 0.5 μm) and flame ionization detector. Samples were subjected to a 140 °C, 15 minute temperature profile. Analyses were conducted on two 10 μL samples withdrawn from the high-throughput reactor wells.

Validation experiments

Results from selected catalysts from the high throughput experiments were validated in a 300 mL Parr reactor with mechanical stirrer and heating jacket. On a bench top, a glass liner for the reactor was charged with the chosen catalyst formulation (*ca.* 5 mol% based on PG), PG (*ca.* 20.7 g, 0.272 mol) and MeCN (*ca.* 31.5 g, 0.767 mol). The reactor top was mounted under a flow of CO_2. After the reactor had been assembled, it was flushed with CO_2 three times, and the temperature was raised to 145 °C under a slight pressure of CO_2 (2–3 bar). Finally, the pressure was raised to *ca.* 50 bar and the CO_2 feed was closed. The reaction was stopped by simply removing the heating jacket and allowing the reactor to cool to room temperature, followed by a slow release of CO_2. The reactor was cleaned with hot isopropanol and dried between runs. Product distributions were determined by the established ^{1}H NMR method.

Parameter screening for $Zn(OTf)_2$

The reactor was dried at 100 °C under vacuum for a minimum of one hour and cooled to <40 °C before use. The reactor was then charged with anhydrous $Zn(OTf)_2$ and 10 mL (191 mmol) of dry MeCN stored under Ar. 60 mmol PG and a magnetic stirring bar was subsequently added. The reactor was then sealed, pressurized to 20 bar with CO_2 and heated to the desired temperature. Once this temperature had been reached, the CO_2 pressure was increased to the desired value. After 16 hours, the reactor was cooled in a bath of liquid nitrogen or dry ice/acetone until the internal temperature dropped below 30 °C, at which time the reactor was depressurized. Yields and selectivities were determined from integration of the ^{1}H NMR spectra of the products.

Conclusions

Due to its thermodynamic stability and kinetic sluggishness, CO_2 needs catalysts in order to be transformed into useful chemical products. In this study, we have endeavored to apply high throughput techniques so as to more rapidly screen a host of different metal salts as catalysts for the carboxylation of propylene glycol to propylene carbonate. Our initial pre-screening results found an improved catalyst system – 5 mol% $Zn(OAc)_2$/*p*-chlorobenzene sulfonic acid – that showed an improvement in PC yield over data reported in the literature. With this benchmark catalyst system in hand, 81 different metal salts and metal salt combinations were screened. This resulted in the discovery of two new catalysts for this reaction, $Zn(OTf)_2$ and $Zn(Tos)_2\cdot H_2O$, the activities of which were validated using a bench-top reactor. The $Zn(OTf)_2$ system underwent further bench-top screening, in order to establish the optimal reaction parameters. The $Zn(OTf)_2$ catalyst showed an overall PC yield at least twice as large as the best catalyst system hitherto reported in the literature, using MeCN as a water trap. The activity of the new catalyst is still inferior to that obtained with CeO_2 and 2-cyanopyridine.[6]

We are continuing our studies into these new Zn catalyst systems, and variations on the basic $Zn(OTf)_2$ catalyst and investigations into the mechanism of the reaction will be reported in due course.

Acknowledgements

The authors would like to thank the European Union Seventh Framework program (FP7/2007–2013) project "CyclicCO2R", under grant agreement number 309497, for financial support. R.H.H. and E.T. would like to thank Kari Anne Andreassen, Anne Andersen, Aud M. Bougza, Dr Silje F. Håkonsen and Ruth Elisabeth Stensrød for experimental assistance.

Notes and references

1 G. M. Bond, J. Stringer, D. K. Brandvold, F. A. Simsek, M.-G. Medina and G. Egeland, *Energy Fuels*, 2001, **15**, 309–316.

2 J. Langanke, A. Wolf, J. Hofmann, K. Böhm, M. A. Subhani, T. E. Müller, W. Leitner and C. Gürtler, *Green Chem.*, 2014, **16**, 1865–1870; A. M. Chapman, C. Keyworth, M. R. Kember, A. J. J. Lennox and C. K. Williams, *ACS Catal.*, 2015, **5**, 1581–1588.

3 For a recent review, see C. Martin, G. Fiorani and A. Kleij, *ACS Catal.*, 2015, **5**, 1353–1370.

4 The standard Gibbs energy of reaction in the ideal gas state ΔRgIG is calculated to be $\approx$+37 kJ mol^{-1} for synthesis of open carbonates such as DMC from CO_2 and 2 equiv. of an alcohol. K. Müller, L. Mokrushina and W. Arlt, *Chem. Ing. Tech.*, 2014, **86**, 497–503.

5 K. Tomishige, H. Yasuda, Y. Yoshida, M. Nurunnabi, B. Li and K. Kunimori, *Green Chem.*, 2004, **6**, 206–214.

6 M. Honda, M. Tamura, K. Nakao, K. Suzuki, Y. Nakagawa and K. Tomishige, *ACS Catal.*, 2014, **4**, 1893–1896.

7 M. Honda, M. Tamura, Y. Nakagawa and K. Tomishige, *Catal. Sci. Technol.*, 2014, **4**, 2830–2845.

8 X. Zhao, N. Sun, S. Wang, F. Li and Y. Wang, *Ind. Eng. Chem. Res.*, 2008, **47**, 1365–1369. After 6 h reaction, the yield of PC had increased to 19%, while the selectivity had decreased to 51%.

9 S. Huang, S. Liu, J. Li, N. Zhao, W. Wei and Y. Sun, *J. Fuel Chem. Technol.*, 2007, **35**, 701–705.
10 S. Huang, S. Liu, J. Li, N. Zhao, W. Wei and Y. Sun, *Catal. Lett.*, 2006, **112**, 187–191.
11 S. Huang, J. Ma, J. Li, N. Zhao, W. Wei and Y. Sun, *Catal. Commun.*, 2008, **9**, 276–280.
12 S. Huang, S. Liu, J. Li, N. Zhao, W. Wei and Y. Sun, *Catal. Lett.*, 2007, **118**, 290–294.
13 J. Meléndez, M. North and R. Pasquale, *Eur. J. Inorg. Chem.*, 2007, 3323–3326; M. North and R. Pasquale, *Angew. Chem., Int. Ed.*, 2009, **48**, 2946–2948; J. A. Castro-Osma, M. North and X. Wu, *Chem.–Eur. J.*, 2014, **20**, 15005–15008.
14 J. R. Ochoa-Gomez, O. Gomez-Jimenez-Aberasturi, B. Maestro-Madurga, A. Pesquera-Rodriguez, C. Ramirez-Lopez, L. Lorenzo-Ibarreta, J. Torrecilla-Soria and M. C. Villaran-Velasco, *Appl. Catal., A*, 2009, **366**, 315–324; A. Kumar, K. Iwatani, S. Nishimura, A. Takagaki and K. Ebitani, *Catal. Today*, 2012, **185**, 241–246.
15 M. Fuming, *J. Mol. Catal. A: Chem.*, 2002, **184**, 465–468.
16 H. Lee, S. J. Kim, B. S. Ahn, W. K. Lee and H. S. Kim, *Catal. Today*, 2003, **87**, 139–144.
17 A. Kütt, I. Leito, I. Kaljurand, L. Sooväli, V. M. Vlasov, L. M. Yagupolskii and I. A. Koppel, *J. Org. Chem.*, 2006, **71**, 2829–2838; F. Eckert, I. Leito, I. Kaljurand, A. Kütt, A. Klamt and M. Diedenhofen, *J. Comput. Chem.*, 2009, **30**, 799–810. The pK_a's for *p*-CBSA and *p*-TSA in MeCN are 7.16 and 8.01.
18 F. M. Mei, E. X. Chen and G. X. Li, *Kinet. Catal.*, 2009, **50**, 666–670.
19 N. A. M. Razali, K. T. Lee, S. Bhatia and A. R. Mohamed, *Renewable Sustainable Energy Rev.*, 2012, **16**, 4951–4964.

Faraday Discussions

PAPER

A MALDI-TOF MS analysis study of the binding of 4-(*N*,*N*-dimethylamino)pyridine to amine-bis(phenolate) chromium(III) chloride complexes: mechanistic insight into differences in catalytic activity for CO_2/epoxide copolymerization

Christopher M. Kozak,* April M. Woods, Christina S. Bottaro, Katalin Devaine-Pressing and Kaijie Ni

Received 1st May 2015, Accepted 5th May 2015
DOI: 10.1039/c5fd00046g

Amine-bis(phenolato)chromium(III) chloride complexes, [LCrCl], are capable of catalyzing the copolymerization of cyclohexene oxide with carbon dioxide to give poly(cyclohexane) carbonate. When combined with 4-(*N*,*N*-dimethylamino)pyridine (DMAP) these catalyst systems yield low molecular weight polymers with moderately narrow polydispersities. The coordination chemistry of DMAP with five amine-bis(phenolato)chromium(III) chloride complexes was studied by matrix assisted laser desorption/ionization time-of-flight mass spectrometry (MALDI-TOF MS). The amine-bis(phenolato) ligands were varied in the nature of their neutral pendant donor-group and include oxygen-containing tetrahydrofurfuryl and methoxyethyl moieties, or nitrogen-containing *N*,*N*-dimethylaminoethyl or 2-pyridyl moieties. The relative abundance of mono and bis(DMAP) adducts, as well as DMAP-free ions is compared under various DMAP : Cr complex ratios. The $[LCr]^+$ cations show the ability to bind two DMAP molecules to form six-coordinate complex ions in all cases, except when the pendant group is *N*,*N*-dimethylaminoethyl (compound **3**). Even in the presence of a 4 : 1 ratio of DMAP to Cr, no ions corresponding to $[L^3Cr(DMAP)_2]^+$ were observed for the complex containing the tertiary sp^3-hybridized amino donor in the pendant arm. The difference in DMAP-binding ability of these compounds results in differences in catalytic activity for alternating copolymerization of CO_2 and cyclohexene oxide. Kinetic investigations by infrared spectroscopy of compounds **2** and **3** show that polycarbonate formation by **3** is twice as fast as that of compound **2** and that no initiation time is observed.

Department of Chemistry, Memorial University of Newfoundland, St. John's, Newfoundland, A1B 3X7, Canada. E-mail: ckozak@mun.ca; Tel: +1-709-864-8082

Introduction

The copolymerization of carbon dioxide with epoxides to yield polycarbonates has become one of the most extensively studied processes that generates potentially valuable materials from CO_2.[1–6] Carbon dioxide is an appealing C1 feedstock because it is widely available, inexpensive, and nontoxic.[7] The variety of metal complexes shown to perform this reaction is impressive and complexes containing metals such as Mg,[8] Al,[9–13] Zn,[14–22] Cr,[23–38] Co[6,39–58] and Fe[59] have been examined for use in the copolymerization of carbon dioxide (CO_2) and epoxides, particularly cyclohexene oxide (CHO) and propylene oxide (PO).

Homogeneous catalysts for epoxide/CO_2 copolymerization have involved various ligand classes including, for example, porphyrin Cr[24,60,61] and Al[13] compounds. Darensbourg and Holtcamp introduced the use of Zn phenoxides for epoxide/CO_2 copolymerization,[62–64] whereas Coates and co-workers explored the use of β-diketiminate (BDI) ligand systems with Zn at lower pressures and temperatures than had previously been reported.[14] Tetravalent group 4 (Ti and Zr) and 14 (Ge and Sn) metals supported by planar trianionic bis(phenolato) ligands reported by Nozaki and co-workers have shown activity towards epoxide and CO_2 copolymerization.[65] By far the most widely studied ligands for epoxide/CO_2 copolymerization have been the salen[3–5] and, more recently, the salan ligands, which have been primarily used with Cr (Fig. 1)[23,25–30,33,36,38,66] and Co.[39,44,45,52,53,67] These catalysts typically require nucleophilic co-catalysts such as chlorides, bromides or azides paired with bulky cations such as PPN (PPN = bis-(triphenylphosphoranylidene)ammonium) or tetrabutylammonium, or neutral bases such as *N*-methylimidazole (*N*-MeIm) or dimethylaminopyridine (DMAP).

Whereas compounds of the salen and salan ligands represent the most investigated homogeneous systems for CO_2/epoxide copolymerization, we are interested in the use of the related tetradentate amine-bis(phenolato) ligand class.[68–71] This ligand offers possibilities for the development of potentially highly active catalysts because of the modifiable nature of the donor sites, their steric and electronic properties, and their geometry, which differs from that exhibited by the salen and salan-based systems. Mechanistic information is, of course, beneficial for the design of highly efficient catalyst systems for CO_2/epoxide copolymerization[17,29,66,72–75] and mass spectrometry has been particularly useful in this regard. Electrospray ionization (ESI) mass spectrometry and collision-induced dissociation (CID) have been employed to study the interactions of DMAP co-catalyst with salen- and salan chromium chloride complexes and how these interactions influence catalyst activity.[66] ESI-MS has also been used to confirm intermediates in CO_2/PO copolymerization by single-site cobalt salen catalysts.[54]

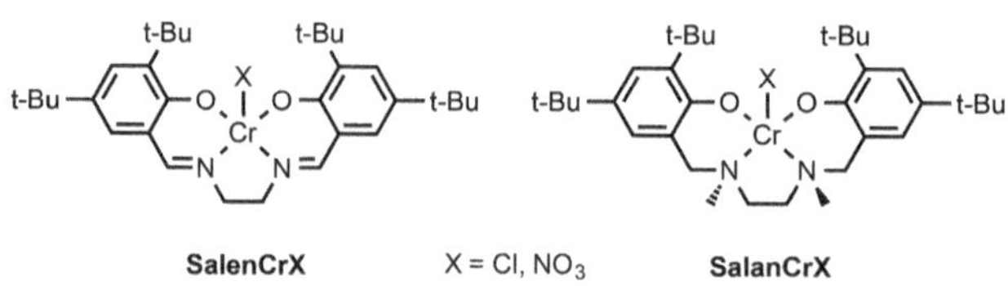

Fig. 1 Salen and salan chromium(III) complexes used in CO_2/epoxide copolymerization.

Matrix assisted laser desorption/ionization time-of-flight (MALDI-TOF) MS has been used by Duchateau and co-workers to investigate zinc catalyzed copolymerization of cyclohexene oxide and CO_2.[76,77]

Darensbourg and co-workers have conducted elegant kinetic studies on epoxide/CO_2 copolymerization catalyzed by binary salenCrX (where X = Cl^- or N_3^-) with Lewis basic (neutral or ionic) co-catalysts, wherein initiation was proposed to occur *via* a bimetallic process and propagation *via* a monometallic enchainment of epoxides.[25,26,78,79] These reports suggest that the binding of the co-catalyst to the metal centre labilizes the *trans*-orientated growing polymer chain. A released "free" alkoxide or carboxylate-terminated polymer fragment promotes insertion of CO_2 into the metal–alkoxide bond. These salen-based complexes also exhibit long induction periods that are not observed in related salan-based systems employing DMAP as the added Lewis base. In this report, we use MALDI-TOF MS to examine the DMAP binding affinity for five amine-bis(phenolato) chromium chloride complexes (Fig. 2). The observed resistance to dimer and bis(DMAP) adduct formation in complex **3** is believed to corroborate with the absence of an induction period, and a faster rate of polycarbonate formation catalyzed by this complex *vs.* the other derivatives. These studies provide a comparative assessment of the mechanistic behaviour of amine-bis(phenolate) chromium-based catalysts for CO_2/epoxide copolymerization against the benchmark of salen and salan chromium(III) systems.

Experimental

Unless otherwise stated, all manipulations were performed under an atmosphere of dry, oxygen-free nitrogen by means of standard Schlenk techniques or using an MBraun Labmaster DP glove box. Anhydrous dichloromethane was obtained by purification using an MBraun Manual Solvent Purification System. Reagents were purchased either from Aldrich or Alfa Aesar and used without further purification. Complexes **1–5** were prepared *via* previously reported methods.[68,80]

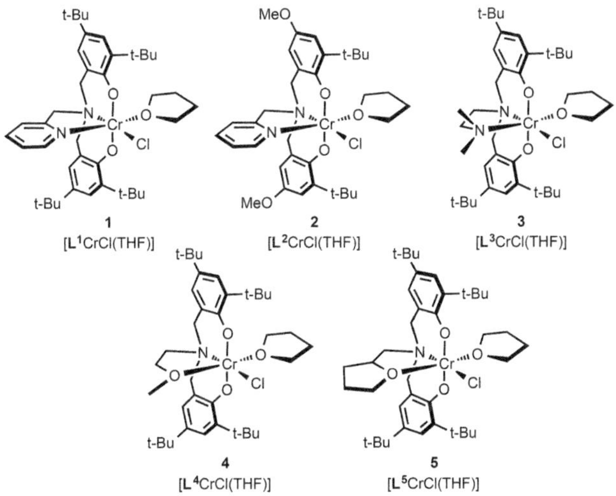

Fig. 2 Chromium(III) complexes used in this study.

MALDI-TOF MS was performed using an Applied Biosystems 4800 MALDI TOF/TOF Analyzer equipped with a reflectron, delayed ion extraction and high performance nitrogen laser (200 Hz operating at 355 nm). Samples were prepared in the glove box and sealed under nitrogen in a Ziploc© bag for transport to the instrument. Anthracene was used as the matrix. The matrix, complex and DMAP were dissolved in dichloromethane at a concentration of approximately 1 mg per 0.1 mL. The complex and DMAP were mixed together in ratios of 1 : 0.5, 1 : 1, 1 : 2 and 1 : 4. Once combined with the matrix, 0.5 μL of this mixture was spotted onto a MALDI plate and left to dry under nitrogen in a glove box. Images of the mass spectra were prepared using mMass™ software (http://www.mmass.org).

Infrared spectra of the reaction progress were obtained using a Mettler-Toledo ReactIR™ 15 spectrometer equipped with silver halide fibre conduit attached to a SiComp Sentinel™ high-pressure window fitted to a 100 mL stainless steel Parr autoclave reactor.

A representative preparation of samples for analysis by MALDI-TOF MS is provided. 0.2094 g of complex **1** (3.323×10^{-4} mol) and 0.0397 g of DMAP (3.249×10^{-4} mol) were placed in a sample vial and just enough dichloromethane was added to dissolve the solids. This solution was filtered through a glass fibre plug into another sample vial and the vial was capped, sealed with Parafilm and stored at −35 °C in the glove box freezer until required.

A stock solution of 8.7 mg of anthracene in 800 μL of dichloromethane was prepared. Samples of complex **1** with their required DMAP loading were dried under vacuum. A measured quantity of each sample was dissolved in 200 μL of CH_2Cl_2 and 80 μL of this solution was transferred to another vial along with 80 μL of the stock anthracene solution. These solutions were spotted onto a MALDI plate and the spectra were obtained.

For MALDI-TOF MS analysis on samples exposed to air, all prepared samples, once analyzed under the air-free preparation described above, were removed from the glove box and exposed to air. The solvents were allowed to evaporate and after two or three days the samples were redissolved in CH_2Cl_2 and analyzed once again.

Results and discussion

The binding of DMAP to five chromium complexes (Fig. 2) was investigated by preparing dichloromethane solutions of each complex with varying ratios of DMAP. The resulting solutions were evaporated until dry and then re-dissolved with the matrix (anthracene). These solutions of analyte and matrix were spotted onto the MALDI plate in air and under air-free conditions. The MALDI-TOF mass spectra of complexes **1**, **2**, **4** and **5** in the absence of DMAP exhibited numerous ions, including ions resulting from dimers and their fragments. The exception was complex **3**, which gave a relatively clean spectrum showing two main ions corresponding to $[\mathbf{L}^3\mathrm{CrCl}]^{+\bullet}$ and $[\mathbf{L}^3\mathrm{Cr}]^{+}$. Ions containing the weakly coordinating THF present in **1–5** are, as anticipated, consistently absent from the mass spectra of these complexes. The MALDI-TOF MS of mixtures of complexes **1–5** and varying amounts of DMAP showed the presence of complicated mixtures of ions. The ions observed or predicted for complex **2** in the presence of DMAP are given in Fig. 3. The labels **F1–F12** also apply to ions arising from complexes **1** and **3–5**, except of course for the nature of the amine-bis(phenolato) ligand present ($\mathbf{L}^1$ to $\mathbf{L}^5$, present

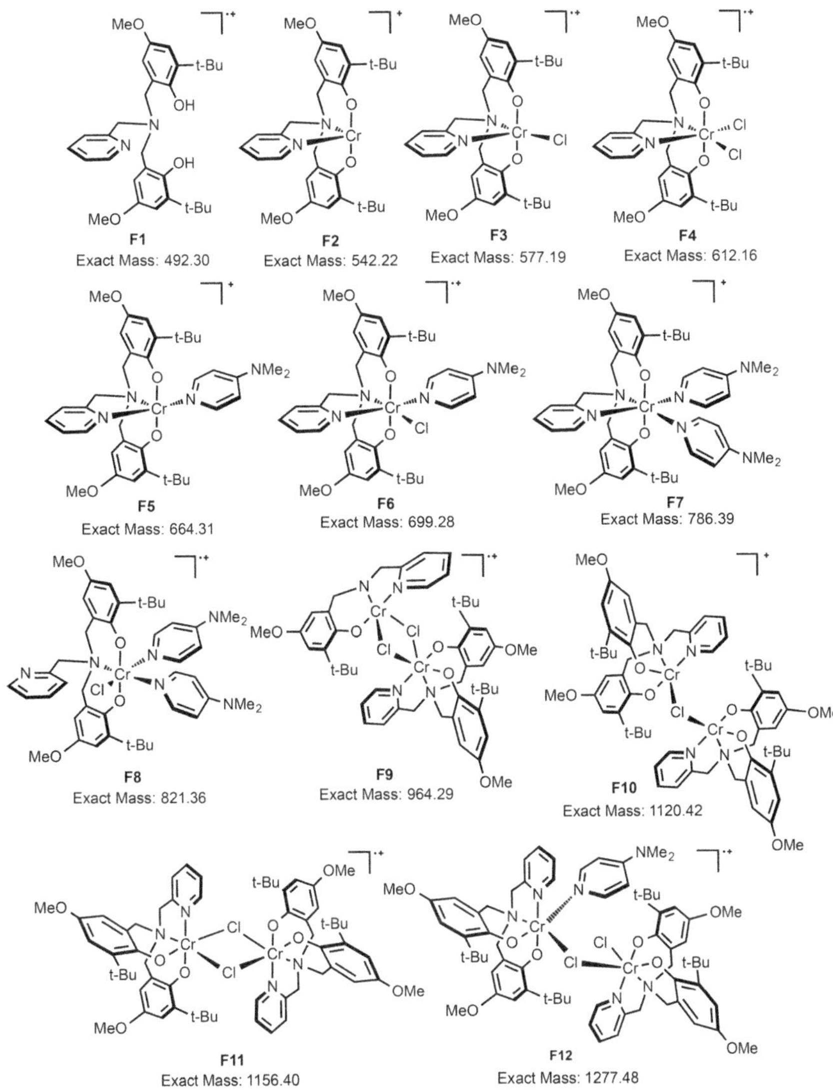

Fig. 3 Anticipated and observed fragments for complex **2**. Fragments of complexes **1** and **3–5** are represented by the analogous structures shown for **F1–F12**.

in complexes **1–5**, respectively). The MALDI-TOF mass spectra for **2**/DMAP and **3**/DMAP mixtures in different molar ratios are shown in Fig. 4. When the molar ratio of **2** to DMAP is 1 : 0.5, a complicated mass spectrum results and many ions could not be readily identified. It is clear, however, that dichromium-containing ions are present along with DMAP-containing fragments, including $[\mathbf{L}^2\mathrm{CrCl(DMAP)}]^{+\bullet}$, **F6**. Furthermore, a significant abundance of ions attributed to an unmetallated ligand, $\mathrm{H_2}\mathbf{L}^2$ (**F1**), are observed. With an increase in the DMAP to Cr ratio of 1 : 1, the mass spectrum simplifies considerably and readily identifiable ions are observed. Under these conditions, the base peak occurs at m/z 577.1, corresponding to $[\mathbf{L}^2\mathrm{CrCl}]^{+\bullet}$ (**F3**), with the next most abundant peak at m/z 699.2,

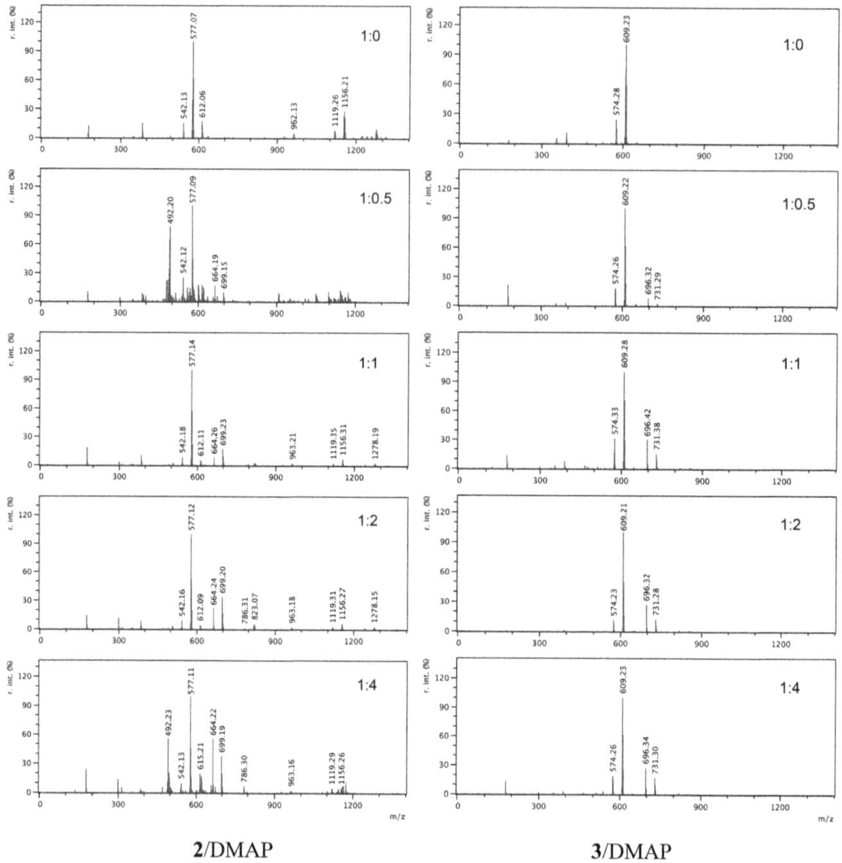

Fig. 4 MALDI-TOF mass spectra of the mixtures resulting from combinations of complex 2 (left) and complex 3 (right) with different molar ratios of DMAP as shown: 1 : 0, 1 : 0.5, 1 : 1, 1 : 2 and 1 : 4.

representing $[L^2CrCl(DMAP)]^{+\bullet}$. Ions at higher *m*/*z* corresponding to dichromium-containing fragments **F9**, **F10**, **F11** and **F12** were also observed. Increasing the loading of DMAP further (**2** : DMAP of 1 : 2 and 1 : 4) resulted in the formation of $[$**2** $-$ THF $-$ Cl + 2DMAP$]^+$ ions, **F7**. The 3/DMAP system demonstrates much less complex mass spectra. The addition of 0.5 equiv. of DMAP per Cr centre causes the appearance of $[L^3Cr(DMAP)]^+$ (**F5**) and $[L^3CrCl(DMAP)]^{+\bullet}$ (**F6**), and increasing the DMAP to Cr ratio (2 and 4 equiv. DMAP to Cr) has little influence on the number of ions observed and there was no evidence of $[L^3Cr(DMAP)_2]^+$ (**F7**) or dichromium-containing ions. The other significant species observed in the **3**/DMAP system is **F4** (*m*/*z* 574.3), which results from chloride loss to give $[L^3Cr]^+$. It is interesting that even with 4 equiv. of DMAP the base peaks for both **2** and **3** correspond to the DMAP-free ion, **F3**, with a *m*/*z* of 577.1 for **2** and 609.2 for **3**, which suggests that this ion exhibits a surprisingly high stability even to loss of chloride. Six-coordinate species of **3** appear to be unfavourable, as shown by the much higher abundance of the four- and five-coordinate ions, **F2**, **F3** and **F5**.

A comparison of the relative abundance of the various fragment ions in compounds **1–5** with different amounts of DMAP added is shown in Fig. 5. For all five complexes, the abundance of the DMAP-coordinated cation, [**L**Cr(DMAP)]$^+$ **F5**, increases expectedly when increasing amounts of DMAP are added. Changing the pendant donor group, however, appears to influence the coordination chemistry of DMAP with these complexes, as well as the stability of the complexes themselves in the gas phase. The nitrogen donor pendant arms, *i.e.* pyridyl in **1** and **2** (abbreviated as BuBuPyr and BuMethPyr in Fig. 5), and dimethylaminoethyl in **3** (abbreviated as $BuBuNMe_2$ in Fig. 5) generate very stable molecular ions **F3**, which comprise the base peaks in their respective mass spectra. Oxygen donor pendant arms, *i.e.* methoxyethyl in **4** (abbreviated as BuBuMe in Fig. 5) and, in particular, tetrahydrofurfuryl in **5** (abbreviated as BuBuTHF in Fig. 5) show a lower stability

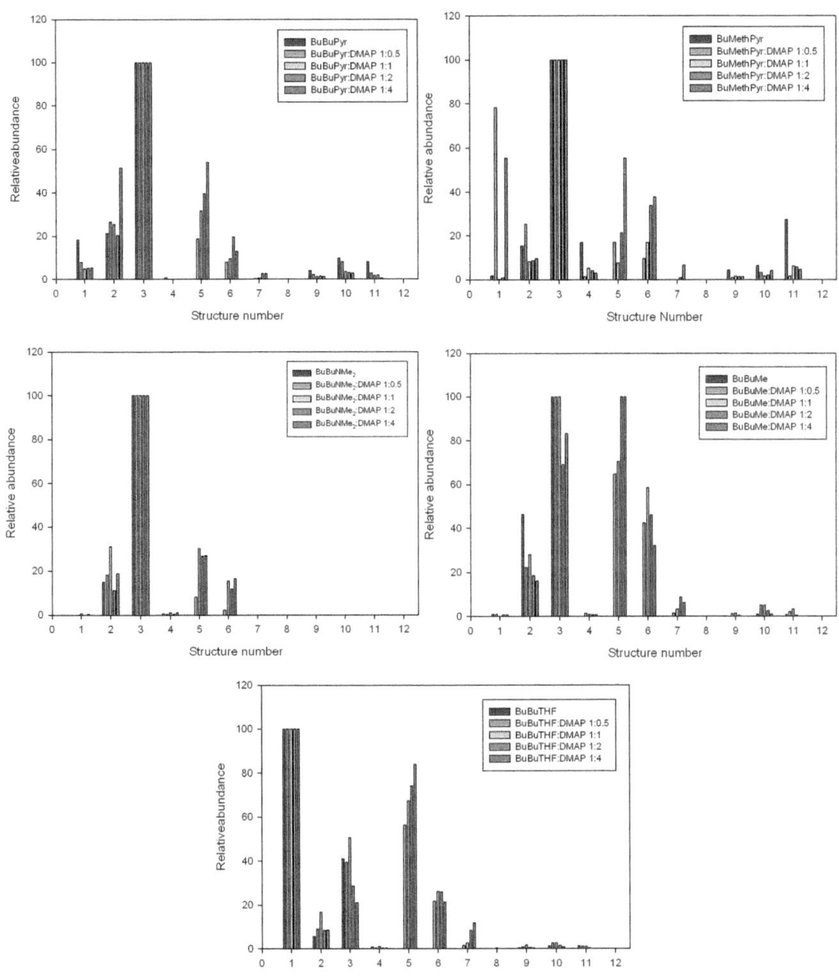

Fig. 5 Comparison of the relative abundance of fragment ions in compounds **1–5** with varying amounts of DMAP, as indicated. Relative abundance is given in % and structure numbers 1–12 correspond to the fragments/ions shown in Fig. 3.

of the molecular ions as judged by the lower relative abundance. As the amount of DMAP is increased compared to **4**, the DMAP adduct accompanied by chloride loss ($[L^4Cr(DMAP)]^+$, **F5**) becomes more abundant as well as the DMAP coordinated radical cation, $[L^4CrCl(DMAP)]^{+\bullet}$ **F6**. Complex **5** stands out as being most readily fragmented under MALDI conditions. The base peak was consistently identified as the unmetallated ligand, H_2L^5 at *m*/*z* 537.4 (the isotopic pattern of this ion is actually a combination of the radical cation of $[H_2L^5]^{+\bullet}$ as well as $[HL^5]^{+\bullet}$ and $[L^5]^{+\bullet}$ resulting from losses of $H^\bullet$ and/or single electrons during laser desorption). This suggests that L^5 is (relatively) poor at stabilizing Cr(III) complexes. Our previous reports of the CO_2/cyclohexene oxide copolymerization catalyzed by **5** with a DMAP co-catalyst showed moderate activities (conversions of 64%, average TONs of 300, $M_n \sim 6000$ g mol^{-1}, $Đ = 1.5$–1.7).[70] By comparison, we observed that the catalyst system comprised of **1**/DMAP under these conditions provided conversions of 80% (TON of 400, $M_n \sim 10\,000$ g mol^{-1}, $Đ = 1.1$ to 1.7).[68]

The difference in ion stability observed between compound **3** and the other four complexes can perhaps be attributed to the steric hindrance created by the dimethylaminoethyl group found in L^3/complex **3**. Furthermore, the presence of two sp^3-hybridized amino donors in **3** reduces the Lewis acidity of the metal centre. Therefore, sterically the Cr centre is more encumbered preventing the formation of chloride-bridged dimers, as previously[68] crystallographically authenticated for $[L^1Cr]_2(\mu\text{-}Cl)_2$ and as postulated based on the presence of *m*/*z* 1260.6 (corresponding to **F11**) in the MALDI-TOF mass spectrum of **1**. Electronically, the more electron-rich Cr centre in **3** compared to **1**, for example, results in a lower affinity for the neutral DMAP than the anionic chloride ligand. A similar trend has been observed in comparing the affinity of salenCrX and salanCrX complexes (Fig. 1) for DMAP.[66] These trends in binding affinity for DMAP are believed to influence the CO_2/epoxide copolymerization activity of the binary catalyst systems of salenCrCl and salanCrCl with DMAP. Rieger and co-workers reported the selectivity for copolymerization of CO_2 with propylene oxide (PO) catalyzed by salenCrCl and DMAP improved considerably with a salenCrCl : DMAP ratio of 1 : 0.5.[27] Therefore, the formation of species containing one DMAP rather than two is likely preferred for initiation of copolymerization.[66]

The mechanism of copolymerization of CO_2 and cyclohexene oxide by binary catalyst systems of salenCrX and N-heterocyclic bases such as DMAP has been investigated in detail by Darensbourg.[29] The proposed mechanism involves the formation of carbamate intermediates resulting from the reaction of DMAP with CO_2 to generate a zwitterion, which was identified by ν_{CO2} bands at 2097 and 2017 cm^{-1}. Furthermore, Darensbourg reports a correlation between the disappearance of "free DMAP" as observed by *in situ* FTIR with the initiation time observed for the formation of polycarbonate. Lu and co-workers studied the initiator role of DMAP in CO_2/propylene oxide copolymerization by ESI-MS and observed species corresponding to $[-OCH(CH_3)CH_2-DMAP^+ + H^+]$ and, with time, the growth of peaks corresponding to the addition of units of PO + CO_2, that is, *m*/*z* [181.1 + *n*(102)]. This was interpreted as evidence that the formation of carbamate zwitterions is not required for initiation of copolymerization. We have conducted detailed end group analysis of poly(cyclohexane)-carbonate[68,70] and poly(propylene)carbonate[69] formed by catalyst systems composed of **1**, **2** and **5** with DMAP. Use of complex **1** or **2** with 0.5 or 1 molar equiv. of DMAP in 500 molar equiv. of neat CHO shows only –Cl and –OH end groups with no initiation by

DMAP observed. When a mixture of co-catalysts is used, the end group properties become more complicated. The polymer obtained using a 1 : 1 : 1 mixture of complex **2**, DMAP and PPNCl (bis(triphenylphosphine)iminium chloride) under the same conditions results in MALDI-TOF mass spectra where –Cl, –DMAP and –OH end groups are observed (Fig. 6). It is likely that the presence of the additional equiv. of nucleophile competes with the binding of DMAP to the chromium centre, hence generating more "free DMAP", which can undergo an intermolecular nucleophilic attack at a chromium-bound epoxide. We also previously reported the MALDI-TOF MS characterization of a polymer obtained using the tetrahydrofurfuryl-functionalized complex **5** and DMAP.[70] Interestingly, the polymer obtained using **5**/DMAP catalyst system *also* showed complicated mass spectra with –Cl, DMAP and –OH end groups despite using equimolar amounts of **5** and DMAP and the apparently stronger bonding affinity of DMAP to the Cr centre in **5** (*i.e.* the appearance of $[\mathbf{L^5}Cr(DMAP)_2]^+$ ions for 1 : 1 rations of **5** : DMAP). However,

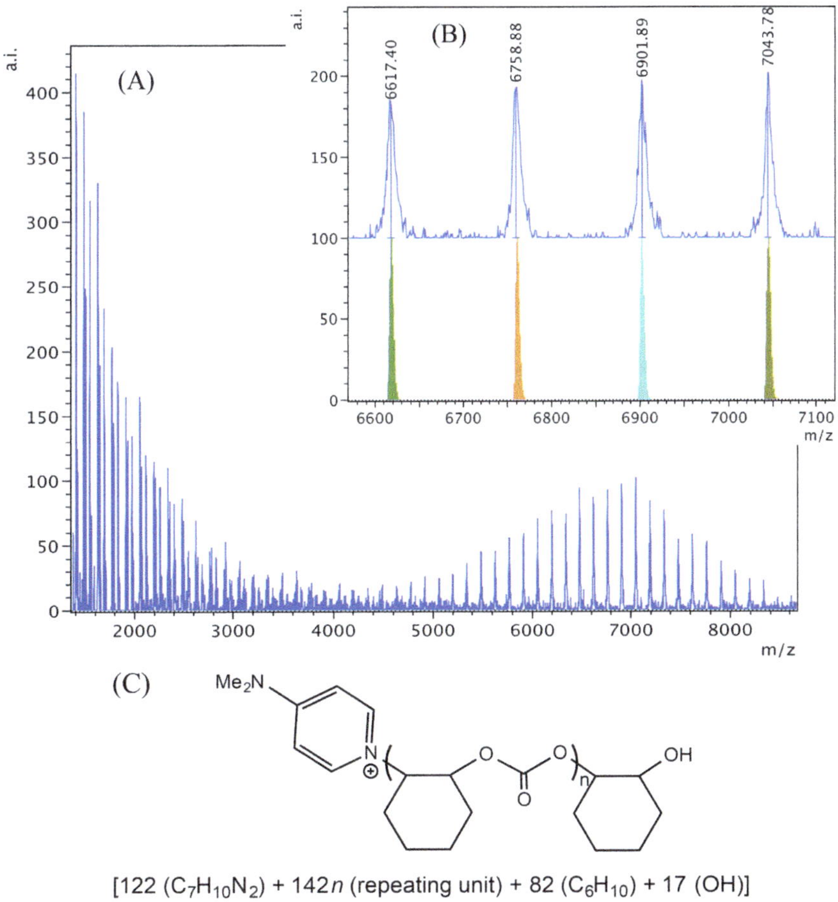

Fig. 6 (A) MALDI-TOF mass spectrum of polycarbonate produced by equimolar **2**/DMAP/PPNCl at 60 °C and 40 bar CO_2. (B) Higher mass region (*m*/*z* 6600–7100, *n* = 45–48) of the spectrum with calculated masses of fragments shown beneath the observed spectrum. (C) Proposed structure of the high mass range polymer.

whereas the base peak for compounds **1–4** correspond to $[LCrCl]^{+\bullet}$ (**F3**) the base peak for **5** is the unmetallated ligand, $[H_2L^5]^{+\bullet}$ (**F1**), regardless of the DMAP concentration. This suggests the L^5 ligand coordinates more weakly to the metal than L^1–L^4, and this results in the poorly controlled polymerization previously observed.

It has been observed that the salenCrX/DMAP system differs considerably from the salanCrX/DMAP system in that it exhibits a lengthy induction period of up to 2 h for the copolymerization of CO_2/PO. The salanCrX/DMAP system, on the other hand, shows propagation of the polymer after a period of 10 min.[66] Also, the salanCrX system displays higher TOFs, selectivity for polymer over cyclic carbonate formation, and gives higher polymer molecular weights with narrow dispersities. The differences in activity were attributed to the differing affinity for DMAP binding to the salan- and salen chromium compounds. Furthermore, as DMAP loading is increased, a decrease in polymer molecular weight is observed for both the salanCrX and salenCrX catalyst systems.[25,27,66,79] The relationship between induction period, reaction rate and DMAP-binding affinity for amine-bis(phenolate)chromium chloride complexes was investigated in order to assess whether a similar trend is observed to that of the salan and salen-based systems. The copolymerization of CHO with CO_2 by complexes **2** and **3** using 1 : 1 DMAP : Cr in 500 molar equiv. of neat epoxide was conducted at 60 °C and 40 bar CO_2. The reaction rates were monitored by infrared spectroscopy following the carbonate region (absorbance at 1750 cm^{-1}) over time (Fig. 7). The time profile for the first hour shows virtually no induction period for **3**/DMAP, and only a short one (approx. 10 min.) for **2**/DMAP. The polymerization rate of **3**/DMAP was also observed to be much faster than that of complex **2** (Fig. 8). These results are in agreement with the MALDI-TOF MS analysis above, namely the propensity of **2** to form stable, six-coordinate adducts and dimers results in a lower activity towards

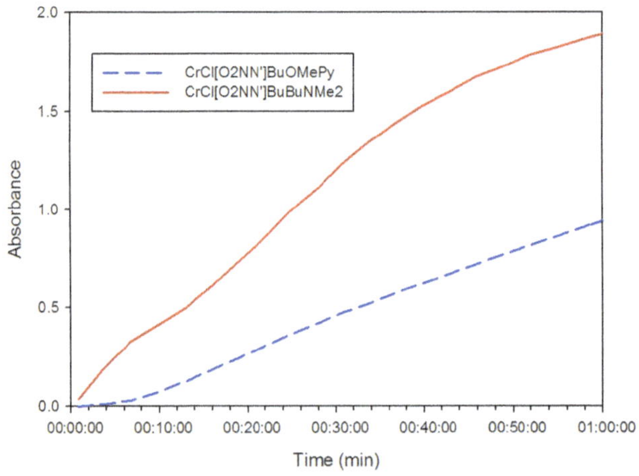

Fig. 7 Time profiles (for the first h) of the absorbance at 1750 cm^{-1} (corresponding to polycarbonate) using **2** (dashed blue line) and **3** (solid red line) as the catalyst. Reaction conditions: [Cr] : [CHO] : [DMAP] = 1 : 500 : 1, 24 h total experiment time, 60 °C, 40 bar CO_2.

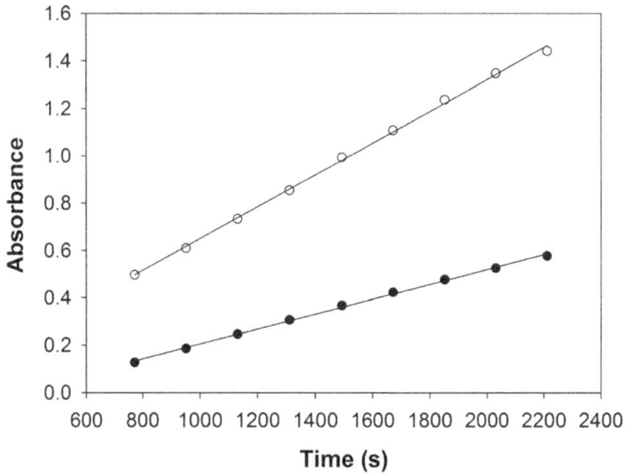

Fig. 8 Representative plots of absorbance *vs.* time for the linear portion of polycarbonate formation for the data presented in Fig. 7. Straight lines represent best fits of the data for compound **2** (●), $y = 3.15 \times 10^{-4}x$ to 1.10×10^{-1}, $R^2 = 9.99 \times 10^{-1}$; and compound **3** (○) $y = 6.72 \times 10^{-4}x$ to 2.09×10^{-2}, $R^2 = 9.99 \times 10^{-1}$.

epoxide/CO_2 copolymerization than complex **3**, which shows no evidence of dimer formation and a relatively low abundance of six-coordinate $[\mathbf{L^3}CrCl(DMAP)]^{+\bullet}$.

Comparisons of the possible mechanisms of action of salenCrX and salanCrX complexes have identified the nature of the binding of the tetradentate ligand as being influential in catalyst activity.[29,35,37,66,81] The crystallographically authenticated salenCrX and salenCrXL complexes (where X is a monodentate anionic ligand, *e.g.* Cl^- or N_3^-, and L is a neutral monodentate ligand, *e.g.* DMAP or O=PCy_3) show the salen ligand exhibits a planar orientation. SalanCrXL complexes, however, with the presence of sp^3-hybridized tertiary amino donors give distorted octahedral structures where the tetradentate salan ligand occupies *cis*-α or *cis*-β orientations, where the two monodentate ligands are found in *cis*-positions rather than the *trans* coordination mode often found in salenCrXL complexes. The amine-bis(phenolate) complexes **1–5** exhibit bonding orientations related to the salan systems. Based on the above comparisons and on the observed end-groups in the polycarbonates obtained under equimolar ratios of DMAP/PPNCl to the Cr compounds, and comparisons of the influence of the ligand on the ability for stable six-coordinate ions to be observed by MALDI-TOF MS, a mechanism can be proposed for CO_2/epoxide copolymerization by **1–5** with DMAP as the co-catalyst. The initiation steps available for the complex **3**/DMAP catalyst system are shown in Scheme 1. The relative abundance of **F3** and **F5** compared to **F6** suggests that the first equilibrium lies to the right, leading to a very short induction period. Also, whereas dimer formation is observed in the MALDI-TOF mass spectrum of **1**, **2**, **4** and **5**, no evidence of dimeric **3** was observed. Dimer formation, therefore, may be a reason for the presence of the induction period observed for **2** but not for complex **3**. As previously proposed by Lu and co-workers,[66] progress along either path A or B may be inhibited by excess DMAP, leading to the formation of $[\mathbf{L}Cr(DMAP)_2]^+$ in the case of path A, or increase the reversibility of the first equilibrium step in path B by hindering

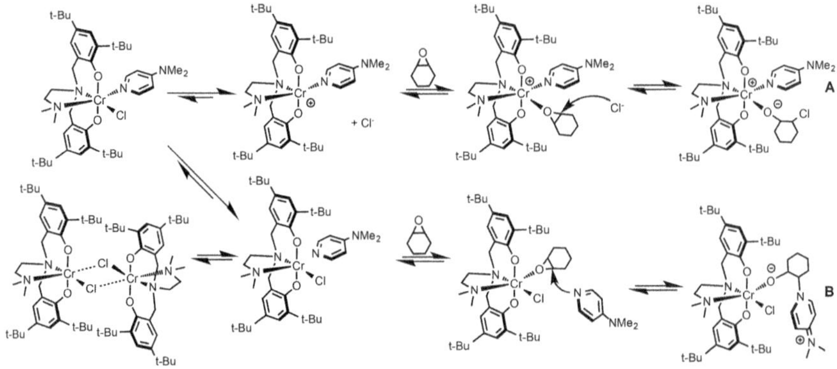

Scheme 1 Two initiation paths *via* an intermolecular nucleophilic attack available for **3**/DMAP. Path A leads to initiation by Cl; path B leads to initiation by DMAP.

DMAP dissociation. Competitive binding of epoxide allows activation of the first monomer fragment, which undergoes nucleophilic attack by either Cl^- or DMAP. For conditions where DMAP end groups are absent (as for 1 : 0.5 or 1 : 1 combinations of complex **1** or **2** with DMAP as discussed above), path A is preferred resulting from the high affinity of DMAP bonding to the metal centre. For complex **3**, which is more electron rich at the Cr centre, DMAP dissociation at lower concentrations of that base is more prevalent, hence "free DMAP" is available to induce ring opening of the epoxide, generating –DMAP terminated polycarbonate (path B). DFT studies have been reported on the mechanism of epoxide/CO_2 copolymerization by metal salen complexes.[82] Rieger and co-workers proposed that the growing polymer chains undergo dissociation from the metal centre during polymerization, with chain-transfer resulting from high co-catalyst loading leading to low molecular weight polymers.[73] Baik and Nguyen studied the role of DMAP in salenCr-catalyzed PO/CO_2 copolymerization.[72] Their findings support the role of the salenCrCl complex as a Lewis acid that activates the epoxide ring, promoting its opening by an external nucleophile, such as DMAP. No evidence for a bimetallic pathway or simultaneous activation of both the epoxide and CO_2 was obtained. Unlike Rieger's proposal, attempts at creating an unsaturated Cr(III) centre through chloride loss or alkoxide dissociation resulted in significant energy penalties. Baik and Nguyen concluded that such steps were not possible in this case, therefore a ring-opening step similar to that shown in Scheme 1B is favoured. If the same case holds for compounds **1–5**, the presence of an epoxide ring opening by chloride (leading to the observed –Cl end groups using MALDI-TOF MS) implies this step must happen by an intramolecular mechanism akin to a migratory insertion between cisoid ligands on the metal centre. Computational studies to investigate this possibility are currently underway.

Conclusions

The MALDI-TOF mass spectra of amine-bis(phenolato)chromium(III) chloride complexes in the presence of DMAP in various amounts demonstrated that the ability of DMAP to bind to the chromium centre was influenced by the nature of the pendant neutral donor group on the tetradentate ligand. Where the donor was an ether (methoxyethyl or tetrahydrofurfuryl) or pyridyl group (complexes **1**, **2**, **4** and **5**), binding of two DMAP molecules was observed. These complexes also demonstrated a propensity to form dimeric species, likely *via* chloride-bridges that we have previously structurally authenticated. For dimethylaminoethyl-functionalized ligands, no evidence for the binding of two DMAP molecules was observed, even at four equiv. of DMAP per Cr. Furthermore, no dimeric species were observed in their mass spectra. This difference in coordination chemistry was found to influence the catalytic activity of these compounds towards the copolymerization of CO_2 with cyclohexene oxide. In the presence of 1 equiv. of DMAP, the activity of the diamino-bis(phenolato)chromium chloride complex **3** was found to be faster than that of complex **2**. End group analysis of the polymers by MALDI-TOF MS suggests that the DMAP-containing end groups occur when the concentration of unbound DMAP is high, suggesting an intermolecular nucleophilic attack. Chloride-containing end-groups were most commonly observed at molar equiv. levels of DMAP loading, suggesting that DMAP is not competitive with chloride in terms of epoxide ring-opening, and that binding to the metal centre is more favourable. Interestingly, increasing the chloride concentration leads to DMAP-containing polycarbonate, which may be explained by the formation of dichloride chromium species and release of free DMAP into the reaction.

Acknowledgements

Financial support was provided by the Canada Foundation for Innovation (CFI), Newfoundland and Labrador Research & Development Corporation, and the Natural Sciences and Engineering Research Council (NSERC) of Canada in the form of grants to C. M. K. and C. S. B., and Memorial University of Newfoundland in the form of graduate studies fellowships to K. D.-P. and K. N. NSERC is also acknowledged for an Undergraduate Student Research Award to A. M. W.

Notes and references

1 D. J. Darensbourg, *Chem. Rev.*, 2007, **107**, 2388–2410.
2 G. W. Coates and D. R. Moore, *Angew. Chem., Int. Ed.*, 2004, **43**, 6618–6639.
3 M. R. Kember, A. Buchard and C. K. Williams, *Chem. Commun.*, 2011, **47**, 141–163.
4 D. J. Darensbourg, *Inorg. Chem.*, 2010, **49**, 10765–10780.
5 S. Klaus, M. W. Lehenmeier, C. E. Anderson and B. Rieger, *Coord. Chem. Rev.*, 2011, **255**, 1460–1479.
6 X.-B. Lu and D. J. Darensbourg, *Chem. Soc. Rev.*, 2012, **41**, 1462–1484.
7 M. Aresta, *Carbon Dioxide as Chemical Feedstock*, Wiley-VCH, Weinheim, 2010.
8 M. R. Kember and C. K. Williams, *J. Am. Chem. Soc.*, 2012, **134**, 15676–15679.
9 D. J. Darensbourg and D. R. Billodeaux, *Inorg. Chem.*, 2005, **44**, 1433–1442.
10 T. Aida and S. Inoue, *Acc. Chem. Res.*, 1996, **29**, 39–48.
11 T. Aida, M. Ishikawa and S. Inoue, *Macromolecules*, 1986, **19**, 8–13.
12 T. Aida and S. Inoue, *J. Am. Chem. Soc.*, 1983, **105**, 1304–1309.
13 F. Kojima, T. Aida and S. Inoue, *J. Am. Chem. Soc.*, 1986, **108**, 391–395.
14 M. Cheng, E. B. Lobkovsky and G. W. Coates, *J. Am. Chem. Soc.*, 1998, **120**, 11018–11019.

15 M. Super, E. Berluche, C. Costello and E. Beckman, *Macromolecules*, 1997, **30**, 368–372.
16 T. Sarbu and E. J. Beckman, *Macromolecules*, 1999, **32**, 6904–6912.
17 M. Cheng, D. R. Moore, J. J. Reczek, B. M. Chamberlain, E. B. Lobkovsky and G. W. Coates, *J. Am. Chem. Soc.*, 2001, **123**, 8738–8749.
18 D. R. Moore, M. Cheng, E. B. Lobkovsky and G. W. Coates, *Angew. Chem., Int. Ed.*, 2002, **41**, 2599–2602.
19 B. Y. Lee, H. Y. Kwon, S. Y. Lee, S. J. Na, S. I. Han, H. S. Yun, H. Lee and Y. W. Park, *J. Am. Chem. Soc.*, 2005, **127**, 3031–3037.
20 M. Kroger, C. Folli, O. Walter and M. Doring, *Adv. Synth. Catal.*, 2005, **347**, 1325–1328.
21 M. R. Kember, P. D. Knight, P. T. R. Reung and C. K. Williams, *Angew. Chem., Int. Ed.*, 2009, **48**, 931–933.
22 Y. L. Xiao, Z. Wang and K. L. Ding, *Chem.–Eur. J.*, 2005, **11**, 3668–3678.
23 R. L. Paddock and S. T. Nguyen, *J. Am. Chem. Soc.*, 2001, **123**, 11498–11499.
24 S. Mang, A. I. Cooper, M. E. Colclough, N. Chauhan and A. B. Holmes, *Macromolecules*, 2000, **33**, 303–308.
25 D. J. Darensbourg and J. C. Yarbrough, *J. Am. Chem. Soc.*, 2002, **124**, 6335–6342.
26 D. J. Darensbourg, J. C. Yarbrough, C. Ortiz and C. C. Fang, *J. Am. Chem. Soc.*, 2003, **125**, 7586–7591.
27 R. Eberhardt, M. Allmendinger and B. Rieger, *Macromol. Rapid Commun.*, 2003, **24**, 194–196.
28 D. J. Darensbourg, R. M. Mackiewicz, J. L. Rodgers, C. C. Fang, D. R. Billodeaux and J. H. Reibenspies, *Inorg. Chem.*, 2004, **43**, 6024–6034.
29 D. J. Darensbourg and R. M. Mackiewicz, *J. Am. Chem. Soc.*, 2005, **127**, 14026–14038.
30 D. J. Darensbourg, R. M. Mackiewicz and D. R. Billodeaux, *Organometallics*, 2005, **24**, 144–148.
31 B. Li, R. Zhang and X. B. Lu, *Macromolecules*, 2007, **40**, 2303–2307.
32 X. Q. Xu, C. M. Wang, H. R. Li, Y. Wang, W. L. Sun and Z. Q. Shen, *Polymer*, 2007, **48**, 3921–3924.
33 B. Li, G. P. Wu, W. M. Ren, Y. M. Wang, D. Y. Rao and X. B. Lu, *J. Polym. Sci., Part A: Polym. Chem.*, 2008, **46**, 6102–6113.
34 D. J. Darensbourg and S. B. Fitch, *Inorg. Chem.*, 2009, **48**, 8668–8677.
35 D. J. Darensbourg, M. Ulusoy, O. Karroonnirum, R. R. Poland, J. H. Reibenspies and B. Çetinkaya, *Macromolecules*, 2009, **42**, 6992–6998.
36 K. Nakano, M. Nakamura and K. Nozaki, *Macromolecules*, 2009, **42**, 6972–6980.
37 D. J. Darensbourg, R. R. Poland and A. L. Strickland, *J. Polym. Sci., Part A: Polym. Chem.*, 2012, **50**, 127–133.
38 L. P. Guo, C. M. Wang, W. J. Zhao, H. R. Li, W. L. Sun and Z. Q. Shen, *Dalton Trans.*, 2009, 5406–5410.
39 C. T. Cohen, T. Chu and G. W. Coates, *J. Am. Chem. Soc.*, 2005, **127**, 10869–10878.
40 C. T. Cohen and G. W. Coates, *J. Polym. Sci., Part A: Polym. Chem.*, 2006, **44**, 5182–5191.
41 M. R. Kember, F. Jutz, A. Buchard, A. J. P. White and C. K. Williams, *Chem. Sci.*, 2012, **3**, 1245–1255.
42 M. R. Kember, A. J. P. White and C. K. Williams, *Macromolecules*, 2010, **43**, 2291–2298.

43 B. Liu, Y. Gao, X. Zhao, W. Yan and X. Wang, *J. Polym. Sci., Part A: Polym. Chem.*, 2010, **48**, 359–365.
44 B. Liu, X. Zhao, H. Guo, Y. Gao, M. Yang and X. Wang, *Polymer*, 2009, **50**, 5071–5075.
45 X. B. Lu and Y. Wang, *Angew. Chem., Int. Ed.*, 2004, **43**, 3574–3577.
46 K. Nakano, S. Hashimoto, M. Nakamura, T. Kamada and K. Nozaki, *Angew. Chem., Int. Ed.*, 2011, **50**, 4868–4871.
47 K. Nakano, S. Hashimoto and K. Nozaki, *Chem. Sci.*, 2010, **1**, 369–373.
48 K. Nakano, T. Kamada and K. Nozaki, *Angew. Chem., Int. Ed.*, 2006, **45**, 7274–7277.
49 Y. S. Niu, H. C. Li, X. S. Chen, W. X. Zhang, X. Zhuang and X. B. Jing, *Macromol. Chem. Phys.*, 2009, **210**, 1224–1229.
50 Y. S. Niu, W. X. Zhang, X. Pang, X. S. Chen, X. L. Zhuang and X. B. Jing, *J. Polym. Sci., Part A: Polym. Chem.*, 2007, **45**, 5050–5056.
51 E. K. Noh, S. J. Na, S. Sujith, S.-W. Kim and B. Y. Lee, *J. Am. Chem. Soc.*, 2007, **129**, 8082–8083.
52 R. L. Paddock and S. T. Nguyen, *Macromolecules*, 2005, **38**, 6251–6253.
53 Z. Q. Qin, C. M. Thomas, S. Lee and G. W. Coates, *J. Am. Chem. Soc.*, 2003, **42**, 5484–5487.
54 W.-M. Ren, Z.-W. Liu, Y.-Q. Wen, R. Zhang and X.-B. Lu, *J. Am. Chem. Soc.*, 2009, **131**, 11509–11518.
55 W. M. Ren, X. Zhang, Y. Liu, J. F. Li, H. Wang and X. B. Lu, *Macromolecules*, 2010, **43**, 1396–1402.
56 J. E. Seong, S. J. Na, A. Cyriac, B.-W. Kim and B. Y. Lee, *Macromolecules*, 2010, **43**, 903–908.
57 L. Shi, X. B. Lu, R. Zhang, X. J. Peng, C. Q. Zhang, J. F. Li and X. M. Peng, *Macromolecules*, 2006, **39**, 5679–5685.
58 S. Sujith, J. K. Min, J. E. Seong, S. J. Na and B. Y. Lee, *Angew. Chem., Int. Ed.*, 2008, **47**, 7306–7309.
59 A. Buchard, M. R. Kember, K. G. Sandeman and C. K. Williams, *Chem. Commun.*, 2011, **47**, 212–214.
60 W. J. Kruper and D. V. Dellar, *J. Org. Chem.*, 1995, **60**, 725–727.
61 L. M. Stamp, S. A. Mang, A. B. Holmes, K. A. Knights, Y. R. de Miguel and I. F. McConvey, *Chem. Commun.*, 2001, 2502–2503.
62 D. J. Darensbourg and M. W. Holtcamp, *Macromolecules*, 1995, **28**, 7577–7579.
63 D. J. Darensbourg, M. W. Holtcamp, G. E. Struck, M. S. Zimmer, S. A. Niezgoda, P. Rainey, J. B. Robertson, J. D. Draper and J. H. Reibenspies, *J. Am. Chem. Soc.*, 1999, **121**, 107–116.
64 D. J. Darensbourg, J. R. Wildeson, J. C. Yarbrough and J. H. Reibenspies, *J. Am. Chem. Soc.*, 2000, **122**, 12487–12496.
65 K. Nakano, K. Kobayashi and K. Nozaki, *J. Am. Chem. Soc.*, 2011, **133**, 10720–10723.
66 D. Y. Rao, B. Li, R. Zhang, H. Wang and X. B. Lu, *Inorg. Chem.*, 2009, **48**, 2830–2836.
67 X. B. Lu, L. Shi, Y. M. Wang, R. Zhang, Y. J. Zhang, X. J. Peng, Z. C. Zhang and B. Li, *J. Am. Chem. Soc.*, 2006, **128**, 1664–1674.
68 R. K. Dean, L. N. Dawe and C. M. Kozak, *Inorg. Chem.*, 2012, **51**, 9095–9103.
69 R. K. Dean, K. Devaine-Pressing, L. N. Dawe and C. M. Kozak, *Dalton Trans.*, 2013, **42**, 9233–9244.

70 H. Chen, L. N. Dawe and C. M. Kozak, *Catal. Sci. Technol.*, 2014, **4**, 1547–1555.
71 L. N. Saunders, N. Ikpo, C. F. Petten, U. K. Das, L. N. Dawe, C. M. Kozak and F. M. Kerton, *Catal. Commun.*, 2012, **18**, 165–167.
72 D. Adhikari, S. T. Nguyen and M.-H. Baik, *Chem. Commun.*, 2014, **50**, 2676–2678.
73 G. A. Luinstra, G. R. Haas, F. Molnar, V. Bernhart, R. Eberhardt and B. Rieger, *Chem.–Eur. J.*, 2005, **11**, 6298–6314.
74 M. H. Chisholm and Z. P. Zhou, *J. Am. Chem. Soc.*, 2004, **126**, 11030–11039.
75 Y. L. Xiao, Z. Wang and K. L. Ding, *Macromolecules*, 2006, **39**, 128–137.
76 W. J. van Meerendonk, R. Duchateau, C. E. Koning and G. J. M. Gruter, *Macromolecules*, 2005, **38**, 7306–7313.
77 R. Duchateau, W. J. van Meerendonk, L. Yajjou, B. B. P. Staal, C. E. Koning and G.-J. M. Gruter, *Macromolecules*, 2006, **39**, 7900–7908.
78 D. J. Darensbourg and A. D. Yeung, *Polym. Chem.*, 2015, **6**, 1103–1117.
79 D. J. Darensbourg and A. L. Phelps, *Inorg. Chem.*, 2005, **44**, 4622–4629.
80 R. K. Dean, S. L. Granville, L. N. Dawe, A. Decken, K. M. Hattenhauer and C. M. Kozak, *Dalton Trans.*, 2010, **39**, 548–559.
81 D. J. Darensbourg, R. M. Mackiewicz, A. L. Phelps and D. R. Billodeaux, *Acc. Chem. Res.*, 2004, **37**, 836–844.
82 D. J. Darensbourg and A. D. Yeung, *Polym. Chem.*, 2014, **5**, 3949–3962.

Faraday Discussions

PAPER

Kinetic and economic analysis of reactive capture of dilute carbon dioxide with Grignard reagents

G. R. M. Dowson,* I. Dimitriou, R. E. Owen, D. G. Reed, R. W. K. Allen and P. Styring*

Received 1st May 2015, Accepted 18th June 2015
DOI: 10.1039/c5fd00049a

Carbon Dioxide Utilisation (CDU) processes face significant challenges, especially in the energetic cost of carbon capture from flue gas and the uphill energy gradient for CO_2 reduction. Both of these stumbling blocks can be addressed by using alkaline earth metal compounds, such as Grignard reagents, as sacrificial capture agents. We have investigated the performance of these reagents in their ability to both capture and activate CO_2 directly from dried flue gas (essentially avoiding the costly capture process entirely) at room temperature and ambient pressures with high yield and selectivity. Naturally, to make the process sustainable, these reagents must then be recycled and regenerated. This would potentially be carried out using existing industrial processes and renewable electricity. This offers the possibility of creating a closed loop system whereby alcohols and certain hydrocarbons may be carboxylated with CO_2 and renewable electricity to create higher-value products containing captured carbon. A preliminary Techno-Economic Analysis (TEA) of an example looped process has been carried out to identify the electrical and raw material supply demands and hence determine production costs. These have compared broadly favourably with existing market values.

Introduction

With the looming threats posed by climate change, a key plank in carbon dioxide reduction strategies is to develop new pathways to mitigate and avoid emissions by the production of low carbon, carbon neutral and even carbon negative alternatives to common and bulk chemicals.[1] One route by which this can be accomplished is to incorporate carbon dioxide, that would otherwise be emitted, into the desired product, thus delaying or avoiding its release into the atmosphere. In this way, when the product is consumed or degraded, there is a reduced

UK Centre for Carbon Dioxide Utilization, Department of Chemical and Biological Engineering, The University of Sheffield, Sir Robert Hadfield Building, Sheffield, S1 3JD, UK. E-mail: p.styring@sheffield.ac.uk; g.dowson@sheffield.ac.uk

change in atmospheric carbon dioxide levels associated with that product, ideally rendering the product itself carbon neutral. Naturally, if the product has a prolonged life or is recycled effectively this method represents an overall sequestering of the original carbon dioxide from the atmosphere to the solid state in a form of long term storage which would complement geological storage aspirations.[2]

This strategy of Carbon Dioxide Utilisation (CDU) is known by a variety of names and acronyms, but in all cases involves the use of carbon sources, including mixed and dilute ones such as flue gas and biogas, to generate products.[3] The resulting products therefore have a reduced carbon footprint leading to an overall net emissions reduction. While ambitious, if CDU processes can be implemented with atmospheric carbon dioxide (also known as "direct air capture", DAC) this would represent both a key technique to form a sustainable carbon product cycle, analogous to the natural carbon cycles, and potentially allow the creation of useful products irrespective of geographical location.

This generation of products using carbon dioxide is what sets CDU apart from Carbon Capture and Sequestration (CCS) where carbon dioxide is instead treated as a waste to be dumped in geological storage sites, with only the environmental benefit of net emissions reduction as a motivation. The major benefit of CDU, in contrast, is the addition of value by the carbon dioxide to the product in addition to the environmental benefit associated with shrinking, or eliminating, its carbon footprint.

The extent of the environmental benefit will of course be product-dependent, however recent publications have shown that even modest targets for CDU allow amounts of CO_2 to be mitigated that match or even exceed current CCS targets.[4] Indeed, while it predates this argument, global production of urea from CO_2 and ammonia (which is essentially a CDU process) utilises *ca.* 140 million tonnes of carbon dioxide annually.[1] Admittedly, when the production of ammonia and the breakdown of urea as a fertilizer are taken into consideration, the net emissions of the overall process are greater than unity. However, urea production acts as an indicator of the scale that CDU processes could hope to achieve, dwarfing even optimistic CCS targets of 102 million tonnes total stored by 2020.[5]

The fact that CDU treats carbon dioxide as a resource rather than a waste also allows traditional economic driving forces to foster innovation. This has already resulted in several examples of economically successful CDU processes.[6] These include the Bayer Dream process,[7] where CO_2 is incorporated into polyurethane plastics, Sunfire's synthetic diesel[8] and Carbon Recycling International, where cheap Icelandic geothermal energy is used to turn CO_2 into methanol.[9]

Importance of techno-economic analysis in CDU processes

Since CDU primarily relies upon the addition of value through using the CO_2, it would therefore seem essential to combine any serious attempt at the development of a CDU process with at least a preliminary or generalised Techno-Economic Analysis/Assessment (TEA) to demonstrate what value, if any, can be added by the process in question.[10]

This is because the likelihood for implementation of existing and new technologies depends on the interest and motivation of private and public investors. Many researchers who develop new technologies have limited knowledge of their economic potentials and pitfalls. Therefore, understanding the investors' needs and the related economic questions enables technology developers to focus their work on the promising options and avoid expending effort in technologies that are fundamentally not economically viable. TEA studies of new technologies and processes, such as CDU systems, allow us to better understand the relationship between process performance (*e.g.* conversion) and costs (*e.g.* production) to help identify the most promising process designs which can provide the highest financial returns for the potential investors. Therefore, in the case of CDU, TEA studies are an invaluable tool which can support policy makers and businesses in their decision-making by establishing whether the production of chemicals and fuels from waste CO_2 is economically feasible; and identifying the modifications and conditions required to improve the economic competitiveness of CDU technologies.

What has already been revealed by these analyses and assessments is that for CDU and CCS strategies that rely on separation of carbon dioxide from dilute sources such as flue gas, this "capture" step is a major energy sink and therefore carries significant costs.[11]

Cost of capture of CO_2

Separating carbon dioxide from other gases on a large scale is one of the biggest challenges for CDU and also CCS. This might be surprising as thermodynamically the cost of separation is not large (Table 1). Even the most dilute source of CO_2, such as that which is found in the atmosphere, requires only 521.7 MJ per tonne for separation (approximated from binary $CO_2 : N_2$ mixtures).

Despite this, calculated capture energy costs using industrial benchmark capture agents such as monoethanolamine (MEA) and other amines frequently exceed 3000–4000 MJ per tonne CO_2 even when starting from relatively concentrated CO_2 sources. This translates into both large capital and large operational costs for carbon capture from power plants, with the cost of the capture plant often exceeding the base plant costs (see Fig. 1).[12]

The cause of this discrepancy in energy cost, in the case of MEA and other amines, is that the capture process is driven by a chemisorption reaction loop involving the creation and then decomposition of carbamates.[13] The high energy cost of this loop is indicative of the relatively low reactivity of carbon dioxide at standard temperatures and pressures. However, while this reactive capture chemisorption approach has the major advantage of ensuring high carbon

Table 1 Minimum thermodynamic energy cost of separating 1 tonne of CO_2 from N_2/CO_2 mixtures at STP

CO_2 concentration in N_2 (vol%)	Minimum entropic demixing cost per tonne CO_2 (MJ)
Atmospheric (400 ppm)	521.7
1%	340.2
5%	248.0
10%	206.9
15%	182.0
20%	163.6

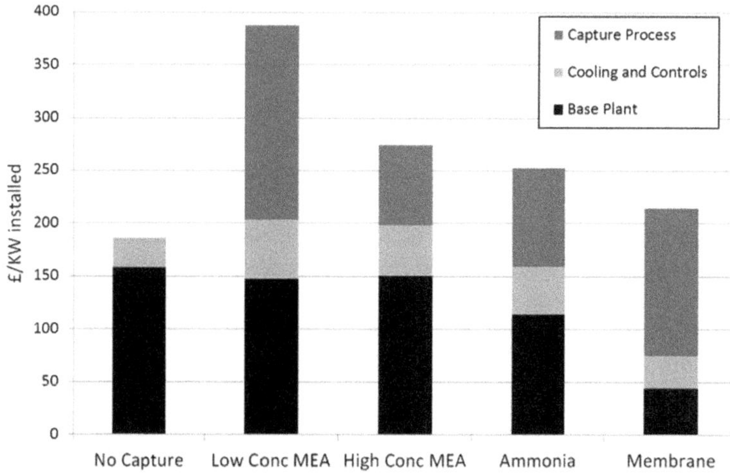

Fig. 1 Breakdown of annual operation and maintenance (O&M) costs per unit electricity by capture processes.

dioxide selectivity over the other main component gases, nitrogen and methane (in flue gas and biogas respectively), it actually represents a waste of energy from CDU perspectives as the product carbamate is itself a CDU product. As a result, the decomposition of the post-capture carbamate to create purified CO_2 only to react that CO_2 again to make another product can be seen as an energetic wild goose chase where the carbon dioxide is essentially utilised twice!

Additionally, the chemisorption approach will often allow some of the trace gases in these feed gas streams to react with the sorbent. At best, this is disadvantageous due to the displacement of CO_2 in the capture step, but at worst the trace gas poisons the sorbent permanently. This is especially problematic when dealing with gas supplies contaminated with sulfur compounds.

While for MEA the carbamate product is a mere intermediate for CO_2 purification, if an analogous process where a valuable carbon–carbon bond is formed instead were developed, it would represent a potentially attractive pathway for direct conversion of flue gas into products without a distinct gas separation step. Naturally the value of this process would be decided by the economic value of the product but it would certainly avoid much of the wasted energetic costs of the carbamate loop.

Any such process would therefore require a stoichiometric reactant of some sort with which the CO_2 must react, however this is already the case when using any capture process for the generation of CDU products. Considering the reactants that could be used, one prominent example is the Grignard reaction, involving organomagnesium halide reagents.

A fresh look at Grignard chemistry in the context of carbon dioxide reactive capture

The classic schoolroom carbon–carbon bond formation process using carbon dioxide is the reaction of a Grignard reagent with a pellet of dry ice. This standard reaction proceeds by the insertion of the carbon dioxide molecule between the carbon–magnesium bond of an organomagnesium halide, formed by the reaction of metallic magnesium with an organohalide (Scheme 1).[14]

This reaction is exothermic and readily carried out at atmospheric pressure and ambient temperatures. Aside from requiring that all components are stringently dried, due to the rapid and exothermic reaction with water, the reaction also proceeds to typically high yields, dependent on reaction conditions and solvents, with a wide range of possible R-groups. The resulting carboxylic acid product is then readily isolated from the magnesium salt by-product in the majority of cases by solvent extraction, esterification and/or distillation.[15]

The strongly exothermic nature of the reaction is driven by the oxidation of the starting elemental magnesium to the 2+ oxidation state. Coupled with the exothermic nature of the formation of the initial organohalide (*e.g.* −42.2 kJ mol^{-1} for methyl bromide from methanol), the overall reaction of the parent alcohol with CO_2 becomes facile.

While methanol would be the parent alcohol in the example in Fig. 2, the Grignard reaction is well-understood and offers a wide range of potential substrates, allowing a variety of industrially interesting CDU products to potentially be formed, a selection of which is shown in Fig. 3.

The substrates for the formation of the above products could open new pathways of commercial or environmental interest. In particular, the use of CO_2-derived carboxylic acids in polymers has already been targeted for large scale CDU applications.[16] For example, adipic acid, used for the synthesis of Nylon, is currently manufactured from petrochemical phenol, cyclohexene or butadiene, but could instead be created from biological 1,4-butanediol and CO_2 *via* halogenation to 1,4-dihalobutane. Similarly, terephthalic acid which is a key component of polyethyleneterephthalate (PET) plastic, the third most common plastic in use today, is currently derived from *p*-xylene, of which there has been a long-standing shortage and safety concerns, highlighted by the recent explosions in Zhangzhou, China.[17] By utilising CO_2, the terephthalic acid polymer precursor could be derived directly from the relatively plentiful benzene (*via* *p*-dihalobenzene, which is easily synthesised).

Collectively, the products shown in Fig. 3 are manufactured on a scale of approximately 71 million tonnes per annum worldwide and could potentially utilise over 40 million tonnes of CO_2, much of which would be sequestered into various polymers and other long life products.

However, as previously mentioned, Grignard reactions of carbon dioxide would typically be carried out using dry ice pellets rather than gas phase or dilute CO_2 mixtures. A series of tests were therefore performed to determine whether direct utilisation of gas phase CO_2 and even simulated flue gas compositions and atmospheric concentrations were in any way suitable for Grignard chemistry. The reaction conditions and reaction rates for different gas compositions would allow a more complete techno-economic analysis of the process to be completed.

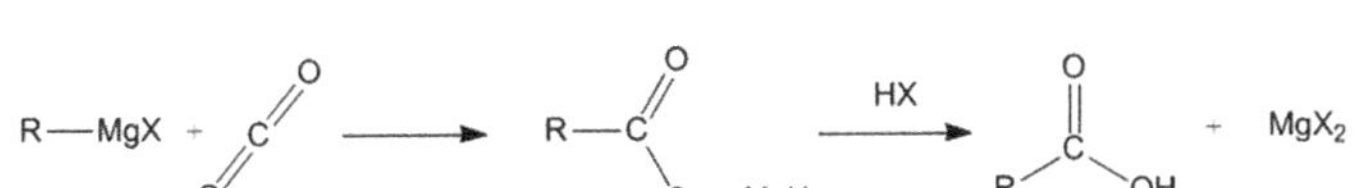

Scheme 1 Generalised reaction of a Grignard reagent with CO_2.

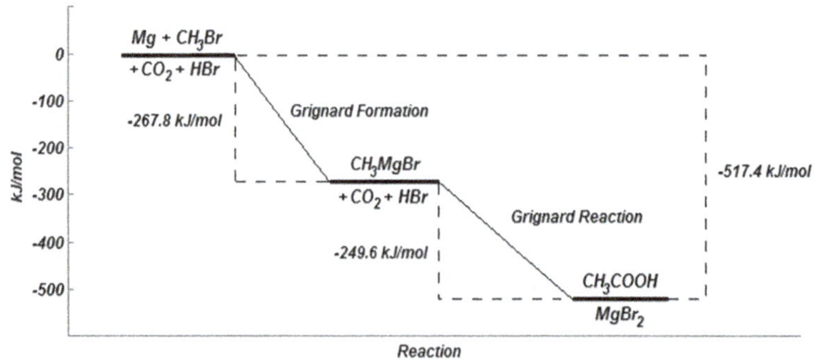

Fig. 2 Reaction enthalpy profile of each step of a Grignard reaction to form acetic acid from CO_2 and methyl bromide.[16]

Grignard reactions can typically be tracked by titration of the remaining active organomagnesium reagent using 1,10-phenanthroline. The phenanthroline forms a coloured, charge transfer complex in a 1 : 1 ratio with the reagent that may then be titrated by using a dry alcohol such as 2-butanol.[18] However, it was found that in the case of the reaction of methylmagnesium chloride and CO_2, no colour change was observed when injected into a solution of 1,10-phenanthroline in dry THF. A possible reason for this is that the intermediate product, a magnesium acetate bromide species, acts as a quenching agent for the charge transfer complex, thereby preventing the colour change needed for titration.

Therefore, in order to determine the rate of reaction of CO_2 with methylmagnesium chloride, a high-accuracy pressure transducer (Omega PX409USB) was used to monitor CO_2 consumption within the reaction apparatus (Fig. 4). In

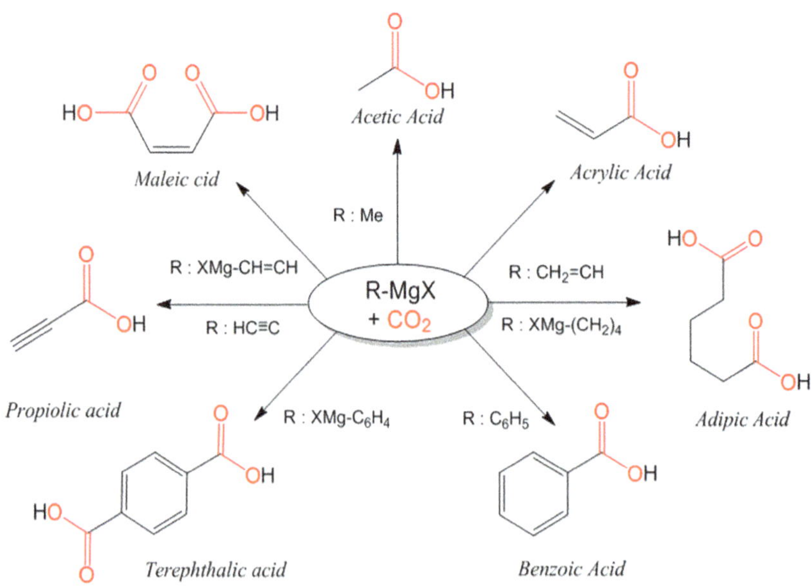

Fig. 3 A series of potential CO_2 Grignard products of potentially industrial significance.

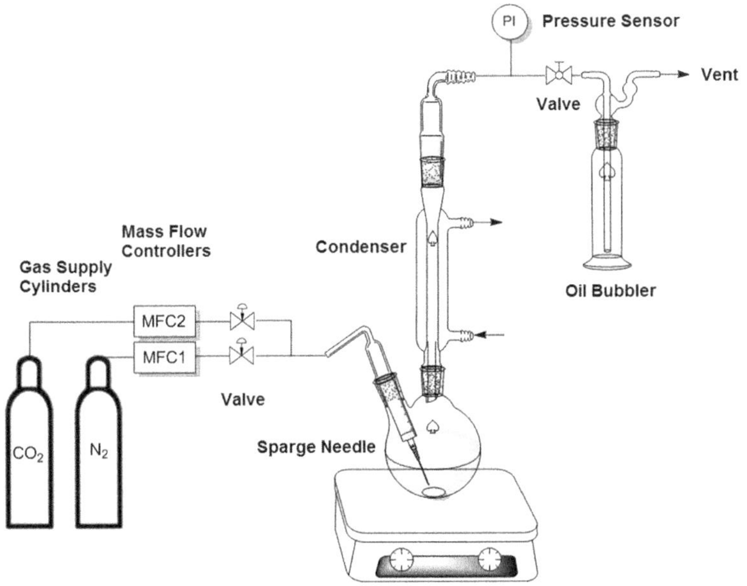

Fig. 4 Schematic of the gas sparging reaction system.

order to maintain a generally stable CO_2 partial pressure, the pressure drop was monitored intermittently between periods of CO_2 sparging.

As shown in Fig. 5, these reactions were found to proceed cleanly when using gas-phase carbon dioxide bubbled through the Grignard reagent solution. Furthermore, this intermittent sparging process demonstrated clear trends in pressure drop rate, despite the exothermic nature of the reaction causing thermal expansion and vapour generation from the ethereal solvent required for Grignard chemistry. This high degree of reactivity with CO_2 prompted tests with increasingly dilute CO_2 mixtures in nitrogen, comparing the rate of change of pressure during the CO_2 interruptions.

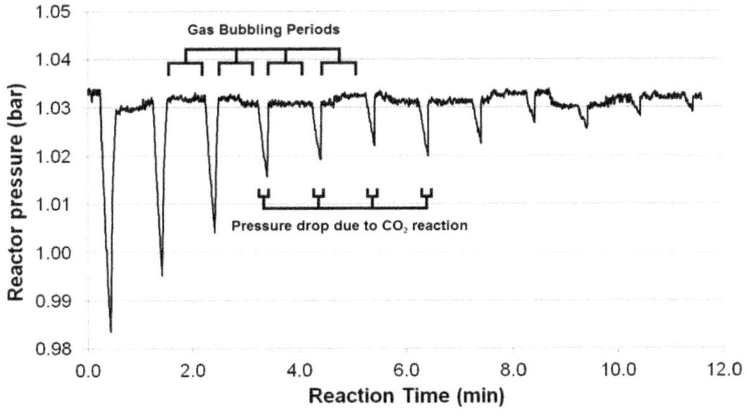

Fig. 5 Typical reaction profile showing rapid pressure drop within the sealed reactor when gas sparging of 50% CO_2/N_2 is paused for 10 seconds every 1 minute.

The calculated rate of CO_2 consumption from the rate of pressure drop seen in Fig. 5 is shown in Fig. 6(a). Fig. 6(b) plots individual natural log plots at varying gas compositions. These appear initially to be pseudo-first order in the presence of a constant supply of CO_2. However, while they are superficially similar to pseudo-first order kinetics, where a large and essentially constant surplus of one reagent is present, the change in rate constant when varying CO_2 concentration demonstrates that the reaction system is more complex. This is because although CO_2 concentration is static (as found in pseudo-first order reactions), it is not in excess, limiting each rate "constant" to the maximum rate that can be achieved at the given CO_2 concentration under these conditions. As a result, there is a decrease in reaction rate "constant" at lower concentrations. Overall combination

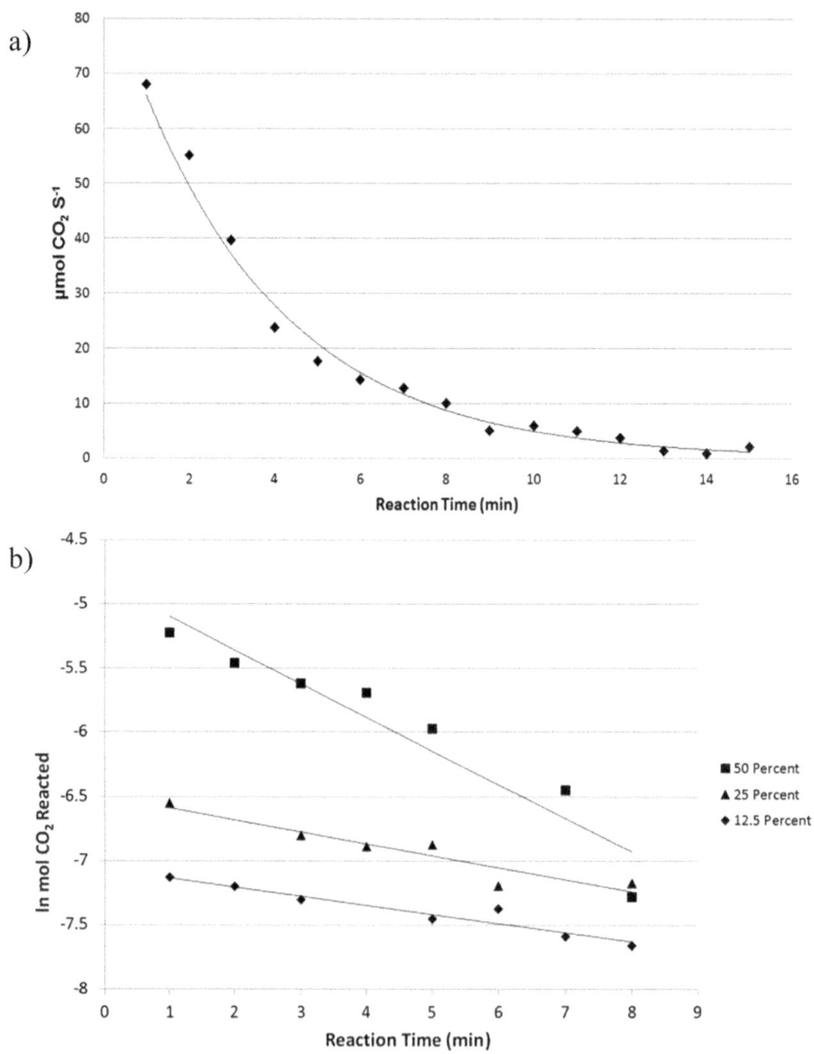

Fig. 6 (a) CO_2 consumption in the Grignard reaction of 1 M MeMgCl with 100% CO_2 by pressure drop detection. (b) Comparison of reaction rate at decreasing CO_2 compositions by dilution in N_2 with total gas flow rate of 100 mL min^{-1}.

of these effects strongly indicates that the rate has a dependence on the concentration of both reagents, appearing to be first order with respect to Grignard.

Table 3 shows Grignard reaction yields for a selection of alkyl and aryl substituents using different volumetric CO_2 concentrations. Grignard reactions are often dramatically affected by the reaction solvent and conditions, so the results shown in Table 3 may not be representative of those that *could* be achieved in a dedicated process where optimisation had been carried out. However, yields in almost all cases were high, regardless of the CO_2 concentration used and no other product was detected in work up mixtures other than the desired carboxylate salt (with the exception of the di-Grignard reagents in entries 5 and 14), indicating the expected extremely high selectivity, although longer reaction times naturally were required at lower CO_2 concentrations. Using lower concentrations of CO_2 for the reaction tended to have somewhat reduced yields, however this is likely due to additional exposure to potential contaminants owing to the longer reaction period. Generation of terephthalic acid from the di-Grignard reagent derived from 1,4-dibromobenzene proceeded in disappointingly modest yield with approximately equimolar generation of the mono-substituted product, 4-bromobenzoic acid. This can be explained by the deactivation of the *para*-position on generation of the mono-substituted version of the starting Grignard reagent. Using more forcing conditions during the synthesis of the Grignard reagent and more reactive magnesium precursors has been found to improve yields of reactions of this type in the past.[19] Once again, optimisation of the reaction conditions and precursor formation will be essential for high yields to be realised.

Potential side reactions in a capture context

With the exception of the last entry in Table 3, all reactions were carried out in the absence of oxygen. However, in a genuine reactive capture scenario, non-negligible quantities of oxygen would be present and would likely represent the largest threat of unwanted side reactions when using a dried flue gas stream. Commonly quoted figures for such streams usually cite O_2 concentrations of around 3% by volume although this figure can vary dependent on combustion process and fuel.[20]

Oxygen is thought to react with Grignard reagents *via* the formation of peroxides, which disproportionate to form the corresponding alcohols after quenching with aqueous acid (Scheme 2).[21]

Due to their sensitivity and explosive nature, the formation of any organic hydroperoxides, potentially by quenching the product from the first reaction shown above, would be very troublesome to a potential Grignard process. Thankfully, prior investigation of these species has shown that the second reaction in Scheme 2 is extremely rapid in comparison with the first, keeping peroxide concentrations negligible and preventing their isolation except when the reaction mixture is both saturated with oxygen and very cold.[22] Uncatalysed, the overall reaction of Grignard with oxygen at room temperature is generally slow and has a poor yield in comparison with that with CO_2. This is exacerbated when the Grignard reaction is carried out in a volatile ethereal solution such as diethyl ether. The ether is thought to form a protective blanket of solvent vapour at ambient or elevated temperatures.[23] However, this blanket effect may actually be

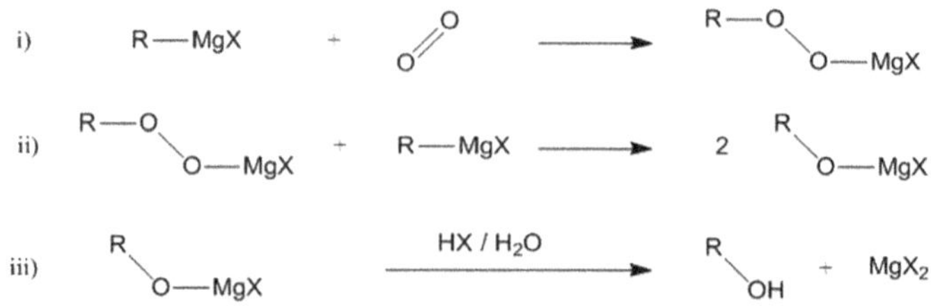

Scheme 2 Generalised reaction of a Grignard reagent with oxygen.

an over-simplification of a more complex process than first thought since, in entry 17, the dry air was sparged through the reaction mixture in the same fashion as the other runs.

Previous literature measurements of reaction rates of simple diethyl ether-solvated alkyl Grignards at room temperature with pure oxygen atmospheres demonstrated that over the course of an hour up to 43% of the initial Grignard

Table 2 Reaction rate constants of methylmagnesium chloride with CO_2/N_2 gas mixtures

Gas composition (vol% CO_2)	Pseudo rate constants (s^{-1})
100	5.49×10^{-3}
50	4.36×10^{-3}
25	1.56×10^{-3}
12.5	1.19×10^{-3}

Table 3 Comparison of yields using pure CO_2 and dilute CO_2/N_2 mixtures with a selection of Grignard agents

Entry	R	CO_2 (vol%)	Yield[a] (%)	Time (min)
1	Me–	100	82.5	20
2	CH_3CH_2–	100	91.8	20
3	iPr–	100	78.1	20
4	Ph–	100	92.4	20
5	XMg–Ph–	100	41.2[b]	20
6	$H_2C{=}CH$–	100	95.2	20
7	$H_2C{=}C(CH_3)$–	100	89.7	20
8	Me–	50	81.2	25
9	Me–	25	79.2	45
10	Me–	12.5	80.1	60
11	CH_3CH_2–	12.5	88.9	60
12	iPr–	12.5	74.2	60
13	Ph–	12.5	88.4	60
14	XMg–Ph–	12.5	42.1[b]	60
15	$H_2C{=}CH$–	12.5	95.0	60
16	$H_2C{=}C(CH_3)$–	12.5	81.1	60
17	Me–	0.05[c]	0.09	360

[a] Yield of isolated sodium carboxylate salt following NaOH quenching of the product. [b] Monocarboxylated product, *p*-bromobenzoic acid isolated in equimolar yield with terephthalic acid. [c] Dry compressed air used.

reagent would react.[23] These results have shown that, with excess oxygen present, this reaction has pseudo-zeroth order characteristics, reacting at a steady rate of 6.67×10^{-7} mol s^{-1}. For comparison, the corresponding pure CO_2 atmosphere experiments demonstrate an initial reaction rate of 0.1087 mol s^{-1}, and an average reaction rate over the course of the reaction of approximately 5.9×10^{-3} mol s^{-1} some 8850 times faster. Optimised CO_2–liquid contacting would be likely to improve this reaction rate further.

The reactions carried out here, shown in Fig. 6(b) and Table 2, have demonstrated that even when using 12.5% volume CO_2 gas, the addition of CO_2 is comparatively rapid compared with the literature values for O_2. The discrepancy in the reactivity of the two species is further demonstrated by entry 17 in Table 3, where compressed air, passed through a Drierite™ column, was used as the CO_2 source. In dry air there is more than 500 times more oxygen than carbon dioxide by volume, yet a small yield of the acetate product was isolated. While this by no means demonstrates viability of atmospheric CO_2 utilisation by Grignard reagents, it confirms that under the right reaction conditions, the yield reduction caused by oxidation could be minimised, especially in oxygen-poor environments.

Other trace gases, especially the sulfur oxides, if dry, will also interfere with the reaction, generating the corresponding sulfinic acids (Scheme 3).[24]

While this process consumes the reagent and would diminish efficiency and selectivity, unlike with some reactive capture agents such as MEA, there is no irreversible loss of the capture agent by interaction with the sulfur.[25] In the case being explored here, the magnesium chloride product is the same as that found in the reaction with CO_2.

The other major acidic gases, nitrogen oxides, are not known to react particularly with Grignard reagents without the presence of a transition metal catalyst. With these catalysts present, nitrous oxide, and also oxygen as previously mentioned, promote homocoupling of the alkyl or aryl moieties present or form other compounds such as hydrazones.[26]

In general, side reactions of this type are obstacles for all flue gas capture processes. However, as discussed above, the high reactivity of the Grignard reagents throws this into sharp focus. The question remains: in an optimistic scenario where these problems have been overcome or mitigated, what would be the cost of a Grignard CDU process?

Calculating costs of a Grignard CDU process

Naturally, the fact that a stoichiometric quantity of the magnesium reagent is required for the capture of the carbon dioxide could lead to the instant conclusion that Grignard reagents are totally unsuitable for CDU from an environmental perspective. However, further consideration of how the Grignard agent may be regenerated by electrolysis, thereby forming an electrolytic magnesium cycle, gives rise to a potentially sustainable, if high-energy, CDU process (Fig. 7). Note that any electricity used in such a process must be derived from low-carbon sources or any carbon sequestration potential for the process is immediately reversed as more carbon must be emitted from the electricity production than would be stored in the resultant products.

The net reaction of such a process would be the dehydration–carboxylation of alcohols using CO_2 and H_2. However, it should be noted that the organohalide

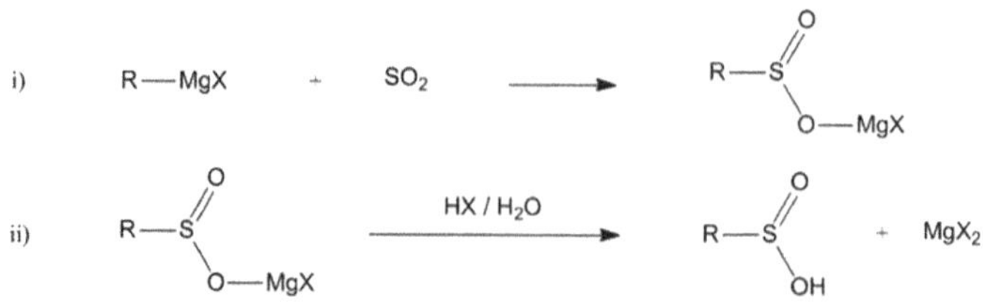

Scheme 3 Generalised reactions of a Grignard reagent with sulfur dioxide.

reagent required to make the starting Grignard reagent need not be sourced from alcohols. If instead hydrocarbons such as methane or benzene were used, no hydrogen would be required in this process as halogenation of both can be carried out directly from the elemental halide under suitable conditions.

Magnesium and halide regeneration

All steps in such a process are readily achievable and exothermic with the exception of the electrolysis step to regenerate the starting magnesium and halogen. Magnesium regeneration from magnesium dihalide, specifically magnesium dichloride, is already a fully established industrial process, and one of the two main ways that metallic magnesium is manufactured.[27]

In this process, the magnesium dihalide (typically dichloride) is heated until molten, at 750–800 °C and then electrolysed requiring a molar electrode potential of 3.74 V. Typically, industrial magnesium production processes are reported to require 10.5–13.2 kW h per kg, including the drying and purification of the starting magnesium dichloride, which is a significantly energy-intensive part of the overall process.[28] In our proposed process, this step could be avoided by using either gaseous or alcoholic HCl (which is available as a by-product of the formation of the initial alkyl chloride from the parent alcohol) to quench the Grignard product. The resulting magnesium chloride would then be much more readily dried, reducing electrical costs and opening a reaction pathway to direct formation of potentially valuable esters.

Furthermore, recent advances in the area of magnesium electrolysis particularly focused on the rapid removal of the chlorine gas by-product and have been able to demonstrate a reduction in the energy cost to just 7.0 kW h per kg.[29] This is very close to the theoretical minimum energy cost of this process of 6.2 kW h per

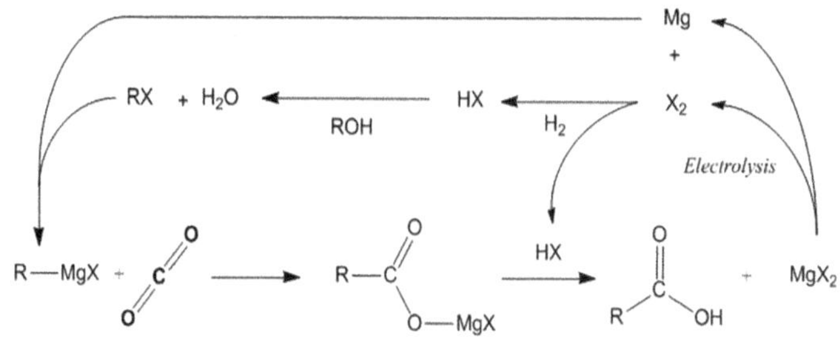

Fig. 7 Grignard reaction with regeneration.

kg magnesium or 542 kJ mol^{-1}, and represents an electrical efficiency of nearly 89%.

Even taking the more standard magnesium production process energy consumption figures, magnesium potentially represents an efficient energy vector for the net reduction of carbon dioxide by electricity.

The other product of magnesium electrolysis is naturally the elemental halide, which when using alcohols as reagents must be reacted with hydrogen to regenerate the acid for both the alkyl halide formation and the post reaction quenching. This reaction is also very exothermic. Typical hydrogen chlorine burners reach 2000–2500 °C, with up to half of the combustion energy being recoverable.[30] It is easy to imagine that this high temperature reaction would further aid the magnesium chloride electrolysis by supplying the temperatures needed to melt the starting magnesium halide salt.

Aspen Plus® modelling

The model shown in Fig. 8 was constructed and used to derive a preliminary techno economic analysis of the reaction of methanol with CO_2 and H_2 *via* methyl magnesium chloride to generate acetic acid. This example was chosen for its simplicity and to be used as a basis to show just how expensive or otherwise a Grignard-based CDU system would be.

The model assumes a magnesium regeneration energy cost of 10.5 kW h kg and a quantitative yield of HCl from the reaction of H_2 and Cl_2, comparable with industrial processes. The extremely exothermic nature of the latter reaction is modelled as supplying all the heat requirements of the plant, but no further energy recovery beyond this is included in the model.

The methanol chlorination reaction is assumed to have a 94% yield, in line with industry performance, however the 5–6% by-product of this reaction, dimethyl ether (DME), is not given a value. The 5–6% of the HCl that would therefore be expected to remain unreacted from the methanol chlorination step, primarily due to dilution by the by-product water and incomplete chlorination, can still be potentially utilised for quenching the Grignard reaction and so HCl utilization would be expected to be quantitative.

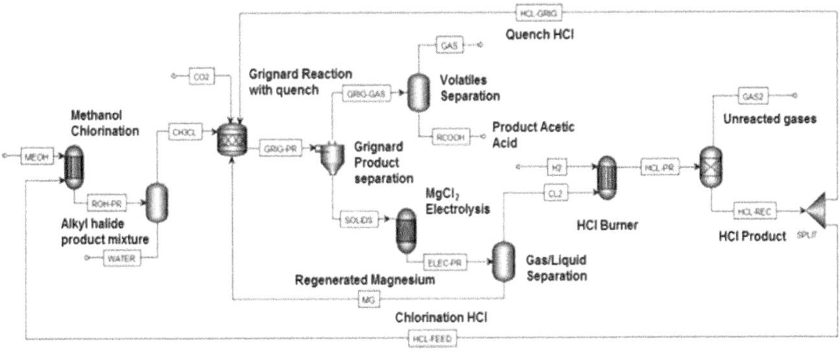

Fig. 8 Aspen Plus® model used for preliminary techno economic assessment of acetic acid production from Grignard reaction of methanol with CO_2.

For simplicity, both the formation of the Grignard reagent and the subsequent reaction with CO_2 was assumed to also be quantitative. While this may seem to be rather optimistic, Grignard formation reactions, under appropriate conditions regularly reach extremely high yields.[21] Likewise, optimisation of the CO_2 reaction and workup did not therefore seem to be wholly unreasonable. Furthermore, the absence of any additional heat recovery from the essentially entirely exothermic reaction pathway and overlooking the valuation of the DME stream will go some way to offset the ignored costs of a true Grignard-based system, which will naturally have some yield limitations, especially if an oxygen-containing gas stream is used.

Tables 4 and 5 summarise the outputs and base conditions of the model shown in Fig. 8. At first glance, perhaps unsurprisingly, the production cost of acetic acid using such a process is significantly higher than that of the market value of the product. However, this is primarily an effect of the small production scale of the model, the scale of which is based on large existing Grignard plants in the US and UK.[31] From the point of view of acetic acid, this is an extremely small scale plant, with commercial acetic acid production facilities reaching a scale exceeding 500 000 tonnes per year. This point is emphasised by noting that the labour costs exceed that of the electricity and raw materials costs in Table 4.

Approximating the effect of increasing plant scale on production price from these results was then carried out using the "six-tenths rule".[32] As shown in Fig. 9, this demonstrates that dramatic reductions in the production cost to levels comparable with current acetic acid market value could be achieved by increasing reaction scale. For example, a tripling of the reactor linear dimensions (which would increase reactor volumes and therefore production by around a factor of 27) would drop production cost to around €865 per tonne from the base case of over €3200. Further size increases may even allow for profitable production of acetic acid in this way, however caution must be taken when dealing with extended extrapolations of this sort.

Sensitivity analysis was also carried out to determine what effect both raw material and electricity price would have on the production cost of the acetic acid (Fig. 10). Here, we can see that by far the most important cost is that of the electricity, as might be expected for an electrochemical cycle, with each change in the price per megawatt hour causing more than a four-fold change in production cost. The electricity figure used in the base case model in Table 5 is a relatively high one, representing the requisite renewable electricity, but this figure could

Table 4 Aspen Plus® model outputs

Model output	
Total capital costs	€4 835 230
Annual cost of capital	€567 944
Annual labour costs	€482 130
Annual electricity costs	€431 161
Annual raw materials costs	€250 444
Total operating costs	€1 367 964
Production cost (per tonne)	€3217
Current European acetic acid value (per tonne)	€525

Table 5 Model calculation notes

Base case conditions, model notes and assumptions	
Methanol	350 € per tonne
Hydrogen	4700 € per tonne
Carbon dioxide	12 € per tonne
Electricity supply	17 C€ kW^{-1} h^{-1}: 170 € MW^{-1} h^{-1}
Plant lifetime	20 years
Interest rate	10%
Production scale	602 tonne per a

easily be halved by more developed lower-cost low carbon sources. Dropping electricity costs in this way to €85 per MW h, within reported ranges for onshore wind and large scale photovoltaic sources, would then decrease production costs by over €360 per tonne.[33] This effect would proportionally complement any drop in production cost by increasing the reaction scale.

Other potential products

As previously mentioned, acetic acid was chosen due to its simplicity; however that same simplicity and acetic acid's market as a bulk chemical makes it an especially difficult target for this process from an economic perspective due to the relatively low value of the product. In comparison, many of the other carboxylic acids shown previously in Fig. 3 have a significantly higher value. This would mean that the value added by the addition of CO_2 *via* Grignard chemistry would be larger, making the overall process far more economically enticing, even if the process cost is high.

This effect is especially apparent in the polymer precursors; adipic, acrylic and terephthalic acids where the cost differential between the starting material and

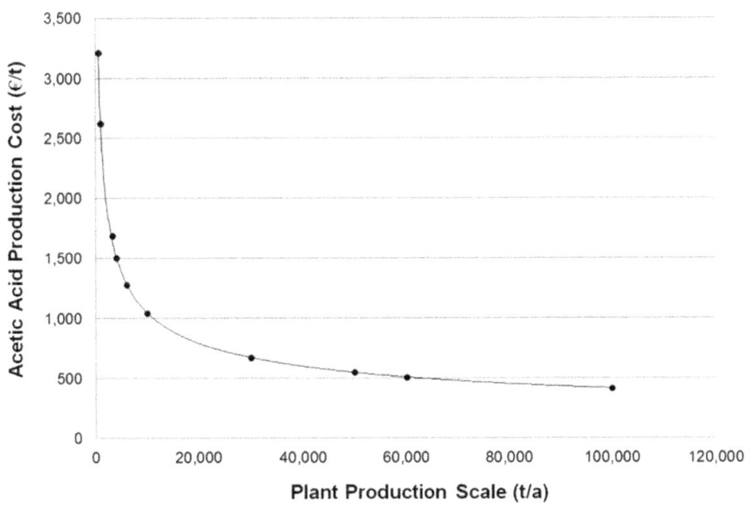

Fig. 9 Projected effect of process scale on production cost.

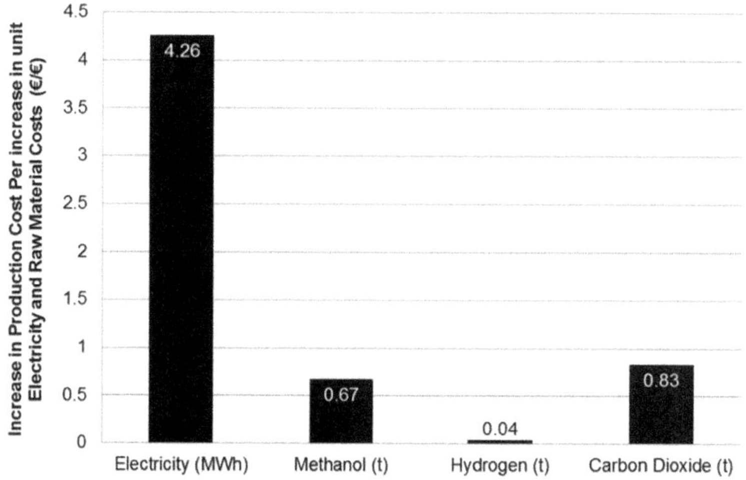

Fig. 10 Model electrical and raw material price sensitivity measured as production cost increase per increase in unit price.

product in a putative Grignard process is significantly larger than that of acetic acid from methanol (Fig. 11).

However, it should be noted that no direct comparison between these putative processes and that of the acetic acid model should be carried out due to stark differences in reaction profile, energetics and potential yields.

Experimental

With the exception of the di-Grignard product of 1,4-dibromobenzoic acid, all Grignard reagents were purchased in THF solution from Sigma Aldrich and used

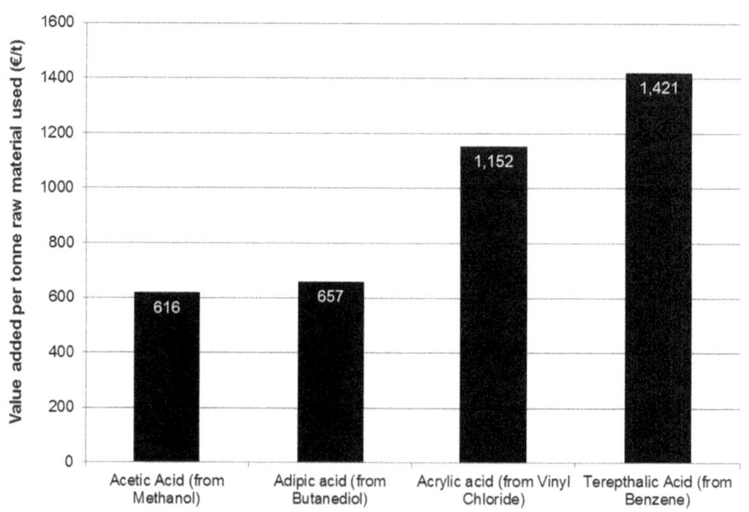

Fig. 11 Value added by addition of CO_2 to starting compound (in brackets). Average 2013 prices used.

without further purification. Reactions were performed using Schlenk techniques under inert (N_2) atmosphere unless otherwise indicated. HPLC grade THF was dried by storage over freshly-regenerated 3 Å molecular sieves *ca.* 30% by volume. CO_2, N_2 and compressed air were supplied by BOC-Linde. CO_2 and N_2 mixture compositions were achieved using a pair of Bronkhorst 100 mL min^{-1} mass flow controllers. Compressed air was dried using a Drierite™ 8 Mesh Laboratory Drying Unit. ^{1}H NMR spectra were recorded using a Bruker Avance 400 MHz spectrometer with D_2O (99.9% D). Pressure data were recorded using an Omega PX409USB High Accuracy Pressure Transducer. Grignard concentrations were verified by titration with 2-butanol (99+%) and 1,10-phenanthroline (99.5%). Modelling was conducted using Aspen Plus® on AspenTech.

General procedure for kinetic investigation of Grignard reaction with dilute CO_2

An oven-dried 2-neck 100 mL round bottom flask with attached empty cold-finger condenser was charged with THF (20 mL dried over 3 Å molecular sieve) under nitrogen using Schlenk techniques. Using a needle adapter the flask was then connected to a sparging needle and a flow of 100 mL min^{-1} of selected CO_2 gas mixture, venting through a silicone oil bubbler. To avoid any amount of water condensation within the reactor during set-up, gas flow was carried out for 20 minutes before a freezing salt-ice slurry was added to the cold finger. A normal ice bath was then placed around the reaction vessel and an additional 30 minutes was given to allow the temperature to equilibrate.

A measured volume of 3 M MeMgCl sufficient to achieve 1 M overall concentration of the Grignard solution (10 mL) was then swiftly added to the vessel under positive nitrogen pressure, and reaction timing initiated. Every 60 seconds, the CO_2-containing gas flow in and out of the reactor was paused for a set period of time (10–20 s) before being resumed, as pressure readings were logged every 0.2 seconds. The reaction was deemed complete when no pressure drop was observed during a prolonged gas flow pause. Aqueous HCl (3.5 M) was then slowly added to the reaction mixture until the mixture clarified and gave an acidic reaction on indicator paper. To this mixture aqueous NaOH (1 M) was then added until the reaction mixture became basic, and a cloudy suspension of salts formed. The basic mixture was then evaporated until dry by rotary evaporation. An accurately weighed sample of the dry salt mixture was then dissolved in D_2O (2 mL) with the addition of DMSO (10 μL) as an internal reference for NMR spectroscopy.

^{1}H NMR (400 MHz D_2O) δ (ppm): 2.81 (s, 6H, DMSO), 1.94 (s, 3H, CH_3COO). Integration of the individual peaks is then used to determine carboxylate yield.

Conclusions

In summary, a series of experiments were carried out to determine the reaction kinetics and feasibility of the reaction of Grignard reagents with dilute sources of CO_2 in N_2. From these results, it was determined that high yield reactive capture of dilute CO_2 is possible with Grignard agents. Although oxygen concentration in any gas stream will likely hamper yields, the previously-measured rate of reaction with oxygen has been found to be nearly negligible in comparison of that with CO_2.

A preliminary techno-economic analysis was also carried out, to determine the economic feasibility of such a capture process to be used for production of acetic acid from methanol, using industrially-relevant data and a simplified reaction process model. It determined that when scaling factors were taken into account, the production costs of acetic acid by this method were comparable with existing market prices.

While a more complete costing and detailed assessment could very possibly demonstrate the economic infeasibility of the described process under less favourable conditions, the assessment carried out here suggests that the use of a Grignard reaction cycle to generate acetic acid from methanol is not perhaps as absurd as would be expected.

Acknowledgements

We would like to thank EPSRC for funding from the 4CU Programme Grant (EP/K001329/1), at the University of Sheffield, which has supported this work, G. R. M. D. and I. D., and for the DTA Studentship (D. G. R.). We would also like to thank Innovate UK and BBSRC (BB/M011917/1) for funding R. E. O. We also thank EPSRC for their funding of the CO_2Chem Grand Challenge Network in CDU (EP/K007947/1 & EP/H035702/1).

Notes and references

1 M. Aresta, A. Dibenedetto and A. Angelini, *J. CO_2 Util.*, 2013, **3–4**, 65–73.

2 M. Perez-Fortez, A. Bocin-Dumitriu and E. Tzimas, *Energy Procedia*, 2014, **63**, 7968–7975.

3 P. Styring, D. Jansen, H. de Coninck, H. Reith and K. Armstrong, *Carbon Capture and Utilisation in the Green Economy*, CO_2Chem Media & Publishing, Sheffield, UK, 2011, ISBN: 978-0-9572588-0-8.

4 P. Styring and K. Armstrong, *Front. Energy Res.*, 2015, **3**, 1–9.

5 Global CCS Institute, *The Global Status of CCS 2014*, Victoria, Australia, 2014.

6 K. Armstrong, *Carbon Dioxide Utilisation: Closing the Carbon Cycle*, ed. P. Strying, E. Quadrelli and K. Armstrong, Elsevier, Cambridge, UK, 2014, pp. 237–249, ISBN: 978-0-4446274-6-9.

7 Bayer Dream Process: J. Langanke, A. Wolf, J. Hofmann, K. Böhm, M. A. Subhani, T. E. Müller, W. Leitner and C. Gürtler, *Green Chem.*, 2014, **16**, 1865–1870.

8 Sunfire, *Sunfire Now Produces Synthetic Fuel from Air, Water and Green Electrical Energy*, Dresden, Germany, 2015.

9 Carbon Recycling International, *Emissions to Liquid Fuels*, Reykjavik, Iceland, 2011.

10 N. von der Assen, J. Jung and A. Bardow, *Energy Environ. Sci.*, 2013, **6**, 2721–2734 and Green Chem.

11 R. M. Cuellar-Franca and A. Azapagic, *J. CO_2 Util.*, 2015, **9**, 82–102.

12 E. Catalanotti and M. Pourkashanian, *Carbon Capture: A Technology Assessment*, Leeds, UK, 2012, pp. 15–22.

13 P. Styring, *Carbon Dioxide Utilisation: Closing the Carbon Cycle*, ed. P. Strying, E. Quadrelli and K. Armstrong, Elsevier, Cambridge, UK, 2014, pp. 19–31, ISBN: 978-0-4446274-6-9.

14 V. Grignard, *C. R. Hebd. Seances Acad. Sci.*, 1900, **130**, 1322–1325.

15 Thermodynamic data for reactions, intermediates and products was sourced from: US national Institutes of Standards and Technology (NIST), Chemistry Webbook, http://webbook.nist.gov/chemistry/, accessed 23/04/2015.

16 E. A. Quadrelli, G. Centi, J.-L. Duplan and S. Perathoner, *ChemSusChem*, 2011, **4**, 1194–1215.

17 E. Stoye, *Chemistry World News*, RSC, UK, 8 April 2015.

18 J. Tammiku, P. Burk and A. Tuulmets, *J. Phys. Chem. A*, 2001, **105**, 8554–8561.

19 R. Rieke, *Top. Curr. Chem.*, 1975, **59**, 1–13.

20 R. Zevenhoven and P. Kilpinen, *Control of Pollutants in Flue Gases and Fluel Gases*, Helsinki, Finland, 2001, pp. 2.1-2.11.

21 G. S. Silverman and P. E. Rakita, *Handbook of Grignard Reagents, Marcel Dekker*, New York, USA, 1996, pp. 9–22.

22 C. S. Foote, J. Valentine, A. Greenberg and J. F. Liebman, *Active Oxygen In Chemistry*, Springer, Houten, Netherlands, 1995, pp. 55–58; C. Walling and S. A. Buckler, *J. Am. Chem. Soc.*, 1955, **77**, 6032–6038.

23 M. Goebel and C. Marvel, *J. Am. Chem. Soc.*, 1933, **55**, 1693–1696.

24 P. Allen, W. Rehl and P. Fuchs, *J. Org. Chem.*, 1955, **20**, 1237–1239.

25 B. R. Strazisar, R. R. Anderson and C. M. White, *Energy Fuels*, 2003, **17**, 1034–1039.

26 For homo and cross-coupling of Grignard reagents with Nitrous oxide see; G. Kiefer, L. Jeanbourquin and K. Severin, *Angew. Chem., Int. Ed.*, 2013, **52**, 6302–6305. For the same with oxygen see; G. Cahiez, C. Duplais and J. Buendia, *Angew. Chem.*, 2009, **121**, 6859–6862. For formation of hydrazones see; A. Tskhovrebov, E. Solari, R. Scopelliti and K. Severin, *Organometallics*, 2014, **33**, 2405–2408.

27 I. J. Polmear, *Magnesium and Magnesium Alloys*, ed. M. M. Avedesian and H. Baker, ASM International, OH, USA, 1999, pp. 2–6.

28 I. J. Polmear, *Magnesium and Magnesium Alloys*, ed. M. M. Avedesian and H. Baker, ASM International, OH, USA, 1999, pp. 2–6.

29 G. Demirci and I. Karakaya, *Magnesium Technology*, ed. S. N. Mathaudhu, W. H. Sillekens, N. R. Neelameggham and N. Hort, John Wiley & Sons, Hoboken, NJ, USA, 2012, pp. 59–62.

30 SGL Group, *Systems – HCl synthesis*, Meitingen, Germany, 2014.

31 D. Welter, *Grignard Manufacturing at WeylChem*, Elgin, USA, 2014.

32 J. Sweeting, *Project Cost Estimating: Principles and Practice*, IChemE, 1997, pp. 157–159.

33 C. Kost, J. N. Mayer, J. Thomsen, N. Hartmann, C. Senkpiel, S. Philipps, S. Nold, S. Lude, N. Saad and T. Schlegl, *Levelized Cost of Electricity, Renewable Energy Technologies*, Fraunhofer Institute ISE, Freiburg, Germany, 2013.

Faraday Discussions

PAPER

Carbon dioxide capture and utilization: using dinuclear catalysts to prepare polycarbonates†

N. Yi, J. Unruangsri, J. Shaw and C. K. Williams*

Received 12th May 2015, Accepted 18th May 2015
DOI: 10.1039/c5fd00073d

The copolymerization of epoxides, including cyclohexene oxide and vinyl-cyclohexene oxide with carbon dioxide are presented. These processes are catalyzed using a homogeneous di-zinc complex that shows good activity and very high selectivities for polycarbonate polyol formation. The polymerizations are investigated in the presence of different amounts of exogenous reagents, including water, diols and diamines, as models for common contaminants in any carbon dioxide capture and utilization scenario.

Introduction

Using carbon dioxide to make useful products, including chemicals and fuels, is an attractive way to add value to a common waste gas and could be a useful method to reduce pollution.[1–4] From a practical, implementation focused perspective, there are two primary considerations for any carbon capture and usage process: (1) the overall carbon dioxide emission must be reduced, which sounds obvious but requires careful consideration of the overall energies of both the process and embedded in the preparation of co-reagents; and (2) the product must be needed and viable, both in terms of the potential market volume and value.

Amongst the myriad of possible carbon dioxide transformations, the production of polymers offers significant potential.[5–10] Such polymers are prepared by the ring-opening copolymerization (ROCOP) of carbon dioxide and epoxides (Scheme 1), a process which requires a catalyst. The reaction has attracted considerable attention, with a number of different catalysts and processes being reported by both academic and industrial researchers worldwide.[5–18]

The products of the reaction are aliphatic polycarbonates; at low molecular weights these polymers are suitable replacements for petrochemical polyols used in polyurethane manufacturing.[19,20] In the context of carbon dioxide reduction, the replacement of polyether polyols by poly(ether carbonate) polyols leads to

Department of Chemistry, Imperial College London, London, SW7 2AZ, UK. E-mail: c.k.williams@imperial.ac.uk
† Electronic supplementary information (ESI) available. See DOI: 10.1039/c5fd00073d

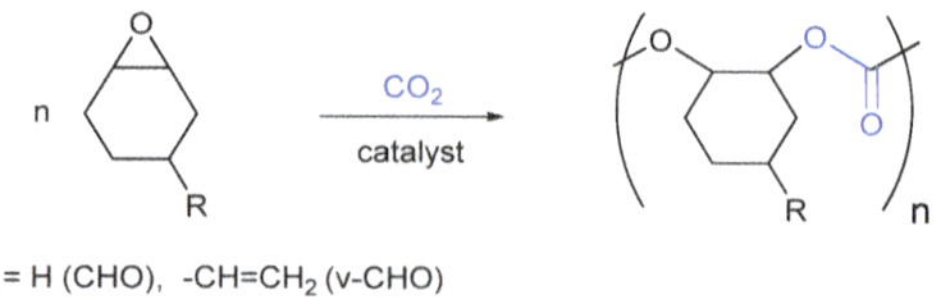

Scheme 1 Illustrates the ring opening copolymerizations using cyclohexene oxide (CHO) and vinyl-cyclohexene oxide (v-CHO).

significant reductions in greenhouse (GHG) emissions (~20%).[21] By selecting the correct catalyst, it is possible to substitute >40% of the epoxide with carbon dioxide. Such a substitution is attractive both economically and environmentally as it may further reduce GHG emissions. Thus, the first criterion can be met by the production of polymers from carbon dioxide.

The second criterion, that the products are valuable and needed/used on a large-scale, is also met by polymer production. Considering only polyols, there are currently >8 Mt of polyurethanes produced globally requiring 3–4 Mt of polyols in their manufacture.[19,20] Therefore, the production of polymers from CO_2 has the potential to offer a viable industrial route to add value to waste emissions. The process is critically dependent on the selection of the catalyst, with a range of very promising heterogeneous and homogeneous catalysts having been reported.[5–18] However, so far there remain far fewer studies on the tolerance of such catalysts to a range of impurities, which are likely to be present in any carbon capture and usage scenario.[22,23] We recently reported that our dinuclear metal catalysts are tolerant to a wide range of impurities found in captured carbon dioxide, including nitrogen, oxygen, oxides of carbon, NO_x and SO_x.[23] Furthermore, using di-magnesium catalysts a wide-ranging study of common contaminants was undertaken which revealed remarkable catalyst stability, even in the presence of large amounts of contaminants.[23] In that study, carbon dioxide captured from a UK power station was applied and showed near-equivalent performance compared to pure carbon dioxide.[23]

Our group have reported a series of homogeneous catalysts for CO_2/epoxide copolymerization, based on dinuclear complexes of Zn(II), Mg(II), Co(II/III) and Fe(III).[24–35] This discussion paper will focus on a di-zinc bis(trifluoroacetate) complex[28] and the influence of the addition of known quantities of water, alcohols and amines to the polymerizations. The additives and monomers are investigated both to test tolerance of the catalyst and also to target particular polymer chain functionalities (end-groups, side-chain substituents) so as to widen the scope, and ultimately the scale, of future applications.

Results and discussion

The preparation and structure of catalyst **1** is shown in Scheme 2.[28]

The catalyst was synthesized from the ancillary ligand, H_2L, which can itself be prepared in high yield (82%) from commercial materials in 3 steps. Its purity was assessed by ^1H NMR spectroscopy (Fig. S1†) and by elemental analysis, which showed results in line with the literature.[28]

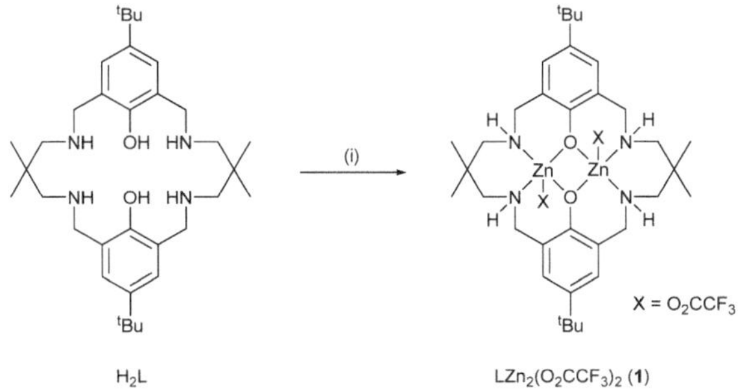

Scheme 2 Synthesis of the dizinc complex ($LZn_2(O_2CCF_3)_2$, **1**). Reagents and conditions: (i) 2 equiv. $Zn(O_2CCF_3)_2\cdot xH_2O$, MeOH, 25 °C, 16 h.

Catalyst **1** was applied in the copolymerization of cyclohexene oxide (CHO) or vinyl cyclohexene oxide (vCHO) and carbon dioxide (Table 1 and Scheme 2). It showed a good performance using both monomers, with the TOF for CHO being 25 h^{-1} and for v-CHO being 24 h^{-1}, at 1 bar CO_2 pressure, 0.1 mol% catalyst loading and 80 °C. The activity for CHO is in-line with the value reported in the literature.[28] The activity using v-CHO has not previously been reported and is more difficult to quantify precisely due to significant overlap of resonances associated with monomer and polymer in the crude ^1H NMR spectrum (Fig. S2 and S3†). For both monomers, the polycarbonates produced show a very high selectivity for carbonate linkages (≥99%) and there is ≤1% formation of a cyclic carbonate by-product.

The alternating polymers show moderate molecular weights, with values typically being lower than 10 000 g mol^{-1}, and narrow dispersities ($Đ < 1.20$). Copolymerizations using **1** show a linear evolution of molecular weight *vs.* CHO conversion and narrow dispersities: both are features of well-controlled/living polymerizations. Such behaviour is indeed typical of this class of dinuclear catalyst and indicates good polymerization control.[30]

Controlled polymerizations are defined as exhibiting rapid and quantitative initiation leading to all chains propagating at the same rate and limited/no

Table 1 Epoxide/CO_2 copolymerization using **1**[a]

Entry	Monomer	Catalyst	Temp. (°C)	TON[b]	TOF[c] (h^{-1})	% carbonate[d]	% cyclic carbonate[d]	M_n[e] (g mol^{-1})	M_w/M_n[e] (Đ)
1[f]	CHO	1	80	608	25	99	3	1500	1.08
2	CHO	1	80	502	25	99	1	6600	1.18
3[g]	v-CHO	1	80	429	20	>99	<1	5900	1.30
4	v-CHO	1	80	389	24	>99	<1	6700	1.18

[a] Polymerization conditions: 0.1 mol% catalyst, in neat epoxide, for 6–24 h, 1 bar research grade CO_2 unless stated otherwise. [b] TON = number of epoxide consumed per mole of catalyst. [c] TOF = TON per hour. [d] Determined by comparing the integrals of resonances at 4.65 ppm (PCHC) or 4.75–4.90 ppm (PVCHC), at 4.0 ppm (cyclic carbonate by-product) and at 3.45 ppm (ether linkages). [e] Determined by SEC, in THF using narrow M_w polystyrene standards. [f] Unpurified CHO. [g] Unpurified v-CHO.

termination or chain transfer reaction. In such a polymerization, the molecular weight of the polymer should be predictable from the monomer conversion and catalyst/initiator concentration. However, the polymerizations using catalyst **1** result in the production of polymers that have molecular weights, determined by size exclusion chromatography, which are substantially lower than predicted; indeed the values obtained suggest that chain transfer reactions occur, and that approximately 8 chains grow per equivalent of catalyst. Thus, the polymerizations appear to be occurring under immortal conditions, whereby chains are rapidly being exchanged or transferred between the catalyst and a chain transfer agent (often an alcohol). Fig. 1 illustrates the key processes occurring during polymerization. It is proposed that 1,2-diols are present in the polymerizations and function as the chain transfer agents. Such chain transfer processes are commonly observed in this field using a range of different catalysts.[14,27,30,36–38]

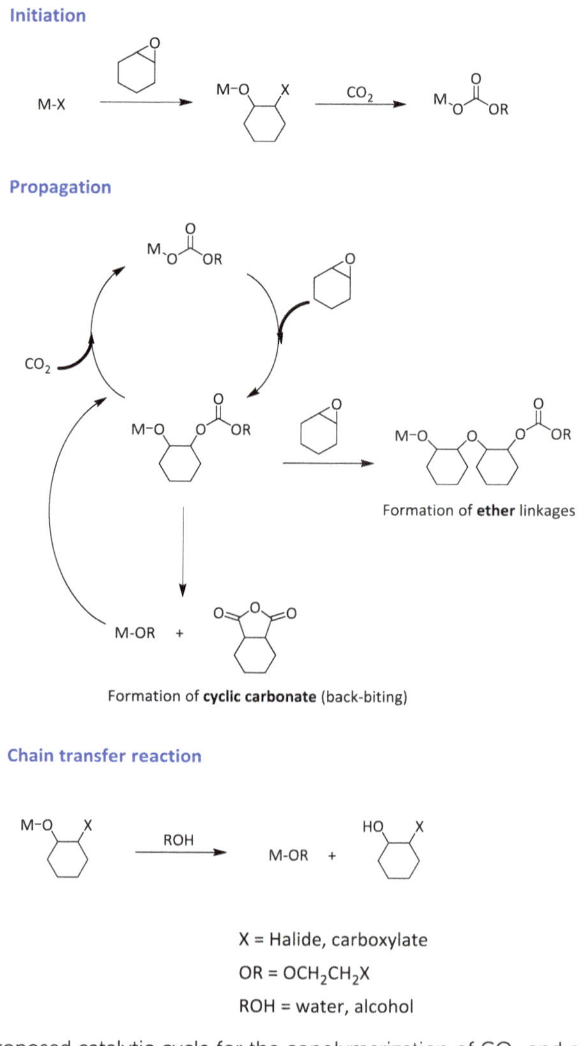

Fig. 1 The proposed catalytic cycle for the copolymerization of CO_2 and epoxides.[8]

A further consideration is that bimodal molecular weight distributions are commonly observed, but seldom explained, for a wide range of catalysts in the literature.[27,29,30,37,39] Indeed, catalyst **1** also shows a distinctly bimodal M_w distribution for both PCHC and PvCHC (Fig. 2, blue trace, Fig. S4 and S5†). In order to study the observed bimodality, two fractions of PCHC containing the higher and lower M_n distributions were isolated by SEC, according to a preparative separation by elution time (see Experimental section). These fractions were analysed by ^{1}H NMR spectroscopy, MALDI-ToF and SEC. For both fractions, the ^{1}H NMR spectra are identical, indicating PCHC end-capped with hydroxyl groups (Fig. S6†). Furthermore ^{19}F NMR spectroscopy does not show any signals consistent with a lack of trifluoroacetate end-groups on the polymer chains. The MALDI-ToF spectra show that both fractions contain a single series attributed to polycarbonates having di-hydroxyl end groups. SEC analyses showed a clear separation into two molecular weight distributions: the lower fraction has M_n of 3900 (*Đ*: 1.14), whilst the higher fraction has M_n of 9000 (*Đ*: 1.05). Based on these observations, it appears that catalyst **1** leads to the production of PCHC with narrow but bimodal molecular weight distributions, where all detectable chains are di-hydroxyl terminated (Fig. 3).

We have previously observed bimodal M_w distributions, due to chains end-capped by the initiating group (lower M_n series) and by the di-hydroxyl groups (high M_n series, Fig. S3†).[30] Based on this previous data, it would be expected that catalyst **1** would show chains end-capped by both trifluoroacetate (from the

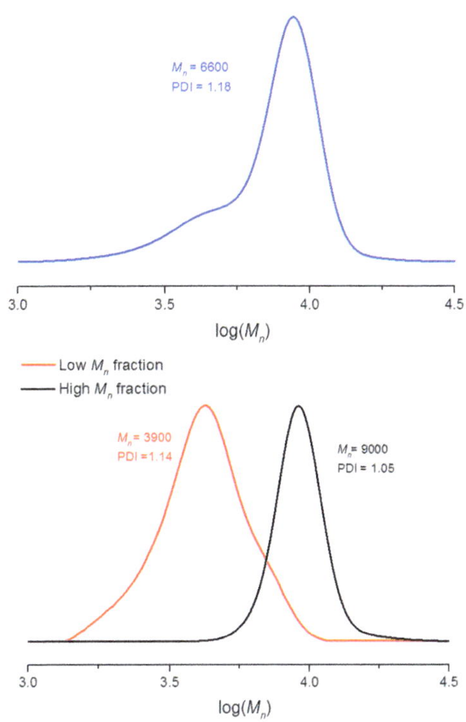

Fig. 2 Top: molecular weight distribution of the isolated (crude) PCHC polymer (Table 1 entry 2) and bottom: the separation of PCHC into two fractions of low and high M_w.

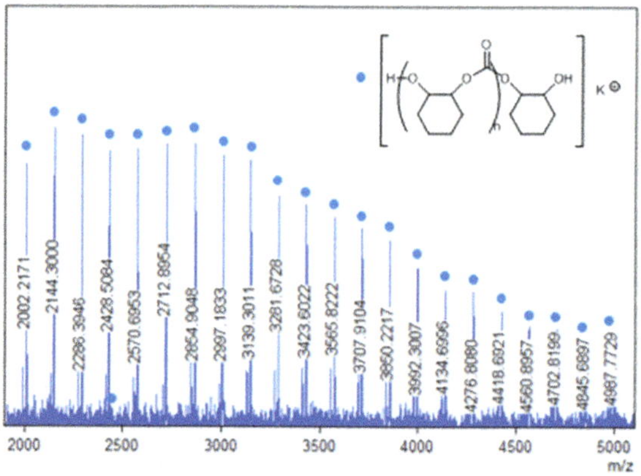

Fig. 3 MALDI-ToF spectrum of the low M_n fraction of PC (Table 1 entry 2), a single series is observed corresponding to: $[HO(C_7H_{10}O_3)_nC_6H_{11}O_2]K^+ = [(142.15)_n + 116.16 + 39.1]$.

catalyst) and di-hydroxyl groups (from 1,2-diols). As the polymerizations are well-controlled, it is expected that the chains propagate at the same rate, therefore the molecular weight of the polymer from the telechelic diol would be expected to be double that of polymers initiated from the trifluoroacetate group. Thus, the bimodality can be generally attributed to the two types of initiating group. In order to rationalize the lack of a trifluoroacetate end-group (by spectroscopy and MALDI-ToF), it is proposed that those chains are susceptible to end-group (trifluoroacetate) hydrolysis during work-up due to the strongly electron-withdrawing substituents on the carboxylate group. The hydrolysis would lead to a bimodal M_w distribution where both series of chains are end-capped by hydroxyl groups, in line with the experimental observations. Darensbourg and co-workers observed that salen-chromium(trifluoroacetate) catalysts also lead to di-hydroxyl terminated polycarbonates.[40] They confirmed that the trifluoroacetate end-group was present during polymerization, using anhydrous electrospray mass spectrometry, but was hydrolysed during termination/work-up to selectively produce the polyol.[40]

The important outcome is that catalyst **1** leads to the highly selective formation of dihydroxyl end-capped polymers or polyols. This is potentially beneficial for further application in polyurethane manufacture or in the production of higher polymers. We have previously demonstrated that such polyols are viable initiators for lactide ring-opening polymerizations, upon addition of a ROP catalyst, so as to produce ABA type triblock copoly(ester–carbonate–esters).[28] Given that selective production of polyols is feasible using catalyst **1**, it is of interest to investigate its tolerance to the presence of various other chain transfer agents and also the potential to prepare mono-modal M_w distributions.

Polymerizations with added water

Chain transfer reactions are known to occur during these polymerizations if protic compounds are present. These reactions serve to 'swap' the growing polymer chain on/off the metal centre and result in more than one polymer chain growing per equivalent of catalyst (Fig. 1). Given that there is good control of the

M_w and the dispersities are narrow, it is proposed that the chain transfer reactions occur more rapidly than propagation. Thus, it should be possible to control the molecular weights *via* the addition of chain transfer agents, the simplest and most cost-effective of these is probably water. Furthermore, water is of interest as it is a common contaminant of carbon dioxide and would be expected to be present in any gases captured from industrial processes. In order to investigate the tolerance of catalyst **1** to water, a series of copolymerizations were run with increasing quantities of water being added (Table 2).

In all cases where water was added, the polymerizations were nearly as active and selective as polymerizations in the absence of any additive (Table 2). It is feasible to add large quantities of water to the polymerizations, including from ~90–5700 ppm by weight, and still maintain effective polymerization. This is particularly notable as typical pipeline specifications for captured CO_2 water content are approximately 100–1000 ppm.[23]

For the polymerizations using water as the chain transfer agent, the polymer MWs reduce as increasing quantities of water are added. Furthermore, monomodal M_w distributions are observed when excess water is added (Fig. 4). A representative MALDI-ToF spectrum (Fig. 4) confirms that only a single series of polycarbonate polyol chains is formed in all cases.

Polymerizations with added alcohols

It has been proposed that water reacts with epoxides to produce cyclohexane-1,2-diol (CHD), such a process likely occurs immediately and may be metal catalyzed.[27–30] Cyclohexane-1,2-diol would be expected to function as a chain transfer agent (CTA), leading to propagation from both of its secondary hydroxyl groups.[41] Thus, it is of interest to investigate polymerizations in the presence of various equivalents of exogenous CHD (Table 3, entry 1–4). Using either CHO or vCHO resulted in successful polymerizations, and all the polymers contained high carbonate linkage contents (>99%) and low quantities of cyclic carbonates (~5% for PCHC and <1% for PvCHC). Using a 5-fold excess of CHD, *vs.* catalyst, with CHO or v-CHO led to polymerizations occurring at near equivalent activity to those run in the absence of additives (Table 1, entries 2 and 4). Increasing the

Table 2 Data obtained for polymers produced by the ring opening copolymerization (ROCOP) of CHO and CO_2 with the addition of water as a chain transfer agent (CTA)[a]

Entry	Amount of water added (mol equiv.)	ppm H_2O^b (by weight)	$M_n{}^c$ (g mol^{-1})	$M_w/M_n{}^c$ ($Đ$)
1	0.5	91	5700	1.17
2	1	182	5200	1.15
3	2	364	4600	1.12
4	4	728	3300	1.11
5	8	888	2500	1.10
6	16	2904	1800	1.10
7	32	5792	800	1.09

[a] Polymerization conditions: 0.1 mol% of 1, 18 h reaction time, 1 bar research grade CO_2, 80 °C. [b] The quantity of water (as ppm weight fraction) added, based on the overall quantities of catalyst 1, epoxide and additive. [c] Determined by SEC, in THF, using narrow molecular weight polystyrene standards to calibrate the instrument.

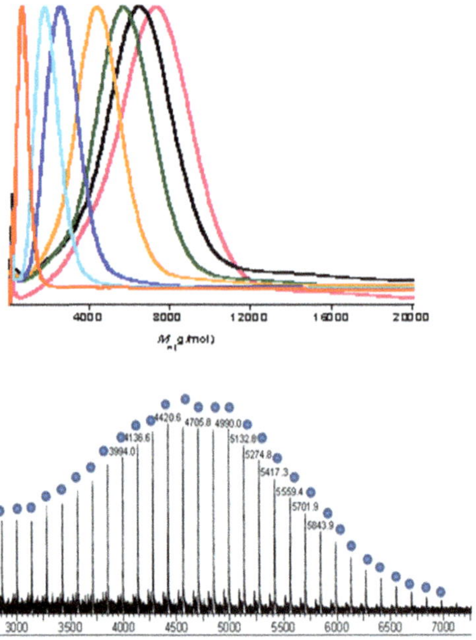

Fig. 4 Top: SEC traces for the PCHC formed in Table 2. Colour code: entry 1: pink, 2: black, 3: green, 4: yellow, 5: dark blue, 6: pale blue, 7: red. Bottom: the MALDI-ToF spectrum obtained from Table 2, entry 2.

Table 3 Shows the data for polymerizations conducted with addition of various additives (mono-, bi- and tri-functional)[a]

Entry	Monomer	Additive[b]	Amount of additive added[c] (mol equiv.)	ppm additives[d] (by weight)	Conversion[e] (%)	% selectivity[f]	M_n[g] ($g\ mol^{-1}$)	M_w/M_n[g] ($Đ$)
1	CHO	CHD	5	5831	53	>99	4400	1.16
2	CHO	CHD	10	11 598	53	>99	2100	1.15
3	v-CHO	CHD	5	4623	40	>99	7500	1.21
4	v-CHO	CHD	10	9231	59	>99	5700	1.17
5	CHO	TEA	20	2905	49	>99	1500	1.23
6	CHO	H_2NBn	20	2116	22	>99	950	1.10
7	CHO	$HNBn_2$	20	3834	33	>99	3700	1.18
8	CHO	MEA	10	6068	65	>99	2200	1.21
9	CHO	EN	10	6008	61	>99	2900	1.21

[a] Polymerization conditions: 0.1 mol% of **1**, in neat epoxide, for 16–72 h, 1 bar research grade CO_2, 80 °C. [b] Weighed directly into reaction vessel. [c] Based on **1**. [d] The quantity of additives (as ppm weight fraction) added, based on the overall quantities of catalyst **1**, epoxide and additive. [e] Determined by ^1H NMR spectroscopy, based on comparing integration values for the polymers (4.55–4.70 ppm, PCHC or 2.45 ppm, PvCHC) *vs.* epoxides (3.10 ppm, CHO or 5.62–5.80 ppm, vCHO). [f] Selectivity for polycarbonate *vs.* polyether, determined by comparing the integrals of resonances in the ^1H NMR spectra at 4.65 ppm (PCHC) or 4.75–4.90 ppm (PVCHC) (corresponding to polymer carbonate linkages), and at 3.45 ppm (corresponding to polymer ether linkages). [g] Determined by SEC, in THF using narrow M_w polystyrene as standard to calibrate the instrument.

amount of CHD to a 10-fold excess also enabled successful polymerization with the selectivity remaining very high.

As would be expected, increased quantities of CHD led to a decrease in polycarbonate M_n due to multiple polymer chains resulting per metal centre. As stated earlier, bimodal molecular weight distributions were observed for both PCHC and PvCHC. In contrast, the addition of an excess of CHD suppressed this bimodality and produced polycarbonates with monomodal molecular weight distributions (Fig. 5 and S8†).

Based on the successful results using CHD (a bifunctional alcohol) as a chain transfer agent, it was of interest to investigate the addition of a tri-functional alcohol to produce branched polycarbonates (Table 3). The efficient copolymerization of CO_2 and CHO was undertaken in the presence of 20 mol equiv. *vs.* catalyst **1**, of triethanolamine ($N(CH_2CH_2OH)_3$, TEA) (Table 3, entry 5). The polycarbonate was analysed by ^1H and $^{31}P\{^1H\}$ NMR spectroscopy, MALDI-ToF and SEC; all of which confirmed the formation of a perfectly alternating copolymer. The synthesis of tri-functional, or branched, PCHC was confirmed from the ^1H NMR spectrum (Fig. S9†), where the ethylene protons of the NCH_2 and OCH_2 groups of the TEA segment were observed at 2.86 and 4.17 ppm, respectively.

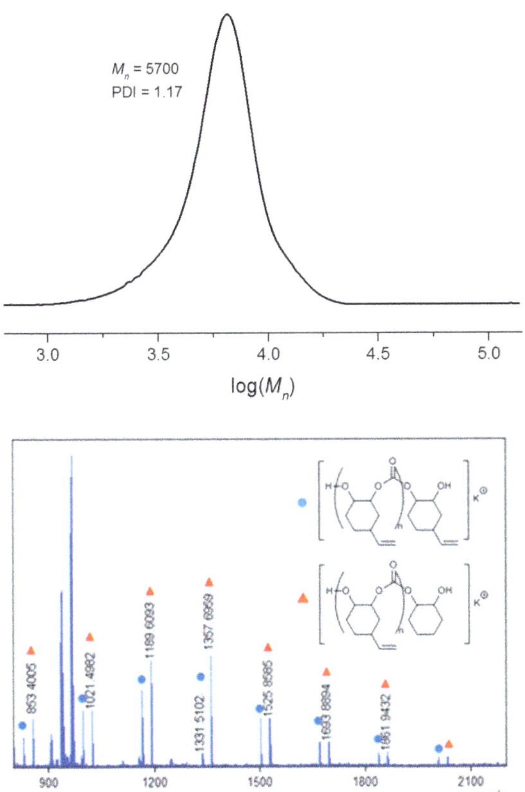

Fig. 5 Top: the monomodal molecular weight distribution of PvCHC (Table 3 entry 4), as determined by SEC, using 10 equiv. of CHD in the polymerization. Bottom: MALDI-ToF spectrum of the PvCHC produced in Table 3 entry 4, using 10 equiv. of CHD.

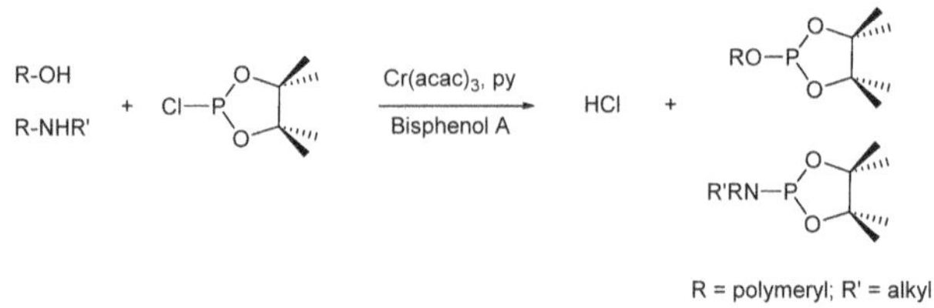

Scheme 3 The reaction between alcohol or amine and 2-chloro-4,4,5,5-tetramethyl dioxaphospholane.

Furthermore, only secondary alcohol groups (145.5 ppm) were identified using a $^{31}P\{^{1}H\}$ NMR spectroscopic analysis method (Fig. S10†). The method involves reacting a sample of the polymer with 2-chloro-4,4,5,5-tetramethyl dioxaphospholane, using bis(phenol)A as an internal standard (Scheme 3).[42–44] It indicates that, within the detection limits of the technique, all polymerizations were successfully initiated from all the hydroxyl functionalities of TEA as all the primary alcohol groups were consumed and only secondary hydroxyl groups, due to cyclohexanol end groups were observed. The MALDI-ToF spectrum (Fig. S11†) showed a major series corresponding to the dihydroxyl end-capped PCHC. The absence of the expected series, corresponding to PCHC initiated from TEA, is proposed to be due to difficulties in branched chains. In this case, the rate of polymerization was relatively slower (TOF $= 7\ h^{-1}$), yielding a polymer with M_n of 1500 g mol^{-1} ($Đ = 1.23$).

Polymerizations with added amines

Amine based post combustion capture of CO_2 has recently received much attention as a promising method for the reduction of atmospheric CO_2 emissions.[45] Approximately, 0.2 ppm of amine, namely methanolamine (MEA) and methylamine, in conjunction with 20 ppm ammonia are found as common contaminants. Denitrification is required to remove the amine-based contaminants in the production of reclaimed CO_2.[46,47]

It has already been established that the successful ROCOP of CHO and reclaimed CO_2 occurs using analogous a di-magnesium catalyst (Table 1, entry 6).[23] It was, therefore, of interest to establish whether $LZn_2(O_2CCF_3)_2$ (**1**) can successfully catalyze ROCOP in the presence of amine additives.

The polymerizations of CHO and CO_2 with primary (H_2NBn) and secondary ($HNBn_2$) amine additives occurred successfully (Table 3, entries 6 and 7, respectively). Slightly lower polymerization rates (TOF $< 20\ h^{-1}$) were obtained in both cases compared to the polymerizations in the absence of any additives (TOF $= 25\ h^{-1}$, Table 1, entry 2), whereas in all cases the polymerization selectivities were high. The ^{1}H NMR spectra of the isolated PCHC clearly indicate the incorporation of the amine reagents into the polymer chains. A representative ^{1}H NMR spectrum of the PCHC from Table 3, entry 7 is shown in Fig. 6, illustrating the key resonances assigned to the benzyl protons at 7.32–7.22 and 4.37 ppm, respectively. In addition, SEC analysis using a UV detector also served to confirm the incorporation of the benzyl units as end-groups to the polymers (Fig. S12†).

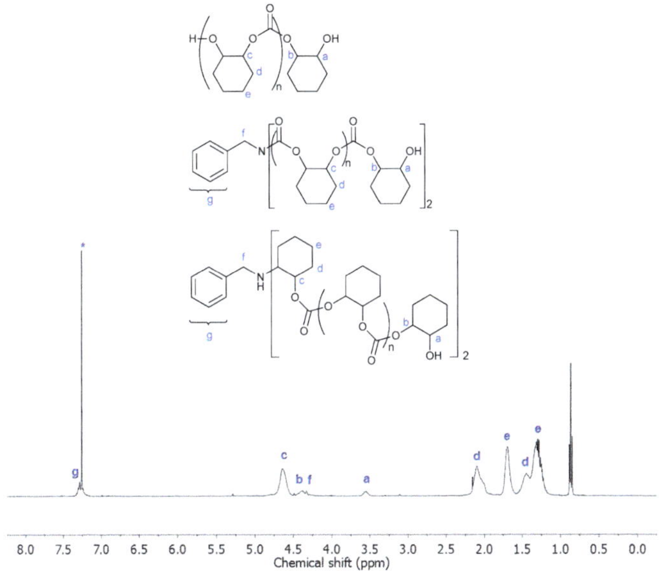

Fig. 6 ^1H NMR spectrum (400.0 MHz, $CDCl_3$-d_1, 298 K) of the isolated PCHC produced in the polymerization with H_2NBn additive (Table 3, entry 7). The asterisk denotes the residual protio-solvent.

The MALDI-ToF mass spectra showed three series attributed to polycarbonates end-capped by: (1) dihydroxy groups, (2) amine/hydroxyl groups and (3) carbamate/hydroxyl groups (Fig. 7).

The presence of both amine and amide end groups must be due to the formation of both the carbamate alcohol (a, Scheme 4) and the amino alcohol (b, Scheme 4) from the reaction between amine, CHO and CO_2, in the presence of

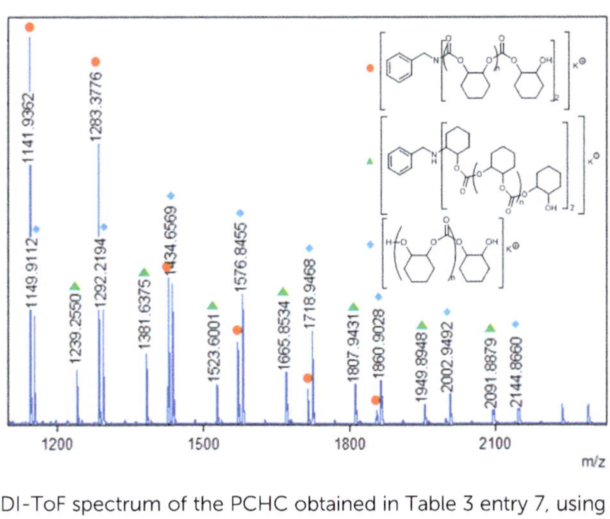

Fig. 7 MALDI-ToF spectrum of the PCHC obtained in Table 3 entry 7, using 20 equiv. of H_2NBn.

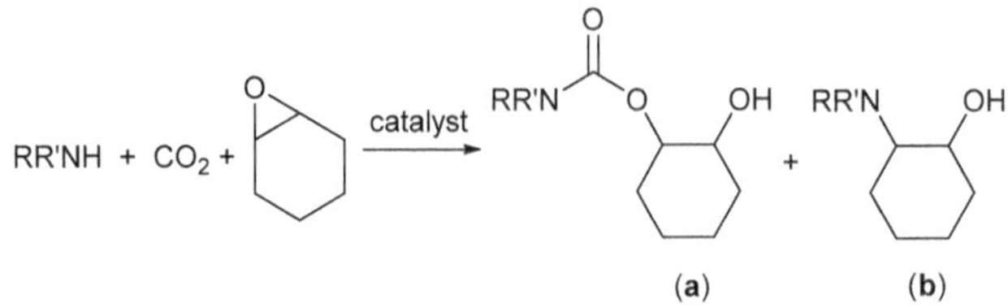

Scheme 4 The synthesis of carbamate alcohol and amino alcohol.

catalyst **1**. Other researchers have reported the reactions of amines, carbon dioxide and epoxides can lead to the formation of both products, although it is notable that in these cases more forcing conditions were generally required.[48–50]

A control experiment was carried out to examine the behaviour of the amine reagents under the polymerization conditions which showed that polymerization only occurred in the presence of the catalyst, *i.e.* the amines themselves are not able to 'catalyze' polymerization. This observation reinforces the proposed metal-mediated ROCOP pathway, whereby monomers are activated through the coordination to the metal centres prior to reactions with the amines. The carbamate alcohol and amino alcohol intermediates are expected to act as chain transfer reagents in the polymerizations. These polymerizations occurred with slightly slower rates than in the absence of additives, this may be due to inhibition of monomer binding by the amines and/or slower initiation processes. The polymers were analyzed using MALDI-ToF MS and by $^{31}P\{^{1}H\}$ NMR spectroscopy, after reaction with 2-chloro-4,4,5,5-tetramethyl dioxaphospholane. The resonances were characteristic of polycarbonates with secondary hydroxyl end-groups, with peaks at 145.5 ppm in $^{31}P\{^{1}H\}$ NMR spectra (C and D, Fig. S10†). Whereas the reaction between dibenzylamine and the dioxaphospholane resulted in a peak at 142.5 ppm in the $^{31}P\{^{1}H\}$ NMR spectrum (B, Fig. S10†). Thus supporting the involvement of the NH groups in the polymerizations, as illustrated in Scheme 4.

The scope of polymerizations in the presence of amines was expanded to include bi-functional reagents, such as, monoethanolamine (MEA) and ethylenediamine (EN). The formation of a white suspension was observed during the initial stages of polymerizations using these additives (Table 3, entries 10 and 11, respectively). Such observations are consistent with the formation of ammonium salts, as part of equilibria between amines, carbon dioxide and carbamic acid.[48–50] The white suspension dissolved as the polymerization proceeded, and the reactions resulted in the formation of low molecular weight PCHC, with a slightly broader dispersity ($M_n = 2200$ g mol^{-1}, $Đ = 1.21$ for MEA and $M_n = 2900$ g mol^{-1}, $Đ = 1.21$ for EN) compared to that obtained from the control experiment ($M_n =$ 6600 g mol^{-1}, $Đ = 1.18$). Once again, the polymerization activities were significantly lower (TOF = 9 h^{-1}) than that of the control experiment. The polymer chain end group analysis, using the reaction with 2-chloro-4,4,5,5-tetramethyl dioxaphospholane, showed the formation of polycarbonates with secondary alcohol end-groups (E and F, Fig. S10†), however, the MALDI-ToF spectra were highly complex and as yet cannot be fully assigned. These findings indicate that although successful polymerizations did occur in the presence of primary amines, the precise polymer structures may be complicated by carbamate formation at the chain ends.

Experimental section

Materials and methods

All manipulations were carried out using standard Schlenk line or dry-box techniques under an inert atmosphere of nitrogen. All glassware was dried at 160 °C for 20 h and cooled under vacuum prior to use. Solvents were dried by passing through a column of appropriate drying agent, degassed and stored under nitrogen atmosphere. Chloroform-d_3 was dried over CaH_2, distilled under reduced pressure and stored over 4 Å molecular sieves, under nitrogen atmosphere.

Ligand H_2L and catalyst **1** were synthesised according to the previously published literature. Cyclohexene oxide and vinyl cyclohexene oxide were purchased from Acros and Sigma-Aldrich, respectively, and fractionally distilled from CaH_2. Cyclohexene diol was recrystallised from ethyl acetate and stored under nitrogen atmosphere. Benzylamine (99%), dibenzylamine (97%), monoethanolamine (99.5%), ethylenediamine (99.5%) and triethanolamine (98%) were purchased from Sigma-Aldrich, degassed and stored under nitrogen atmosphere. High-purity CO_2 was obtained by passing research grade CO_2 (99.99995%) purchased from BOC (Linde Gas) through a high performance purifier (Drierite column).

^{1}H, $^{13}C\{^{1}H\}$, $^{31}P\{^{1}H\}$ NMR spectra were recorded using a Bruker AV400 MHz spectrometer at ambient temperature (unless stated otherwise). ^{1}H, $^{13}C\{^{1}H\}$ NMR spectra were referenced internally to residue proteo-solvent (^{1}H) or solvent (^{13}C) resonances, and are reported relative to tetramethylsilane. SEC data were determined by a Shimadzu LC-20AD instrument using MALLS detector (Wyatt Dawn 8+), with THF as the eluent, at a flow rate of 1.0 mL min^{-1} at 30 °C. Two Mixed Bed PSS SDV linear S columns were used in series. The MALLS detector was calibrated using a polystyrene standard. MALDI-ToF MS experiments were carried out on Waters/Micromass MALDI micro MX spectrometer, using a dithranol matrix in THF at a 1 : 1 loading with potassium trifluoroacetate (KO_2CCF_3) as the cationizing agent. Elemental analyses were carried out using the Elemental Analysis Service at London Metropolitan University.

Synthetic procedures

General polymerization protocol

Copolymerization of cyclohexene oxide or vinyl cyclohexene oxide and carbon dioxide. Cyclohexene oxide (1.67 mL, 17 mmol) or vinyl cyclohexene oxide (2.15 mL, 17 mmol) and $LZn_2(O_2CCF_3)$ (**1**, 15.0 mg, 0.017 mmol) were added into a Schlenk tube. The epoxide was degassed, before being left stirring under 1 atm CO_2 at 80 °C for a pre-determined time period. The crude reaction mixture was dried *in vacuo* to remove unreacted epoxide. The polymer was purified by repeat precipitations from methylene chloride using methanol.

Separation of poly (cyclohexene carbonate) (PCHC) with a bimodal molecular weight distribution. 100 mg of PCHC in THF (1.5 mL) was analysed using the previously stated SEC equipment. The delay time of the instrument was calculated based on the flow rate and the column length. Different fractions were collected manually according to the pre-determined time (elution and delay time).

Polymerizaitons in the presence of chain transfer agents. The general procedure stated above was followed, whereby multi-equivalents of chain transfer agents were added to the reaction mixture prior to the degassing process.

Conclusions

In conclusion, a di-zinc trifluoroacetate catalyst was investigated for the copolymerization of cyclohexene oxide or vinyl-cyclohexene oxide and carbon dioxide. The catalyst shows moderate/good activity and productivity and very high selectivity for carbonate linkage formation. It also selectively produces polycarbonate polyols, with bimodal molecular weight distributions. The catalyst functions well in the presence of additives, including water, diols and amines. These added reagents lead to good catalytic performances and enable control over the molecular weight of the polymers and the production of polyols with monomodal molecular weight distributions. The catalyst shows a high tolerance to such contaminants, including the ability to selectively prepare polyols in the presence of >5000 ppm of water.

Acknowledgements

The authors acknowledge the Engineering and Physical Sciences Research Council (EPSRC grants: EP/K035274/1, EP/L017393/1, EP/H046380) for funding.

Notes and references

1 P. Markewitz, W. Kuckshinrichs, W. Leitner, J. Linssen, P. Zapp, R. Bongartz, A. Schreiber and T. E. Muller, *Energy Environ. Sci.*, 2012, **5**, 7281–7305.
2 N. MacDowell, N. Florin, A. Buchard, J. Hallett, A. Galindo, G. Jackson, C. S. Adjiman, C. K. Williams, N. Shah and P. Fennell, *Energy Environ. Sci.*, 2010, **3**, 1645–1669.
3 M. Mikkelsen, M. Jorgensen and F. C. Krebs, *Energy Environ. Sci.*, 2010, **3**, 43–81.
4 D. J. Darensbourg, *Inorg. Chem.*, 2010, **49**, 10765–10780.
5 D. J. Darensbourg and A. D. Yeung, *Polym. Chem.*, 2014, **5**, 3949–3962.
6 M. I. Childers, J. M. Longo, N. J. Van Zee, A. M. LaPointe and G. W. Coates, *Chem. Rev.*, 2014, **114**, 8129–8152.
7 X.-B. Lu and D. J. Darensbourg, *Chem. Soc. Rev.*, 2012, **41**, 1462–1484.
8 M. R. Kember, A. Buchard and C. K. Williams, *Chem. Commun.*, 2011, **47**, 141–163.
9 D. J. Darensbourg, *Chem. Rev.*, 2007, **107**, 2388–2410.
10 S. Paul, Y. Zhu, C. Romain, R. Brooks, P. K. Saini and C. K. Williams, *Chem. Commun.*, 2015, **51**, 6459–6479.
11 K. Nakano, K. Kobayashi, T. Ohkawara, H. Imoto and K. Nozaki, *J. Am. Chem. Soc.*, 2013, **135**, 8456–8459.
12 M. W. Lehenmeier, S. Kissling, P. T. Altenbuchner, C. Bruckmeier, P. Deglmann, A.-K. Brym and B. Rieger, *Angew. Chem., Int. Ed.*, 2013, **52**, 9821–9826.
13 J. K. Varghese, D. S. Park, J. Y. Jeon and B. Y. Lee, *J. Polym. Sci., Part A: Polym. Chem.*, 2013, **51**, 4811–4818.
14 W. C. Ellis, Y. Jung, M. Mulzer, R. Di Girolamo, E. B. Lobkovsky and G. W. Coates, *Chem. Sci.*, 2014, **5**, 4004–4011.
15 G.-P. Wu, D. J. Darensbourg and X.-B. Lu, *J. Am. Chem. Soc.*, 2012, **134**, 17739–17745.

16 Y. Liu, W. M. Ren, M. Wang, C. Liu and X. B. Lu, *Angew. Chem., Int. Ed.*, 2015, **54**, 2241–2244.
17 L. Pena Carrodeguas, J. Gonzalez-Fabra, F. Castro-Gomez, C. Bo and A. W. Kleij, *Chem.–Eur. J.*, 2015, **21**, 6115–6122.
18 P. K. Saini, C. Romain and C. K. Williams, *Chem. Commun.*, 2014, **50**, 4164–4167.
19 S. H. Lee, A. Cyriac, J. Y. Jeon and B. Y. Lee, *Polym. Chem.*, 2012, **3**, 1215–1220.
20 J. Langanke, A. Wolf, J. Hofmann, K. Bohm, M. A. Subhani, T. E. Muller, W. Leitner and C. Gurtler, *Green Chem.*, 2014, **16**, 1865–1870.
21 N. Von der Assen and A. Bardow, *Green Chem.*, 2014, **16**, 3272–3280.
22 D. J. Darensbourg, W.-C. Chung, K. Wang and H.-C. Zhou, *ACS Catal.*, 2014, **4**, 1511–1515.
23 A. C. Chapman, C. Keyworth, M. R. Kember, A. J. J. Lennox and C. K. Williams, *ACS Catal.*, 2015, 1581–1588.
24 M. Winkler, C. Romain, M. A. R. Meier and C. K. Williams, *Green Chem.*, 2015, **17**, 300–306.
25 C. Romain, A. Thevenon, P. K. Saini and C. K. Williams, in *Topics in Organometallic Chemistry: Carbon Dioxide and Organometallics*, ed. X. B. Lu, Springer DE, 2015.
26 P. K. Saini, C. Romain, Y. Zhu and C. K. Williams, *Polym. Chem.*, 2014, **5**, 6068–6075.
27 M. R. Kember and C. K. Williams, *J. Am. Chem. Soc.*, 2012, **134**, 15676–15679.
28 M. R. Kember, J. Copley, A. Buchard and C. K. Williams, *Polym. Chem.*, 2012, **3**, 1196–1201.
29 A. Buchard, F. Jutz, M. R. Kember, A. J. P. White, H. S. Rzepa and C. K. Williams, *Macromolecules*, 2012, **45**, 6781–6795.
30 F. Jutz, A. Buchard, M. R. Kember, S. B. Fredrickson and C. K. Williams, *J. Am. Chem. Soc.*, 2011, **133**, 17395–17405.
31 A. Buchard, M. R. Kember, K. G. Sandeman and C. K. Williams, *Chem. Commun.*, 2011, **47**, 212–214.
32 M. R. Kember, A. J. P. White and C. K. Williams, *Macromolecules*, 2010, **43**, 2291–2298.
33 M. R. Kember, A. J. P. White and C. K. Williams, *Inorg. Chem.*, 2009, **48**, 9535–9542.
34 M. R. Kember, P. D. Knight, P. T. R. Reung and C. K. Williams, *Angew. Chem., Int. Ed.*, 2009, **48**, 931–933.
35 M. R. Kember, F. Jutz, A. Buchard, A. J. P. White and C. K. Williams, *Chem. Sci.*, 2012, **3**, 1245–1255.
36 Y. Liu, W. M. Ren, K. K. He and X. B. Lu, *Nat. Commun.*, 2014, **5**, 1–7.
37 R. K. Dean, L. N. Dawe and C. M. Kozak, *Inorg. Chem.*, 2012, **51**, 9095–9103.
38 M. W. Lehenmeier, C. Bruckmeier, S. Klaus, J. E. Dengler, P. Deglmann, A.-K. Ott and B. Rieger, *Chem. – Eur. J.*, 2011, **17**, 8858–8869.
39 K. Nakano, M. Nakamura and K. Nozaki, *Macromolecules*, 2009, **42**, 6972–6980.
40 D. J. Darensbourg and G.-P. Wu, *Angew. Chem.*, 2013, 10796–10800.
41 F. Jutz, A. Buchard, M. R. Kember, S. B. Fredriksen and C. K. Williams, *J. Am. Chem. Soc.*, 2011, **133**, 17395–17405.
42 Y. Zhu, C. Romain, V. Poirier and C. K. Williams, *Macromolecules*, 2015, **48**, 2407–2416.
43 E. Hatzakis, E. Archavlis and P. Dais, *J. Am. Oil Chem. Soc.*, 2007, **84**, 615–619.

44 A. Spyros, D. S. Argyropoulos and R. H. Marchessault, *Macromolecules*, 1997, **30**, 327–329.
45 N. MacDowell, N. Florin, A. Buchard, J. Hallett, A. Galindo, G. Jackson, C. S. Adjiman, C. K. Williams, N. Shah and P. Fennell, *Energy Environ. Sci.*, 2010, **3**, 1645–1669.
46 IEAGHG, Gaseous emissions from amine based PCC processes and their deep removal, May, 2012.
47 I. Hauser, A. Einbu, H. F. Stevendsen and K. Ostgaard, *J. Exp. Clin. Med.*, 2014, **1**, 1–5.
48 Y. Yoshida and S. Inoue, *J. Chem. Soc., Perkin Trans. 1*, 1979, 3146–3150.
49 Y. Yoshida and S. Inoue, *Chem. Lett.*, 1978, **7**, 139–140.
50 D. Chaturvedi, *Tetrahedron*, 2012, **68**, 15–45.

Faraday Discussions

PAPER

Thioether-triphenolate bimetallic iron(III) complexes as robust and highly efficient catalysts for cycloaddition of carbon dioxide to epoxides†

Antonio Buonerba,[ab] Francesco Della Monica,[a] Assunta De Nisi,[ac] Ermanno Luciano,[a] Stefano Milione,[ab] Alfonso Grassi,[ab] Carmine Capacchione*[ab] and Bernhard Rieger[d]

Received 11th May 2015, Accepted 21st May 2015
DOI: 10.1039/c5fd00070j

The selective and effective synthesis of organic carbonates under mild conditions, starting from carbon dioxide and oxiranes, catalyzed by metal complexes is currently a focus of interest for both industrial and academic researchers. We recently developed a novel thioether-triphenolate iron(III) catalyst (**C**$_{t\text{-Bu}}$) that has proven to be highly active for the coupling of CO_2 with epoxides, resulting in cyclic organic carbonates under solvent-free conditions. In the current work, the properties of this novel class of catalysts were extensively investigated. In particular, the steric properties of the ligand were modulated by changing the substituents of the aromatic rings in order to obtain a deeper knowledge of the relationship between the complex structure and catalytic performance/selectivity for these iron complexes. Notably, the less steric demanding iron(III) **C**$_{H}$ complex synthesized shows, when activated by *n*-tetrabutylammonium bromide, an impressive turnover frequency (TOF) of 3800 h^{-1} for the formation of propylene carbonate and glycerol carbonate which are, by far, the highest reported for an iron based catalyst and compares well with the most active catalyst based on other metals.

Introduction

The possibility of converting carbon dioxide into useful chemicals has become an important area of research in both academic and industrial environments.[1] For

[a]*Dipartimento di Chimica e Biologia "Adolfo Zambelli", Università degli Studi di Salerno, via Giovanni Paolo II 132, 84084 Fisciano (SA), Italy. E-mail: ccapacchione@unisa.it*
[b]*CIRCC, Interuniversity Consortium Chemical Reactivity and Catalysis, via Celso Ulpiani 27, 70126 Bari, Italy*
[c]*Dipartimento di Chimica G. Ciamician, Alma Mater Studiorum, Università di Bologna, via Selmi 2, Bologna, Italy*
[d]*WACKER Lehrstuhl für Makromolekulare Chemie, Technische Universität München, Lichtenbergstrasse 4, 85748 Garching, Germany*

† Electronic supplementary information (ESI) available: Further details of NMR, ESI-MS, FT-IR and UV-Vis spectroscopic characterizations. See DOI: 10.1039/c5fd00070j

example, the reduction of CO_2 by reaction with hydrogen or water to produce formic acid,[2] formaldehyde and methanol[3] has been intensively studied for the intrinsic value of these molecules as chemical intermediates or fuel.[4] Like other industrially relevant chemical processes, the key to overcoming the chemical sluggishness of CO_2 is the development of properly designed catalytic systems able to enhance the reactivity and selectivity of such chemical transformations. In this scenario, a prominent role is played by the coupling of CO_2 with epoxides to form cyclic organic carbonates (COCs)[5] and polycarbonates.[6] In particular, COCs, due their high molecular dipole moments, dielectric constants and boiling temperatures, are employed as *green* highly polar aprotic solvents and as ion-carriers for lithium-based batteries.[7] Furthermore, they find application as valuable intermediates in organic synthesis and as monomers in the ring opening polymerization to form polycarbonates. As a result of the growing interest in using COCs in chemical synthesis, the development of efficient, selective, non-toxic and inexpensive catalytic systems able to promote the coupling between CO_2 and oxiranes or oxetanes has become a focus of the field of CO_2 chemistry.

Generally, the most active catalytic systems are based on the combination of a transition metal complex acting as a Lewis acid and a suitable nucleophile (ammonium salts and phosphonium salts).[8] The widely accepted mechanism involves Lewis acid activation of the epoxide toward the ring opening by nucleophilic attack on the less substituted carbon atom (see Scheme 1). This reaction is followed by CO_2 insertion into the formed metal-alkoxide intermediate leading to an alkyl emi-carbonato unit which can undergo the ring closure, *via* a backbiting mechanism producing the cyclic carbonate (**A**), or by inserting additional epoxide/CO_2 molecules producing polycarbonates (**B**). The fortuitous insertion of two consecutive epoxide molecules also leads to the formation of polyether segments (**C**). Consequently, the choice of metal complex, the ratio between the Lewis acid and the nucleophile and the nature of the co-catalyst, the solvent, and the pressure and temperature are of pivotal importance for the selectivity of the reaction.[5a]

We have recently developed a new catalytic system[9] based on iron,[10] an abundant and non-toxic metal. In particular we have reported that the dinuclear Fe(III) ($\mathbf{C}_{t\text{-}\mathbf{Bu}}$) (see Scheme 2) complex coordinated by dithioether-triphenolate-based ligands in combination with tetrabutylammonium bromide (TBAB) is an active and selective catalyst for the synthesis of various COCs. Notably, the

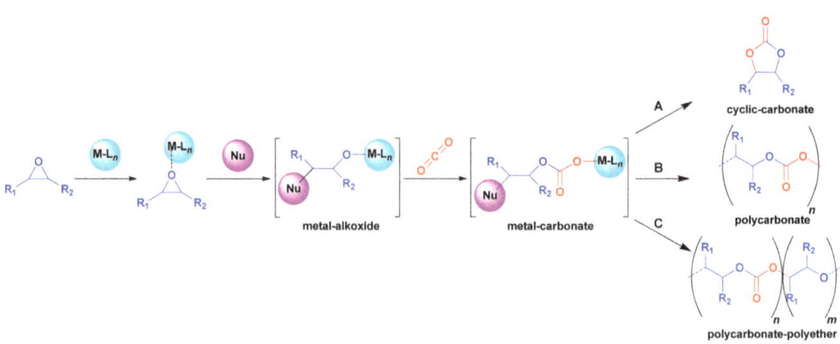

Scheme 1 Metal catalysed carbon dioxide–epoxide coupling.

coupling of propylene oxide with CO_2, in the presence of the aforementioned catalytic system, gives the corresponding carbonate with the highest TOF (580 h^{-1}) so far reported for an iron based catalyst.[11]

In order to establish a more refined catalyst structure–reactivity relationship for this new Fe(III) catalyst family, we report here the synthesis of two new iron(III) complexes, $\mathbf{C_H}$ and $\mathbf{C_{Cum}}$. We report on their influence on the catalytic performance, in terms of activity and selectivity, in the CO_2/epoxide coupling reaction of the substituents on the aromatic rings in the ligands $\mathbf{L_H}$–$\mathbf{L_{Cum}}$.

Experimental

All manipulation involving air- and/or moisture-sensitive compounds were performed under nitrogen atmosphere using a standard Schlenk technique and a MBraun glovebox. Toluene (99.5%; Carlo Erba) and THF (99%; Sigma-Aldrich) were used as received or refluxed for 48 h over sodium and distilled before use for the moisture- and oxygen-sensitive reactions. All other reagents were used as received (Sigma-Aldrich) or distilled under reduced pressure over calcium hydride. The ligand precursor 6,6′-(((5-(*tert*-butyl)-2-hydroxy-1,3-phenylene)bis-(methylene))bis(sulfanediyl))-bis(2,4-di-*tert*-butylphenol) ($\mathbf{L}_{t\text{-}\mathbf{Bu}}$, Scheme 2) and the corresponding iron(III) complex ($\mathbf{C}_{t\text{-}\mathbf{Bu}}$, Scheme 2) were synthesized according to the reported procedure.[9] Deuterated solvents were purchased from Euriso-Top or Sigma-Aldrich and used as received.

NMR spectra were collected on Bruker Avance spectrometers (600, 400, 300 or 250 MHz for ^1H), and the chemical shifts were referenced to tetramethylsilane (TMS) as the external reference, using the residual protio signal of the deuterated solvents. Measurements of effective magnetic moments were performed on a Bruker Avance 600 MHz spectrometer at 25 °C in CD_3CN using a 5 mm Wilmad coaxial insert NMR tube. Solutions of the complex $\mathbf{C_H}$ (1.7 mM) and $\mathbf{C_{Cum}}$ (2.1 mM) in CD_3CN with 1% of TMS were prepared under a nitrogen atmosphere. The effective magnetic moment (μ_{eff}) was calculated from $\mu_{\text{eff}} = 8\chi_g M_w T$, where χ_g ($cm^3\ g^{-1}$) is the corrected molar susceptibility derived from $\chi_g = 3\Delta_f/4\pi f_o C M_w + \chi_o$,[12] Δ_f is the shift in frequency (Hz) of the signal of the solvent in the presence of the complex from the value of the pure solvent, C is the concentration of the complex (mol cm^{-3}), M_w is the molecular weight of the complex (g mol^{-1}), f_o is the operating frequency of the spectrometer (Hz), and χ_o is the mass susceptibility of the pure solvent (-0.524×10^{-6} $cm^3\ g^{-1}$ for CD_3CN). The $4\pi/3$ is the shape factor

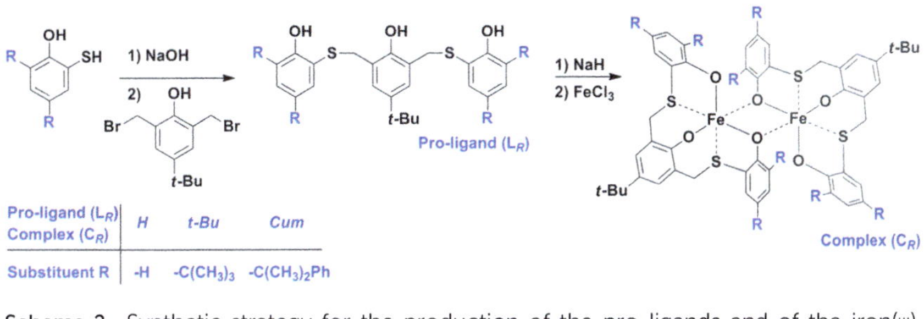

Scheme 2 Synthetic strategy for the production of the pro-ligands and of the iron(III) complexes.

for a cylindrical sample in a superconducting magnet. Elemental analysis was performed on a CHNS Thermo Scientific Flash EA 1112 equipped with a thermal conductivity detector. ESI-MS spectra were acquired on a Quattro *micro*™ API triple quadrupole mass spectrometer from Waters equipped with an electrospray ion source, using acetonitrile as the solvent. FT-IR measurements were carried out on a Bruker Vertex 70 spectrometer equipped with a DTGS detector and a Ge/KBr beam splitter. The samples were analysed as chloroform solutions or in the form of KBr disks. UV-Vis acquisitions were collected on a PerkinElmer Lambda EZ 201.

Synthesis of the pro-ligand 2,2′-(((5-(*tert*-butyl)-2-hydroxy-1,3-phenylene) bis(methylene))bis(sulfanediyl))-bisphenol (L_H, Scheme 2)

A 500 mL two-neck round-bottom flask equipped with a condenser and a magnetic stirring bar was charged, under a nitrogen atmosphere, with 5.00 g of 2-mercaptophenol (35.7 mmol) dissolved in 160 mL of ethanol, and 1.43 g of NaOH (35.7 mmol); the mixture was refluxed for about 1 hour until complete dissolution of the hydroxide. 6.00 g of 2,6-dibromomethyl-4-*t*-butylphenol (17.83 mmol) dissolved in 50 mL of ethanol were slowly added at 0 °C and the mixture was heated to the reflux of the solvent that was kept overnight. The solvent was distilled off, water was added until dissolution of the NaBr by-product and the aqueous phase was extracted twice with methylene chloride. The combined organic phases were dried with $MgSO_4$ and, after evaporation of the solvent, the product was recovered as a white solid. Yield: 7.36 g, 96.8%. EA for $C_{24}H_{26}O_3S_2$ calc.: C, 67.57; H, 6.14; S 15.03; found: C, 67.27; H, 6.07; S, 14.93. Mass spectrum: 449.3 *m*/*z* (MNa^+). Spectroscopic data: ^{1}H-NMR (400 MHz, CD_2Cl_2, 25 °C): δ 1.02 (9H, s); δ 3.93 (4H, s); δ 5.82 (1H, s); δ 6.61 (2H, s); δ 6.75 (1H, s); δ 6.80 (2H, t); δ 6.95 (2H, d); δ 7.26 (4H, t). ^{13}C-NMR (250 MHz, CD_2Cl_2, 25 °C): δ 31.21; δ 34.01; δ 37.02; δ 115.21; δ 118.12; δ 121.05; δ 123.78; δ 127.69; δ 131.78; δ 134.94; δ 143.34; δ 150.00; δ 157.66.

Synthesis of the iron(III) complex C_H (Scheme 2)

A 500 mL two-neck round-bottom flask, equipped with a magnetic stirring bar, was charged with 0.51 g of sodium hydride (21.0 mmol) suspended in 130 mL of THF. A solution of 2.72 g of the ligand precursor L_H (6.37 mmol) dissolved in 180 mL of THF was slowly added at room temperature under a protective nitrogen atmosphere, and allowed to react overnight. The resulting suspension was filtered through celite and slowly added at room temperature to 1.00 g of anhydrous iron(III) chloride (6.24 mmol) dissolved in 100 mL of THF. The rapid colour change to a deep purple was observed and the reaction was kept overnight. The mixture was then filtered through celite and the solvent removed under reduced pressure affording a deep purple crystalline solid. Yield: 2.80 g, 94%. EA for $C_{48}H_{46}Fe_2O_6S_4$ calc.: C, 60.13; H, 4.84; S, 13.38; found: C, 60.33 H, 4.47; S, 13.28. Mass spectrum: 981.2 *m*/*z* (MNa^+). UV-Vis: $\varepsilon_{470} = 2804$ L mol^{-1} cm^{-1}.

Synthesis of the pro-ligand 2,2′-(((5-(*tert*-butyl)-2-hydroxy-1,3-phenylene) bis(methylene))bis(sulfanediyl))-bis(2,4-di-(α,α′-dimethylbenzyl)phenol) (L_{Cum}, Scheme 2)

A 50 mL two-neck round-bottom flask equipped with a condenser and a magnetic stirring bar was charged, under a nitrogen atmosphere, with 2.02 g of 2-mercapto-4,6-bis-(α,α-dimethylbenzyl)-2-phenol (5.56 mmol) dissolved in 25 mL of ethanol,

0.30 g of NaOH (22.2 mmol) and the mixture was refluxed for about 1 hour until complete dissolution of the hydroxide. 0.94 g of 2,6-dibromomethyl-4-*t*-butylphenol (2.8 mmol) dissolved in 10 mL of ethanol was slowly added at 0 °C and the mixture was heated to the reflux of the solvent that was kept overnight. The solvent was distilled off, water was added until dissolution of the NaBr by-product, and the aqueous phase was extracted twice with methylene chloride. The combined organic phases were dried with $MgSO_4$ and, after evaporation of the solvent, the resulting crude solid was purified by column chromatography (light petroleum/ethyl acetate = 95/5). Yield: 0.81 g, 33%. EA for $C_{60}H_{66}O_3S_2$ calc.: C, 80.13; H, 7.40; S 7.13; found: C, 79.94; H, 7.27; S, 7.03. Mass spectrum: 899.2 *m*/*z* (M^+), 922.2 *m*/*z* (MNa^+). Spectroscopic data: ^{1}H-NMR (400 MHz, CD_2Cl_2, 25 °C): δ 1.13 (9H, s); δ 1.54 (12H, s); δ 1.59 (12H, s); δ 3.80 (4H, s); δ 5.97 (1H, s); δ 6.44 (2H, s); δ 6.76 (2H, s); δ 7.10–7.24 (22H, Ar-H). ^{13}C-NMR (400 MHz, CD_2Cl_2, 25 °C): δ 29.61; δ 31.10; δ 31.58; δ 34.16; δ 36.80; δ 42.71; δ 42.83; δ 118.46; δ 123.25; δ 125.93; δ 126.04; δ 127.04; δ 127.61; δ 128.09; δ 128.31; δ 128.40; δ 132.53; δ 134.83; δ 142.28; δ 143.33; δ 150.51; δ 150.91; δ 151.06; δ 153.19.

Synthesis of the iron(III) complex $\mathbf{C_{Cum}}$ (Scheme 2)

A 50 mL two-neck round-bottom flask, equipped with a magnetic stirring bar was charged with 0.073 g of sodium hydride (2.60 mmol) suspended in 20 mL of THF, and a solution of 0.78 g of the ligand precursor $\mathbf{L_{Cum}}$ (0.87 mmol) dissolved in 25 mL of THF was slowly added at room temperature under a protective nitrogen atmosphere, and allowed to react overnight. The resulting suspension was filtered through celite and slowly added at room temperature to 0.138 g of anhydrous iron(III) chloride (0.85 mmol) dissolved in 20 mL of THF. A rapid colour change to deep blue was observed and the reaction was kept overnight. The mixture was then filtered through celite and the solvent removed under reduced pressure affording a deep blue crystalline solid. Yield: 0.79 g, 97.7%. EA for $C_{120}H_{126}Fe_2O_6S_4$ calc.: C, 75.69; H, 6.67; S, 6.74; found: C, 75.42; H, 6.58; S, 6.67. Mass spectrum: 1904 *m*/*z* (M^+), 1927 *m*/*z* (MNa^+). UV-Vis: $\varepsilon_{585} = 7188$ L mol^{-1} cm^{-1}.

Typical procedure for CO_2/epoxide coupling to cyclic carbonates catalyzed by C_H/TBAB (referred to entry 3, Table 1)

A 60 mL stainless steel pressure reactor equipped with a magnetic stirring bar was charged, under CO_2 atmosphere, with 17.1 mg of catalyst $\mathbf{C_H}$ (1.79×10^{-5} mol) and 11.5 mg of TBAB (3.57×10^{-5} mol) dissolved in 5.0 mL of PO (7.15×10^{-2} mol). The reaction mixture was pressurized with CO_2 at 2 MPa and stirred at 120 °C for 1 h. The reactor was cooled with ice, the CO_2 was released, and mesitylene was added as an internal standard before the mixture was analysed by ^{1}H-NMR spectroscopy using CD_2Cl_2 as the solvent. Yield 49.2%.

Results and discussion

The ligands $\mathbf{L_H}$–$\mathbf{L_{Cum}}$ were synthesized using the same synthetic strategy reported for the synthesis of ligand $\mathbf{L_{\mathit{t}\text{-}Bu}}$ (see Scheme 2).[9] The new ligands were purified by recrystallization or by column chromatography, and fully characterized using NMR, elemental analysis, MS, and FT-IR (see ESI†).

These ligands were chosen in order to study the influence of the substituents on the aromatic rings on the catalytic activity resulting by diminishing ($\mathbf{L_H}$) and increasing ($\mathbf{L_{Cum}}$) the steric bulk. The complexes $\mathbf{C_H}$ and $\mathbf{C_{Cum}}$ were obtained by reaction of the corresponding sodium salt obtained by reaction with 3 equivalents of NaH in THF with an equimolar amount of anhydrous $FeCl_3$ in analogy to the procedure used for the synthesis complex $\mathbf{C_{\textit{t}\text{-}Bu}}$.[9]

The complexes, $\mathbf{C_H}$ and $\mathbf{C_{Cum}}$, were recovered by crystallization in tetrahydrofuran as small needle-shaped crystals with a red-purplish and deep-blue colour respectively. Unfortunately, in both cases, the crystals were not suitable for structural resolution by means of single crystal X-ray diffraction. Indeed the formation of the adduct iron-ligand was initially confirmed using elemental analysis and FT-IR spectroscopy, which revealed the disappearance of the vibration due to the hydroxyl of the pro-ligand as a result of the coordination of the corresponding phenolate to the iron (see Fig. S11 and S17†). The ESI-MS spectra of $\mathbf{C_H}$ and $\mathbf{C_{Cum}}$ show molecular ion peaks at 899 and 1927 *m*/*z* respectively, which are consistent with the dinuclear structure observed in the case of complex $\mathbf{C_{\textit{t}\text{-}Bu}}$. The Evans method[12] was applied for the determination of the solution magnetic susceptibility and a value of 8.1 μB for $\mathbf{C_H}$ and 7.2 μB for $\mathbf{C_{Cum}}$ resulted at 25 °C, which are close to the values calculated for two isolated high spin (HS) iron(III) centres (8.37 μB)[13] and to that observed previously for $\mathbf{C_{\textit{t}\text{-}Bu}}$,[9] confirming the dimeric nature of the complexes and indicating some degree of ferromagnetic coupling between the iron atoms being both in a HS ($s = 5/2$) state. Consequently, the UV-Vis spectra, as expected for two HS iron(III) centres in an octahedral environment, showed no d–d transitions but only a strong ligand-to-metal charge transfer absorption at 470 nm (21 276 cm^{-1}; $\varepsilon_{470} = 2804$ L mol^{-1} cm^{-1}) for $\mathbf{C_H}$ and at 585 nm (21 276 cm^{-1}; $\varepsilon_{585} = 7188$ L mol^{-1} cm^{-1}) for $\mathbf{C_{Cum}}$. In addition, the coordination of the sulfur atoms, necessary to ensure the octahedral geometry for the iron centre, was confirmed by the shift of the vibration bands in the infrared spectra of the complex to respect that of the free pro-ligand, respectively in the spectral regions: 1450–1415 cm^{-1}, 1270–1220 cm^{-1} and 680–600 cm^{-1}, diagnostics for the alkyl-sulphide moiety (see Fig. S12–S14 and S18–S20†).[14] In particular the shift to the red, observed for the CH_2–S–C deformation (spectral region 1450–1415 cm^{-1}; Fig. S12 and S18†) and for the C–S stretching (spectral region 680–600 cm^{-1}; Fig. S14 and S20†), accounts for a weakening of the C–S bond as a result of the coordination of the sulphur to the iron, whereas the blue shift for the CH_2–S–C wagging (spectral region 1270–1220 cm^{-1}; Fig. S13 and S19†) indicates a reduced conformational mobility of the moieties after coordination to the metal.

In order to check the catalytic performances of the iron(III) complexes, $\mathbf{C_H}$ and $\mathbf{C_{Cum}}$, for the coupling of (±)-propylene oxide (PO) with carbon dioxide and directly compare the results with the previously reported catalyst $\mathbf{C_{\textit{t}\text{-}Bu}}$, we conducted several catalytic tests for the three complexes ($\mathbf{C_H}$, $\mathbf{C_{\textit{t}\text{-}Bu}}$, $\mathbf{C_{Cum}}$) under the same reaction conditions. The main results are summarized in Table 1. Due to the encouraging preliminarily results obtained with catalyst $\mathbf{C_{\textit{t}\text{-}Bu}}$ we initially decided to use a low catalyst loading (0.025% mol) and as co-catalyst the *n*-tetrabutylammonium bromide (TBAB) (with a co-catalyst/catalyst molar ratio of 2) under a CO_2 pressure of 2 MPa and solvent-free conditions. The reaction time was reduced to 1 hour in order to better evaluate the turnover frequencies (TOF) of the catalytic systems. In the temperature range 80–140 °C, an increase of activity by increasing the temperature was observed for all the titled catalysts, with the maximum

Table 1 Synthesis of propylene carbonate through CO_2/propylene oxide coupling catalyzed by Fe(III) catalysts

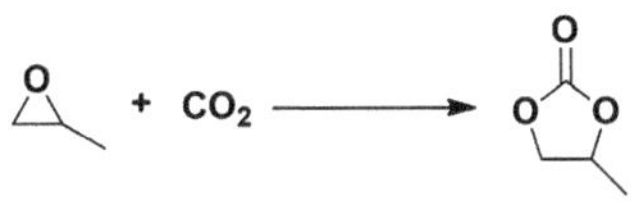

Entry[a]	Catalyst	Co-catalyst	(mol%)	PO/co-cat./cat. (molar ratio)	Solvent	T (°C)	P_{CO_2} (MPa)	Conversion[bc] (%)	TOF[d] (h^{-1})
1	$\mathbf{C_H}$	TBAB	0.05	4000/2/1	—	80	2	2.4	96
2	$\mathbf{C_H}$	TBAB	0.05	4000/2/1	—	100	2	16.9	680
3	$\mathbf{C_H}$	TBAB	0.05	4000/2/1	—	120	2	49.2	1970
4	$\mathbf{C_H}$	TBAB	0.05	4000/2/1	—	140	2	60.5	2420
5	$\mathbf{C_{\mathit{t}\text{-}Bu}}$	TBAB	0.05	4000/2/1	—	80	2	4.6	185
6	$\mathbf{C_{\mathit{t}\text{-}Bu}}$	TBAB	0.05	4000/2/1	—	100	2	16.8	675
7	$\mathbf{C_{\mathit{t}\text{-}Bu}}$	TBAB	0.05	4000/2/1	—	120	2	39.8	1595
8	$\mathbf{C_{\mathit{t}\text{-}Bu}}$	TBAB	0.05	4000/2/1	—	140	2	57.4	2300
9	$\mathbf{C_{Cum}}$	TBAB	0.05	4000/2/1	—	80	2	4.6	185
10	$\mathbf{C_{Cum}}$	TBAB	0.05	4000/2/1	—	100	2	20.1	805
11	$\mathbf{C_{Cum}}$	TBAB	0.05	4000/2/1	—	120	2	29.4	1175
12	$\mathbf{C_{Cum}}$	TBAB	0.05	4000/2/1	—	140	2	47.5	1900
13	$\mathbf{C_H}$	TBAB	0.05	4000/2/1	MEK	100	2	8.9	355
14	$\mathbf{C_H}$	TBAB	0.05	4000/2/1	MEK	120	2	31.1	1245
15	$\mathbf{C_{\mathit{t}\text{-}Bu}}$	TBAB	0.05	4000/2/1	MEK	100	2	7.8	310
16	$\mathbf{C_{\mathit{t}\text{-}Bu}}$	TBAB	0.05	4000/2/1	MEK	120	2	20.3	810
17	$\mathbf{C_{Cum}}$	TBAB	0.05	4000/2/1	MEK	100	2	8.3	330
18	$\mathbf{C_{Cum}}$	TBAB	0.05	4000/2/1	MEK	120	2	21.4	855
19	$\mathbf{C_H}$	TBAB	0.05	4000/2/1	—	120	0.5	35.6	1425
20	$\mathbf{C_H}$	TBAB	0.05	4000/2/1	—	120	4	55.7	2230
21	$\mathbf{C_H}$	TBAI	0.05	4000/2/1	—	120	2	42.3	1690
22	$\mathbf{C_H}$	[PPN]Cl	0.05	4000/2/1	—	120	2	38.9	1555
23	$\mathbf{C_H}$	DMAP	0.05	4000/2/1	—	120	2	0.4	16
24	$\mathbf{C_H}$	TBAB	0.1	4000/4/1	—	120	2	82.4	3300
25	$\mathbf{C_H}$	TBAB	0.25	4000/10/1	—	120	2	95.2	3810
26	$\mathbf{C_{\mathit{t}\text{-}Bu}}$	TBAB	0.25	4000/10/1	—	120	2	84.6	3385
27	$\mathbf{C_{Cum}}$	TBAB	0.25	4000/10/1	—	120	2	63.2	2530
28[e]	—	TBAB	0.05	4000/2/0	—	120	2	32.6	—

[a] Basic reaction conditions: propylene oxide (5 mL; 7.15×10^{-2} mol), catalyst loading = 0.025 mol% (1.79×10^{-5} mol), 1 h. [b] Determined by NMR (using mesitylene as an internal standard). [c] The selectivity for the formation of propylene carbonate was found to be >99%. [d] Turnover frequency (mol_{PC} $mol_{catalyst}^{-1}$ reaction $time^{-1}$). [e] Control experiment in absence of catalyst: reaction time = 24 h.

activity shown by $\mathbf{C_H}$ which reached a conversion of 61% with a TOF value of 2420 h^{-1} at 140 °C (entry 4), showing that a less encumbered metal centre is more accessible to the substrate. Notably, at a lower temperature (80–100 °C) the catalytic performances of $\mathbf{C_H}$ are sensibly scarcer (entries 1 and 2) compared to the more sterically demanding catalysts $\mathbf{C_{\mathit{t}\text{-}Bu}}$ and $\mathbf{C_{Cum}}$ (entries 5, 6 and 9, 10 respectively; see also Fig. 1). This apparent contradiction is probably due to the lower solubility of complex $\mathbf{C_H}$ in the reaction medium compared to the more soluble catalysts $\mathbf{C_{\mathit{t}\text{-}Bu}}$ and $\mathbf{C_{Cum}}$. In order to support our hypothesis, we performed the reactions using methyl ethyl ketone (MEK) as the solvent with the aim of suppressing the negative effects on the catalytic activity due to the low solubility of catalyst $\mathbf{C_H}$. As a matter of fact, the results obtained at both 100 and 120 °C clearly show that the catalytic activity of the more open catalyst $\mathbf{C_H}$, is comparable at 100 °C to those of catalysts $\mathbf{C_{\mathit{t}\text{-Bu}}}$ and $\mathbf{C_{Cum}}$ (entries 13, 15 and 17) and is sensibly higher at 120 °C (entries 14, 16 and 18), indicating a crucial role for temperature in determining the catalytic activity for this class of catalysts.

For the more active catalyst $\mathbf{C_H}$, we also explored the effect of CO_2 pressure on the catalytic activity showing that halving or doubling carbon dioxide pressure has a small effect on the catalytic performances (compare entry 3 with entries 19 and 20).

In analogy to the results obtained in the presence of the complex $\mathbf{C_{\mathit{t}\text{-Bu}}}$, the use of other commonly employed co-catalysts such as *n*-tetrabutylammonium iodide (TBAI), bis(triphenylphosphine)iminium chloride (PPNCl), and 4-dimethylaminopyridine (DMAP) resulted in a lowering of the catalytic activity (entries 21–23). In particular, while the two halogen containing co-catalysts (TBAI and PPNCl) show an activity comparable to that of TBAB, the use of DMAP results in a dramatic reduction of the catalytic activity. Moreover, the absence of the cocatalyst (TBAB) or of the iron(III) catalyst $\mathbf{C_H}$ resulted in a complete loss of the catalytic activity in the first case and in a conversion of only 33% (entry 28) in the second case even after 24 h, highlighting a synergic effect between the catalyst and the co-catalyst for carbonate formation.

Intriguingly, by increasing the co-catalyst/catalyst molar ratio to higher values (4 for entry 24 and 10 for entry 25) a further increase of the catalytic activity was observed with a conversion of 95% in 1 hour and a TOF of 3800, which is not only the highest reported so far for an iron based catalyst but compares well with the more active catalysts based on other metals.[5a] It is worth noting that under these conditions the order of reactivity is $\mathbf{C_H} > \mathbf{C_{\mathit{t}\text{-Bu}}} > \mathbf{C_{Cum}}$ (entries 25–27) confirming that a less sterically congested metal centre is more reactive in the CO_2/epoxide coupling.

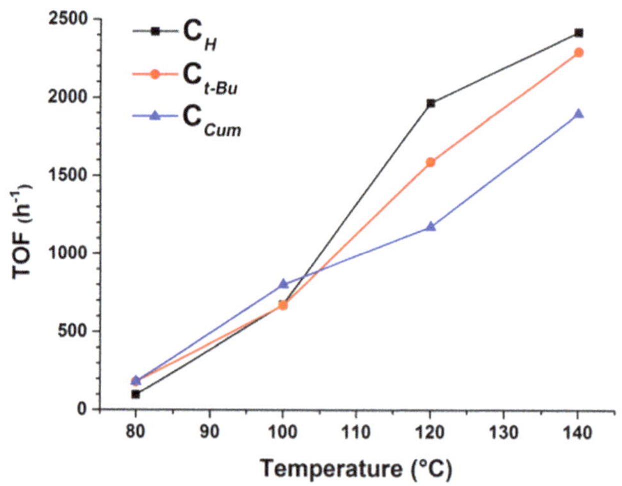

Fig. 1 TOF of the iron(III) complexes as a function of the reaction temperature (entries 1–12, Table 1).

Table 2 Synthesis of organic cyclic carbonate through CO_2/epoxide coupling catalyzed by Fe(III) catalysts

Entry[a]	Catalyst	(mol%)	TBAB (mol%)	Epox./co-cat./cat. (molar ratio)	Substrate	Product[bc]	Time (h)	Conversion[bc] (%)	TOF[d] (h^{-1})
29	C_H	0.025	0.05	4000/2/1	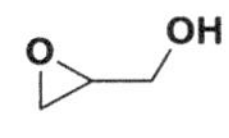	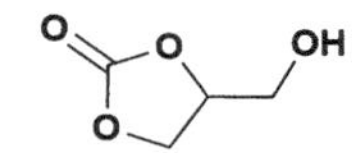	1	95.7	3830
30	C_H	0.025	0.05	4000/2/1	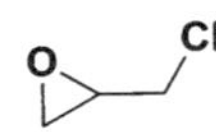	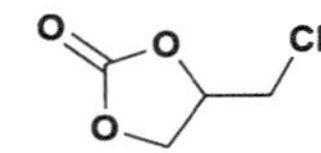	1	40.2	1610
31	C_H	0.025	0.05	4000/2/1	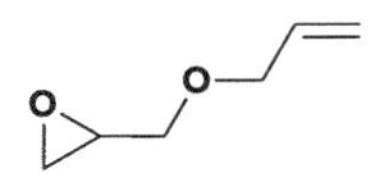	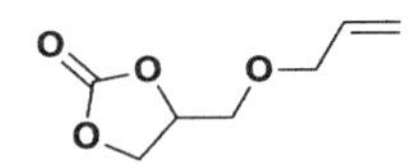	1	40	1600
32	C_H	0.025	0.05	4000/2/1	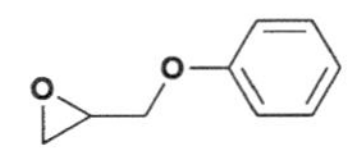	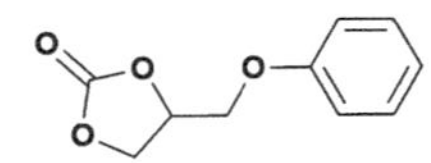	1	18.6	745
33	C_H	0.025	0.05	4000/2/1	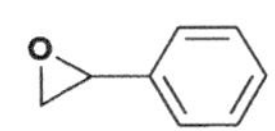	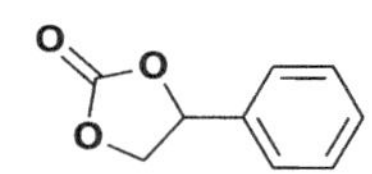	1	16.4	655
34[e]	C_H	0.2	0.4	500/2/1	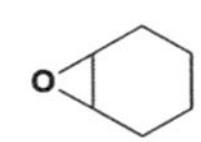	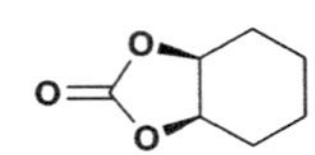	3	62.8	105

Table 2 (*Contd.*)

Entry[a]	Catalyst	(mol%)	TBAB (mol%)	Epox./co-cat./cat. (molar ratio)	Substrate	Product[bc]	Time (h)	Conversion[bc] (%)	TOF[d] (h^{-1})
35[e]	$C_{t\text{-Bu}}$	0.2	0.4	500/2/1	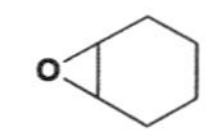	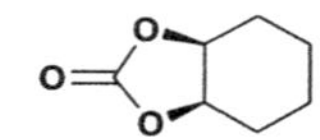	3	23.9	40
36[f]	C_{Cum}	0.2	0.4	500/2/1	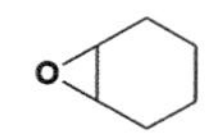	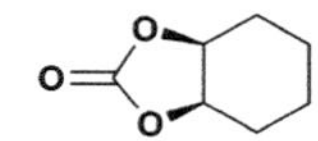	3	6.8	11

[a] Basic reaction conditions: epoxide = 7.15×10^{-2} mol, temperature = 120 °C, carbon dioxide pressure = 2 MPa. [b] Determined by NMR (using mesitylene as an internal standard). [c] The selectivity for the formation of cyclic carbonate was found to be >99%. [d] Turnover frequency ($mol_{COC}\ mol_{catalyst}^{-1}$ reaction $time^{-1}$). [e] A selectivity >99% for the formation of the *cis* isomer of the carbonate was observed. [f] Traces of polycarbonate were found.

Having established that catalyst $\mathbf{C}_H$ is a highly active and selective catalyst for the formation of cyclic carbonates, in order to expand the scope of this catalyst we explored the cycloaddition of CO_2 to a variety of functionalized terminal epoxides providing organic cyclic carbonates, under similar catalytic conditions as used in the case of propylene oxide (reaction condition of entry 3 of Table 1). The main results are reported in Table 2. All the substrates were conveniently converted to the corresponding cyclic carbonates with good yields even at relatively short reaction times (1 h). Notably, glycerol carbonate[15] which is a very interesting target molecule due to its properties, is easily obtained from glycidol in excellent yield (96%) with a TOF of 3830 h^{-1}. These results are even superior to those observed for propylene oxide under the same reaction conditions (compare entry 3 of Table 1 with entry 29 of Table 2) and, also in this case, it is the highest observed TOF reported for this substrate.

Furthermore, we explored the synthesis of cyclic carbonates starting from the more challenging substrate cyclohexene oxide (CHO). Actually, the formation of the cyclic carbonate from CHO is disfavored with respect to the formation of the corresponding polycarbonate because of the geometric strain due to the formation of two interconnected six- and five-membered cycles. In our case, we obtained good conversion (63%) within 3 h with a complete selectivity toward the cyclic product in the case of complex $\mathbf{C}_\mathbf{H}$. In this case the role of the steric bulkiness of the substituents on the aromatic rings is even more pronounced than in the case of PO, with the catalytic activity decreasing with steric bulk of the substituents: $\mathbf{C}_H$ > $\mathbf{C}_{t\text{-}\mathbf{Bu}}$ > $\mathbf{C}_\mathbf{Cum}$ (see entries 34–36).

It is worth mentioning that, as already observed for other iron based catalytic systems, the cyclic carbonates from CHO were produced selectively with *cis*-stereochemistry, as pointed out by NMR and FT-IR analysis, that revealed strong adsorption bands at 1802 cm^{-1}, previously assigned to the stretching of the carbonyl, respectively for the *cis*-cyclohexene carbonate.[11i]

Conclusions

In conclusion, we reported on the synthesis and characterization of two new, air stable, dimeric iron(III) complexes bearing thioether-triphenolate ligands. When activated by tetrabutylammonium bromide the title iron(III) complexes are very effective in the coupling of CO_2 with epoxides, giving excellent results in terms of activity, and chemo- and stereo- selectivity under solvent-free conditions. In particular, the less sterically demanding $\mathbf{C}_\mathbf{H}$ complex shows the best catalytic performances in terms of activity and selectivity for a wide number of substrates with TOF values up to 3830 h^{-1}. These results demonstrate, for this class of catalysts, that a more accessible metal centre is crucial to increase the activity in the cycloaddition of CO_2 to epoxide. Finally, these outcomes clearly show that this class of iron(III) based catalysts, bearing two sulfur atoms in the ligand skeleton, are competitive in term of activity, selectivity and substrate scope respect to other catalytic systems based on more toxic and/or expensive metals, for the formation of cyclic carbonates.

Acknowledgements

We gratefully acknowledge the Ministero dell'Istruzione, dell'Università e della Ricerca (MIUR, Roma, Italy) for FARB 2014 and PRIN 2010–2011, the Regione

Campania (POR FSE, project: "MAteriali e STrutture Intelligenti", MASTRI, Code 4-17-3, CUPB25B09000010007) and the Centro di Tecnologie Integrate per la Salute (Project PONa3_00138) for the 600 MHz NMR instrument time funding. Alexander von Humboldt Foundation is deeply acknowledged for a renewal fellowship (C.C.). Dr Patrizia Oliva, Dr Patrizia Iannece and Dr Ivano Immediata from Università degli Studi di Salerno are acknowledged for technical assistance.

Notes and references

1 (*a*) M. Aresta and A. Dibenedetto, *Dalton Trans.*, 2007, 2975–2992; (*b*) M. Aresta, A. Dibenedetto and A. Angelini, *Chem. Rev.*, 2013, **114**, 1709–1742; (*c*) D. J. Darensbourg, *Inorg. Chem.*, 2010, **49**, 10765–10780; (*d*) I. Omae, *Coord. Chem. Rev.*, 2012, **256**, 1384–1405; (*e*) E. A. Quadrelli, G. Centi, J.-L. Duplan and S. Perathoner, *ChemSusChem*, 2011, **4**, 1194–1215; (*f*) *Activation of Carbon Dioxide*, ed. S. L. Suib, Elsevier, Amsterdam, 2013.

2 (*a*) M. Grasemann and G. Laurenczy, *Energy Environ. Sci.*, 2012, **5**, 8171–8181; (*b*) S. Moret, P. J. Dyson and G. Laurenczy, *Nat. Commun.*, 2014, **5**, 4017.

3 F. Studt, I. Sharafutdinov, F. Abild-Pedersen, C. F. Elkjær, J. S. Hummelshøj, S. Dahl, I. Chorkendorff and J. K. Nørskov, *Nat. Chem.*, 2014, **6**, 320–324.

4 C. Maeda, Y. Miyazaki and T. Ema, *Catal. Sci. Technol.*, 2014, **4**, 1482–1497.

5 (*a*) C. Martín, G. Fiorani and A. W. Kleij, *ACS Catal.*, 2015, **5**, 1353–1370; (*b*) J. W. Comerford, I. D. V. Ingram, M. North and X. Wu, *Green Chem.*, 2015, **17**, 1966–1987; (*c*) N. Kielland, C. J. Whiteoak and A. W. Kleij, *Adv. Synth. Catal.*, 2013, **355**, 2115–2138; (*d*) M. North, in *New and Future Developments in Catalysis*, ed. S. L. Suib, Elsevier, Amsterdam, 2013, pp. 379–413; (*e*) M. Cokoja, C. Bruckmeier, B. Rieger, W. A. Herrmann and F. E. Kühn, *Angew. Chem., Int. Ed.*, 2011, **50**, 8510–8537; (*f*) R. Li, X. Tong, X. Li and C. Hu, *Pure Appl. Chem.*, 2012, **84**, 621–636; (*g*) A. Decortes, A. M. Castilla and A. W. Kleij, *Angew. Chem., Int. Ed.*, 2010, **49**, 9822–9837.

6 (*a*) M. I. Childers, J. M. Longo, N. J. Van Zee, A. M. LaPointe and G. W. Coates, *Chem. Rev.*, 2014, **114**, 8129–8152; (*b*) M. Taherimehr and P. P. Pescarmona, *J. Appl. Polym. Sci.*, 2014, **131**, 41141; (*c*) D. J. Darensbourg and S. J. Wilson, *Green Chem.*, 2012, **14**, 2665–2671; (*d*) S. Klaus, M. W. Lehenmeier, C. E. Anderson and B. Rieger, *Coord. Chem. Rev.*, 2011, **255**, 1460–1479.

7 (*a*) B. Schäffner, F. Schäffner, S. P. Verevkin and A. Börner, *Chem. Rev.*, 2010, **110**, 4554–4581; (*b*) A.-A. G. Shaikh and S. Sivaram, *Chem. Rev.*, 1996, **96**, 951–976.

8 (*a*) D. J. Darensbourg and A. D. Yeung, *Polym. Chem.*, 2014, **5**, 3949–3962; (*b*) T.-T. Wang, Y. Xie and W.-Q. Deng, *J. Phys. Chem. A*, 2014, **118**, 9239–9243.

9 A. Buonerba, A. De Nisi, A. Grassi, S. Milione, C. Capacchione, S. Vagin and B. Rieger, *Catal. Sci.Technol.*, 2015, **5**, 118–123.

10 (*a*) I. Bauer and H.-J. Knölker, *Chem. Rev.*, 2015, **115**, 3170–3387; (*b*) *Iron Catalysis Fundamentals and Applications*, ed. B. Plietker, Springer, 2011; (*c*) C.-L. Sun, B.-J. Li and Z.-J. Shi, *Chem. Rev.*, 2010, **111**, 1293–1314; (*d*) A. Correa, O. Garcia Mancheno and C. Bolm, *Chem. Soc. Rev.*, 2008, **37**, 1108–1117.

11 (*a*) M. Taherimehr, J. P. C. C. Sertã, A. W. Kleij, C. J. Whiteoak and P. P. Pescarmona, *ChemSusChem*, 2015, **8**, 1034–1042; (*b*) X. Sheng, L. Qiao, Y. Qin, X. Wang and F. Wang, *Polyhedron*, 2014, **74**, 129–133; (*c*) M. Adolph,

T. A. Zevaco, C. Altesleben, O. Walter and E. Dinjus, *Dalton Trans.*, 2014, 3285–3296; (*d*) M. A. Fuchs, T. A. Zevaco, E. Ember, O. Walter, I. Held, E. Dinjus and M. Doring, *Dalton Trans.*, 2013, 5322–5329; (*e*) M. Taherimehr, S. M. Al-Amsyar, C. J. Whiteoak, A. W. Kleij and P. P. Pescarmona, *Green Chem.*, 2013, **15**, 3083–3090; (*f*) C. J. Whiteoak, E. Martin, E. Escudero-Adán and A. W. Kleij, *Adv. Synth. Catal.*, 2013, **355**, 2233–2239; (*g*) C. J. Whiteoak, E. Martin, M. M. Belmonte, J. Benet-Buchholz and A. W. Kleij, *Adv. Synth. Catal.*, 2012, **354**, 469–476; (*h*) C. J. Whiteoak, B. Gjoka, E. Martin, M. M. Belmonte, E. C. Escudero-Adán, C. Zonta, G. Licini and A. W. Kleij, *Inorg. Chem.*, 2012, **51**, 10639–10649; (*i*) A. Buchard, M. R. Kember, K. G. Sandeman and C. K. Williams, *Chem. Commun.*, 2011, **47**, 212–214; (*j*) J. E. Dengler, M. W. Lehenmeier, S. Klaus, C. E. Anderson, E. Herdtweck and B. Rieger, *Eur. J. Inorg. Chem.*, 2011, 336–343.

12 *Physical Methods in Bioinorganic Chemistry*, ed. J. J. J. Y. Girerd, University Science Books, Sausalito, CA, 2000.

13 *Advanced Inorganic Chemistry*, ed. F. A. Cotton, G. Wilkinson, C. A. Murillo and E. M. Bochmann, Wiley, 6th edn, 1999.

14 (*a*) *Interpreting Infrared, Raman, and Nuclear Magnetic Resonance Spectra*, ed. R. A. Nyquist, Academic Press, San Diego, 2001, pp. 65–83; (*b*) *The Sadtler handbook of infrared spectra*, ed. W. W. Simons, Sadtler Research Laboratories, Philadelphia, 1978.

15 M. O. Sonnati, S. Amigoni, E. P. Taffin de Givenchy, T. Darmanin, O. Choulet and F. Guittard, *Green Chem.*, 2013, **15**, 283–306.

Faraday Discussions

DISCUSSIONS

Atom efficiency in small molecule and macromolecule synthesis: general discussion

Antoine Buchard, Michael North, Christopher Kozak, Eryk Remiezowicz, Pedro Abrantes, André Bardow, Jennifer Dodson, Charlotte Williams, Jonathan Albo, Ian Ingram, Gonçalo V. S. M. Carrera, Peter Styring, Martyn Poliakoff, Katie Lamb, Elsje Alessandra Quadrelli, Niall MacDowell, George Dowson, Michele Aresta, Richard Heyn, José A. Castro-Osma, William Webb, Renata Silva, Aryane Marciniak, Christopher Jones, Cafer Tayyar Yavuz, Carmine Capacchione, Anthony Coogan and Michael Priestnall

DOI: 10.1039/c5fd90079d

Peter Styring opened a general discussion of the introductory lecture by Martyn Poliakoff: You say that the CO_2 needs to be purified before it can be used in chemical reactions. Do you have an idea of the degree of purity and is this the case for all reactions?

Martyn Poliakoff replied: I think that it obviously depends on the reaction. However, most reactions will probably work with the levels of purification currently suggested in draft specifications for captured CO_2 that is to be transported by pipelines.

Peter Styring asked: The question about closing down power stations at times of low demand is an interesting one. In reality that is not economic. In fact power stations continue to generate power, albeit at a lower load, because of start up time and costs. In Europe this is then exported across national boundaries, but how would this operate for an island nation such as the UK?

Michele Aresta answered: One of the not-solved problems today is the efficient storage of electricity. Batteries are a way, but they have some limitations. The conversion of electricity into chemicals by converting carbon dioxide is an interesting approach (I have been proposing it for over three decades) that can be implemented with good efficiency today: at the end the produced chemicals/fuels can enter the existing market and fuels can be used to generate energy. This will allow the existing infrastructures in transport to be maintained.

Christopher Kozak responded: Britain currently does buy electricity from France and the Netherlands (and to a much lesser extent sells surplus electricity)

via the HVDC Cross-Channel interconnector, so its island location is not (really) an issue. Storage of surplus electricity is an issue, however, and we are seeing similar problems with the current glut of oil available on the market. Some oil and gas storage facilities in North America are nearing capacity and hoping for a rise in prices (for example, see http://www.reuters.com/article/2015/04/21/us-oil-storage-cushing-idUSKBN0NC1XS20150421). Regardless of what form future energy sources/fuels take, storage will continue to be an issue.

Michael Priestnall replied: In a recent visit to Little Barton Power Station (gas) near St Neots, I learned they had ~46 shut downs in 2014 in response to fluctuating grid demand. Electricity spot prices can swing wildly from +£600/M_wh to −£30/M_wh and the background load of 14M_w for the power station when shut down can be very expensive to maintain. This is not economically viable and capacity payments are essential to subsidise operation.

Michele Aresta queried: The chemical and fermentation industry release *ca.* 3300 Mt/y of CO_2. The purity of such streams may be very high and the streams may not contain NO_x nor SO_y. A part of such emissions (NH_3 industry) is used (for the synthesis of urea). Wouldn't it be better to target the use of such emissions for CCU instead of the power stations flue gases? If we were able to use the entire quantity of industrial CO_2, CCU would contribute a *ca.* 10% cut of the emissions. LCA (life cycle assessment) may say if the transportation of clean CO_2 is more expensive than cleaning dirty CO_2.

Martyn Poliakoff responded: Your point is valid. However it is not exactly the same as what I was trying to say in my lecture. I was pointing out that CO_2 from CCS activities will be compressed. Therefore there is an opportunity, at least in principle, to obtain already compressed CO_2 from power plants for use as high pressure or supercritical solvents for chemical reactions without the need to pay the full compression costs. As far as I am aware, CO_2 from fermentation is not compressed to the necessary pressures.

Eryk Remiezowicz asked: Carbon dioxide usage processes often assume supply of renewable energy. However, with large scale penetration of renewable energy, there will be less CO_2 emitted, which will decrease the need for CO_2 usage and will decrease CO_2 availability. Why develop carbon dioxide usage processes?

Martyn Poliakoff answered: If only you were right! I cannot see renewable energy becoming so predominant in the short term that there will be a shortage of CO_2. Sadly, a shortage of water is much more likely.

Niall MacDowell remarked: You mentioned that CCU options can afford to be picky and to opt to use CO_2 that is obtained from high purity sources such as gas sweetening operations, for example. My question is, what do you think is a better use for that CO_2, from a global perspective: its chemical conversion into a product or its permanent geological sequestration?

Martyn Poliakoff responded: You are conflating two different issues. The purpose of making chemicals is not primarily to sequester CO_2. It is to make use of a source of carbon that is not directly from crude oil. Therefore, we need to source carbon from somewhere and CO_2 is the most obvious, if not the only source. Having said that, I should reiterate what I said in my lecture, namely that the release of anthropogenic CO_2 is orders of magnitude larger than the potential need for CO_2 to make chemicals. So there will also be a need to sequester CO_2 as well as making chemicals.

Jonathan Albo opened the discussion of the paper by Richard Heyn: An overall propylene carbonate yield greater than 50% (with 69% selectivity) in 24 h is a promising result. What are the expected mechanisms and other by-products from the reaction? Do you have a hypothesis for the improved catalytic activity of Zn-based catalysts in comparison to other transition metals?

Richard Heyn replied: The other by-products from the reaction are the three isomers of acetoxylpropanol, the monosubstituted 1- and 2-isomers, and the disubstituted version. These arise *via* the formation of acetamide, from the reaction of water with acetonitrile, and then reaction of the acetamide or acetic acid (from the reaction of water with acetamide) and then reaction with propylene glycol.

We do not have much information on the mechanism at the current time. The reaction presumably proceeds *via* formation of Zn-glycolate, followed by CO_2 insertion into the Zn–O bond. We have been working on the synthesis of Zn-glycolate species that could function as model compounds, but we have thus far been unsuccessful. We have some evidence that the acetoxypropanols undergo a redistribution reaction under the reaction conditions, thus providing more propylene glycol at longer reaction times.

As far as a hypothesis for the generalized better performance of Zn salts *vs.* other transition metals, it is difficult to answer this question, since we know very little about the active catalyst species (or even resting states). Our best hypothesis at the moment is that Zn^{2+} salts are better Lewis acids than the other divalent transition metal salts. This could then promote deprotonation of a coordinated propylene glycol, although this is mere speculation. I think it is relevant that Nature uses Zn in the enzyme carbonic anhydrase; replacement of Zn with other transition metals reduces the enzyme activity.

Michele Aresta queried: The use of organic water traps adds complexity to the system in terms of post reaction separation. As a matter of fact, the water trap is used in excess and the hydrated product is formed in stoichiometric amounts with the cyclic carbonate. Moreover, the formed amide, in the case of a cyanide as a water trap, may react with the carbonate. Have you estimated the real benefit of moving from a 6% thermodynamic yield of carbonate from a diol and CO_2 and 100% selectivity, to the use of water traps which as you have shown may afford 40% conversion of the diol with formation of the amide and other by-products?

Richard Heyn answered: No, I have not estimated this benefit, and I would be surprised if our system with the higher yield and lower selectivity would win such an estimation. My view on the use of the acetonitrile in these reactions is that it is a necessary evil required to provide robust product yields that permit a proper comparison of catalytic activities. In particular, we are primarily interested in the time needed to reach equilibrium, so a comparison of product yields at short reaction times is the most valid approach. This is hard to accomplish at the very low equilibrium yields of these reactions without water traps.

Aryane Marciniak communicated: In your study, you use acetonitrile as a chemical water trap, but acetonitrile is a toxic solvent and many by-products, acetamide, acetic acid and acetylated glycols, are formed from the water-trapping reaction. This is a problem in this kind the reaction. So, I would like suggest the use of a chemical water trap can be recycled and no waste is formed. For example, in my studies, for the synthesis of dimethyl carbonate (DMC) from CO_2 and methanol, I use methyl esters as a chemical water trap and the carboxylic acid formed upon the ester hydrolysis could be recycled to the methyl ester after reaction. The most principal point is no by-products is formed. This could be something for you to think about.

Richard Heyn communicated in reply: I must admit that I am unfamiliar with this method for removing water. If the recycling step is not particularly energy intensive, I could see the benefits of this approach.

Renata Silva communicated: In the first instance I would like to congratulate you on your work. It's very nice! You talked in the paper about the use of MeCN (acetonitrile) as a water trap, but acetonitrile is very toxic. So, what other solvent could you use as a water trap to make be reaction greener? I saw that you also used methanol, but it did not work very well. Did you test some other greener solvents or are there others you want to test?

Richard Heyn communicated in reply: Thank you for the kind words. Our ultimate goal is to find a catalyst that is fast enough such that water removal could be designed within the process. My general view is that any reaction to an organic carbonate that needs a water trap to obtain viable yields, immediately loses on the basis of overall sustainability. Having said that, coupling of this reaction with an industrially relevant hydrolysis may be an interesting route to test.

Gonçalo V. S. M. Carrera remarked: Besides the coordination of the ligands to the metal center (Zn^{2+}), one of the reasons behind the improved performances associated with trifluoromethanesulfonate and tosylate as ligands could be a possible improved solubility of these complexes in a relative high pressure, CO_2 rich environment, which is related with the known affinity of organofluorides and aromatics with CO_2. One possible study to complement the work performed could be a correlation between the catalytic activity of the multiple complexes checked and its solubility in the reactional media. Another possible anion with associated poor coordinations and high CO_2 affinity is bistriflamide (on the framework of ionic liquids), in this context this could be a target ligand to test. What's your perspective regarding these issues?

Richard Heyn responded: Your suggestion is very valid. I cannot comment on the homogeneity of the majority of catalytic species, since the high-throughput reactor wells are literally black boxes. Batch scale reactions of $Zn(OAc)_2$ + *para*-chlorobenzene sulfonic acid, $Zn(OTf)_2$ and $Zn(Tos)_2$ run with sample removal during the reaction show only homogeneous solutions. I think a complementary study investigating the number of phases in the reaction media and the solubility of different catalysts in these phases would be very worthwhile.

Antoine Buchard asked: In Table 1, entry 20, the addition of 13X molecular sieves does not improve the yield of the reaction and even inhibits it, despite the presence of acetonitrile to trap water. What is your explanation? Have you observed the same phenomenon when using $Zn(OTf)_2$ as a catalyst?

Richard Heyn replied: Molecular sieves do not improve the yield of the reaction because sieves are not very effective water absorbers at the temperature of the reaction. One hypothesis for the lower yield in the presence of the sieves could be preferential absorption of the reaction products. We have not used molecular sieves in any of the reactions with $Zn(OTf)_2$.

George Dowson continued: With regards to the loss of activity seen when using molecular sieves within the reaction mixture, molecular sieves in homogeneous reactions can sometimes act as traps not only for water but also for dissolved reactive species and intermediates, thus preventing further reactivity. This behaviour of the molecular sieves could be an indication of the homogeneous nature of the catalyst system being discussed, however other methods of testing homogeneity would need to be used for confirmation. Suspending the molecular sieves above the liquid medium during the reaction should avoid this reaction trapping process and help demonstrate the importance of water removal to drive the reaction onwards.

Richard Heyn noted: Suspension of molecular sieves over the reaction media has been used with some success in a few studies.[1]

1 Aresta, M., *et.al*, *J. Mol. Catal.*, 2006, **257**, 149–153.

Michael North commented: There was a suggestion from the floor that molecular sieves would be helpful as dehydrating agents during cyclic carbonate synthesis from diols and CO_2. At the temperature the reaction was being carried out in our work (DOI: 10.1039/c5fd00061k), the molecular sieves would not be effective dehydrating agents.

Michele Aresta remarked: It has been proven that molecular sieves are effective as water traps at low temperature (240 K), that means that the reaction mixture must be cooled from 410 K (reaction temperature) to 240 K and then reheated. This is not an efficient approach because heat recovery in that range of temperature is not efficient and the does not find industrial applications. Modified molecular sieves may be more effective at the reaction temperature: the issue is the formation of acidic H-bonds that may convert back the carbonate. Other techniques for water trapping may be more efficient.

Eryk Remiezowicz questioned: Have you tried to relate the parameters of your reaction (yield, speed, atom efficiency *etc.*) with the existing industrial processes for the production of propylene carbonate?

Richard Heyn answered: No, we have not done that analysis. In my opinion, the reaction and catalyst development are, at the moment, not developed enough for such a comparison. There is still much work to be done to improve the reaction.

Michael North opened the discussion of the paper by Christopher Kozak: There has been previous work published on the interaction of DMAP with CO_2, how does the formation of this complex affect your results? Your mechanism involves chloride acting as a nucleophile to ring open the epoxide, what is the effect of adding silver salts to remove the chloride from solution?

Christopher Kozak responded: Yes, D. J. Darensbourg[1] observed the formation of a DMAP-CO_2 adduct giving a carbamate ligand based on *in situ* IR studies (ν_{CO_2} bands at 2097 and 2018 cm^{-1}). We have looked for bands in that region of the IR spectra for our systems but have not conclusively found anything that would suggest this occurs. Regarding the removal of chloride, although we have not added external agents such as silver salts to remove the Cl^- *in situ*, we have prepared a number of derivatives that replace the chloride with ligands such as acetate, nitrate, azide and 2,4-dinitrophenolate. The catalytic studies of these systems are in their early stages, but Cr complexes bearing these groups do show activity for poly(cyclohexenecarbonate) formation. Analagous Co(III) complexes, however, show lower tendencies to give the polymer and instead exhibit cyclic carbonate formation.

1 Donald J. Darensbourg, Ryan Mackiewicz and Jody L. Rodgers, *J. Am. Chem. Soc.*, 2005, **127**, 14026–14038 and subsequent papers.

Charlotte Williams commented: The paper presents a very interesting study of the influence of the Lewis base co-catalyst on the structure of the Cr(III) complexes used in catalysis. I have two questions:

1) Would it be possible to expand the method so as to be able to characterize the products of the reaction between the Cr(III) complex, co-catalyst and epoxide? If such a method could be developed it would be very useful to understand the position of the equilibrium between catalyst and monomer.

2) Given that some dimeric complexes were isolated by mass spectrometry for the catalysts, what is the order in Cr in the rate law? Is there any evidence that such dimeric species are implicated (*i.e* non-integral orders)?

Christopher Kozak answered: Regarding the first question, a technical issue arises with MALDI-TOF MS in that the sample is loaded into the instrument and then placed under a very high vacuum prior to laser desorption/ionization. Volatile components such as cyclohexene oxide are lost readily under these conditions. Any attempts by us to incorporate CHO into the matrix and analyte mixture have been unsuccessful. LC-MS techniques have proven more useful in this regard and we plan on performing studies using this technique in the future, akin to the work of Xiao-Bing Lu.[1] For the second question, we have preliminary data that are consistent with first order behaviour in Cr concentration, but further studies are needed to confirm this.

1 Dun-Yan Rao, Bo Li, Rong Zhang, Hui Wang and Xiao-Bing Lu, *Inorg. Chem.*, 2009, **48**, 2830–2836

Martyn Poliakoff asked: In these current times when people are worrying about chromium in the environment, how can you justify the use of chromium in an industrial process?

Christopher Kozak responded: I agree that chromium is not the optimum choice of metal for a large scale process. Although the Phillips Catalyst (using Cr(VI) oxide supported on silica) is still widely used for the industrial synthesis of polyethylene, there is a need for more benign metals in industrial catalysis. The Cr(III) in our catalysts is considered non-toxic due to its insolubility in water, unlike Cr(VI), which is highly toxic. Furthermore, we see no evidence of oxidation of our catalysts to Cr(VI), but this can happen in the environment. Nonetheless, we are currently studying iron and zinc-based catalysts for CO_2/epoxide copolymerization in order to address this problem, but the chromium catalysts in our hands work so well by comparison that for mechanistic studies they are a valuable system from which we can learn.

André Bardow opened the discussion of the paper by George Dowson: I believe that it is a great idea to avoid CO_2 capture and purification since pure CO_2 might not be required for CO_2 utilization. With respect to your specific approach, I have two questions:

1) Do you actually fulfil the full vision to avoiding CO_2 capture? In essence, you are using a capture agent and you are recycling it. I would suggest using a reactant leading to a valuable product directly as the capture agent.

2) The capture agent used in your work is recycled by electrolysis. You correctly point out that magnesium electrolysis is an established and well optimized process running close to the thermodynamic limit. But still, the required energy is large; in fact, much larger than the energy required for CO_2 capture by amines. In addition, you require electric energy while amines can be recycled using heat with much lower energy content. Why do you believe that the magnesium route is sensible? I would challenge you by claiming that the conventional CO_2 capture route would be more efficient if it could use the same amount of renewable energy the magnesium route is using.

George Dowson replied: In answer to your first question, that is exactly what we are doing. The Grignard is acting as a reactant directly leading to a valuable product, bypassing an independent capture step.

In answer to your second question, capture by magnesium compounds is definitely more energetically costly than using amines, but unlike with amines, the capture process directly generates a higher-value product through one capture and regeneration cycle. This valuable product is what should be compared with the energy cost, not the capture process itself. To emphasise; direct reaction of flue gas in this way is not directly comparable with amine capture, unless the cost of the transformation of the post-amine-capture CO_2 into a value-added product is included.

Richard Heyn queried: For your concept to be a useful process, the CO_2 must be dried, otherwise water vapor would react with the Grignard reagent. As well, it is indicated in Fig. 7 that water is produced during alcoholysis of the acid, and the water must be separated from the alkyl halide prior to its reaction with Mg. Have you included the costs of these drying steps in your techno-economic assessment?

George Dowson answered: For a Grignard reaction, water is one of the chief concerns and indeed dry reagents must be used for the process to work.

For the first question, in the creation of the alkyl halide, as in the case of methyl chloride for the worked example, the drying process would simply be one of condensation as methyl chloride is a highly volatile gas with a boiling point below −20 °C. For other products, especially those involving alkyl halides with higher boiling points, dehumidification may be a larger issue, and for others water may cause no issues at all. For our techno-economic assessment, as a purely preliminary study, drying of the methyl chloride was not considered.

Similarly, the CO_2 that was modelled within the techno-economic assessment was assumed to be dry and pure and was costed using Aspen model standards. It was not intended to be a direct simulation of flue gas capture but more an outline of the economic concept.

Martyn Poliakoff stated: For the two chemicals you mentioned (terephthalic acid and methyl methacrylate) – both used to make polymers where colour is important – trace impurities at a low level are very important. Therefore, the targets you have chosen are possibly ones where you're making it harder for yourself because the products have to be exceptionally pure.

George Dowson remarked: Firstly thank you, I welcome the comment (and am a big admirer!). Selection of appropriate products for any CDU process intended for commercialisation is of course extremely important. However, I would suggest that the difficulty of the targets is actually a potential benefit to the attraction of such a process to industry. This is because there is the potential for these difficult compounds to be made or purified more easily, although this is certainly not guaranteed! For example, in the case of terephthalic acid one of the common yellowing impurities is 4-carboxybenzaldehyde, formed by incomplete oxidation of the *p*-xylene starting material. Starting instead from *p*-dibromobenzene *via* Grignard chemistry and carbon dioxide, it seems unlikely this problematic aldehyde can form. Naturally, the Grignard process will instead leave us with other side products and by-products (such as *p*-bromobenzoic acid) to deal with instead, but these may be more readily removed and extracted than their oxidation process counterparts, allowing the desired exceptional purity to be attained more readily, and all this at room temperature and atmospheric pressures.

Only after a full and thorough investigation of how these more challenging targets could be made and purified, can we make any comparisons with existing technology.

Niall MacDowell continued: I am concerned that the techno-economic analysis presented in the paper has a number of potential weaknesses. Namely, that you appear to assume – captured CO_2 (pure, and high pressure) is available at no cost – that hydrogen and methanol are available at current market prices, or in other words, these are products derived from fossil fuels, and thus not derived from renewable energy and should be associated with a commensurate CO_2 footprint. High purity CO_2 is unlikely to be available at a zero cost – if we consider other instances where anthropogenic CO_2 was used for, *e.g.*, EOR, there is always a cost associated. Similarly, renewable H_2/methanol will tend to cost substantially more than fossil derived H_2/methanol. Both of these facts should be taken into account in this analysis, with reference to, as a minimum, the existing literature on CO_2 EOR for the costs of CO_2 and renewable H_2/methanol.[1–4]

The availability of renewable energy in this paper is an important point. In the UK, onshore wind has a capacity factor of 28% and offshore wind has a capacity factor of 38% on average. It would be necessary to account for the intermittent character if the renewable energy by either reflecting this in the number of hours per year your facility could operate or explicitly consider the added cost of additional deployed renewable capacity and energy storage.

If you're planning on considering the utilisation of raw exhaust gas, it must be recognised that drying the gas stream is itself costly, and this must be accounted for in your analyses.

Finally, given that you have an extant model of one of these processes, it is not necessary to simply assume that the process scales based on an assumed 0.6 scaling factor – surely you can simulate the process at different scales, and provide insight into bottlenecks, and the likely correct scaling factor?

Could you please provide some insight into how the above factors affect your conclusions?

1 Mignard and Pritchard, *Trans IChemE, Part A, Chemical Engineering Research and Design*, 2006, **84**, 828–836.
2 Ayers *et al.*, *ECS Transactions*, 2010, **33**, 3–15.
3 van der Ham, *Chem Eng Trans*, 2012, **29**, 181–186.
4 Van-Dal, *J Cleaner Production*, 2013, **57**, 38–45.

George Dowson responded: As stated in the paper, this techno-economic assessment was carried out in only a preliminary fashion and some of the weaknesses in its execution are to be expected. Additionally, due to the time constraints of a Faraday discussion presentation, multiple techno-economic assessments at different scales were not possible so a scale comparable with existing large Grignard plants was chosen to at least reflect the cost at a reasonable scale. This cost, while high, demonstrates that recycling of magnesium species to generate products is not as outlandish as might first be assumed.

The sources of the CO_2 and hydrogen prices came from the works of Becker[1] and Dimitriou[2] respectively. The methanol price used was that of the upper edge of the European methanol commercial price range at the time of writing. This figure was chosen with renewable methanol, such as that generated by Carbon Recycling International in mind, however even if fossil methanol and hydrogen are used, while this does impact the carbon footprint of the product, we are still sequestering one molecule of carbon dioxide per molecule of product. However, the life cycle assessment of this process was not being carried out for this process so estimations of carbon footprints of our preliminary assessment is inappropriate.

To the drying costs of the input gases, it was hard to find this data to be included although I would caution that while drying costs of captured CO_2 from amine capture processes are very high, this is primarily due to this CO_2 being 100% humid upon passing through the capture stream. Direct flue gas is much drier and depending on the nature of the CO_2 source and the ESP system it passes through, the level of moisture can easily be in the single digit percentage, greatly reducing the drying cost. Furthermore, I think it is clear that the electrolysis cost of the magnesium salts is going to be much larger than any gas drying cost.

Therefore for a simple preliminary assessment, ignoring the drying cost should not greatly affect the product price although, as you correctly identify, it is an added cost that has not been considered.

Finally to the availability of the renewable energy: production of 500,000 tonnes of acetic acid a year (matching the production of BP Saltend and thus the largest acetic acid plant in Europe) by this process would consume at least 200,000 tonnes of magnesium. Using the electrolysis values of magnesium produced industrially, this would require approximately 2 TW h^{-1} of electricity per year, equalling around 5% of the current renewable energy output in the UK. Frankly, using enough energy to affect renewable capacity would be a nice problem to have!

1) W.L. Becker, R.J. Braun, M. Penev, M. Melaina, Production of Fischer-Tropsch liquid fuels from high temperature solid oxide co-electrolysis units, *Energy*, 2012, **47**, 99–115.

2) I. Dimitriou, P. García-Gutiérrez, R. H. Elder, R. M. Cuéllar-Franca, A. Azapagic, R. W. K. Allen, Carbon dioxide utilisation for production of transport fuels: process and economic analysis, *Energy and Environmental Science 8*, 2015, 1775–1789.

Peter Styring followed: As George Dowson has stated, this is a preliminary techno-economic analysis and has made a number of assumptions for various scenarios. As this is a laboratory scale process at present we are using the Aspen Plus model to look at scaling. This is an on-going process with a number of scenarios planned. The 2/3 scaling rule was applied as a standard factor consistent with many process development scenarios. The TEA presented should be viewed as a worst case scenario taking upper end values for hydrogen and methanol.

In the TEAs that will follow we will look at CO_2 derived methanol (such as Vulcanol from CRI) and hydrogen from recycled energy through, for example, PEM electrolysis of water. The costs at present represent a large scale synthesis of acetic acid. What we plan to do is look at the smaller scale synthesis of high value acids and esters, particularly in communities isolated from CO_2 producing power generation sources or sequestration sites. Therefore the full LCA and TEA will need to include relative transportation costs (with or without pipelines). The issue of the CO_2 source is very interesting. Pricing is difficult to assess until a carbon price is agreed. Whether or not the CO_2 is free or not really depends on the source. If there is no capture facility the emitter will have three options: emit and pay the tax, install a capture facility (assuming there is a storage location available) or allow a utilisation facility to take the CO_2 free of charge (thereby removing disposal or taxation costs) to transform it into a valuable commodity chemical, and maybe even receiving a share in the profit.

So, our TEA is preliminary but also conservative and will be refined in the forthcoming years as real plant data become available.

Elsje Alessandra Quadrelli stated: Fig. 3 reports a series of industrially-relevant carboxylic acids and diacids that could be potentially obtained by reacting CO_2 and Grignard reagents. To the best of my knowledge no such route has led to an economically viable processes. Could you share with us your analysis of the current hurdles underlying the non-existence of such industrial routes and to what extent your techno-economic analysis on Grignard-based capture routes—which have the compounding difficulty – and connected purification costs – of handling presumably wet and variously contaminated CO_2 sources—integrate these current hurdles?

George Dowson replied: The major current hurdle of any Grignard process is scaling. Grignard chemistry is seen as somewhat unpredictable and risky at a large scale by industry. Whether this perception is actually true remains to be seen and I would expect further investigation is warranted find the truth.

The other point is that we are specifically looking at a process where the magnesium by-product of the Grignard process is recycled. From those who have experience of industrial Grignard chemistry that I have spoken to, this has not previously been considered. This is likely because Grignard chemistry has been confined to the production of fine chemicals with values high enough to afford the "loss" of a stoichiometric mass of magnesium on production. Looking at bulk chemicals this is not the case, and I am certain that without recycling there is no economic case (and certainly no environmental case) whatsoever for the Grignard route. Once recycling and electrolysis is included however, these bulk chemicals *via* Grignard chemistry might start to become feasible, as indicated in the paper, and at the very least a sustainable route for CO_2 addition is revealed.

Michael North asked: You have focused on the use of magnesium which is not the most abundant metal. Have you chosen the wrong metal? Would it have been better to consider aluminium which is much more abundant? There is plenty of waste aluminium available.

George Dowson answered: While aluminium is more plentiful than magnesium, magnesium remains the 8th most common element in the Earth's crust and is also available in appreciable amounts in seawater (1.3 g per litre), so availability of the magnesium is not likely to be a limitation for this process, especially considering that the magnesium is recycled.

However, alternative metals for this process should indeed be investigated including candidates such as aluminium, which may have advantageous electrolysis properties if Grignard-type chemistry can be made to occur, although the very high energy nature of aluminium electrolysis would be a concern (aluminium chloride would have to be oxidised to bauxite first from what I can determine).

Jennifer Dodson commented: I like the vision of this paper, studying the direct use of the CO_2 stream rather than initially purifying it. However, there are several challenges related to the scale-up, production and use of Grignard reagents, particularly due to the use of flammable solvents during a highly exothermic reaction. They are also moisture and air sensitive. Can you comment on why you chose to focus on this reaction considering these green chemistry problems and if you are going to move on to look at other reactions?

George Dowson responded: Thank you for your question. Grignard chemistry on a large scale faces a multitude of challenges indeed, however it is one of the simpler and more accessible reactions involving CO_2 to generate value-added products that can be investigated. The reason for this is the same reason that

Grignards can be difficult at scale; Grignard reagents are very high-energy and their reactions with CO_2 are highly exothermic. This is in stark contrast to many other reactions of CO_2 where large energy barriers have to be overcome to generate the CO_2-incorporated product, such as in Fischer–Tropsch. Therefore it seemed sensible to get some idea of the cost of using a higher-energy intermediate (in this case, Grignard reagents) to benchmark against other processes that use lower energy reagents, but endothermic reaction conditions. Now that it has been shown that using Grignards may not be completely outlandish, we may well look at other high-energy reagents that have better environmental and economic credentials.

Michele Aresta questioned: When you use Mg(0) for making the Grignard it is oxidized to Mg(II), while CO_2 is reduced by $1e^-$ from C^{4+} to C^{3+}. Therefore, there is a penalty as oxidized Mg should be reduced for making a new Grignard. The penalty is even higher as Mg is used in excess with respect to the organic halide for making the Grignard and the excess Mg is often passivated. Do you have an energy balance for the reaction, based on real conditions? Moreover, not all the halides promptly react to afford the relevant Grignard, the yield may be lower than 100% and not all halides of the parent compounds of the acids you have shown are available. I guess that such a reaction may be valid for a selected number of cases, considering those syntheses that today are performed in very unfriendly conditions.

George Dowson replied: During the overall process, you are correct in observing that magnesium is oxidised from Mg(0) to Mg(II). While only one electron of these two ends up in the carbon dioxide, it should be noted that the other overall reduction acts upon the methanol carbon, so this second electron is not wasted in the synthesis of acetic acid.

This magnesium oxidation is also the driving force for the reaction, allowing the capture and reaction of CO_2 to occur at room temperature and atmospheric pressure. These conditions are a benefit that must be weighted against the higher energy cost of the magnesium-based process.

For simplicity we have not used real energy balances, however I do not believe they would differ greatly from those of the more idealised scenarios. A multitude of papers and reports detailing quantitative conversion, even using the most challenging of Grignard syntheses, strongly suggest to me that there is no reason to expect yields significantly lower than 100% from the majority of Grignard processes, although exceptions may indeed exist. I believe all the chemicals shown in Fig. 3, with the exception of the propriolic acid, have available halides. The alkynyl grignards can indeed only be prepared through the substitution reaction with another Grignard.

I believe that you are entirely correct in estimating that this process may be only appropriate in specific cases where harsh conditions are usually required.

Martyn Poliakoff opened a general discussion of the papers by Richard Heyn, Christopher Kozak and George Dowson by commenting: The current mid-century process for making terephthalic acid by aerobic oxidation of *para*-xylene is an exceptionally efficient one compared to most manufacture of bulk chemicals. It delivers products with very high purity and high yields. Therefore it seems unlikely that a route from CO_2 will be able to compete commercially.

Michele Aresta responded: *p*-Xylene to terephthalic acid is a particularly efficient process, but other oxidations of organics to acids are not so C-efficient and in such cases the direct carboxylation of organic substrates may find useful application. The barrier is the C-H activation.

Christopher Kozak remarked: Another concern with terephthalic acid as a target molecule for an industrial process using CO_2 is that a major use for it is in making PET. The phthalates in PET have been linked to health issues (they are suspected endocrine disruptors). The EU intends to restrict phthalates under RoHS 2, so regardless of how they are made, their market may be limited.

Cafer Tayyar Yavuz proposed: Martyn, I did some reading and came across some work by Li *et al.*[1] where they claim they have better solutions. What do you say about this paper?

1 Meng Li, Thomas Ruddy, Darryl Fahey, Daryle H. Busch and Bala Subramanian, *ACS Sustainable Chem. Eng.*, 2014, **2**(4), 823–835.

Peter Styring replied: The efficiency of the process will be determined by the availability of the pre-cursors. If fossil derived starting materials are to be avoided, then CO_2-derived processes may provide replacements

Pedro Abrantes questioned: This is a general question to all lecturers When you work on a solution it's probably to become a valuable industrial deployment. In your presentations, nobody spoke about the perspective of his solution. None of you have given us a vision of the potential or the position confronting the existing solutions. Can you tell us about the volumes of CO_2 you try to manage? Or the volume you will need to run your processes? A vision for the future maybe?

Michele Aresta answered: We have carried out a study for a USA catalyst company and have shown that, considering the existing scenarios of expansion of the market of chemicals that can be derived from CO_2 and of technologies which use but not convert CO_2 , the amount of used CO_2 may jump from *ca.* 200 Mt/y to 330 Mt/y in 2030. It is obvious that this figure does not give a great perspective to CCU. Also if we consider the fact that such new technologies will be more efficient in energy and the "avoided" CO_2 may be as much as 2 times the used CO_2, it will nevertheless be only with the implementation of large use of perennial energies (solar, wind) and with larger exploitation of biomass (microalgae) that we may reach a significant target in CO_2 avoidance. My forecast is that exploiting at their best chemical processes, perennial energies as the primary energy source in CO_2 conversion processes and biomass (terrestrial and aquatic) as the raw material a 10% cut of the CO_2 emissions may be reached in 15–20 years from now.

Christopher Kozak continued: The current focus of my research is on the fundamental science of catalyst design and optimization of conditions *via* mechanistic studies. Our product, as it were, is to increase the number of tools available to chemists by understanding the process of our reactions. The amount of CO_2 consumed by manufacturing polymers or even solvents from CO_2 will be marginal compared with the amount of CO_2 produced. Our goal is not to mitigate large volumes of CO_2 so much as to demonstrate that it should be treated as a feedstock in lieu of (or in conjunction with) petroleum feedstocks.

Richard Heyn responded: This is an oversimplification of the necessary chemical engineering, but in general any organic carbonate can be made from another *via* a transesterification reaction. Therefore, if only one organic carbonate (linear or cyclic) could be made from CO_2 and the appropriate alcohol or diol, then all organic carbonates could be obtained from CO_2. After transesterification, the original alcohol or diol could then be recarboxylated. I have illustrated this with ethylene glycol and methanol to give dimethylcarbonate as shown in Fig. 1.

Note that this is currently done today, but that the starting point is propylene oxide. The propylene glycol thus formed after the transesterification with methanol to dimethylcarbonate is just a by-product. If there is a market for the by-product and the economics are OK, then this works to some degree, but I doubt this is a general or sustainable approach.

I do not have the necessary numbers to comment on volumes of CO_2, but one of my visions is that a viable process for making only one organic carbonate from CO_2 will permit the production of a wide range of carbonates from CO_2.

Eryk Remiezowicz continued the discussion on the future vision by asking: What is your vision to further develop your work? Where do you think your work will develop? Will you search for different chemistry or will you try to go for pilot scale?

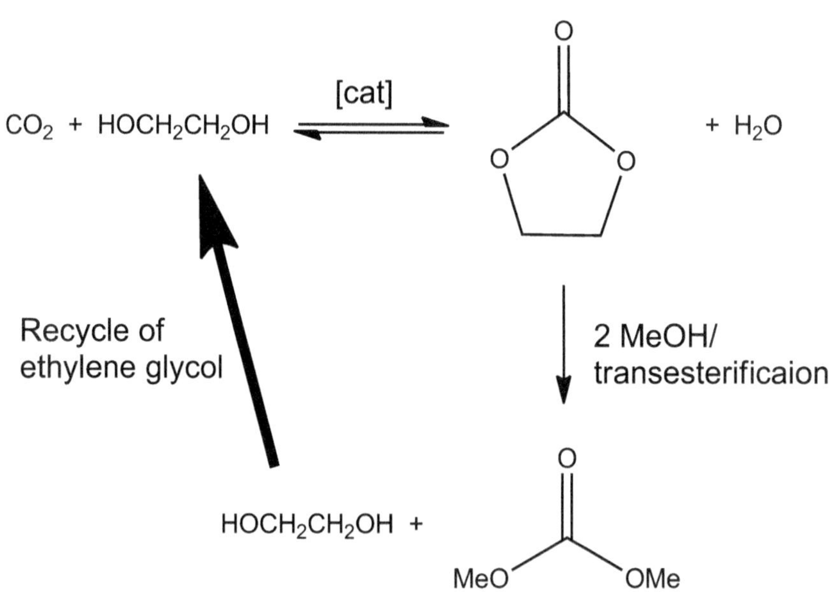

Fig. 1 Formation of dimethylcarbonate via initial reaction of CO_2 and ethylene glycol, followed by transesterification with methanol.

George Dowson responded: Thank you for your question. This is a hard one to answer at this time as a more complete and scaled techno-economic assessment and life cycle assessment must be carried out to fully understand the costs and implications of such a process. In my view I don't believe this process will be viable for a cheap bulk chemical like acetic acid, but may well be viable for higher value carboxylic acid targets which are commonly found in polyamide and polyester plastics. More importantly perhaps, I think we have demonstrated that using sacrificial agents such as magnesium species is not as big a roadblock to economically creating a CO_2-containing products as one might think and that indeed some different options than magnesium may yield routes that are even more economic.

Ian Ingram stated: Several commenters had asserted that in order to be economically feasible, environmentally sustainable products need to be at the same price or cheaper than their conventional equivalents. The success of brands like Ecover demonstrates that at least some people are willing to pay significantly more for a product because it is marketed as being more sustainable, at least in the consumer sector. Whilst this market is not necessarily huge, in many cases it might allow sufficient investment to gain economies of scale or further develop the technology and ultimately reduce the price.

Christopher Jones remarked: I would argue that narrowing the price-gap between sustainable and unsustainable product options is important to garnering mass-market appeal. Through my own research (into fashion and energy purchasing) I have found that people would often be willing to pay a little more for sustainable options but only really if the price difference was fairly narrow and the product quality equally good/desirable.

The price of products is a strong driver of purchasing behaviour and while it is not the only predictor of consumer choice (*e.g.* normative considerations about what is 'appropriate' and hedonic considerations as to what is 'desirable' do influence consumer choice) it remains a prominent consideration for many people. Of course there will always be people there that buck the trend and will be willing to pay substantially more for more sustainable products (*e.g.* due to having particularly strong environmental values); however, I would argue that for the majority of people – particularly in a world where there are so many competing priorities and goals – the relative affordability of sustainable products will be a major consideration. What I would say, however, is that we need to bear in mind that the nature of the product being produced/retailed will also have a bearing on people's willingness to pay for it. Where a product is seen to be a right or necessity (*e.g.* energy/water), people tend to be quite reluctant to pay over the odds for it. Where a product is deemed to be more of a luxury item or indulgence (or perhaps a symbolic/status item) then perhaps people would be willing to 'loosen the purse strings' a little more.

Martyn Poliakoff replied: I think that you have a point. However, Ecover is making products containing mixtures of chemicals and the public might respond less favourably to paying more for individual chemicals, especially as they do not buy chemicals directly from producers.

Pedro Abrantes responded: The challenge in front of us, is to handle giga tons of CO_2. Of course we can find several companies able to buy an expensive CO_2 ton. A best price to valuate sustainability action but the volume to handle is far away from this exceptional position

Richard Heyn answered: Rather than individuals, one should also consider the marketing strategies of service industries. As an example, perhaps a hotel chain puts only "CO_2-containing" mattresses in their hotel beds. While the mattress may be more expensive, the increase in costs to only one consumer would likely be insignificant, and the hotel chain could promote how "green" they are.

Martyn Poliakoff said: In response to the comment on consumers wanting eco-friendly goods, I think supermarkets rather than consumers will make the difference because supermarkets have sufficient commercial power to force suppliers to take notice. Of course, the supermarkets may be influenced by consumers but consumers alone are less likely to get things changed.

Christopher Kozak replied: I think supermarkets will only make a change if they see the consumer demand for it. Shelf space is expensive and there are many, many variants of products such as personal care, toiletries, household cleaners, *etc.* Supermarkets won't carry something that doesn't sell, but if the consumers call for it they will. In Canada there was a public backlash on the fashion of "microbeads", small bits of plastic in shower gels and personal items used as exfoliants and mild abrasives. Environmentalists, through the popular media, did a good job illustrating the long term problems with environmental accumulation of plastic from these sources and the public ceased buying these products. Supermarkets caught on and used this as a marketing tool by promising to not carry any such products on their shelves. If consumers are informed and vocal about their choices, the market will follow.

Elsje Alessandra Quadrelli commented: I think there is a point about relevance of CCS (carbon capture and storage) in the context of CDU (carbon dioxide utilization) that is not mentioned often enough: we have to be careful when the overall goal is green chemistry about how CCS is brought into play. If the goal is fossil fuel avoidance, and/or switching from a fossil-based economy to one where renewable energies have a larger contribution, CCS is sometimes being used a Trojan horse to help "green" fossil-fuel focused industries or even to derail efforts to advance (chemical) CDU in the context of climate mitigation. We need to draw the lines surrounding green technologies carefully: yes CCS has a role to play in the context of carbon mitigation, yes enhanced oil recovery (EOR) is a CDU technology, yes EOR is economically viable to recover more fossil fuels using (in the best case) less energy than the incumbent technology to get this new oil out of the ground and finally, yes EOR can be used to learn how to perform large scale CCS experiments; but no, EOR is not directly a technology that will help the transition to renewable energies use : it is not because an industry uses CO_2 that it directly serves the purpose of a circular carbon economy. There are technologies that use CO_2 that have a substantial potential to help a circular carbon economy, but not all technologies that use CO_2 necessarily fulfil that goal. Caution should be exercised to separate the wheat from the chaff.

Anthony Coogan stated: I have a suggestion of how CDU can assist CCS. Is it a case of having a second bite at the cherry? There appears to be a perception that CCS and CDU are going forward independently. Perhaps they are even competing for funding?

We may have missed a potentially large opportunity: CDU schemes may sometimes be unsuccessful, often for economic reasons. The considerable expenditure by CCS may be reduced by using the CDU idea. Carbon dioxide utilisation takes in CO_2 but primarily not other carbon compounds; on the other hand, carbon capture and storage could include methane, graphite or almost any other compound containing carbon. Assuming that both processes can use the same capture process, then much of the excess costs of CCS will involve the carbon storage process.

Costs of transporting carbon dioxide gas to the coast and storing it below the sea bed must be enormous. These costs include piping and pumping costs, research into finding undersea locations where the gas can be stored safely *etc.* Here is our opportunity: at our conference, we have heard how polycarbonates, polyurethane, chemical feedstocks, magnesium carbonate (or bicarbonate) and other substances are made profitably by utilising carbon dioxide, but suppose the worst should happen? After spending time, effort and cash, a certain project happens to fail. Are all those resources to be written off? No! Let us continue to make a certain amount of product, at a small loss, and store it at ground level. Although the CDU business will lose money, the overall CCS and CDU system will save money in potentially taking carbon out of the atmosphere and save the CCS costs outlined above.

For example, take magnesium carbonate – with apologies to Priestnall *et al* (DOI: 10.1039/c5fd00047e). Playing devil's advocate, suppose a business failed through no fault of their management: it could continue to manufacture $MgCO_3$ and store it in a warehouse at ground level, with benefits to the environment and a net saving to the CCS organisation. Undersea storage may be unsafe, but storing $MgCO_3$ would be safe. Another example is what happened in Japan. In 1998, a process manufactured graphite fibre (and water), from methane and CO_2 at a profit, but then the competitive acrylonitrile process took over. It would have been too costly for that process to continue, but if solid carbon could be stored in warehouses at ground level, CCS costs could be reduced. With warnings of global water shortages, the process could be revived in arid areas where natural gas was cheap.

Niall MacDowell asked: Which would be preferred, enhanced oil recovery or other unconventional oil sources such as fracking or tar sands?

Michael North answered: The question is hypothetical as what would happen in practice is that if allowed the oil industry would do the cheapest first, then all the others in order of increasing cost, resulting in a further increase in atmospheric CO_2 levels.

Niall MacDowell responded: This is wrong. It is not a hypothetical question – both EOR and producing oil from so called 'unconventional sources' have been relatively common practice in recent decades. The key driver is the combination of oil price and demand. Once demand drives the oil price sufficiently high, oil derived from CO_2-EOR or tar sands becomes commercially viable. The CO_2

footprint associated with oil derived from tar sands is much greater than that derived from CO_2-EOR, so from the perspective of minimising CO_2 emissions to the atmosphere, oil derived from EOR is preferable to oil derived from tar sands.

The point is one of displacement, which is slightly more subtle than the stepwise process that you have outlined above. If there is sufficient demand for oil, and there is no CO_2-EOR activity, then that demand will be met using tar sands, resulting in a greater rate of CO_2 emission. Therefore, CO_2-EOR allows the avoidance of CO_2 emissions and the deployment of CCS infrastructure. It is, of course, possible to capture the CO_2 arising from a tar sands upgrader process, similarly to the Shell QUEST project in Edmonton, but it is less clear how this leads to the deployment of larger-scale CCS infrastructure.

Katie Lamb opened the discussion of the paper by Carmine Capacchione: I would like to query some results in terms of catalyst and co-catalyst screening that you carried out for your iron(III) catalysts. In Table 1 where you quote your screening results, entry 25 quotes the best conversion obtained overall for the study, is around 95%. I noticed that in this run you use a 10 : 1 molar ratio of TBAB : iron(III) catalyst in the reaction. In your TBAB control test quoted in this paper (entry 28, Table 1), you state that TBAB was tested over a 24 h period and only obtained 32% conversion. In this run however you only used a molar ratio of 2 with respect to TBAB. I was therefore wondering if you have run a control test using a TBAB molar ratio of 10 in this reaction, using the same reaction conditions as entry 25 except with no iron(III) catalyst, and compared this to your result when using TBAB and catalyst together, to check that the high conversion obtained in entry 25 is not due to the presence of excess TBAB?

Carmine Capacchione replied: In entry 25 Table 1, we reported a conversion of 95% after one hour. We have run a control test using TBAB under the same conditions (molar ratio of 10 and with no catalyst), obtaining a conversion of 10% with the same reaction time. From these data it is clear that the conversion obtained is due to the synergic effect between the catalyst and TBAB.

Michael North enquired: Can your catalyst or ligand be recovered and recycled?

Carmine Capacchione answered: The possibility to recover the catalyst has been taken into account. Preliminary results showed that the catalyst can be reused at least two times, *i.e.*, using propylene oxide as the substrate and complex CH as the catalyst, under the same conditions mentioned in Table 1 entry 3, we observed conversions equal to: 56, 38 and 17%; respectively for the first, second and third run. Despite this not being a complete recycling of the catalyst, we are working in order to improve the efficiency of the recovery procedure.

Antoine Buchard asked: What is the catalytic activity of iron trichloride in the coupling of CO_2 and propylene oxide under your reaction conditions, alone or in combination with TBAB?

Carmine Capacchione responded: We have done control experiments to evaluate the activity of iron(III) chloride in our reaction conditions. In order to compare the results with entry 25 Table 1, we used the same reaction conditions. Since the complexes are dinuclear we use twice the number of moles of iron(III) chloride with respect to the catalyst, resulting in a propylene oxide : TBAB : $FeCl_3$ molar ratio equal to 2000 : 5 : 1. In the absence of TBAB, after 1 h we observed a conversion of about 9% with the exclusive formation of poly(propylene oxide). In the presence of TBAB the conversion reach the 65%, with the formation of propylene carbonate and traces of poly(propylene oxide).

Richard Heyn remarked: Each Fe center in the catalysts are 6-coordinate, so there is no apparent, readily available coordination site for the epoxide. I might expect that the catalysts with the bulkier R groups would be more prone to dissociate into monomers, which would provide more sites for epoxide coordination and thus higher activities, but this is not supported by your data. Is there a dimer–monomer equilibrium that provides a coordination site for the epoxide? If not, how is the epoxide activated by the Fe_2 catalyst?

Carmine Capacchione replied: From our experimental data we have no evidence of a monomer–dimer equilibrium, even in the case of the bulkier cumyl groups. We suppose that, during the reaction, the dimer is able to reorganise in order to coordinate an epoxide or carbon dioxide molecule. Our hypothesis is supported by the labile nature of the sulphur–metal coordination bond. Indeed we have already reported a similar behaviour in a previous work[1] also regarding a reaction with a coordination-insertion mechanism like the ring-opening polymerisation.

1 Francesco Della Monica, Ermanno Luciano, Giuseppina Roviello, Alfonso Grassi, Stefano Milione and Carmine Capacchione, *Macromolecules*, 2014, **47**, 2830–2841

Peter Styring said: From the table of data it is clear that there is no conversion unless TBAB is present. It strikes me that the iron catalyst is actually the co-catalyst while TBAB is the actual catalyst. The same is true where the organic species possess halide ions. Could you comment on the role of TBAB in the reaction mechanism as it was not included in the scheme?

Carmine Capacchione answered: In our system the role of TBAB is essential to introduce a nucleophile in the reaction medium. Comparing the conversion obtained in entry 25 Table 1 (95%) with those obtained in the same reaction conditions in the absence of the catalyst CH (10%; see question 200) it is clear that the Lewis acidity of the iron centres plays a dominant role in the activation of the reactants respect to the ammonium ion.

Christopher Kozak commented: Most of your reactions are conducted at elevated temperatures. Have you done any low temperature experiments to change the selectivity from cyclic to polycarbonates?

Carmine Capacchione responded: Using the most active catalyst CH, we have done low temperature experiments using propylene oxide and cyclohexene oxide as substrates, respectively under the conditions of entry 25 Table 1 and entry 34 Table 2. In both cases, by decreasing the temperature to 35 °C, we only observed a drop in the activity and no changes in terms of selectivity towards cyclic carbonate, with no appreciable formation of polycabonates.

Peter Styring asked: It is not clear in the paper if the pressures you refer to in the reaction conditions refer to the gauge pressure before or after heating to the reaction temperature. Could you please confirm which it is? If it is the pressure at reaction temperature how is it controlled to the accuracy presented in the paper?

Carmine Capacchione replied: In our reaction procedure, we adjust the pressure soon after the equilibration of the reactor at the desired temperature and the pressure during the reaction is kept constant by a common pressure control valve.

Ian Ingram queried: In our own laboratory we often observe a significant loss of unreacted volatile epoxides, such as propylene oxide, when the CO_2 pressure is released at the end of a reaction, even if the vessel is first cooled with liquid nitrogen. For us this has sometimes made it more difficult to determine an accurate conversion by NMR, although I appreciate that you have used an internal standard which should solve this problem. Did you check the mass remaining at the end of your reactions, and was any epoxide lost?

Carmine Capacchione answered: Yes, we checked the mass of the reaction mixture remaining at the end or our reactions and we observed no significant loss in epoxide with respect to the quantity determined by NMR analysis using an internal standard.

José A. Castro-Osma continued: Firstly, I would like to congratulate you for the nice work you have presented at the Faraday Discussion. My question is regarding the use of internal epoxides as it has been reported that these substrates are more challenging and there are not many catalysts that can convert internal epoxides into disubstituted cyclic carbonates. Have you studied the catalytic activity of your catalyst using internal epoxides? If not, are you planning to do it?

Carmine Capacchione stated: As reported in the manuscript, we performed the cycloaddition of carbon dioxide with cyclohexene oxide (Table 2, entry 34–36), obtaining interesting results in the case of the less hindered catalyst CH. Starting from these results, at the moment we are evaluating the scope of the catalytic system on a series of internal epoxides

Michael North questioned: In addition to carrying out reactions under solvent free conditions, you have done reactions in methyl ethyl ketone. Why does this solvent work when most others fail?

Carmine Capacchione responded: For this kind of reaction, the beneficial effect of a polar aprotic solvent is often observed in the literature.[1] In our case a polar solvent like acetonitrile gave poor results in terms of activity, likely due to the competition between the solvent and the reactants for the coordination site. Methyl ethyl ketone results in a reasonable compromise between polarity, in order to have a good carbon dioxide solubility, and coordination capability, so the slowing down of the reaction using this solvent is probably only due to a dilution effect.

1 (a) C. Martín, G. Fiorani and A. W. Kleij, *ACS Catal.*, 2015, **5**, 1353–1370; (b) J. W. Comerford, I. D. V. Ingram, M. North and X. Wu, *Green Chem.*, 2015, **17**, 1966–1987; (c) N. Kielland, C. J. Whiteoak and A. W. Kleij, *Adv. Synth. Catal.*, 2013, **355**, 2115–2138; (d) M. North, in *New and Future Developments in Catalysis*, ed. S. L. Suib, Elsevier, Amsterdam, 2013, pp. 379–413; (e) M. Cokoja, C. Bruckmeier, B. Rieger, W. A. Herrmann and F. E. Kühn, *Angew. Chem., Int. Ed.*, 2011, **50**, 8510–8537; (f) R. Li, X. Tong, X. Li and C. Hu, *Pure Appl. Chem.*, 2012, **84**, 621–636; (g) A. Decortes, A. M. Castilla and A. W. Kleij, *Angew. Chem., Int. Ed.*, 2010, **49**, 9822–9837.

William Webb opened a discussion of the papers by Carmine Capacchione and Charlotte Williams by communicating: I was curious to know what effect the substrate size and the distance of the metal centres from one another had on the reactions, I understand that in the case of the metal centres the vacant sites are considered to be *trans* from one another so the metal–metal distance may not be important but wanted to know if you had considered this? For the substrate, does the size of the substrate have any effect on rate of reaction, could size affect the access to the active sites?

Charlotte Williams communicated in reply: A good question. If the epoxides are changed, there are a number of affects which should be considered. The first is that for each epoxide there are different barriers to polymerization *vs.* cyclic carbonate formation. Thus, the most obvious effect noted on changing epoxide (size or electronic properties) is a difference in polymerization rate and selectivity (cyclic *vs.* polymer formation). Using larger/more substituted epoxides, it is notable in the literature that there are fewer examples of active catalysts. For example, there are only two reports of catalysts able to copolymerize limonene oxide/CO_2. The metal–metal distance is important. Recent studies using a range of catalysts and throughout the literature have highlighted the importance of bimetallic/dinuclear active sites. It has also been shown that the distance between such sites can be quite variable, from 3–7 Å. There are also proposals that having 'flexible' ancillary ligands which allow moderation of the metal–metal distance can improve activity. However, such studies are still early and more data and a wider range of catalysts need to be studied.

Carmine Capacchione also communicated in reply: With regard to the metal–metal distance, at the moment we are not able to evaluate the influence of this parameter on the reaction because we don't know how it changes from one complex to another, or during the reaction. About the influence of the substrate size on the reaction, we can look at the data listed in Table 2. It is clear, from the data obtained in the case of a bulkier substrate like styrene oxide (entry 33) and cyclohexene oxide (entry 34–36), that the activity decreases by increasing the size of the substrate.

Michael North asked: How well can the catalyst be removed from your polycarbonate polyols and for polyurethane synthesis does this actually matter?

Charlotte Williams answered: There are several strategies published in the academic literature to remove catalysts from polymers in this type of polymerization.

These include using silica plugs, biphasic systems and heterogenization of the catalyst. These have been shown to be effective in removing, and in some cases allowing recycling of the catalysts. It is important to remove the catalyst in order to improve the polymer thermal stability which may have implications for processes/engineering aspects. In the case of PU synthesis it may be less important, from the point of view of allowing the polyol-isocyanate coupling, but I am not aware of any published studies describing the influence of catalyst residues on that reaction. A further consideration relating to catalyst removal, is the loading at which catalyst can be applied. We have demonstrated in previous publications that lower loadings are feasible using this class of catalyst. Furthermore, in the broader literature it is already established that viable loadings, including at 1 : 100 000 (catalyst : monomer, m/m) are feasible. At such lower loadings, the issue of catalyst removal may be obviated and it would be expected that the influence of residual catalyst would be significantly decreased.

Christopher Kozak asked: The turnover numbers suggest 50% of the cyclohexene oxide is consumed, yet the polymer molecular weights are rather low. How do you reconcile the difference between low observed M_n values and the number of consumed monomer? Was SEC conducted on purified polymer or on the crude material? What is the dispersity of the crude product?

Charlotte Williams replied: The SEC was conducted on crude polymers. The crude products either show mono-modal distributions (when chain transfer agents are present) or bimodal distributions (when trifluoroacetate is the initiating group). The dispersities in M_w values are typically very narrow. This paper specifically sets out to examine the influence of various chain transfer agents so as to target low M_w polyols for PU synthesis. The question of polymer M_w is an important question and key to understanding the field as a whole. In this field of catalysis, particularly using cyclohexene oxide as the epoxide, it is commonly observed that although turn-over-numbers can be very high, the polymer molecular weights are typically in the polyol regime. That is, the molecular weights are typically lower than would be expected on the basis of catalyst loading. Such calculations can be made because the polymerizations exhibit features of control. In fact, these polymerizations are likely occurring under immortal conditions, *i.e.* where there are chain transfer agents present. It appears to be quite a general observation, including using a range of catalyst world-wide. Therefore, it is proposed, that under these immortal conditions the presence of protic compounds (either added as chain transfer agents or present due to reactions between epoxide and water) controls the M_w of the product. This is demonstrated in our paper where we use chain transfer agents to prepare monomodal distribution polyols.

Richard Heyn questioned: Do you observe different initiation rates depending on the type or amount of chain transfer agent added to the polymerization?

Charlotte Williams replied: This is also a very good question. In these polymerizations the catalysts are operating under so-called immortal regimes. This immortal polymerization was first described by Inoue (the pioneer of CO_2/epoxide copolymerization) and refers to well controlled catalysts which are operating in the presence of chain transfer agents. For a very good review of immortal polymerization see the work of Inoue.[1]

In an immortal polymerization system it is proposed that the rate of chain transfer occurs significantly more rapidly than any other reaction (*e.g.* than propagation). Thus, by using protic reagent chains are 'swapped' on and off metals very rapidly. There have been model studies of such chain transfer reactions for metal alkoxides and these substantiate the proposal that chain transfer occurs rapidly and reversibly.[2] Thus in an immortal polymerization the features are that dispersity is narrow, M_w is predicted on the basis of the total concentration of hydroxyl groups, M_w increases linearly with conversion, the rates of initiation are faster than propagation, there are limited termination reactions and block copolymers can be prepared. The polymerizations described in this paper exhibit the features of immortal polymerizaiton. In particular the M_w values correspond to the total concentrations of hydroxyl/protic groups which are present. The question is whether there is an effect on initiation by using different chain transfer agents. It is a very good question but one that cannot be answered fully on the basis of the current experimental data. It would be necessary to analyse in detail the polymerization kinetics so as to quantify the rates of initation *vs.* propagation. It is possible to observe experimentally that the M_w distributions are very narrow for the polymers and that the polymerizations exhibit the typical features of rapid chain transfer (M_w predictable from OH concentration, M_w increases linearly with conversion, narrow PDI, end group control). Thus, it is likely that chain transfer is still a rapid and reversible process. We can also note that in most cases the rate (TOF) is not affected significantly by the nature of the chain transfer agent. The exception to this are the amines (discussed already). In the case of hydroxyl groups as chain transfer agents then there is not a significant effect on TOF on varying the nature of the chain transfer agent or in varying the amount of chain transfer agent. Thus, it would be expected (but cannot yet be proven) that the chain transfer agent is not significantly affecting the rate of propagation. This question needs to be answered by a more detailed kinetic study.

1 S. J. Inoue, *Polym. Sci, Part A: Polym. Chem.*, 2000, **38**, 2861–2871.
2 M. H. Chisholm, E. E. Delbridge, J. C. Gallucci, *New. J. Chem.*, 2004, **6**, 145.

Gonçalo V. S. M. Carrera commented: The decrease on the activity observed when amines are used as additives on the preparation of polycarbonates, could be related with the formation of carbamates leading to a possible acidification of the reactional medium and a consequent change on the protonation state of the ligand, affecting the coordination with the metal centers and a possible influence on the solubility of the complex (decrease) contributing to the poorer performances observed when these additives are added. What's your perspective regarding this issue?

Charlotte Williams answered: This is a good question and relates to a minor decrease in activity observed using amines as the chain transfer agent. Firstly, it is necessary to qualify that although the values for turn-over-frequency are marginally lower using amines, caution should be applied to over-interpret such data. It is well documented that TOF values are point kinetic measurements and as such do not represent the true picture of whether activity is compromised. For example, if there was a slower initiation period, but overall propagation occurred at the same rate, an analysis of TOF would not be sufficient to detect such subtle effects. Nevertheless, it does appear that under these conditions and using some amines there are side-reactions occurring. It is documented in our paper that using diamines we observe precipitate formation and we tentatively propose that this is due to side reactions occurring between the amine and carbon dioxide. Thus, exactly as you comment, it is very well known that amines can reversibly interact with carbon dioxide to form carbamate species. It is not known the extent to which such carbamates can under-go chain transfer reactions. Thus, one interpretation of the slight decrease in TOF may be that the carbamate formation is slowing overall propagation and/or affecting rates of chain transfer.

Christopher Kozak said: You mentioned trifluoroacetate (TFA) as being a good initiator. I wonder if it's acting as a labile bridging group, or if being a potentially bidentate ligand, it can remain bound to one metal site while also being a nucleophile to an epoxide bound to the other metal centre? Do you have any evidence of bridging? Do you think that increasing the pressure affects the rate in this system? What is really the special influence of TFA for the polymerization?

Charlotte Williams replied: In this paper we study the trifluoroacetate ligand which is the initiating site during polymerization. It is important to note that using this ligand obviates the need for any Lewis base/ionic co-catalysts. Thus, the catalysts described are used without any additives. The paper describes that using the di-zinc bis(trifluoroaceate) complex the polymer produced shows a bimodal M_w distribution, in the absence of any chain transfer agents, but both distributions are due to dihydroxyl end-capped polymer chains. This is established by using SEC to separate the two M_w distributions and subsequent analysis of each distribution using NMR spectroscopy/MALDI-ToF. Thus, the trifluoroacetate ligand is special, compared to for example acetate, because it produces distributions where both M_ws are polyols. This has also been observed by other researchers, using different catalysts, for example Donald Darensbourg using metal salen catalysts with trifluoroacetate initiating groups. In Darensbourg's study he showed, using *in situ* mass spectrometry, that the trifluoroacetate group is present at the chain end during propagation but is hydrolysed on exposure of the polymer to work-up. We propose that a similar explanation is feasible to rationalise the experimental observation that chains show bimodal distributions, where both series are polyol. If chain transfer agents are added, the paper establishes that monomodal, narrow M_w distributions of polyols can be produced (using water, diol or triol as CTA). It is an interesting suggestion to investigate its coordination modes at the metal and is part of on-going catalyst characterization studies. Increasing the pressure does not affect the rate using dizinc catalysts. These catalysts are active at 1 bar pressure. A detailed kinetic study[1] has shown that there is a zero order dependence of the rate on the pressure of carbon dioxide, over the range 1–40 bar.

1 Jutz *et al.*, *J. Am. Chem. Soc.*, 2011, **133**, 17395–17405.

Peter Styring asked: Your catalyst gave moderate/low molecular weights but narrow polydispersities. There are data in the literature for other catalysts that show high molecular weights but broad polysispersities. Is there a way to get a high molecular weight and low polydispersity?

Charlotte Williams responded: This is a good question and there are certainly literature reports using other epoxides (*e.g.* propylene oxide) and other catalysts to allow a tuning of molecular weight.

Indeed, in our own previous publications we also report that high molecular weight, narrow dispersity polycarbonates can be produced and we provide evidence that the polymerizations are well controlled.[1,2] The reason it is possible to prepare high M_w, narrow dispersity polymers is because such single site catalysts are well controlled. Thus, is it possible to predict the M_w based on the amount of catalyst applied. Therefore, by using lower loadings of catalysts it is possible to access (at higher conversions) high M_w polymers. If ill-defined catalysts or heterogeneous catalysts are applied it is not generally possible to access high M_w, narrow dispersity polymers because such catalysts have multiple initiating sites and thus cannot allow polymerization control. The major current application for polycarbonates is to use low M_w polyols and so less focus is currently placed on high M_w materials.

1 A. Buchard, M. R. Kember, K. G. Sandeman, C. K. Williams, *Chem. Commun.*, 2011, **47**, 212–214.
2 A. M. Chapman, C. Keyworth, M. R. Kember, A. J. J. Lennox, C. K. Williams, *ACS Catal.*, 2015, **5**, 1581–1588.

Christopher Kozak made a general comment relating to selectivity for higher/lower molecular weight polymers: My students found that decreasing the epoxide : catalyst ratio, they were often getting higher M_w polycarbonate. As Charlotte said, impurities generate more chain transfer events, and the impurities come from the epoxide. So, for example, the less CHO that is present, the fewer chain transfer events because there are fewer impurities per catalyst, hence leading to higher molecular weights.

Michael North noted: Some catalysts make cyclic carbonate whilst others which are very similar make polycarbonate. I ask an open question to everyone on how we could predict in advance which product a particular catalyst would produce?

Charlotte Williams answered: This is an important consideration. It is important to note that cyclic carbonates are the thermodynamic products of the reaction between carbon dioxide and epoxides. Thus, at sufficiently forcing conditions all catalysts will be able to prepare cyclic carbonate. The key point to note is that only a few catalysts can currently prepare polycarbonate (the kinetic product). We have previously shown in studies using di-zinc catalysts, that it is possible to quantify the difference in experimental activation barriers between the polymer and cyclic product. In the case of di-zinc catalysts, using cyclohexene oxide, the barrier for polymer production is significantly lower, explaining the high selectivity for polymer.[1] Therefore, one strategy to predict the product (polymer *vs.* cyclic) produced by a catalyst which is effective for polymerization is to experimentally determine the energy barriers and then use this to select the optimum conditions (temperature) at which to carry out the reaction so as to maximise the desired product. It may also be possible to calculate the energy barriers, using DFT.[2] It is also generally observed in the literature that ionic complexes or complexes with a good leaving group have a much stronger tendency to form cyclic carbonates and to fail to produce a polymer. Another consideration is that if the polymer chain is 'off-metal' for a significant period of time then there is a greater tendency to form the cyclic carbonate. This may be the reason that the salen catalysts incorporating the co-ligand into the scaffold are highly effective (such systems were pioneered by Nozaki, Lee and Lu). Finally, there is emerging evidence that dinuclear catalysts shown high activity as polymerization catalysts. We have recently reviewed this area.[3]

As a final observation, I would note that currently the range of different ligands/metal complexes observed active for CO_2/epoxide polymerization is still narrow. Thus, it is rather difficult to make 'rules' about what the design features for an effective catalyst would be. It will be important to expand the range and scope of ligands/metals so as to garner further knowledge on the catalysis.

1 F. Jutz, A. Buchard, M. R. Kember, S. W. Fredriksen, C. K. Williams, *J. Am. Chem. Soc.*, 2011, **133**, 17395–17405.
2 A. Buchard, F. Jutz, M. R. Kember, A. J. P. White, H. S. Rzepa, C. K. Williams, *Macromolecules*, 2012, **45**, 6781–6795.
3 C. Romain, A. Thevenon, P. K. Saini, C. K. Williams, in *Topics in Organometallic Chemistry: Carbon Dioxide and Organometallics*, ed. X. B. Lu, Springer DE, 2015.

Christopher Kozak commented: We looked at these tripodal amine-bis(phenolate) ligands because most of the previous ligands were planar, giving vacant sites that are *trans* opposed. We wanted to generate a forced *cis* arrangement of vacant sites to promote intramolecular nucleophilic attack. Some of the most active catalysts for polycarbonate synthesis are planar, however we see that this is not necessarily a prerequisite.

Michael North remarked: There are two routes by which cyclic carbonates could be formed, directly or by back biting of a polycarbonate. The two processes have different stereochemistries and polymer chemists are generally good at limiting back biting.

Eryk Remiezowicz enquired: Are you trying to design the chemical structure of polyol-polycarbonate products, having in mind properties of the final isocyanate? Are you cooperating in that subject with isocyanate producers?

Charlotte Williams responded: This is a very important question as the properties of the resulting polyurethanes must be suitable for existing applications. We are certainly actively investigating the applications and opportunities for polycarbonate polyols with respect to polyurethane applications. The major current industrial interest lies with propylene or ethylene oxides and in using the polycarbonates as substitutes for existing polyether/ester/carbonate polyols. The aliphatic epoxides have been shown to allow the production of polycarbonate or polyether carbonate polyols which are suitable for applications in PU, including as foams and adhesives. The general class of bimetallic catalysts are active using such epoxides and can be used to prepare polyols efficiently. There are a number of on-going collaborations with PU suppliers worldwide.

Michael North continued: It appears that the cyclic *versus* polycarbonate selectivity was not determined by the metal, ligand or substrate and I'm therefore running out of ideas!

Michele Aresta answered: In our experience in this field (we are mostly interested in monomeric compounds) the coupling of cyclic ethers with CO_2 is a multiparameter driven reaction. Not only the metal and its coordination and geometry (it has been suggested that the coupling is a "single site" reaction and two free-*cis* positions are not necessary), but also the ligand (are they innocent?), the solvent (how do they assist in ring-opening or CO_2 activation?), the (accidental) presence of nucleophiles which will promote ring opening, or electrophiles (which may attack carbon of CO_2 or ethers), and so on.

It is evident that the alternation in insertion of CO_2 and epoxide is the key factor for a polymer with good properties. But then, why not polymerise cyclic carbonates? There are not too many studies in this direction.

Peter Styring remarked: In our homogeneous catalyst systems, aluminium complexes give only cyclic carbonates while zinc catalysts give only polycarbonates. We have control over the metal geometry by using ligands that allow variations away from planarity.

Michael North concluded: To conclude the discussion of cyclic *versus* polycarbonate synthesis it seems that we need a major multi laboratory international grant to investigate this further.

Faraday Discussions

PAPER

Electrocatalytic conversion of CO_2 to produce solar fuels in electrolyte or electrolyte-less configurations of PEC cells

C. Ampelli, C. Genovese, B. C. Marepally,† G. Papanikolaou, S. Perathoner* and G. Centi

Received 10th May 2015, Accepted 10th June 2015

DOI: 10.1039/c5fd00069f

The electrocatalytic reduction of CO_2 is studied on a series of electrodes (based on Cu, Co, Fe and Pt metal nanoparticles deposited on carbon nanotubes or carbon black and then placed at the interface between a Nafion membrane and a gas-diffusion-layer electrode) on two types of cells: one operating in the presence of a liquid bulk electrolyte and the other in the absence of the electrolyte (electrolyte-less conditions). The results evidence how the latter conditions allow productivity of about one order of magnitude higher and how to change the type of products formed. Under electrolyte-less conditions, the formation of >C2 products such as acetone and isopropanol is observed, but not in liquid-phase cell operations on the same electrodes. The relative order of productivity in CO_2 electrocatalytic reduction in the series of electrodes investigated is also different between the two types of cells. The implications of these results in terms of possible differences in the reaction mechanism are commented on, as well as in terms of the design of photoelectrocatalytic (PEC) solar cells.

Introduction

There is an increasing interest in the literature on solar fuels,[1–10] although different concepts are often associated to this definition. Solar fuels are indicated here as those having a carbon-negative footprint, because they utilize emitted CO_2 to produce energy carriers or chemicals through the use of renewable energy (RE) sources. The net effect is thus of introducing RE into the energy or chemical production cycle. The type of products obtained in this process is relevant, because they should preferably be *drop-in* liquid-fuels according to both economic and sustainability perspectives. In fact, it is often forgotten that there are many problems (economic, normative, authorization *etc.*) associated with the

Dept. DIECII, Section Industrial Chemistry, University of Messina, CASPE/INSTM and ERIC aisbl, V.le F. Stagno D'Alcontres 31, 98166 Messina, Italy. E-mail: centi@unime.it; perathon@unime.it

† B. C. Marepally currently at C2P2 UMR 5265 (CNRS, CP2 Lyon Université de Lyon), Villeurbanne Cedex (France).

introduction of new products requiring a novel infrastructure. Being on such a large scale as the problem of energy is, a "sustainable" transition to a low-carbon economy is only possible by the development of energy vectors and raw materials for chemistry, which smoothly integrate in the actual system and infrastructure, minimizing the above issues. Liquid products are necessary to allow energy-efficient and low cost transport over long distances and storage for a long time, one of the actual main advantages of oil and derivate energy carriers.

We previously discussed the above aspects in more detail, in relation to the need to develop a sustainable low-carbon energy and chemical production method.[11–14] In this vision, CO_2, biomass and solar energy are the three key elements around which future sustainable scenarios for energy and chemical production should be developed.[12,15] Critical factors to enable this scenario are the development of new approaches for the electrochemical conversion of CO_2, a key step towards artificial leaf-type devices.[7,16] Many reviews have recently discussed the electrocatalytic conversion of CO_2,[17–23] as well as artificial leaf-type or photosynthetic devices,[7,10,24–29] although only a few of them are cited.

A common aspect is the presence of a bulk electrolyte used to close the electric circuit in the electrocatalytic cell or photoelectrocatalytic (PEC) devices. We have shown, however, that a different approach is possible, indicated as electrolyte-less (EL), because a bulk liquid electrolyte is absent.[30] The ionic conduction is realized through a membrane having the photo- and electro-catalysts on the two opposite faces of the membrane, while the electronic conduction is realized through an external wire.[31] This EL approach allows the conversion of CO_2 electrocatalytically to >C1 products,[32–34] while essentially C1 products, with traces of C2 products are detected in the presence of an electrolyte.[35–40]

EL design for PEC solar cells shows some advantages: (i) it is possible to operate at higher temperatures, allowing process intensification and the use of solar concentrators; (ii) the problems related to CO_2 adsorption/diffusion in the liquid phase are eliminated; (iii) the sealing and design of the cell is greatly simplified; (iv) gas-cap formation on the electrode (which may largely decrease the performance) is avoided; (v) operations under pressure are simplified; (vi) light scattering and absorption by the liquid are avoided; (vii) corrosion of the electrode (a main issue in conventional PEC cells) is virtually absent. However, the greatest advantage is the more effective (from cost and energy perspectives) recovery of the liquid products of CO_2 reduction, because they can be collected from the gas stream leaving the EL-PEC cell and not from the liquid electrolyte, requiring distillation or other costly separation procedures. However, the electrodes in a EL-PEC cell, or the hemi-cell for the electrocatalytic reduction of CO_2, should be different from the conventional ones utilized in PEC or electrochemical cells. In fact, together with good electron conductivity, the electrode for the electro-reduction of CO_2 should guarantee a good transport of the protons coming from the membrane to the active centers for CO_2 reduction.

There is thus an intriguing question as to whether this difference in electrode characteristics is responsible for the different type of products observed experimentally during the electrocatalytic reduction of CO_2 or instead there are intrinsic differences in terms of reactivity and reaction mechanism deriving from the presence or not of a bulk liquid electrolyte. The electric double layer, which determines the characteristics of the interface at the catalytic centers, depends on the electrolyte. In PEC cells, concentrated electrolyte solutions are used to minimize internal resistance, but in a EL-PEC cell approach the electrolyte is virtually absent. However, a thin liquid film may be present over the electrocatalysts. Although the ionic transport (to close the circuit) can be realized by surface transport over the electrocatalysts and then through the membrane (without thus the need of a liquid film over the electrode), a surface aqueous acid thin layer may eventually be present as well as the occurrence of capillary condensation in micropores, depending on the temperature and pressure of operations. Nevertheless, the interface between the electrocatalysts and a bulk electrolyte is clearly different from that when the electrocatalysts are in direct contact with the gas phase (CO_2), even when a thin aqueous film may be present. The concentration of CO_2 at the surface of the electrocatalyst is thus expected to be different. However, the different type of electrodes used in conventional and EL hemi-cells for the electrocatalytic reduction of CO_2 do not allow for clarification of above question or to obtain better insight into the motivation for the different behaviours between conventional and EL-cells in the electrocatalytic conversion of CO_2.

The aim of this contribution is thus to utilize the same type of electrodes and reaction conditions, except for the presence of the liquid bulk electrolyte, in order to compare conventional and EL approaches, determine more precisely the differences in the productivity and type of products, and obtain indications about possible differences in the reaction mechanism.

Electrochemical cells for the reduction of CO_2

In electrochemistry, attention is typically focused on the electrode and operative conditions, while cell design is often not considered, except in engineering terms for the realization of optimal charge transport, homogeneous flow and electron distribution in large cells, *etc.* The possibility to operate without a liquid electrolyte is not typically considered for the electrocatalytic reduction of CO_2, although PEM (proton exchange membrane) fuel cells may be considered an example of EL operations. We will compare here two different approaches for the electrochemical reduction of CO_2: (1) gas phase (EL-cell approach) and (2) liquid phase, *e.g.* the conventional one, although using the same type of electrocatalysts as the gas-phase approach. The two devices show common aspects:

(i) the two electrodic compartments are separated by a proton selective membrane, which allows the migration of protons from the anode to the cathode side;

(ii) the electrocatalyst (metal nanoparticles supported over a functionalized carbon nanotube – CNT) is located at the cathode side, supported over a conductive net (carbon paper) and in close contact with the membrane, while the external part, which is in contact with the gas or liquid phase, is coated with Teflon to realize a gas diffusion electrode (GDE) (Fig. 1) as in PEM fuel cells;

(iii) CO_2 continuously flows to the cathode of the electrocatalytic cell as 100% CO_2 at a rate of 10–20 ml min^{-1};

(iv) the currents/potentials applied are similar (1–2 V; 10–100 mA).

These cells are the CO_2 reduction hemi-cells of PEC solar cells, where a photoanode is also present. The latter is a photo-active material able to absorb sunlight to create charge-separation, with the holes utilized at suitable catalytic centers for the reaction of water oxidation to produce O_2 and protons, the latter being transported to the electrocatalyst through the membrane. The electrons are

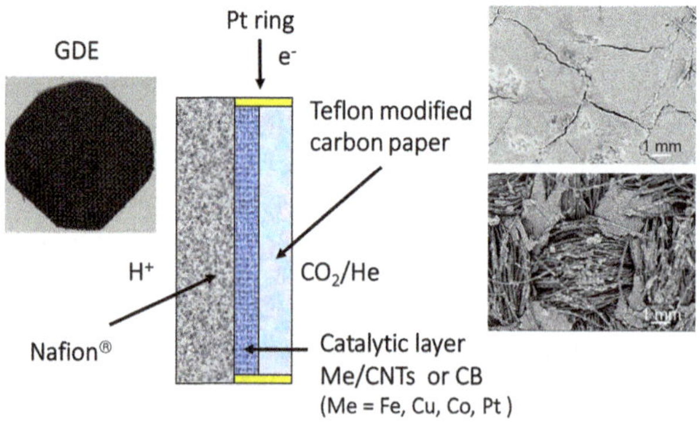

Fig. 1 Schematic illustration of the GDE-type electrodes utilized for CO_2 electrochemical reduction. The image on the left shows the full gas-diffusion electrode (GDE), while the images on the right are scanning electron microscopy images of the side in contact with the membrane and of the side in contact with the gas phase (CO_2) or the electrolyte in which the CO_2 is bubbled.

instead collected and transported externally (through a wire) to the conductive electrocatalyst. We have studied the photo-anode behaviour by preparing TiO_2 nanostructured electrodes,[33] doped with noble (Pt, Au)[41] and non-noble nanoparticles (Cu)[42] to enhance their visible response. Electrolyte and EL operations for the photoanode side (indicated as liquid and gas phase operations, respectively) were both investigated. In water photoelectrolysis and ethanol dehydrogenation processes, higher performances (in terms of H_2 production) of the gas phase with respect to liquid-phase configuration were observed.[42,43] Therefore, a full EL-PEC cell may be developed, both for the anodic and cathodic parts. However, discussion will be focused here only on the electrocatalytic reduction of CO_2, thus on the cathodic part of the PEC cell.

Experimental

Synthesis of the electrode materials

The working electrode for the electrochemical cells for CO_2 reduction consists of a carbon substrate on which the metal nanoparticles are deposited. This electrocatalyst is then deposited on a gas diffusion layer (GDL 25 BC Sigracet®), on the side not modified with Teflon. The carbon substrates are commercial carbon black (CB, Vulcan® XC-72) and carbon nanotubes (CNTs, PR-24-XT-PS Pyrograf®).

PR-24-XT-PS CNTs have an average diameter of about 100 nanometers. The inner part shows well-ordered graphitic layers aligned along the main axis, but the external surface displays a turbographic structure. The CNTs were pyrolyzed at 750 °C to remove polyaromatic hydrocarbons from their surface. Due to the turbographic structure, these CNTs offer a large amount of sites for functionalization of the external surface.

The nature of the functional groups on the carbon surface plays a key role in the catalytic activity of the electrocatalysts. Thus, CNTs were functionalized by direct oxidative treatment in concentrated HNO_3, introducing oxygen functionalities on the carbon surface. In detail, 1 g of CNTs (or CB) was suspended in 50 ml

HNO_3 (65% Sigma Aldrich) and treated in reflux at 100 °C for 3 h, followed by rinsing until at a neutral pH, filtering, and drying overnight. Different types of oxygen functionalities were introduced by this treatment. The total quantity and relative distribution can vary as a function of the annealing post-treatment in an inert atmosphere, as shown from synchrotron radiation XPS data.[23]

The main properties of a commercial GDL and CNTs are reported in Table 1.

Before depositing the carbon substrates on the GDL, metal (M = Cu, Co, Fe or Pt) nanoparticles (NPs) were deposited on CB and CNTs by an incipient wetness impregnation method using an ethanolic solution containing the proper metal precursor [$Cu(NO_3)_2\cdot3H_2O$; $Co(NO_3)_2\cdot6H_2O$; $Fe(NO_3)_3\cdot9H_2O$; $H_2PtCl_6\cdot xH_2O$]. After drying at 60 °C for 24 h, the samples were reduced for 2 h at 350 °C under a slow H_2 flow. The total amount of metal loaded onto the carbon substrate was 10 wt%. This amount was chosen in order to have an amount comparable to the metal loading in the electrocatalysts for PEM fuel cells (usually 10–20 wt%), which corresponds to a small metal loading in the final catalyst (about 0.5 mg cm^{-2}).

The as-prepared carbon substrates with the deposited nanoparticles were then deposited on the GDL using a similar impregnation in anhydrous ethanol and after joining the GDL with the Nafion membrane, the samples were tested as working electrodes both in liquid and gas phase cell configurations. While in the liquid-phase cell the electrode is in contact with the electrolyte solution saturated with CO_2, in the gas-phase cell the electrode is absent and ionic conductivity is guaranteed from the contact with the proton-conducting membrane (Nafion®). Before use, the Nafion® membrane was pre-treated with hydrogen peroxide to eliminate organic impurities and finally activated with H_2SO_4.

Characterization

Surface area of the samples (BET method) was determined by the physical adsorption of N_2 at liquid nitrogen temperatures by using a Micrometrics ASAP 2010 system.

Table 1 Properties of the as-produced commercial materials used to prepare the electrocatalyst

Properties of GDL 25 BC Sigracet®	
Thickness	235 µm
Areal weight	86 g m^{-2}
Porosity	80%
Air permeability	1.0 cm^3 (cm^2 s^{-1})
Electrical resistance (through plane)	<12 mΩ cm^2
Properties of CNTs PR-24-XT-PS Pyrograf®	
Fiber diameter (average)	100 nm
Surface area	45 m^2 g^{-1}
Dispersive surface energy	85 mJ m^{-2}
Moisture	<5 wt%
Iron	<14 000 ppm
Polyaromatic hydrocarbons	<1 mg PAH per g fiber

Transmission electron microscopy (TEM) images were acquired by using a Philips CM12 microscope (resolution 0.2 nm) with an accelerating voltage of 120 kV, while SEM images were recorded with a Philips XL-30-FEG scanning electron microscope.

Liquid-phase cell

The electrochemical cell, made in Plexiglas® to allow visual inspection, has a three-electrode configuration and is schematically illustrated in Fig. 2. The working electrode (about 3 cm^2) is located at the cathode side, at a small distance (0.5 cm) from a saturated Ag/AgCl electrode (working as the reference electrode) to reduce the solution resistance. The electric contact with the working electrode is maintained with a Pt wire. The counter-electrode is a commercial Pt rod (Amel) immersed in the anode compartment. A potentiostat/galvanostat (Amel mod. 2049A) is employed to supply a constant current/bias between the electrodes.

The anode compartment is physically separated from the cathode side by a proton-conducting membrane (Nafion® 117, Ion Power). A 0.5 M aqueous solution of $KHCO_3$ was used as the electrolyte solution in both the cathode and anode compartments. The volume of the electrolyte solution at the anode was about 9–10 ml. The electrochemical cell was designed in order to have a large surface area of the electrode and to minimize the electrolyte solution in direct contact with the electrocatalyst. A continuous flow of pure CO_2 (10 ml min^{-1}) was introduced into an external reservoir to saturate the electrolyte solution. This prevents interference from gas bubbles striking the electrode surface in the cathode compartment. The electrolyte solution is continuously circulated between the cathode compartment and the external container by using a peristaltic pump. The total amount of solution (cathode + external container) was 20 ml.

The liquid products were analysed by sampling the liquid in the external container and determining the composition of the solution using ^1H nuclear magnetic resonance spectroscopy (NMR, Varian NMR 500) and gas chromatography-mass spectrometry (GC-MS, Thermo Trace 1310, ISQ Single Quadrupole MS, column Stabilwax MS). The gas products were detected by sampling the

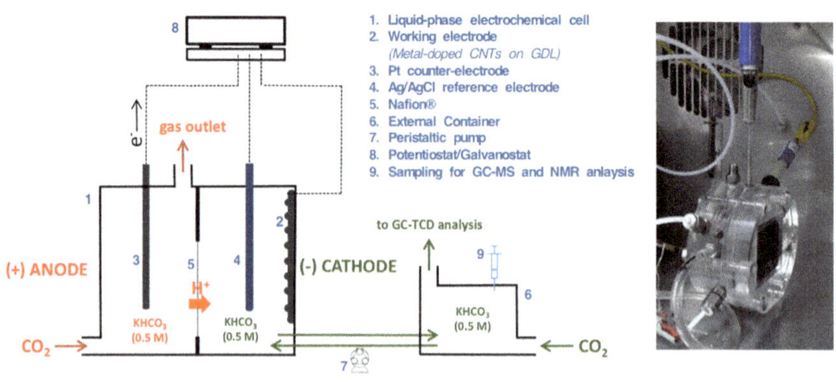

Fig. 2 Schematic drawing of the experimental apparatus for electrocatalytic tests of CO_2 reduction in liquid phase, *e.g.* in the presence of a liquid electrolyte.

gaseous stream leaving the external container at regular intervals and analysing using gas-chromatography (GC-TCD, Agilent 7890A, column 5A Plot).

Gas-phase cell (EL conditions)

The electrochemical cell for CO_2 reduction working in the gas-phase (electrolyte-less – EL – conditions) was designed to operate under a continuous flow of CO_2 diffusing through the GDL to arrive at the surface of the electrocatalyst. The cell was made in Plexiglas® and it is located within an oven to work eventually at higher temperatures, up to about 70–80 °C, although all the tests reported here were carried out at room temperature to be comparable to those in the liquid cell, which do not allow operations at temperatures above about 40 °C. Fig. 3 reports a schematic drawing of the experimental apparatus.

The cell has a three-electrode configuration, with a Pt wire as the counter-electrode and a saturated Ag/AgCl electrode as the reference electrode, both immersed in the anode compartment. A membrane electrode assembly (MEA) separates the two cell compartments. The MEA consists of: (i) a proton-conducting membrane (Nafion® 117) and (ii) the electrocatalyst deposited on the GDL. These two samples were assembled together by hot pressing at 80 atm and 130 °C for 90 s. The electrocatalyst is located at the interface with the membrane. The anode side, to be more comparable with the liquid cell, works in the liquid phase and is used to provide the protons (through the Nafion membrane) needed for the CO_2 reduction process. The anode side is filled with an electrolyte aqueous solution (KCl 0.5 M) and it is in direct contact with one side of the Nafion® membrane. A gas flow of pure CO_2 (10 ml min^{-1}) is continuously fluxed into the cathodic compartment.

A potentiostat/galvanostat (Amel mod. 2049A) was used to supply a constant current (10–20 mA) between the electrodes. The voltage increased as a function of time-on-stream during the first hour stabilizing to a value of around 1.4 to 1.5 V. At the end of the reaction (typically 1 h), the current polarity was inverted to facilitate desorption of the products from the working electrode. The liquid products were collected in a cold trap from the flux leaving the cathodic part of the cell. They were analyzed using a gas chromatograph equipped with a mass detector (Thermo Scientific GC Trace 1310 – ISQ MS).

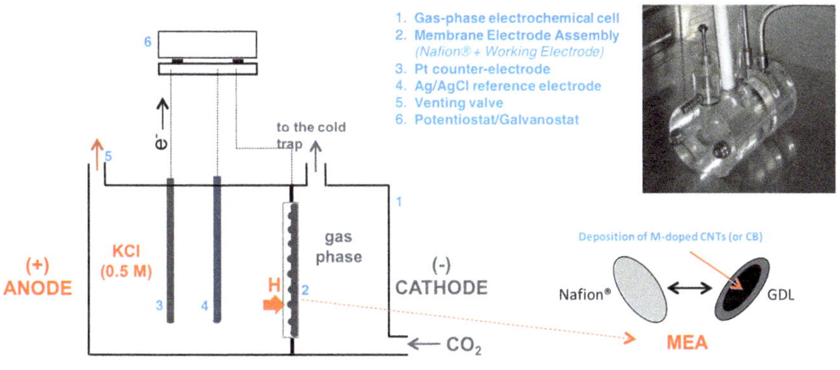

Fig. 3 Schematic drawing of the experimental apparatus for electrocatalytic tests of CO_2 reduction in gas phase, *e.g.* in electrolyte-less (EL) conditions.

Results and discussion

Characterization of the electrodes

The BET characterization of the electrocatalysts is reported in Table 2. An increase in the surface area and pore volume after the addition of Cu, Co, Fe nanoparticles is observed, but not for Pt, although no relevant changes in the average pore volume were observed in all cases. The effect may thus be interpreted as a reduced sticking between the CNTs, due to the change in the surface properties of functionalized CNTs during the process of addition of the metallic nanoparticles. On the contrary, the samples based on CB (carbon black) as the support (only an iron sample is reported in Table 2, with the results for the other samples being quite similar) have a significantly larger BET surface area and pore volume.

Fig. 4 reports an image by TEM of a Fe-CNT sample. It may be observed that there is presence of a good dispersion of iron particles, preferentially located on the external surface of the CNT. The average size is between 3 and 5 nm, with a relatively narrow distribution. A similar distribution of the metal nanoparticles is observed in the other samples. Pt allows for obtaining a slightly lower average size for the metal particles, centred around 1–3 nm. A good dispersion is also observed in the samples based on a CB support, without determining relevant differences in terms of dispersion and average metal nanoparticle size. It may thus be concluded that in spite of the different surface areas, both $CNTs_{ox}$ and CB allow a good dispersion of the metal nanoparticles and an analogous average size to be obtained.

CO_2 electrocatalytic reduction in a liquid phase cell

The main products detected in the liquid electrolyte on the cathode side are reported in Table 3 for all the electrocatalysts tested in the liquid-phase cell. As a reference, (i) metal Cu foil and (ii) CNT_{ox} alone (*e.g.* without metal particles, but deposed on the GDL) were analysed.

The main products detected in the liquid electrolyte solution were: formic acid, acetic acid and methyl formate. The pure Cu foil electrode also formed methanol in traces. The main other product of the reaction, detected in the gas stream leaving the cell (see Fig. 2) is H_2, while CO, CH_4 or C2 hydrocarbons were not detected in all the samples.

The first interesting observation is that CNT_{ox} itself is able to form H_2 (from protons and electrons), as well as some products of CO_2 conversion. In the

Table 2 Characterization of the textural characteristics of the electrocatalysts using a BET method

Electro-catalyst	a_{BET}, $m^2\ g^{-1}$	V_m, $cm^3\ g^{-1}$	Average pore diameter, nm
CNT_{ox}	23, 1	5, 3	11, 4
Cu-CNT_{ox}	48, 3	11, 1	10, 8
Co-CNT_{ox}	72, 6	16, 7	14, 9
Fe-CNT_{ox}	63, 8	14, 6	9, 2
Fe-CB	221, 2	50, 8	12, 2
Pt-CNT_{ox}	23, 6	5, 4	11, 3

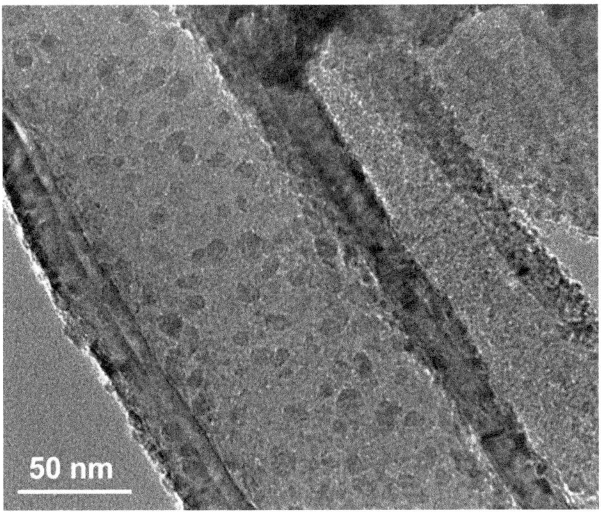

Fig. 4 Transmission electron microscopy (TEM) image of a Fe-CNT sample. Image provided by M Laurent Veyre and E.A. Quadrelli (CPE Lyon, France) who are acknowledged.

absence of CNTs, the products of CO_2 conversion were instead not detected. Thus, the CNT itself (without metal particles) is able to convert CO_2 electrocatalytically. It is worth noting that without the oxidative pretreatment, the CNT is inactive, thus the behaviour observed cannot be attributed to residual traces of metal utilized for the synthesis of the CNT itself (by catalytic chemical vapour deposition). The activity has to be related to the oxygen functional groups created during the CNT pretreatment (see the Experimental part).

Formic acid formation may be explained as deriving from the reaction of H_2 (formed by the catalytic electroreduction of the protons diffusing through the membrane) with the CO_2 present in the electrolyte. However, bubbling together H_2 and CO_2 in the absence of a potential/current applied to the electrode did not result in the formation of products of CO_2 conversion. Reasonably, the functional groups present on CNT_{ox}, likely chetonic groups as observed for other catalytic reactions,[44] are able to reduce CO_2 to formic acid according to a mechanism tentatively outlined in Scheme 1.

Table 3 Main products formed in 1 h of reaction in the CO_2 electrochemical reduction in the liquid phase

Electrode	Formic acid [M]	Acetic acid [M]	Methyl formate [M]
Cu foil	6.1×10^{-4}	4.0×10^{-4}	—
Cu-CB/GDL	3.6×10^{-5}	1.4×10^{-5}	1.7×10^{-4}
Fe-CB/GDL	4.6×10^{-5}	5.4×10^{-5}	6.0×10^{-6}
Co-CB/GDL	8.1×10^{-5}	9.5×10^{-5}	7×10^{-6}
Cu-CNT_{ox}/GDL	1.2×10^{-4}	8.4×10^{-5}	1.5×10^{-5}
Fe-CNT_{ox}/GDL	1.8×10^{-5}	2.1×10^{-5}	1.2×10^{-5}
Pt-CNT_{ox}/GDL	2.3×10^{-4}	1.8×10^{-4}	—
CNT_{ox}/GDL	2.2×10^{-5}	1.4×10^{-5}	—

Chetonic groups present at the edges or defects of CNTs may act as trapping sites for electrons, forming resonance species as outline in Scheme 1. The presence of an electron localized on the carbon activates the nearlying C–C bond, becoming able to coordinate the oxygen in CO_2, breaking the molecule from linearity. The activation of nearlying C–C bonds, making it susceptible for O_2 activation, is the mechanism proposed for N-doped CNTs active in the oxygen reduction reaction (ORR).[44,45]

Acetic acid is the other main product observed using a CNT_{ox}/GDL electrocatalyst (Table 3). It is not easy to explain the formation of this product. A preliminary hypothesis is that the further reduction of formic acid, with a mechanism similar to that outlined in Scheme 1, leads to the formation of formaldehyde. Formaldehyde may selectively transform with even 100% selectivity to acetic acid in relatively mild conditions (100 °C, 500 psi CO) in the presence of a strong acid (HI).[46] Formic acid in strong acid conditions (as present near the electrocatalyst surface, being located close to the Nafion membrane) decomposes to CO and H_2O. Carbon monoxide may react with formaldehyde in the presence of a strong acid to selectively give acetic acid (a variation of the Kock–Haaf reaction).[46] It is reasonable that a similar mechanism occurs in our case during the electrochemical reduction of CO_2:

$$CO_2 \xrightarrow{2e^-,\ 2H^+} HCOOH \tag{1}$$

$$HCOOH \xrightarrow{\text{strong acid catal.}} CO + H_2 \tag{2a}$$

$$HCOOH \xrightarrow{2e^-,\ 2H^+} HCHO + H_2O \tag{2b}$$

$$HCHO + CO \xrightarrow{\text{strong acid, } H^+} CH_3COOH \tag{3}$$

This may also explain why methanol, the further product of formaldehyde reduction, is not observed.

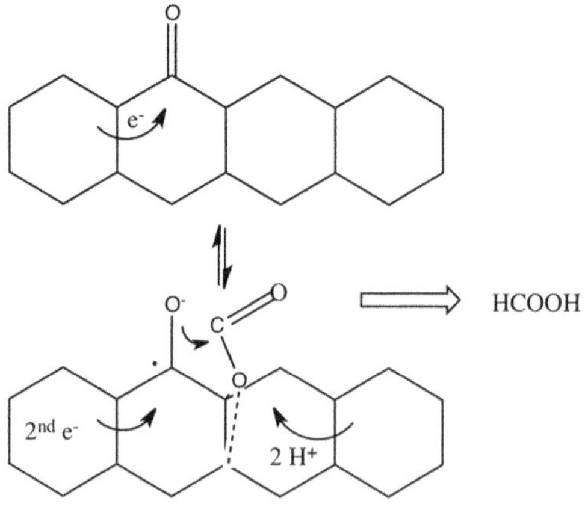

Scheme 1 Tentative reaction mechanism of CO_2 at chetonic groups present on the surface of CNT_{ox}.

The presence of a metal nanoparticle leads to an increase in the productivity in the conversion of CO_2 and the presence of methyl formate in some cases (Table 3). The latter reasonably derives from the reaction of formic acid esterification with methanol under acid conditions:

$$HCOOH + CH_3OH \xrightarrow{\text{strong acid}} HCOOCH_3 + H_2O \quad (4)$$

Methyl formate may thus be associated with the formation of methanol and its fast conversion under the reaction conditions present in our experiments.

Various other observations can be made on the results evidenced in Table 3. Copper-based electrodes have been tested in three types of forms: as Cu foil (as a reference, being one of the type of electrodes often utilized in the electrocatalytic reduction of CO_2), and in the form of copper nanoparticles supported over CNT_{ox} or CB. It may be observed that methyl formate is observed in both cases using CB or CNT_{ox} as a support for copper nanoparticles, but not when copper foil is used. This result indicates that the presence of copper metal nanoparticles supported on carbon materials rather than copper itself (as suggested often in literature) promotes the formation of methanol. In terms of productivity, a Cu-CB/GDL has a formation of methyl formate nearly one order of magnitude higher than a Cu-CNT_{ox}/GDL, although the copper amount and size of the copper nanoparticles is analogous. The higher surface area of CB with respect to CNT_{ox} (Table 2) may thus not be the reason for the different behaviour, as the dispersion of copper in the two supports is relatively similar. We may also note in Table 3 that formic and acetic acid formation is instead higher in a Cu-CNT_{ox}/GDL in comparison to a Cu-CB/GDL. Thus the main difference between the two samples is in the rate of the further reduction of formaldehyde to methanol:

$$HCHO \xrightarrow{2e^-,\ 2H^+} CH_3OH \quad (5)$$

with methanol then further reacting according to eqn (4). CB (Vulcan XC-72) contains about 0.3% S, present as sulphonic groups ($-SO_3H$) on the surface. These acid groups, stronger than the –COOH groups present on CNT_{ox} as a consequence of the oxidative pretreatment, reasonably favour a better surface transport of protons coming from the membrane. Probably the difference in the behaviour observed between the CNT_{ox}/GDL and Cu-CB/GDL is associated with the different concentration of protons at the electrocatalytic centers, although this hypothesis should have better proof. A higher proton concentration favours the further reduction up to methanol, reducing the rate of side reactions. In fact, Table 3 shows that productivity to acetic acid (which can be considered a side reaction with respect to methanol formation) is about six times higher in the CNT_{ox}/GDL with respect to the Cu-CB/GDL, which parallels the strong decrease in methyl formate productivity.

A comparison of the results of the Cu-foil electrode with those of the Cu-CB/GDL and Cu-CNTox/GDL electrodes provides some further interesting indications. To estimate the amount of copper present in the two types of samples, it may be considered that in the Cu-CB/GDL or Cu-CNT_{ox}/GDL the total amount of copper in the 3 cm^2 electrode is about 1.5 mg. To estimate the amount of copper active in copper foil, it is not possible to just have the weight of the foils, but it may be assumed that a foil of the same electrode size (3 cm^2) having a 10 nm thickness would have a comparable thickness to the size of copper nanoparticles in the samples over the carbon support.

Considering for copper a density of 8.9 g cm^{-3}, this estimated amount of "surface" copper in the copper foil is about 0.03 mg, thus significantly lower than that present in the "3D-like" electrodes based on a carbon support. Alternatively, it may be considered as the surface area of metal copper. For copper-foil, it is simply the geometrical electrode area of the copper foil, *e.g.* 3 cm^2. For copper nanoparticles, the surface area can be estimated assuming round-shaped copper nanoparticles of 10 nm size. With this simplification, the surface area for a total amount of 1.5 mg of copper in the electrode results to be about 0.5×10^6 cm^2, *e.g.* much larger than that of the geometrical surface of the copper foil electrode. It may be argued that the real surface of the copper foil is not flat, but some roughness is present as well as some porosity. Also for metal nanoparticles, the part of the surface in contact with the carbon is not accessible (lowering the electrocatalytic active surface), but also particles smaller than 10 nm are present, increasing the overall metal surface. However, the differences estimated for the two types of electrodes (copper foil and copper nanoparticles on a carbon support) are so large indicating that the issue is not related to a too simple model to calculate the active surface area of copper.

Therefore, the electrocatalytic active copper is much lower in copper-foil than in samples supported over carbon. Nevertheless, the productivity in conversion of CO_2 is about five times higher in Cu-foil than that of the Cu-CB/GDL sample and even more with respect to the Cu-CNT_{ox}/GDL (Table 3). This indicates that transport of protons/electrons as well as CO_2 diffusion to the active electrocatalytic centers is dominating the behaviour and productivity is not related to the metal active surface area in our experimental conditions.

This conclusion may also explain why notwithstanding some differences, the behaviour of other metal nanoparticles, namely iron and cobalt, is not significantly different from that observed for copper nanoparticles (Table 3). However, copper, particularly when on a CB support, allows significantly higher methyl formate formation with respect to iron nanoparticles on the same support. Pt nanoparticles, on the contrary, do not form methyl formate but, as for Cu foil, they give larger quantities of formic acid as well as acetic acid.

The change of production rate with time during liquid-phase cell experiments is shown in Fig. 5 for Cu-CB/GDL and Cu-CNTs/GDL electrodes. Formic acid formation decreases after 3 h of reaction, while the concentration of acetic acid increases in both the cases in agreement with the reaction mechanism discussed before. The methyl formate concentration in the liquid phase is higher after 1 h for the Cu-CB/GDL, strongly decreasing after 3 h, due to its high volatility (methyl formate boiling point = 32 °C). Traces of methyl formate were found in the gas outlet from the external container.

It should also be commented that pH changes greatly influence the performance, in agreement with the discussion on the possible reaction mechanism. The aqueous electrolyte used for liquid phase experiments is $KHCO_3$ (0.5 M). When CO_2 is bubbled into the hemi-cell compartments, an acid–base equilibrium between CO_2 and HCO_3^- establishes in solution:

$$CO_2 + H_2O \rightleftharpoons HCO_3^- + H^+ \quad (6)$$

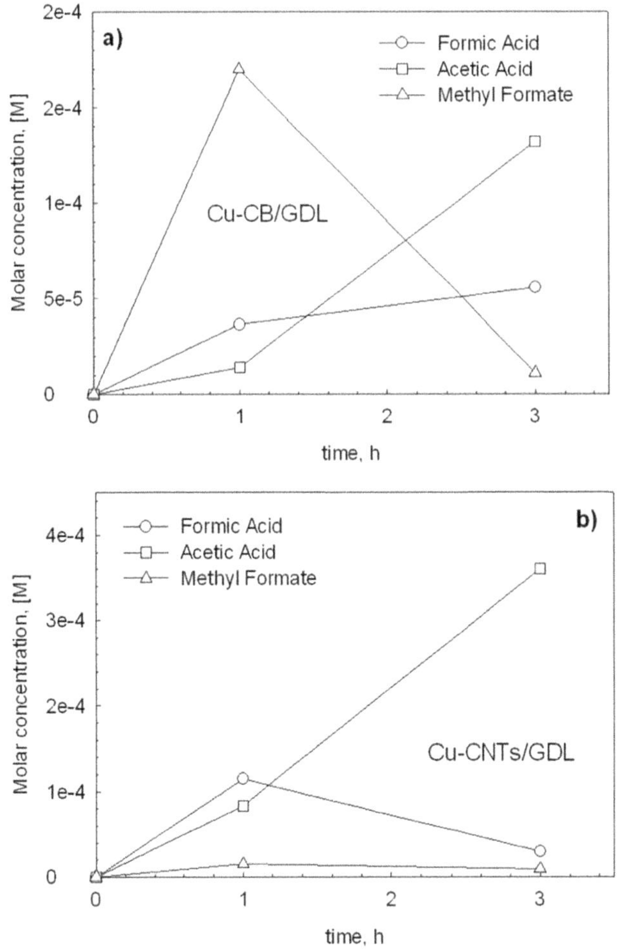

Fig. 5 Time dependence of formic acid, acetic acid and methyl formate formation for liquid-phase testing experiments with (a) Cu-CB/GDL and (b) Cu-CNT_{ox}/GDL electrocatalysts.

The pH of the fresh electrolyte is about 12, but decreases to about 9 when CO_2 is bubbled in it. A small flow of pure CO_2 (10 ml min^{-1}) was also fluxed directly into the anode compartment to favour desorption of O_2 and oxygen species on the Pt counter-electrode surface which may increase the overpotential of the cell.[47] The flux of CO_2 in both compartments of the cell also avoids the creation of pH gradients related to the solubilisation of CO_2, as indicated in eqn (6). However, the pH in the anodic section varies as a consequence of the reaction. The pH of the electrolyte solution at the anode compartment is initially 6. It maintains stable during the first 30–60 min, depending on the electrocatalyst in the cathodic part. During this initial time, the voltage and current given to the electrochemical cell remain stable (Fig. 6). Then, a sharp decrease of both the current and potential is observed, due to the following reaction at the anode, which decreases the pH inhibiting the half-reaction at the anode:

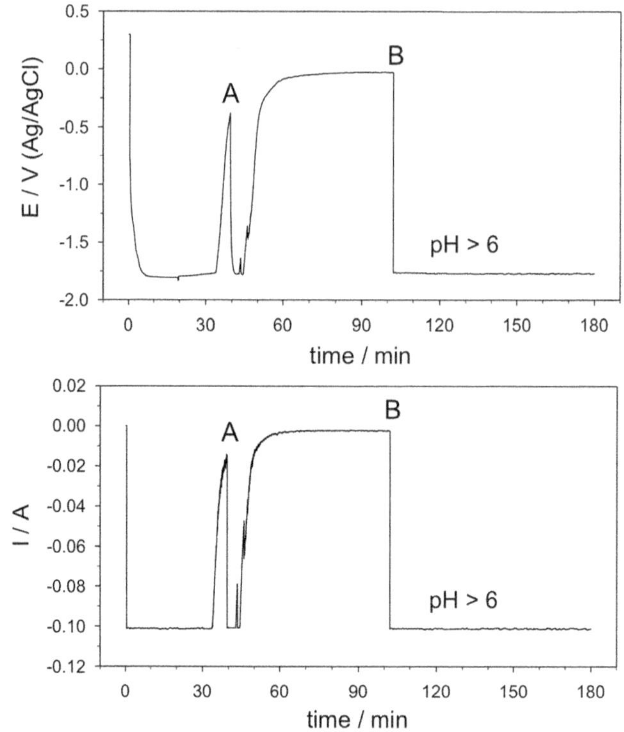

Fig. 6 Profiles of voltage (*E*) and current (*I*) as a function of time in experiments in the liquid-phase. (A) addition of some drops of fresh electrolyte solution (0.5 M $KHCO_3$); (B) addition of some drops of 1 M NaOH aqueous solution to maintain pH in the anode part at pH > 6.

$$H_2O \rightleftharpoons 1/2O_2 + 2H^+ + 2e^- \quad (7)$$

By adding some drops of 1 M NaOH aqueous solution, the pH increased to values above 6 and the current rapidly returned back to the set-point value. In order to have stable performances, it is thus necessary to maintain a pH > 6 in the anodic part by adding a base.

This corroborates that the transport of protons across the membrane is the rate limiting process in our case, because otherwise they should be transported to the cathodic part where they are consumed to generate H_2 and reduce CO_2.

The comparison of these results with those reported in the literature in terms of TOF (turnover frequency) is not easy, because the cell/electrodes and cell operation conditions are different and TOF values are typically not reported in the literature for the electrocatalytic reduction of CO_2. Kuhl *et al.*[35] have not only made a quite detailed investigation of this reaction, but have also reported the TOF values for each of the products determined (in the supplementary part). Due to the different types of electrode (a copper foil mechanically polished) and reaction conditions, they observed different types of products in the reduction of CO_2 (CH_4, formate, CO, methanol, ethylene, ethanol, glycolaldehyde, acetaldehyde, acetate, ethylenglycol, *n*-propanol, allyl alcohol, acetone). It is thus possible to compare the total TOF in products of CO_2 reduction, rather than for the single species. The results are reported in Table 4. A potential of −1.1 V (*vs.* RHE), corresponding to the maximum formation of products of CO_2 reduction in the Kuhl *et al.*[35] results, was chosen for this comparison. TOF data are reported as μmol s^{-1} cm^{-2}, *e.g.* the same unit used by Kuhl *et al.*[35] They also reported in parallel the TOF values expressed as molecules s^{-1} surface Cu $atoms^{-1}$. However, this estimate requires a series of assumptions on the number of Cu atoms which are even less valid in our case, due to the different types of electrodes. As Kuhl *et al.*[35] used only a Cu foil electrode, we report the TOF data in Table 4 only for the Cu-based electrode, for a more homogeneous comparison. TOF results may be easily derived for the other electrodes, being proportional to the total amount of product formed. Having not detected in our case CO formation, the results of Kuhl *et al.*[35] are differentiated to or not to include the amount of CO formed. However, the differences are limited.

As shown in Table 4, TOF data in our electrodes are from three times higher (for dispersed Cu nanoparticles on a carbon support) to 13 times higher (for Cu foil).

CO_2 electrocatalytic reduction in a gas phase cell (EL operations)

Experiments made in the gas-phase electrochemical cell on the same electrodes used for the liquid-phase cell show relevant differences, regarding both the types of products formed and the productivity. However, a common aspect in both cells is the relevant formation of H_2. Hydrogen is an undesired product, because it reduces the Faradaic efficiency to the products of CO_2 reduction. This indicates that in our experimental conditions the use of electrons/protons for the reduction of CO_2 is a slower process with respect their recombination to form H_2. Inhibiting the last reaction is thus a requirement to improve the performances in the electrocatalytic reduction of CO_2. On the other hand, these results also show that the formation of H_2 is a facile reaction not specifically requiring dedicated catalysts.

There are main differences in the type of products of reaction between liquid- and gas-phase cell operations, *e.g.* the presence or absence of the bulk electrolyte. In the gas outlet stream leaving the cathodic part of the cell, together with H_2, CO and in a small amount CH_4 are also detected in the case of EL operations. The two last products are not detected in liquid phase operations. Although CO may form from formic acid decomposition (eqn (2a)) in liquid-phase operations, it immediately reacts and CO is not detected in the product stream leaving the cell, at least up to the detection limit. On the contrary, CO is a main product of CO_2

Table 4 TOF values (μmol s^{-1} cm^{-2}) on copper-based electrodes, in comparison with TOF values reported by Kuhl *et al.*[35] (*) for Cu foil electrode

Electrode	TOF (all products)	TOF (all products, except CO)	TOF (all products)
Cu foil	9.35×10^{-2}	—	—
Cu-CB/GDL	2.04×10^{-2}	—	—
Cu-CNT_{ox}/GDL	2.03×10^{-2}	—	—
Cu foil (*)		6.66×10^{-3}	6.96×10^{-3}

electrocatalytic reduction in gas-phase cell operations, indicating a different mechanism of formation, with probably CO being the primary product of CO_2 reduction, rather than a secondary product, as observed in liquid-phase operations. This may explain why methane, deriving from the catalytic reduction of CO on copper, iron and cobalt metal nanoparticles, is observed even if in small amounts in experiments with gas-phase cells, but not in those with liquid-phase cells.

The type of liquid products formed, detected in the electrolyte or condensed in the cold trap from the gaseous stream leaving the gas-phase cell, are also different in the two cases. Table 5 summarizes the behaviour of the different types of tested electrodes in gas-phase cells (EL conditions). A first observation is that different types of products are observed:

(i) methanol, rather than formic acid as the C1 main product (together with CO);

(ii) acetaldehyde and ethanol together with acetic acid as C2 products, while only the latter was observed together with methyl formate in some electrocatalysts;

(iii) acetone and isopropanol as C3 products, while no C3 products were detected in liquid phase operations.

This different type of products clearly suggests a different type of mechanism of CO_2 electrocatalytic conversion, in agreement with what is commented earlier. Table 6 summarizes the difference observed in the type of products of reduction of CO_2 between gas-phase and liquid-phase cell operations.

It may be observed that not only is the productivity different between gas-phase (Table 5) and liquid-phase (Table 3) operations, but also the relative order of activity. In gas-phase cells, the best productivity to products of CO_2 reduction is shown by a Fe-CNT_{ox}/GDL (around 4.8×10^{-4} mmol h^{-1}). The same electrode in liquid phase operations, considering the total volume of electrolyte, has a productivity of about 0.3×10^{-4} mmol h^{-1}, *e.g.* about one order of magnitude lower. For the Cu-CTN_{ox}/GDL catalyst, productivity in gas-phase operations is about 1.4×10^{-4} mmol h^{-1}, while in liquid-phase operations about 0.15×10^{-4} mmol h^{-1}. Thus again a difference of about one order of magnitude is observed.

Table 5 Products obtained in the CO_2 electrochemical reduction in gas-phase

Electrode	Methanol, mmol h^{-1}	Acetaldehyde, mmol h^{-1}	Ethanol, mmol h^{-1}	Acetone, mmol h^{-1}	Isopropanol, mmol h^{-1}	Acetic acid, mmol h^{-1}	H_2, mmol h^{-1}	CO, mmol h^{-1}
Co-CNT_{ox}/GDL	1.2×10^{-4}	4.5×10^{-5}	2.0×10^{-4}	1.3×10^{-6}	—	3.4×10^{-5}	2.0×10^{-1}	3.2×10^{-2}
Cu-CNT_{ox}/GDL	4.8×10^{-5}	1.1×10^{-5}	6.7×10^{-5}	1.0×10^{-7}	9.3×10^{-6}	—	1.7×10^{-1}	9.6×10^{-3}
Fe-CB/GDL	1.1×10^{-4}	2.9×10^{-5}	1.1×10^{-4}	1.2×10^{-6}	1.1×10^{-6}	1.5×10^{-5}	3.2×10^{-1}	3.6×10^{-2}
Fe-CNT_{ox}/GDL	1.4×10^{-4}	6.7×10^{-5}	9.4×10^{-5}	1.8×10^{-7}	8.4×10^{-5}	9.2×10^{-5}	3.6×10^{-1}	6.5×10^{-2}
Pt-CNT_{ox}/GDL	1.0×10^{-5}	1.7×10^{-5}	3.6×10^{-5}	1.7×10^{-7}	2.4×10^{-5}	1.4×10^{-5}	4.5×10^{-1}	3.5×10^{-2}

Table 6 Comparison of types of products obtained in liquid and gas phase operations in CO_2 electrocatalytic reduction

Gas phase	Liquid phase
Methanol	Formic acid
Acetaldehyde	Acetic acid
Ethanol	Methanol (in traces)
Acetone	
Isopropanol	
Hydrocarbons C4–C9 (in traces)	

Also the relative order of productivity to products of CO_2 reduction is different, being:

$$\text{Fe-CNT}_{\text{ox}}\text{/GDL} > \text{Fe-CB/GDL} > \text{Cu-CNT}_{\text{ox}}\text{/GDL} > \text{Pt-CNT}_{\text{ox}}\text{/GDL}$$

in gas-phase cells (Table 5) while for liquid-phase cells (Table 3):

$$\text{Pt-CNT}_{\text{ox}}\text{/GDL} \approx \text{Fe-CNT}_{\text{ox}}\text{/GDL} > \text{Cu-CNT}_{\text{ox}}\text{/GDL} > \text{Fe-CB/GDL}.$$

This observation further remarks that different aspects may determine the productivity in the reduction of CO_2, and can be reasonably associated to the different reaction mechanisms.

Differences in the reaction mechanism

Clarification of the above question requires a more in depth analysis of the reaction mechanisms, with operando techniques, which, however, are not simple to apply to electrochemical tests under relevant environmental conditions. We may thus advance only some initial considerations, which help in setting the scene for more detailed mechanistic studies.

As earlier commented, the mechanism in the liquid-phase may be associated (on the specific carbon-based electrocatalysts we utilize) with the activation mechanism of CO_2 at specific carbon sites, such as the carbonyl groups formed by oxidation pretreatment. One electron transfer to this site generates a $C–O^-$ center, with the charge localized on the carbon activating the nearlying C–C bond making it susceptible of activating the oxygen in the O═C bond of carbon dioxide, thus generating a $\delta+$ charge on the C of CO_2, facilitating the electron transfer from a $C–O^-$ center. In a different paper on the reactivity of carbon nanotubes (modified by a specific surface nitrogen doping mechanism),[48] we showed that the amount of nitrogen in CNTs influences performance in the electrocatalytic reduction of CO_2. The behaviour linearly correlated with the change of work function in these materials. The latter aspect influences the electronic coupling between the electron donor/acceptor and in turn the rate of electron transfer.

This can provide an explanation of the role of metal nanoparticles in our electrocatalysts. It may be noted, in fact, that there are differences in the behaviour and productivity, but limited with respect to the very relevant change in the type of metal nanoparticles, from Pt to Cu, Fe and Co. If the reaction occurs only at the metal surface, much greater differences are expected. On the other hand, metal nanoparticles may change locally the work function of carbon, thus facilitating the electron transfer. It may be also facilitate the electron transfer to protons, generating H$^{\bullet}$ species, which are more reactive in reacting with oxygen of CO_2 or with the products of its reduction. The metal will thus act as co-catalyst with the functional groups present on activated carbon, rather than as unique catalytic sites. The sites will thus be mainly located at the perimetral edge between carbon and metal particles, in agreement with previous studies using calorimetry on this type of electrocatalysts for the reduction of CO_2.[49] Although clearly further demonstrations are necessary to support this mechanism, it can provide preliminary indications on the aspects investigated, as being different from the actual mechanisms proposed for CO_2 electrocatalytic conversion, which are focused only on the role of the metal surface.

Reasonably the intermediate generated in this mechanism of reduction of CO_2 requires a solvent to be stabilized. On the other hand, metal nanoparticles are easy passivated, because CO_2 dissociation at the electrocatalyst metal surface generates CO and an O species, both remain strongly bound to the metal surface. In the absence of the bulk electrolyte, both the absence of the solvent (electrolyte) and the higher CO_2 concentration at the surface of the electrocatalyst reasonably contribute in inhibiting one side of the mechanism of electron transfer/hydrogenation of CO_2 (*via* formic acid and formaldehyde), and on the other hand, promoting the mechanism of CO_2 dissociation to CO at the metal surface or reasonably at the metal-carbon perimetral region. The further reaction of adsorbed species on the metal surface leads to C–C bond formation and to the products being observed experimentally.

Although speculative, this change in the type of mechanism of CO_2 reduction may explain the differences observed in the productivity, type of products and relative rates of reactions between different electrodes. These observations show how a more complex surface chemistry in the electrocatalytic reduction of CO_2 may exist with respect to what is previously suggested in the literature. A better understanding of these aspects opens new possibilities in controlling the type of products formed and their productivity, as evidenced from the comparison between liquid- and gas-phase operations on the same electrodes. On the other hand, the results also remark that carbon may not only be a support, but also plays a relevant role in understanding the chemistry of the reduction of CO_2. This also opens new possibilities in the design of electrocatalysts for the reduction of CO_2 and of the critical elements to consider for their improvement.

Conclusions

The comparison of the same electrodes for the electrocatalytic reduction of CO_2 in operations in the presence (liquid-phase) or absence (gas-phase) of a bulk electrolyte provide a series of interesting indications both on the limiting steps of the process and on the reaction mechanism.

It is demonstrated that under electrolyte-less (EL) operations, the productivity in the reduction of CO_2 is about one order of magnitude larger, and different types of products are formed. This is related to differences in the reaction mechanism, which were discussed, although further studies are needed to better clarify the reaction mechanism and the influence of the presence of the electrolyte.

In liquid-phase operations it is suggested that the mechanism involves step reduction *via* intermediate formation of formic acid and formaldehyde and finally methanol, although the latter step is observed only in some electrodes. In fact, due to the strong acidic conditions present at the electrocatalyst, located at the interface with Nafion membrane, formic acid decomposes to generate CO, which reacts with formaldehyde to form acetic acid. Methanol also reacts further in these strong acid conditions to form methyl formate.

In gas-phase operations, this mechanism is no longer effective and the conversion of CO_2 involves its dissociation to CO, which remains strongly chemisorbed, giving rise to further transformation reactions and the formation of C–C bonds that produce a different spectrum of products of that observed in liquid-phase operations.

The electrocatalysts utilized here are based on metal nanoparticles (Fe, Cu, Co, Pt) deposited on two types of conductive carbon supports: carbon nanotubes functionalized by oxidation treatment (CNT_{ox}) and Vulcan XC-72 carbon black (CB). $CNTs_{ox}$ without metal particles are active both in producing H_2 from protons/electrons and in the reduction of CO_2. The possible mechanism and the role of carbonyl groups, formed during the oxidative treatment, is outlined. It is commented on why in general and also when metal nanoparticles are present, the carbon does not act only as a support or to transport charges, but has also an active role in the reaction mechanism. It is suggested that the active sites for the electrocatalytic reduction of CO_2 are located at the perimetral edge between metal nanoparticles and the carbon. The nature of the latter, in particular the type of surface functional groups, thus determine considerably the performance, as experimentally observed.

As commented, these results are the start, not the end of the analysis of the mechanism of reaction in the type of electrodes investigated here. However, we believe that these results provide evidence on how a more complex surface chemistry than typically supposed in the literature is present in the electrocatalytic reduction of CO_2 in the type of electrodes we investigated. It should be noted that their performances are better than that of various other electrodes reported in the literature for the conversion of CO_2, even if the different reaction conditions and the way data are reported do not often allow for a precise comparison.

There is thus the need of a better understanding of the performance and reaction mechanism of these electrocatalysts (based on metal nanoparticles on conductive, functionalized carbon supports) for the reduction of CO_2. They may open new possibilities in controlling the type of products formed and their productivity in this challenging reaction.

Acknowledgements

The authors acknowledge the PRIN10-11 project "Mechanisms of activation of CO_2 for the design of new materials for energy and resource efficiency" and the EU project ECO^2CO_2 ("Eco-friendly biorefinery fine chemicals from CO_2 photocatalytic reduction"), which have partially supported this work. C. Marepally also thanks the European Doctoral Programme on Sustainable Industrial Chemistry (SINCHEM) for supporting his PhD programme in collaboration with CPE Lyon

(Prof. E.A. Quadrelli). M Laurent VEYRE (UMR 5265 CNRS, CP2 Lyon Université de Lyon) is acknowledged for TEM images.

Notes and references

1 G. A. Ozin, *Adv. Mater.*, 2015, **27**, 1957.
2 G. Centi and S. Perathoner, in *Green Carbon Dioxide*, ed. G. Centi and S. Perathoner, Wiley & Sons, 2014, ch. 1, p. 1.
3 M. D. Kaerkaes, O. Verho, E. V. Johnston and B. Aakermark, *Chem. Rev.*, 2014, **114**, 11863.
4 S. Protti, A. Albini and N. Serpone, *Phys. Chem. Chem. Phys.*, 2014, **16**, 19790.
5 C. van der Giesen, R. Kleijn and G. J. Kramer, *Environ. Sci. Technol.*, 2014, **48**, 7111.
6 J. M. Thomas, *Energy Environ. Sci.*, 2014, **7**, 19.
7 S. Bensaid, G. Centi, E. Garrone, S. Perathoner and G. Saracco, *ChemSusChem*, 2012, **5**, 50.
8 G. Centi and S. Perathoner, *Greenhouse Gases: Sci. Technol.*, 2011, **1**, 21.
9 G. Centi and S. Perathoner, *ChemSusChem*, 2010, **3**, 195.
10 D. G. Nocera, *Acc. Chem. Res.*, 2012, **45**, 767.
11 C. Ampelli, S. Perathoner and G. Centi, *Philos. Trans. R. Soc., A*, 2015, **373**, 1.
12 P. Lanzafame, G. Centi and S. Perathoner, *Chem. Soc. Rev.*, 2014, **43**, 7562.
13 S. Perathoner and G. Centi, *ChemSusChem*, 2014, **7**, 1274.
14 G. Centi, E. A. Quadrelli and S. Perathoner, *Energy Environ. Sci.*, 2013, **6**, 1711.
15 S. Perathoner and G. Centi, *J. Chin. Chem. Soc.*, 2014, **61**, 719.
16 G. Centi and S. Perathoner, Artificial leaves, in *Kirk-Othmer Encyclopedia of Chemical Technology*, 2014.
17 J. Albo, M. Alvarez-Guerra, P. Castano and A. Irabien, *Green Chem.*, 2015, **17**, 2304.
18 P. Kang, Z. Chen, M. Brookhart and T. J. Meyer, *Top. Catal.*, 2015, **58**, 30.
19 Q. Lu, J. Rosen and F. Jiao, *ChemCatChem*, 2015, **7**, 38.
20 E. S. Rountree, B. D. McCarthy, T. T. Eisenhart and J. L. Dempsey, *Inorg. Chem.*, 2014, **53**, 9983.
21 J. Qiao, Y. Liu, F. Hong and J. Zhang, *Chem. Soc. Rev.*, 2014, **43**, 631.
22 E. V. Kondratenko, G. Mul, J. Baltrusaitis, G. O. Larrazabal and J. Perez-Ramirez, *Energy Environ. Sci.*, 2013, **6**, 3112.
23 C. Genovese, C. Ampelli, S. Perathoner and G. Centi, *J. Energy Chem.*, 2013, **22**, 202.
24 G. Centi and S. Perathoner, in *Chemical Energy Storage*, ed. R. Schlögl, De Gruyter, 2013, p. 379.
25 K. S. Joya, Y. F. Joya, K. Ocakoglu and R. van de Krol, *Angew. Chem., Int. Ed.*, 2013, **52**, 10426.
26 J. Barber and P. D. Tran, *J. R. Soc., Interface*, 2013, **10**, 20120984.
27 M. D. Kaerkaes, O. Verho, E. V. Johnston and B. Aakermark, *Chem. Rev.*, 2014, **114**, 11863.
28 S. Fukuzumi and Y. Yamada, *ChemSusChem*, 2013, **6**, 1834.
29 D. K. Bora, A. Braun and E. C. Constable, *Energy Environ. Sci.*, 2013, **6**, 407.
30 C. Ampelli, G. Centi, R. Passalacqua and S. Perathoner, *Catal. Today*, 2015, DOI: 10.1016/j.cattod.2015.07.020.

31 C. Ampelli, C. Genovese, S. Perathoner, G. Centi, M. Errahali, G. Gatti and L. Marchese, *Chem. Eng. Trans.*, 2014, **41**, 13; C. Ampelli, C. Genovese, R. Passalacqua, S. Perathoner and G. Centi, *Theor. Found. Chem. Eng.*, 2012, **46**, 651.
32 C. Genovese, C. Ampelli, S. Perathoner and G. Centi, *J. Catal.*, 2013, **308**, 237; C. Ampelli, S. Perathoner and G. Centi, *Chin. J. Catal.*, 2014, **35**, 783.
33 C. Ampelli, G. Centi, R. Passalacqua and S. Perathoner, *Energy Environ. Sci.*, 2010, **3**, 292.
34 G. Centi, S. Perathoner, G. Wine and M. Gangeri, *Green Chem.*, 2007, **9**, 671.
35 K. P. Kuhl, E. R. Cave, D. N. Abramc and T. F. Jaramillo, *Energy Environ. Sci.*, 2012, **5**, 7050.
36 P. Hirunsit, W. Soodsawang and J. Limtrakul, *J. Phys. Chem. C*, 2015, **119**, 8238.
37 Q. Shen, Z. Chen, X. Huang, M. Liu and G. Zhao, *Environ. Sci. Technol.*, 2015, **49**, 5828.
38 J. Albo, M. Alvarez-Guerra, P. Castano and A. Irabien, *Green Chem.*, 2015, **17**, 2304.
39 S. Back, H. Kim and Y. Jung, *ACS Catal.*, 2015, **5**, 965.
40 S. Rasul, D. H. Anjum, A. Jedidi, Y. Minenkov, L. Cavallo and K. Takanabe, *Angew. Chem., Int. Ed.*, 2015, **54**, 2146.
41 C. Ampelli, C. Genovese, P. Lanzafame, S. Perathoner and G. Centi, *Chem. Eng. Trans.*, 2014, **39**, 1627.
42 C. Ampelli, R. Passalacqua, C. Genovese, S. Perathoner, G. Centi, T. Montini, V. Gombac and P. Fornasiero, *Chem. Eng. Trans.*, 2013, **35**, 583; C. Ampelli, R. Passalacqua, C. Genovese, S. Perathoner, G. Centi, T. Montini, V. Gombac, J. J. Delgado and P. Fornasiero, *RSC Adv.*, 2013, **3**, 21776.
43 C. Ampelli, C. Genovese, R. Passalacqua, S. Perathoner and G. Centi, *Appl. Therm. Eng.*, 2014, **70**, 1270.
44 D.-W. Wang and D. S. Su, *Energy Environ. Sci.*, 2014, **7**, 576.
45 G. Centi, S. Perathoner and D. S. Su, *Catal. Surv. Asia*, 2014, **18**, 149; D. S. Su, S. Perathoner and G. Centi, *Chem. Rev.*, 2013, **113**, 5782.
46 L. Kaplan, *J. Org. Chem.*, 1985, **50**, 5376.
47 J. K. Nørskov, J. Rossmeisl, A. Logadottir, L. Lindqvist, J. R. Kitchin, T. Bligaard and H. Jónsson, *J. Phys. Chem. B*, 2004, **108**, 17886.
48 J. Xu, R. Huang, X. Sun, B. Zhang, Y. Lin, Q. Li, D. Su and G. Centi, *Nano Energy*, 2015, submitted.
49 R. Arrigo, M. E. Schuster, S. Wrabetz, F. Girgsdies, J.-P. Tessonnier, G. Centi, S. Perathoner, D. S. Su and R. Schlögl, *ChemSusChem*, 2012, **5**, 577.

Faraday Discussions

PAPER

Improving the efficiency of electrochemical CO_2 reduction using immobilized manganese complexes†

James J. Walsh,‡ Charlotte L. Smith,‡ Gaia Neri, George F. S. Whitehead, Craig M. Robertson and Alexander J. Cowan*

Received 11th May 2015, Accepted 10th June 2015
DOI: 10.1039/c5fd00071h

Immobilization of $[Mn(bpy)(CO)_3Br]$, (**1**) and $[Mn(bpy(^tBu)_2)(CO)_3Br]$ (**2**, where $(bpy(^tBu)_2)$ = 4,4′-di-*tert*-butyl-2,2′-bipyridine) in Nafion/multi-walled carbon nanotubes (MWCNT) on glassy carbon yielded highly active electrodes for the reduction of CO_2 to CO in aqueous solutions at pH 7. Films incorporating **2** have significantly improved selectivity towards CO_2, with CO : H_2 ~ 1 at −1.4 V *vs.* SCE, exceeding that for the previously reported **1**/MWCNT/Nafion electrode. Furthermore, we report the synthesis and subsequent electrochemical characterization of two new substituted Mn(I) bipyridine complexes, $[Mn(bpy(COOH)_2)(CO)_3Br]$ (**3**) and $[Mn(bpy(OH)_2)(CO)_3Br]$ (**4**) (where $(bpy(COOH)_2)$ = 4,4′-di-carboxy-2,2′-bipyridine and $(bpy(OH)_2)$ = 4,4′-di-hydroxy-2,2′-bipyridine). Both **3** and **4** were found to have some activity towards CO_2 in acetonitrile solutions; however once immobilized in Nafion membranes CO_2 reduction was found to not occur at significant levels.

1 Introduction

Carbon capture coupled with utilisation is a potentially sustainable route to useful feedstocks and fuels such as CO, HCOOH, CH_3OH and CH_4. Electrocatalytic CO_2 reduction is a particularly attractive utilisation pathway as the energetic input for CO_2 reduction can come from renewable sources of power (*e.g.* wind, solar) with the production of chemical fuels offering a solution to the intermittent nature of such renewables. Whilst the direct one electron reduction of CO_2 is extremely challenging, occurring at −2.15 V *vs.* SCE (pH 7),[1] the

Department of Chemistry, Stephenson Institute for Renewable Energy, The University of Liverpool, L69 7ZD, Liverpool, UK. E-mail: a.j.cowan@liverpool.ac.uk

† Electronic supplementary information (ESI) available: X-ray diffraction details and data, scan rate dependence of **3** and **4** in solution, FTIR spectra showing solvolysis, deconvoluted FTIR-SEC data, the effects of hourly gas re-purging on CPE of **2**/Nafion/MWCNT and temperature dependence on drop drying. CCDC 1061484 and 1061485. For ESI and crystallographic data in CIF or other electronic format see DOI: 10.1039/c5fd00071h

‡ These authors share equal contribution.

multielectron, multiproton reduction of CO_2 can be achieved at more moderate potentials ($CO_2 + 2H^+ + 2e^- \rightarrow CO + H_2O$, −0.77 V *vs.* SCE, pH 7).[2] However in order to achieve high Faradaic efficiencies for CO_2 reduction a suitably selective electrocatalyst for CO_2 is required as the production of H_2 ($2H^+ + 2e^- \rightarrow H_2$, −0.65 V *vs.* SCE, pH 7) often dominates in protic solvents.

Homogenous electrocatalysts based on *fac*-[M(bpy)$(CO)_3$X] (where M = Re, Mn; X = Cl^-, Br^- and bpy = 2,2′-bipyridine) have been widely studied for the reduction of CO_2 to CO due to their high selectivities and ease of synthetic modification.[3,4] Early studies concentrated on [Re(bpy)$(CO)_3$Br] and its derivatives;[5,6] however in 2011 it was shown that the lower cost manganese analogue, [Mn(bpy)$(CO)_3$Br] **1**, is also an excellent homogeneous electrocatalyst for the reduction of CO_2 to CO.[7] Using **1**, CO_2 reduction was reported to occur at a potential (*ca.* −1.4 V *vs.* SCE) significantly more positive than that required using the analogous rhenium complex, providing that a suitable Brønsted acid such as water was present in the organic solvent (typically CH_3CN or DMF). A range of catalysts of the form [Mn^I(L)$(CO)_3$X] (L = bpy,[7] 4,4′ or 6,6′-disubstituted-2,2′-bipyridine)[8,9] have since been reported with the aim of improving both the turnover frequency (TOF) and overpotential for CO_2 catalysis using **1** and amongst the most active is [Mn(bpy(^{t}Bu)$_2$)$(CO)_3$Br] (**2**, where (bpy(^{t}Bu)$_2$) = 4,4′-di-*tert*-butyl-bipyridine). **2** has been reported to have a TOF of 340 s^{-1} at −2.2 V *vs.* SCE for CO production in the presence of 1.4 M 2,2,2-trifluoroethanol (TFE), approximately 300 times higher than that of complex **1**, although it should be noted that the onset for catalysis is *ca.* 200 mV cathodic of **1**.[8] Detailed electrochemical, spectroscopic and theoretical studies have identified several key aspects of the mechanism of CO_2 reduction using **1** and **2**.[8,10–12] It is known that upon initial one-electron reduction, bromide loss and dimerization with another reduced moiety to yield [Mn(bpy)$(CO)_3$]$_2$ rapidly occurs. Reduction of the dimer then leads to the formation of the primary active catalyst, [Mn(bpy)$(CO)_3$]$^-$.[7] Using 2D-EPR measurements it has also been shown that with a closely related complex, [Mn(bpy(Me)$_2$)$(CO)_3$Br] (where (bpy(Me)$_2$) = 4,4′-dimethyl-2,2′-bipyridine), CO_2 reduction can also occur *via* a second pathway involving direct CO_2 and H^+ addition to the dimeric species formed upon reduction.[13] Due to the propensity of the [Mn^I(L)$(CO)_3$X] complexes to dimerize upon reduction, Kubiak *et al.* developed a complex with bulky bipyridine ligands, [Mn(mesbpy)$(CO)_3$Br] (where mesbpy = 6,6′-dimesityl-2,2′-bipyridine),[9] that was too sterically hindered to undergo dimerization, with the aim of modifying the catalytic mechanism. Promisingly, such an approach led to a large enhancement in TOF (*ca.* 5000 s^{-1}, 1.4 M TFE at *ca.* −1.8 V *vs.* SCE), although no improvement in onset potential for catalysis was achieved when compared to **1**. In addition to studies on the modification of the bpy ligand, several groups have also explored the use of N-heterocyclic carbenes,[14,15] nonaromatic α-diimine ligands,[16,17] hydroxyphenol-substituted bpy[18] and pentadentate chelating[19] ligands with Mn(I) centres for use as potential CO_2 reduction catalysts with varying degrees of success.

To date, studies on [Mn^I(L)$(CO)_3$X] derivatives have concentrated on the development of the homogenous catalyst in an aprotic organic solvent doped with a proton source such as water, methanol or TFE. Here we address the development of this class of catalysts for use in an immobilized, recyclable form in aqueous solvents at pH 7. Nafion membranes have been used extensively as catalyst supports as they are chemically stable proton conductors with high permeability towards water. Previous examples of Nafion-supported molecular CO_2 reduction catalysts include [Re(bpy)(CO)$_3$Br],[20] [Ni(cyclam)]$^{2+}$,[21] and a cobalt tetramethylpyridoporphyrazine;[22] however the current densities reported during sustained electrolysis were typically low (<0.1 mA cm^{-2}) due to the low electrical conductivity of the Nafion support. Recently we reported on an immobilized high surface area **1**/MWCNT/Nafion cathode (where MWCNT = multi-walled carbon nanotubes) that achieved CO and H_2 production in ratios suitable for use as syngas (CO : H_2 = 1 : 2) at greatly increased sustained current densities (0.7 mA cm^{-2}) although with a relatively high overpotential (η = 0.63 V, −1.4 V *vs.* SCE) at pH 7.[23] MWCNT have been previously used as supports in a number of elegant studies on molecular catalysts for water oxidation,[24] water reduction,[25] and CO_2 reduction[26] where the catalysts are either directly grafted to, or modified with pyrene groups to enable interaction with, the MWCNT. In contrast, we previously employed a very simple, potentially low-cost approach where MWCNT were added to a polymer/catalyst solution before being drop-cast onto an electrode surface. It has been reported that a weak π–π interaction can occur between bpy ligands and MWCNT,[27,28] and in our co-cast film FTIR spectroscopy indicated the presence of an interaction between the bipyridine ligand of **1** and the high surface area support,[23] although the exact nature of this MWCNT–**1** interaction remains under investigation. Critically the use of the high surface area report led to catalytic current densities an order of magnitude higher than could be achieved for the **1**/Nafion membrane alone. In this contribution, we report a **1**/MWCNT/Nafion electrode that operates at more than twice the current density (1.79 mA cm^{-2}, CO : H_2 ~ 0.3 at −1.4 V *vs.* SCE) of our previous study. We also examine a range of new [Mn(L)(CO)$_3$X]/MWCNT/Nafion films leading to an electrode with a high selectivity towards CO_2 over H^+ reduction even in aqueous solutions.

We also synthesise and study, in solution, two previously unreported [Mn^I(L)(CO)$_3$X] complexes, modified at the 4,4′-position of the bipyridine ligand, [Mn(bpy(COOH)$_2$)(CO)$_3$Br] (**3**) and [Mn(bpy(OH)$_2$)(CO)$_3$Br] (**4**) (where (bpy(COOH)$_2$) = 4,4′-di-carboxy-2,2′-bipyridine and (bpy(OH)$_2$) = 4,4′-di-hydroxy-2,2′-bipyridine), see Fig. 1. Complex **3** is of interest as a route to overcoming the relatively high overpotential for the reduction of CO_2 by **1** (η ~ 580 mV in CH_3CN/H_2O, 95 : 5).[29] The addition of electron-withdrawing groups (EWGs) to the 4,4′-position of the bipyridine ligand has been shown previously to stabilise the first reduction of the analogous rhenium complex,[30] and it is anticipated that a similar anodic shift in the reduction potentials of **3** may occur. Conversely, complex **4** was identified due to the unique ability of the 4,4′-di-hydroxy-2,2′-bipyridine to act as a "push–pull"

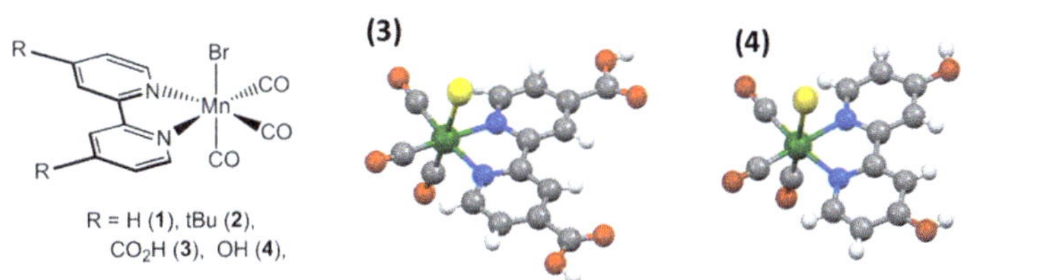

Fig. 1 [Mn^I(L)(CO)$_3$X] complexes under study, and single crystal X-ray structures of (3) [Mn(bpy(COOH)$_2$)(CO)$_3$Br] and (4) [Mn(bpy(OH)$_2$)(CO)$_3$Br], with solvent atoms omitted for clarity (details in the ESI†).

ligand, where deprotonation of the hydroxyl group leads to the formation of an oxyanion bipyridine which has been shown with $[Ru(bpy)_2(bpy(OH)_2)]^{2+}$ to be a strong electron donor.[31] This may be a promising approach as the high catalytic activity of **2** is proposed to be related to the increased electron density at the Mn centre due to the presence of the bulky electron-donating *t*-butyl groups.[8] During the preparation of this manuscript the analogous rhenium complex, $[Re(bpy(OH)_2)(CO)_3Br]$ was reported to have comparable activity to $[Re(bpy(^tBu)_2)(CO)_3Br]$ for the production of CO.[32] Following homogenous studies in aprotic solvents we incorporated both the new complexes (**3** and **4**) and the current benchmark catalysts **1** and **2**, in high surface area $[Mn(L)(CO)_3X]$/MWCNT/Nafion films, leading to the report of both a new **1**/MWCNT/Nafion electrode that operates at improved current densities (1.79 mA cm^{-2}, $CO : H_2 \sim 0.3$ at −1.4 V *vs.* SCE) and a **2**/MWCNT/Nafion electrode with improved selectivity for CO_2 reduction ($CO : H_2 \sim 1$ at −1.4 V *vs.* SCE), compared to our previous report.[23]

2 Experimental

Materials and methods

Bromopentacarbonylmanganese(I), 2,2′-bipyridine, 4,4′-di-hydroxy-2,2′-bipyridine, 4,4′-di-carboxy-2,2′-bipyridine and 4,4′-di-*tert*-butyl-2,2′-bipyridine were purchased from Sigma-Aldrich and used as received. Nafion (5% w/w in lower alcohols, Sigma-Aldrich), sodium phosphate monobasic (≥99%, Sigma-Aldrich), sodium phosphate dibasic (≥99%, Sigma-Aldrich), tetrabutylammonium hexafluorophosphate (TBA PF_6, 99%, electrochemical grade) and multi-walled carbon nanotubes (MWCNT > 98% carbon, 6–13 nm × 2.5–20 μm, Sigma-Aldrich) were used as received. Acetonitrile (chromatographic grade, Sigma-Aldrich) was dried over molecular sieves (3 Å) prior to use and stored under argon. Milli-Q water (18.2 MΩ) was used for all aqueous solutions. CO_2 and argon (CP and pureshield grade respectively) were purchased from BOC.

Synthesis of complexes

All syntheses were carried out in the dark and under an atmosphere of Ar. $[Mn(bpy)(CO)_3Br]$ and $[Mn(bpy(^tBu)_2)(CO)_3Br]$ were synthesized and characterized according to literature methods (details in the ESI†).

$[Mn(bpy(COOH)_2)(CO)_3Br]$, 3. A solution of 4,4′-dicarboxy-2,2′-bipyridine (180 mg, 0.74 mmol) in 8 ml of methanol was added to a solution of $Mn(CO)_5Br$ (200 mg, 0.72 mmol) in toluene (20 ml). The mixture was refluxed for 90 minutes at 60 °C, then kept in the fridge overnight. The orange-red precipitate that formed was filtered off and washed with cold diethyl ether, and the resulting red solution was rotary evaporated to dryness to yield a bright red powder. The final product was recrystallized from hot acetone and hexane, and dried *in vacuo* for 48 hours. Crystals of suitable quality for diffraction studies were grown by slow vapour diffusion of *n*-hexane into concentrated THF solutions of the complex. Yield = 46.8%; ^1H-NMR (300 MHz, DMSO-d_6): δ ppm 7.17 (br. d, J = 1.00 Hz, 2H), 8.09 (br. s, 2H), 9.41 (br. s, 2H), 9.68 (m, 1H); FTIR (CH_3CN/H_2O, 95 : 5): 2030, 1940, 1931 cm^{-1}; anal. calcd for $C_{15}H_8BrMnN_2O_7\cdot H_2O$: C, 37.45; H, 2.10; N, 5.82. Found: C, 37.30; H, 1.81; N, 5.57. *m*/*z* (ESI); 415 ($[M - Br^- + HOMe]^+$), 383 ($[M - Br^-]^+$).

[Mn(bpy(OH)$_2$)(CO)$_3$Br], 4. A solution of 4,4′-dihydroxy-2,2′-bipyridine (140 mg, 0.74 mmol) in 8 ml of methanol was added to a solution of Mn(CO)$_5$Br (200 mg, 0.72 mmol) in toluene (20 ml). The mixture was refluxed for 90 minutes at 60 °C, and then placed in the fridge overnight. The orange-red precipitate that formed was filtered off and washed with cold diethyl ether, and the resulting yellow solution was rotary evaporated to dryness to yield a yellow powder. The final product was recrystallized from hot acetone and hexane, and dried *in vacuo* for 48 hours. Crystals of suitable quality for diffraction studies were grown by slow vapour diffusion of *n*-hexane into concentrated acetone solutions of the complex. Yield = 50%; ^{1}H-NMR (300 MHz, CD$_3$OD): δ ppm 7.05 (dd, J = 1.00 Hz, 2H), 7.70 (dd, J = 1.00 Hz, 2H), 8.86 (dd, J = 1.00 Hz, 2H); FTIR (CH$_3$CN): 2024, 1930, 1915 cm^{-1}; anal. calcd for $C_{13}H_8BrMnN_2O_5$: C, 38.36; H, 1.98; N, 6.88. Found: C, 38.20; H, 1.97; N, 6.78. *m*/*z* (ESI); 359 ([M − Br$^-$ + HOMe]$^+$), 327 ([M − Br$^-$]$^+$).

Electrochemical measurements

A PalmSens[3] potentiostat, glassy carbon disc (A = 0.0717 cm^2, BASi) working electrode and a Pt basket counter electrode were used in all experiments. An Ag/AgCl (3 M KCl, BASi) or a freshly cleaned Ag wire, calibrated *vs.* Fc/Fc$^+$, was used as the reference electrode in water or organic electrolytes, respectively. Bulk electrolysis measurements were carried out in a custom glass H-Cell with the Pt counter electrode separated from the working and reference electrodes by a fine glass frit. All samples were thoroughly purged with either Ar or CO$_2$ for 30 minutes prior to use, and kept in the dark throughout the experiments. Gas chromatography (GC) was performed using an Agilent 6890N employing N6.0 He (BOC) as the carrier gas (5 ml min^{-1}). A 5 Å molecular sieve column (ValcoPLOT, 30 m length, 0.53 mm ID) and a pulsed discharge detector (D-3-I-HP, Valco Vici) were employed. CO and H$_2$ peak areas were quantified by comparison to multiple calibrant gas injections (500 ppm H$_2$ and 200 ppm CO in helium (STG)). NMR spectra were recorded on a Bruker Advance 400 NMR spectrometer operating at 400 MHz (^{1}H) or 101 MHz (^{13}C) or on a Bruker Advance 300 NMR spectrometer operating at 300 MHz (^{1}H) using an amber NMR tube. ESI-MS and elemental analyses were performed by the University of Liverpool analytical services. UV-vis spectra were recorded on a Shimadzu UV-2600 spectrophotometer using 1 cm or 1 mm pathlength quartz cuvettes. FTIR spectra were recorded on a Bruker Vertex instruments either in solution, cast on a CaF$_2$ plate, or using a diamond ATR crystal (Pike). Spectroelectrochemistry (SEC) was carried out using a liquid IR-SEC cell with a pathlength of approximately 0.2 mm that has been reported elsewhere.[33] The working and counter electrodes were Pt minigrids and a silver wire was used as the pseudo-reference. The concentration of the complex was approximately 0.1 mM in dry CH$_3$CN containing 0.1 M TBA PF$_6$ electrolyte, which was recrystallized from absolute ethanol and dried overnight on a Schlenk line at 70 °C prior to use.

Preparation of membrane supported catalyst electrodes

The membrane was prepared from a sample of complex **1–4** dissolved in 0.5 ml of acetonitrile, mixed with 0.5 ml of a Nafion/alcohol solution (5.0% w/w) in a 1 : 1 volume ratio in the dark to make up a 1.4×10^{-2} M solution. In a typical experiment 5 µl of this yellow solution, corresponding to 70 nmol of complex in

2.5% Nafion/CH_3CN 1 : 1, was transferred onto a polished glassy carbon electrode using an autopipette and left to dry in air in the dark for at least 2 hours at a constant temperature (maintained at 25 °C), yielding a continuous pale yellow polymer film that completely coated the working area. It has been observed that the temperature control is critical in obtaining highly active films (see full discussion in the ESI†). Casting solutions for MWCNT/Nafion/complex films were prepared as above with the addition of MWCNT in a 1 : 1 mass ratio with the complex to the casting solution, followed by sonication in the dark for 15 minutes. 5 μl of this darkly coloured casting solution was then transferred onto a polished glassy carbon electrode using an autopipette and left to dry in air in the dark for at least 2 hours. Following bulk electrolysis experiments the aqueous electrolyte was studied using UV/vis spectroscopy to confirm that complex leaching from the membrane into the electrolyte had not occurred. Morphological characterization was performed on films cast on glass slides using an Ambios Technology XP200 profilometer.

3 Results and discussion

Synthesis and characterisation

Complexes **3** and **4** are readily prepared by modification of the existing synthetic route of **1**.[7] It is noted that complex **3** is particularly light sensitive, undergoing rapid decomposition within minutes in solution under ambient light, therefore all electrochemical experiments were carried out in the dark. Both **3** and **4** have been characterised using FTIR spectroscopy which shows three ν(CO) bands corresponding to *fac*-tricarbonyl species (Table 1) in line with the single crystal structures, Fig. 1.

Due to the limited solubility of **3** in acetonitrile all experiments using this complex are carried out in the presence of 5% water. We observe ν(CO) bands at 2030, 1940 and 1931 cm^{-1} for **3** in CH_3CN/H_2O (95 : 5), with the increase in the ν(CO) frequencies compared to complex **1** indicating decreased electron density at the Mn centre due to the introduction of the electron withdrawing carboxylic acid group at the 4,4′-position of the bipyridine ligand.[27] In neat acetonitrile complex **4** has ν(CO) bands at 2024, 1930 and 1915 cm^{-1}, with the decrease in frequency of the stretches *versus* **1** indicating that the introduction of the hydroxyl groups to the bipyridine ligand has increased the electron density at the metal centre in **4**.

Table 1 Infrared carbonyl stretching frequencies and reduction potentials for complexes 1–4 in CH_3CN, and reduction potentials of complexes 1–4 in CH_3CN or CH_3CN/H_2O (95 : 5) containing 0.1 M TBA PF_6 electrolyte

Complex	E 1st (V *vs.* SCE)	E 2nd (V *vs.* SCE)	ν(CO) (cm^{-1})			Solvent
1	−1.24	−1.46	2027	1933	1924	CH_3CN
2	−1.39	−1.57	2025	1930	1921	CH_3CN
3	−1.07	−1.41	2030	1940	1931	CH_3CN/H_2O
4	−1.38[a]	−1.73[a]	2024	1930	1915	CH_3CN

[a] The first reductions of **4** are proposed to be primarily based on the bpy ligand (see Fig. 3 and 4).

The magnitude of the shift of the ν(CO) bands in **4** is very similar to that previously observed, and reproduced here also, for **2**.[8] Solvolysis of the Mn–Br bonds in the case of **4** occurs rapidly in the dark leading to the formation of the solvent complex $[Mn(bpy(OH)_2)(CO)_3(CH_3CN)]^+$ (ν(CO) 2045, 1955 and 1944 cm^{-1}, Fig. S3†), which has been seen previously for this class of complex.[15,34] It has been previously reported that *fac*-$[Mn(bpy)(CO)_3Br]$ can undergo photochemical transformation to *mer*-$[Mn(bpy)(CO)_3Br]$, which has ν(CO) at 2043 (vw), 1948 (s) and 1903 (m) in THF.[35] We rule out the formation of *mer* complexes of **4** due to the relative intensities of the three new ν(CO) stretches formed upon solvation, which are all of similar intensity, indicating the presence of a *fac*-tricarbonyl species.

Electrochemical characterisation of complex 3

The electrochemistry of **1** and **2** in CH_3CN has been reported previously,[7,8] but the cyclic voltammetry of **1** is also shown in Fig. 2(a) to allow comparison to the previously unreported complex **3**, Fig. 2(b). Initial reduction of **1** at −1.24 V in anhydrous CH_3CN results in bromide loss followed by fast dimerization to form $[Mn(bpy)(CO)_3]_2$ which is reduced at −1.46 V to form the catalytically active species, $[Mn(bpy)(CO)_3]^-$. In anhydrous CH_3CN, oxidation of $[Mn(bpy)(CO)_3]^-$ can occur at −1.08 V leading to dimer formation, which is oxidised at −0.21 V.[7] In the presence of water, solvolysis of the Mn–Br bond occurs and the reduction of $[Mn(bpy)(CO)_3(S)]^+$ to form the dimer is also observed at −1.11 V.

Spectroelectrochemical measurements which would be required to confirm the chemical nature of the species produced in analogous experiments with **3**

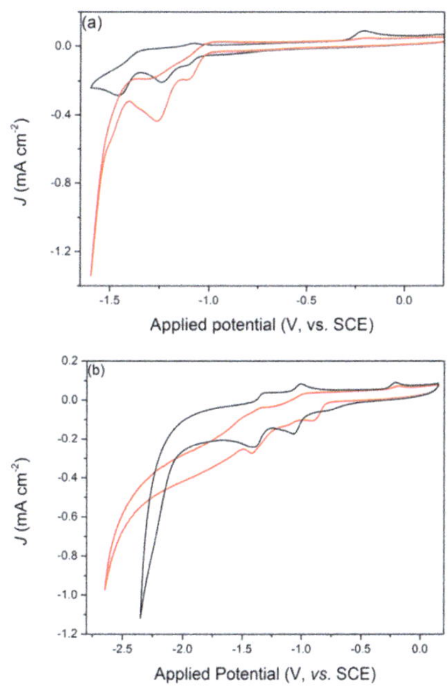

Fig. 2 (a) Solution phase CVs of the parent complex **1** under Ar (black) and CO_2 (red). (b) CVs of complex **3** under Ar (black) and CO_2 (red). The supporting electrolyte was 0.1 M TBA PF_6 in CH_3CN/H_2O (95 : 5), $v = 100$ mV s^{-1}.

were not possible due to excessive H_2 production occurring in the spectroelectrochemical cell (5% H_2O in CH_3CN is required to solubilise **3**). The similarities in the cyclic voltammograms of **1** and **3** allow the following tentative assignments to be made by analogy (Fig. 2(b)). Under argon, the reduction at −1.07 V is assigned to complex **3**, with the shoulder at *ca.* −0.8 V *vs.* SCE in CH_3CN/H_2O due to $[Mn(bpy(COOH)_2)(CO)_3(Solv)]^+$, Fig. 2(b). The significant anodic shift of the reduction potentials of both the cationic solvent complex (−0.80 V) and the bromide complex (−1.07 V) of **3** compared to **1** indicates that the electron withdrawing carboxylic acid groups are able to stabilise the reduced complex as hypothesised. In light of the precedent for similar $[Mn^I(L)(CO)_3X]$ complexes to dimerize following initial reduction, it is proposed that the two-electron reduction of the dimer $[Mn(bpy(COOH)_2)(CO)_3]_2$ occurs at −1.41 V to form two $[Mn(bpy(COOH)_2)(CO)_3]^-$, with re-oxidation of $[Mn(bpy\text{-}COOH)(CO)_3]^-$ and the subsequently formed dimer occurring at −1.00 and −0.21 V, respectively. Our assignment of the formation of the proposed active catalyst $[Mn(bpy(COOH)_2)(CO)_3]^-$ by dimer cleavage at −1.41 V is supported by the enhanced current in the presence of CO_2, which is particularly pronounced at slower scan rates (Fig. S8†). However we are unable to definitively rule out other assignments, including the direct reduction of $[Mn(bpy(COOH)_2)(CO)_3]$ in the absence of dimerization, or the reduction of a manganese hydride species as has been previously proposed by Gobetto *et al.* for a $[Mn^I(L)(CO)_3X]$ complex containing hydroxyl groups.[36] It is notable that, although present, the current enhancement achieved under CO_2 with **3** in solution is small compared to that seen for **1** (Fig. 2), indicating a low catalytic TOF.[8] $[Re(bpy(COOH)_2)(CO)_3Cl]$ has previously been explored as a potential CO_2 reduction electrocatalyst, but no current enhancement in the presence of CO_2 was observed,[30] and it is likely that the reduced nucleophilic nature of the active complex here formed upon reduction of **3**, due to the presence of the EWG, is also a significant factor in decreasing the activity towards CO_2 reduction. Nonetheless, despite the reduced activity the promising anodic shift in reduction potentials of **4** warrants the study of the complex in a polymer supported electrode.

Electrochemical characterisation of complex 4

The cyclic voltammetry of **4** in anhydrous CH_3CN under argon at 100 mV s^{-1} is not typical of that observed for other $[Mn(L)(CO)_3Br]$ complexes.[7,8] The CVs in Fig. 3 show the presence of reductions at −1.38, −2.01 and −2.20 V *vs.* SCE (labelled processes I, III and IV in Fig. 3), in addition to a scan rate dependent feature at −1.73 V *vs.* SCE (II), which becomes increasingly prominent at higher scan rates (Fig. S6†).

Whilst this manuscript was under preparation an IR-SEC study of the related complex $[Re(bpy(OH)_2)(CO)_3Cl]$ was reported, with the modified bpy ligand being able to undergo two reductive deprotonation steps to form $[Re(bpy(OH)(O^-))(CO)_3Cl]^-$ and $[Re(bpy(O^-)_2)(CO)_3]^-$, with assignments based upon DFT calculations and IR analysis of chemically deprotonated complexes.[32] IR-SEC indicates that the behaviour of the equivalent Mn complex (**4**) is also similar, Fig. 4. At open circuit potential (OCP), ν(CO) corresponding to both complex **4** (2024, 1930 and 1915 cm^{-1}) and the solvent complex (2044, 1955, 1944 cm^{-1}) are observed (Fig. 4(a), multi-Lorentzian fittings Fig. S9†). Upon initial

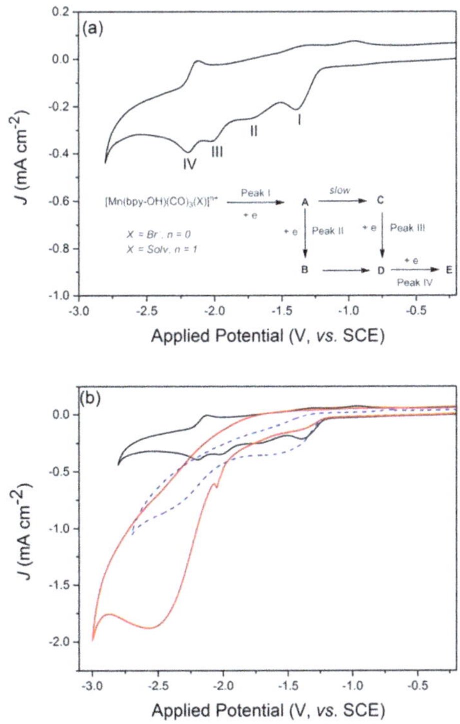

Fig. 3 (a) Cyclic voltammetry of **4** (~1 mM) in Ar-purged CH_3CN. The inset shows the proposed electrochemical scheme. (b) Cyclic voltammetry of **4** (~1 mM) in CH_3CN/Ar (black), CH_3CN/CO_2 (red) and CH_3CN/H_2O (95 : 5)/CO_2 (blue). All scan rates were 100 mV s^{-1}, the supporting electrolyte was 0.1 M TBA PF_6.

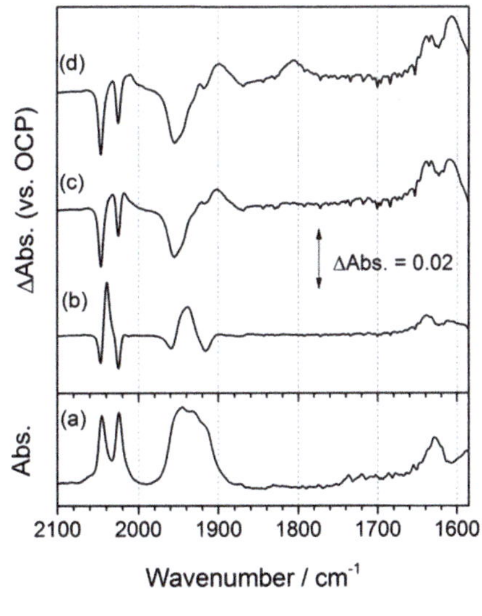

Fig. 4 IR-SEC of **4** in dry CH_3CN/0.1 M TBA PF_6. (a) At OCP, difference spectrum *versus* OCP and (b) at reduction (I), (c) reduction III and (d) reduction IV.

reduction (peak I) the IR difference spectrum shows the loss of **4** and its related solvent complex and the growth of a new species with ν(CO) that heavily overlap with the starting materials, Fig. 4(b). Multi-Lorentzian curve fitting (Fig. S10†) demonstrates that the product formed upon initial reduction has ν(CO) at 2039, 1944 and 1932 cm^{-1} in addition to weaker IR modes assigned to the bipyridine ligand at 1640 and 1610 cm^{-1}. By analogy to the reported spectrum of $[Re(bpy(OH)(O^-))(CO)_3Cl]^-$, which has bpy modes at *ca.* 1630 and 1590 cm^{-1} and ν(CO) at 2012, 1900 and 1881 cm^{-1} corresponding to a shift of −7 to −12 cm^{-1} *vs.* the parent $[Re(bpy(OH)_2)Cl]$ complex,[32] we assign the complex formed during IR-SEC to the singly reductively deprotonated species $[Mn(bpy(OH)(O^-))(CO)_3(Solv)]$. Given the timescales required for bulk reduction, the process labelled II in Fig. 3(a) was not identified during IR-SEC. Further reduction of $[Mn(bpy(OH)(O^-))(CO)_3(Solv)]$ (peak III) leads to the formation of a complex with ν(CO) at 2024, 1927 and 1903 cm^{-1}, in addition to a strong ligand mode at 1607 cm^{-1} (Fig. 4(c) and S11†). This is assigned to the doubly reduced, doubly deprotonated complex $[Mn(bpy(O^-)_2)(CO)_3(Solv)]^-$, due to the excellent agreement with the reported IR frequencies of $[Re(bpy(O^-)_2)(CO)_3]^-$ (2002, 1886, 1865 and 1605 cm^{-1} (str.)).[32] Applying increasingly negative potentials leads to the formation of two new ν(CO) at 1807 cm^{-1} and 1889 cm^{-1}, Fig. 4(d), with the higher frequency ν(CO) again masked by the bleaching of the starting material. The related Mn^0 complex $[Mn(bpy(^tBu)_2)(CO)_3]^-$, formed at *ca.* −1.6 V *vs.* SCE,[8] has ν(CO) at 1907 and 1807 cm^{-1} and here we tentatively assign the complex with ν(CO) at 1889 cm^{-1} and 1807 cm^{-1} to either $[Mn(bpy(OH)(O^-))(CO)_3]^-$ or $[Mn(bpy(O^-)_2)(CO)_3]^{2-}$.

In anhydrous acetonitrile a large current enhancement is seen in the cyclic voltammetry of **4** in the presence of CO_2, with the enhancement onsetting at potentials cathodic of −1.79 V *vs.* SCE (Fig. 3(b)). In contrast to other $[Mn(L)(CO)_3Br]$ complexes where the addition of a proton donor aids CO_2 reduction, here we find that the addition of 5% water actually decreases the current density during cyclic voltammetry. Gobetto *et al.* have reported CO_2 reduction using a hydroxyl-substituted Mn(bpy) complex in the absence of an additional proton source, but do not comment on the role of added water in the process.[36] Bulk electrolysis of 0.2 mM **4** in anhydrous CH_3CN under CO_2 at −2.1 V *vs.* SCE led to CO production with a Faradaic efficiency (FE) of 89% and a turnover number (TON) of 0.6, with no H_2 production detected in the first four hours. Sustained electrolysis (19.5 h) however led to complex degradation and the FE dropped to 5% for CO. In the presence of water (5%) bulk electrolysis of **4** at −2.1 V shows that primarily H_2 (FE = 45%) is produced, with only low levels of CO (FE = 6%). It is therefore apparent that although complex **4** is a poor CO_2 reduction catalyst in CH_3CN it displays markedly different electrochemical behaviour in differing solvent environments, making it of interest to study in the highly acidic environment of the Nafion film.

Polymer supported electrocatalytic reduction of CO_2 in water

To compare the catalytic behaviour of both new complexes (**3** and **4**) and the known benchmark catalysts (**1** and **2**) in a polymer support for use in aqueous electrolyte, complexes **1–4** were cast in Nafion films mixed with MWCNT onto a glassy carbon working electrode and allowed to dry in the dark. We have reported on CO_2 electrocatalysis using **1**/MWCNT/Nafion previously.[23] We have optimised the drying conditions and found that the highest current densities occur when the films are dried at 25 °C and the role of drying temperature is discussed further in the ESI and Fig. S14 and S15.† In brief, there is a broad reduction at approximately −1.2 V *vs.* SCE which has been shown previously to be two overlapping processes: dimer formation followed by dimer reduction, Fig. 5(a) and (b). Dimer formation may seem surprising given that **1** is immobilized, but we have shown previously that clustering on the MWCNT surfaces is likely to enable the formation of bimetallic complexes.[23] The dimer reoxidation peak is observed at −0.4 V. In the presence of CO_2 there is a significant current enhancement for **1** at potentials negative of −1.2 V, indicative of CO_2 reduction (Fig. 5). Due to the optimisation of the film drying conditions we report a large improvement in sustained current densities during CPE measurements (1.8 mA cm^{-2}, *vs.* 0.7 mA cm^{-2} previously reported at −1.4 V),[23] albeit with a slight loss in selectivity towards CO_2 (here 0.28, previous 0.47), see Table 1.

In contrast to electrodes of **1**/MWCNT/Nafion, equivalent experiments using complex **3** showed no significant enhancement in current density under CO_2. Square wave voltammetry (SWV) of a **3**/MWCNT/Nafion electrode under argon did show the presence of two reductions at *ca.* −0.8 and −1.25 V (Fig. 5(b)). The reduction at −0.8 V is present in films of MWCNT/Nafion without complexes **1–4** indicating that it is MWCNT-based. The reduction at −1.25 V is assigned to the reduction of a solvated $[Mn(bpy(COOH)_2)(CO)_3(Solv)]^+$ complex (Solv = CH_3CN, H_2O or Nafion) indicating that a concentration of **3** was electrochemically active within the polymer. However despite the promising solution electrochemical behaviour of **3**, prolonged electrolysis of a **3**/MWCNT/Nafion electrode under CO_2 at −1.4 V in aqueous solutions at pH 7 lead exclusively to H_2 production. In a similar manner **4**/MWCNT/Nafion electrodes under argon in aqueous solutions at pH 7 showed electrochemical behaviour in line with the solution studies in acetonitrile outlined above with two clear complex reductions assigned to the reduction of the solvated complex of **4** $[Mn(bpy(OH)_2)(CO)_3(Solv)]^+$ at −1.15 V and

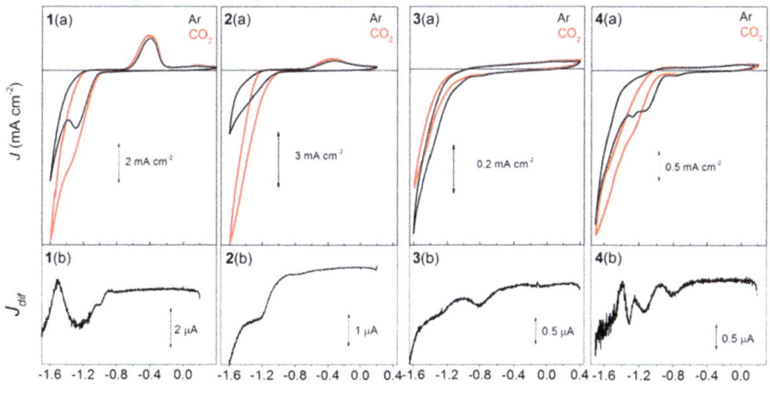

Fig. 5 Cyclic voltammograms (a) and square wave voltammograms (SWV) (b) for complexes **1–4** cast in Nafion/MWCNT on glassy carbon electrodes. Scan rates were 10 mV s^{-1} *vs.* SCE in a 0.1 M phosphate buffer electrolyte. For comparison, CVs of all four complexes in solution are overlaid in Fig. S7.†

of **4** at [Mn(bpy(OH)$_2$)(CO)$_3$(Br)] at −1.3 V, in addition to the MWCNT-based reduction at −0.8 V. In the presence of CO_2 a notable current increase was observed at potentials cathodic of −1.2 V, however prolonged electrolysis at −1.4 V under CO_2 indicated that the primary product formed was H_2, Table 2.

Electrochemical studies of **2**/MWCNT/Nafion electrodes are however very promising. Reductions visible by SWV at −1.28 and −1.53 V (Fig. 5(b)) are assigned to those observed in solution for the initial reduction of **2** and its subsequently formed dimer at −1.27 and −1.48 V *vs.* SCE in CH_3CN, Fig. S12.† A dimer reoxidation peak is also visible on the return sweep of the **2**/MWCNT/Nafion electrode at *ca.* −0.4 V, Fig. 5. Under CO_2 a large increase in current density is measured relative to under argon (−8.8 mA cm^{-2} *vs.* −3.2 mA cm^{-2} at −1.6 V) that onsets at approximately −1.3 V. Bulk electrolysis measurements carried out at −1.4 V demonstrate that **2**/MWCNT/Nafion electrodes are active for CO_2 reduction in pH 7 electrolyte with CO being produced with a selectivity (CO : H_2 ~ 1.04) exceeding that achieved with complex **1** (CO : H_2 ~ 0.28) under the same conditions. Based on analysis of the solution phase electrocatalysis of **2**, we postulated that films of **2**/MWCNT/Nafion would operate efficiently at more negative potentials whilst maintaining high selectivity for CO_2 reduction. CPE of **1**/MWCNT/Nafion at −1.5 V yielded CO with an average current density (J_{avg}) of *ca.* 4.38 mA cm^{-2} and a TON_{CO} of 35.9 after 4 hours, with a decreased selectivity (CO : H_2 = 0.21). Surprisingly CPE at −1.5 V using **2**/MWCNT/Nafion for 2 hours resulted in lower average current densities (1.14 mA cm^{-2}) than for **1**, and after 2 hours the current density decreased dramatically. Accompanying this drop in current density was a concurrent loss in selectivity for CO_2 reduction, which may suggest that the **2**/MWCNT/Nafion electrode is being limited by the available CO_2 concentration. Indeed, re-purging with CO_2 each hour led to recovery of the activity of the system, yielding an improved J_{avg} (2.65 mA cm^{-2}), without a loss in selectivity (CO : H_2 = 0.33) over 4 hours and a total TON_{CO} of over 45 at −1.5 V (Fig. S13†). It is therefore apparent the **2**/MWCNT/Nafion electrode is likely to be limited by the available CO_2 concentration under typical operating conditions. Future work will explore the use of both alternative polymeric materials and electrode configurations to enhance the available CO_2 concentration at the catalytic sites. Recent examples using gas diffusion electrodes coupled to immobilized catalysts have demonstrated marked improvements in catalytic current densities.[26] The relatively weak interaction between the MWCNT and complexes is also likely to be a significant limiting factor and the development of

Table 2 CPE data for Nafion/MWCNT films containing **1**, **2** or **4**

Catalyst	*E* (V, *vs.* SCE)	J_{avg} (mA cm^{-2})	Time (h)	TON_{CO} total	CO : H_2	FE H_2, CO (%)
(**1**)/Nafion/MWCNT	−1.4	1.79	4	18.7	0.28	50, 14
(**1**)/Nafion/MWCNT	−1.5	4.38	4	35.9	0.21	52, 11
(**2**)/Nafion/MWCNT	−1.4	0.2	4	7.3	1.04	44, 46
(**2**)/Nafion/MWCNT[a]	−1.5	2.65	4	46.1	0.32	71, 23
(**2**)/Nafion/MWCNT	−1.5	1.14	2	10.6	0.33	75, 24
(**4**)/Nafion/MWCNT	−1.4	0.83	4	0.3	0.02	40, 1

[a] Cell repurged every hour.

similar complexes with aromatic functionalities better suited to π-stacking on MWCNT, such as pyrene groups, is also likely to yield further improvements in current densities.[37]

4 Conclusions

Two novel Mn complexes incorporating substituted bipyridine ligands, $[Mn(bpy(COOH)_2)(CO)_3Br]$ (**3**) and $[Mn(bpy(OH)_2)(CO)_3Br]$ (**4**), have been synthesized and characterized with the aim of utilisation in a high surface area $[Mn(L)(CO)_3X]$/MWCNT/Nafion electrode for CO_2 reduction. Detailed electrochemical studies of **3** and **4** were carried out and both complexes were found to have some activity towards CO_2 reduction in acetonitrile solutions; however when immobilized for use in aqueous conditions CO_2 reduction to CO was not found to occur in significant quantities. In contrast, immobilization of the benchmark catalyst **2** as a **2**/MWCNT/Nafion electrode was found to be a highly effective approach for enabling CO_2 reduction to CO in aqueous solvents at neutral pH values, with a J_{avg} of 2.65 mA cm^{-2} at −1.5 V and a selectivity towards CO_2 (CO : H_2 = 0.33) being achieved.

Acknowledgements

AC and JJW acknowledge the EPSRC (EP/K006851/1) for a fellowship and funding respectively. GN thanks the University of Liverpool for funding. Thanks to Prof. Dmitry Shchukin (SIRE) for access to the FTIR. Thanks to Prof. Ken Durose and Dr Rob Treharne (SIRE) for access to the profilometer.

References

1 A. J. Morris, G. J. Meyer and E. Fujita, *Acc. Chem. Res.*, 2009, **42**, 1983–1994.
2 A. J. Cowan and J. R. Durrant, *Chem. Soc. Rev.*, 2013, **42**, 2281–2293.
3 C. D. Windle and R. N. Perutz, *Coord. Chem. Rev.*, 2012, **256**, 2562–2570.
4 J.-M. Savéant, *Chem. Rev.*, 2008, **108**, 2348–2478.
5 J. Hawecker, J.-M. Lehn and R. Ziessel, *J. Chem. Soc., Chem. Commun.*, 1983, 536–538.
6 J. Hawecker, J.-M. Lehn and R. Ziessel, *J. Chem. Soc., Chem. Commun.*, 1984, 328–330.
7 M. Bourrez, F. Molton, S. Chardon-Noblat and A. Deronzier, *Angew. Chem., Int. Ed.*, 2011, **50**, 9903–9906.
8 J. M. Smieja, M. D. Sampson, K. A. Grice, E. E. Benson, J. D. Froehlich and C. P. Kubiak, *Inorg. Chem.*, 2013, **52**, 2484–2491.
9 M. D. Sampson, A. D. Nguyen, K. A. Grice, C. E. Moore, A. L. Rheingold and C. P. Kubiak, *J. Am. Chem. Soc.*, 2014, **136**, 5460–5471.
10 C. Riplinger, M. D. Sampson, A. M. Ritzmann, C. P. Kubiak and E. A. Carter, *J. Am. Chem. Soc.*, 2014, **136**, 16285–16298.
11 C. Riplinger and E. A. Carter, *ACS Catal.*, 2015, **5**, 900–908.
12 D. C. Grills, J. A. Farrington, B. H. Layne, S. V. Lymar, B. A. Mello, J. M. Preses and J. F. Wishart, *J. Am. Chem. Soc.*, 2014, **136**, 5563–5566.
13 M. Bourrez, M. Orio, F. Molton, H. Vezin, C. Duboc, A. Deronzier and S. Chardon-Noblat, *Angew. Chem., Int. Ed.*, 2014, **53**, 240–243.

14 J. Agarwal, T. W. Shaw, C. J. Stanton III, G. F. Majetich, A. B. Bocarsly and H. F. Schaefer III, *Angew. Chem., Int. Ed.*, 2014, **53**, 1–5.
15 J. Agarwal, C. J. Stanton III, T. W. Shaw, J. E. Vandezande, G. F. Majetich, A. B. Bocarsly and H. F. Schaefer III, *Dalton Trans.*, 2015, **44**, 2122–2131.
16 Q. Zeng, J. Tory and F. Hartl, *Organometallics*, 2014, **33**, 5002–5008.
17 M. V. Vollmer, C. W. Machan, M. L. Clark, W. E. Antholine, J. Agarwal, H. F. Schaefer III, C. P. Kubiak and J. R. Walensky, *Organometallics*, 2015, **34**, 3–12.
18 J. Agarwal, T. W. Shaw, H. F. Schaefer III and A. B. Bocarsly, *Inorg. Chem.*, 2015, **54**, 5285–5294.
19 T. K. Mukhopadhyay, N. L. MacLean, L. Gan, D. C. Ashley, T. L. Groy, M.-H. Baik, A. K. Jones and R. J. Trovitch, *Inorg. Chem.*, 2015, **54**, 4475–4482.
20 T. Yoshida, K. Tsutsumida, S. Teratani, K. Yasufuku and M. Kaneko, *J. Chem. Soc., Chem. Commun.*, 1993, 631–633.
21 A. Jarzebinska, P. Rowinski, I. Zawisza, R. Bilewicz, L. Siegfried and T. Kaden, *Anal. Chim. Acta*, 1999, **396**, 1–12.
22 J. Zhang, W. J. Pietro and A. B. P. Lever, *J. Electroanal. Chem.*, 1996, **403**, 93–100.
23 J. J. Walsh, G. Neri, C. L. Smith and A. J. Cowan, *Chem. Commun.*, 2014, **50**, 12698–12701.
24 F. Li, B. Zhang, X. Li, Y. Jiang, L. Chen, Y. Li and L. Sun, *Angew. Chem., Int. Ed.*, 2011, **50**, 12276–12279.
25 E. S. Andreiadis, P.-A. Jacques, P. D. Tran, A. Leyris, M. Chavarot-Kerlidou, B. Jousselme, M. Matheron, J. Pécaut, S. Palacin, M. Fontecave and V. Artero, *Nat. Chem.*, 2013, **5**, 48–53.
26 P. Kang, S. Zhang, T. J. Meyer and M. Brookhart, *Angew. Chem., Int. Ed.*, 2014, **53**, 8709–8713.
27 J. Li, Y. Xu, H. Wei, T. Huo and E. Wang, *Anal. Chem.*, 2007, **79**, 5439–5443.
28 F. Liu, X. Yang and S. Sun, *Analyst*, 2011, **136**, 374–378.
29 C. Costentin, S. Drouet, M. Robert and J.-M. Savéant, *Science*, 2012, **338**, 90–94.
30 J. M. Smieja and C. P. Kubiak, *Inorg. Chem.*, 2010, **49**, 9283–9289.
31 S. Klein, W. G. Dougherty, W. S. Kassel, T. J. Dudley and J. J. Paul, *Inorg. Chem.*, 2011, **50**, 2754–2763.
32 G. F. Manbeck, J. T. Muckerman, D. J. Szalda, Y. Himeda and E. Fujita, *J. Phys. Chem. B*, 2015, **119**, 7457–7466.
33 M. Krejčik, M. Daněk and F. Hartl, *J. Electroanal. Chem. Interfacial Electrochem.*, 1991, **317**, 179–187.
34 J.-D. Compain, M. Bourrez, M. Haukka, A. Deronzier and S. Chardon-Noblat, *Chem. Commun.*, 2014, **50**, 2539–2542.
35 G. J. Stor, S. L. Morrison, D. J. Stuf and A. Oskam, *Organometallics*, 1994, **3**, 2641–2650.
36 F. Franco, C. Cometto, F. F. Vallana, F. Sordello, E. Priola, C. Minero, C. Nervi and R. Gobetto, *Chem. Commun.*, 2014, **50**, 14670–14673.
37 J. D. Blakemore, A. Gupta, J. J. Warren, B. S. Brunschwig and H. B. Gray, *J. Am. Chem. Soc.*, 2013, **135**, 18288–18291.

Faraday Discussions

PAPER

Catalytic dehydrogenation of propane by carbon dioxide: a medium-temperature thermochemical process for carbon dioxide utilisation

X. Du,[a] B. Yao,[b] S. Gonzalez-Cortes,[a] V. L. Kuznetsov,[a] Hamid AlMegren,[c] T. Xiao*[a] and P. P. Edwards*[a]

Received 5th May 2015, Accepted 27th May 2015
DOI: 10.1039/c5fd00062a

The dehydrogenation of C_3H_8 in the presence of CO_2 is an attractive catalytic route for C_3H_6 production. In studying the various possibilities to utilise CO_2 to convert hydrocarbons using the sustainable energy source of solar thermal energy, thermodynamic calculations were carried out for the dehydrogenation of C_3H_8 using $CO_2$for the process operating in the temperature range of 300–500 °C. Importantly, the results highlight the enhanced potential of C_3H_8 as compared to its lighter and heavier homologues (C_2H_6 and C_4H_{10}, respectively). To be utilised in this CO_2 utilisation reaction the Gibbs free energy ($\Delta_r G_m^{\theta}$) of each reaction in the modelled, complete reacting system of the dehydrogenation of C_3H_8 in the presence of CO_2 also indicate that further cracking of C_3H_6 will affect the ultimate yield and selectivity of the final products. In a parallel experimental study, catalytic tests of the dehydrogenation of C_3H_8 in the presence of CO_2 over 5 wt%-Cr_2O_3/ZrO_2 catalysts operating at 500 °C, atmospheric pressure, and for various C_3H_8 partial pressures and various overall GHSV (Gas Hourly Space Velocity) values. The results showed that an increase in the C_3H_8 partial pressure produced an inhibition of C_3H_8 conversion but, importantly, a promising enhancement of C_3H_6 selectivity. This phenomenon can be attributed to competitive adsorption on the catalyst between the generated C_3H_6 and inactivated C_3H_8, which inhibits any further cracking effect on C_3H_6 to produce by-products. As a comparison, the increase of the overall GHSV can also decrease the C_3H_8 conversion to a similar extent, but the further cracking of C_3H_6 cannot be limited.

[a]*King Abdulaziz City of Science and Technology (KACST) – Oxford Centre of Excellence in Petrochemicals, Inorganic Chemistry Laboratory, Department of Chemistry, University of Oxford, Oxford, OX1 3QR, UK. E-mail: peter.edwards@chem.ox.ac.uk; Fax: +44(0) 1865 272690; Tel: +44(0) 1865 272660*

[b]*SINOPEC Shanghai Petrochemical Company LTD., Shanghai, 200540, China*

[c]*Petrochemicals Research Institute (PRI), King Abdulaziz City of Science and Technology (KACST), P. O. Box 6086, Riyadh 11442, Saudi Arabia*

1. Introduction

In an attempt to ameliorate the burgeoning growth in greenhouse gas emissions, there are currently intense world-wide efforts aimed at the utilisation of CO_2, particularly in its conversion to fuels and high-value chemical products.[1–7] Chemical processes aimed at the utilisation and conversion of CO_2 are, of course, driven in a thermodynamic sense by the difference in Gibbs free energy between the resulting final chemical products, and CO_2 and the targeted reactants under the relevant experimental conditions. However, CO_2 being a highly stable molecule will require a substantial amount of energy, effective catalysts and effective reaction conditions for any chemical conversion processes into fuels or high-value chemical products.

Thus probably all chemical reactions for CO_2 conversion and utilisation are endothermic and will consume considerable amounts of energy. If such energy is provided by fossil fuels, the net effect based on any well-to-wheels analysis of the process will invariably result in a net production of CO_2; this situation could only be deemed beneficial from a climate mitigation perspective if the necessary input process energy is provided from renewable or sustainable sources.

The motivation behind this work, therefore, is to identify and develop specific chemical utilisation processes for CO_2 that can be achieved by the application of relatively easily-accessible solar thermal energy and associated thermochemical processes, using so-called "Low and Medium Temperature Thermochemical Processes", typically operating for temperature ranges of $T \sim 250$ °C and $T \sim 250$–500 °C, respectively.[8]

Utilising solar thermal energy at these lower temperatures rather than the widely-utilised high temperature process regime[9] ($T \sim 1000$ °C) for thermochemical processes, creates highly interesting and important challenges for catalysis science. Coupling and optimising the catalyst chemistry of CO_2 chemical reactions in the temperature/energy range of *ca.* 300–500 °C with the engineering challenges of cheap, solar thermal collectors may allow the prospect of accessible, sustainable CO_2 utilisation. These coupled challenges therefore provide a high level of opportunity for modern catalysis science and engineering.

As an "improved" reaction of the dehydrogenation of C_3H_8, the dehydrogenation of C_3H_8 in the presence of CO_2 ($C_3H_8 + CO_2 \rightarrow C_3H_6 + H_2O + CO$) has been heavily studied over the last decade.[10–15] In this promising reaction, CO_2 acts as a mild oxidant to combine the dehydrogenation of C_3H_8 with a reverse water gas shift ($H_2 + CO_2 \rightarrow H_2O + CO$), and hence, the equilibrium of the dehydrogenation of C_3H_8 can be shifted to the product side. Moreover, CO_2 may also reduce the coking effect of the catalyst by coke gasification ($C + CO_2 \rightarrow 2CO$). Competitive adsorption among C_3H_8, CO_2 and generated gaseous water can also explain the relatively lower initial conversion of C_3H_8, the reduced coking effect, and the higher stability of catalysts in the CO_2 atmosphere.[16]

In this paper, the advantages of specifically selecting C_3H_8 as the feedstock for CO_2 utilisation are advanced for the lower temperature requirements, as compared to both its lighter and heavier alkane homologues, C_2H_6 and C_4H_{10}, respectively. Furthermore, in seeking the possibility to convert alkanes with the energy source from accessible solar heating technology, the temperature range to be targeted in thermodynamic calculations has been localised in the range 300–

500 °C which can be easily achieved, for example, by high pressure steam from a solar heating system.

The shift of the equilibrium state by CO_2 in the dehydrogenation of C_3H_8 has been identified from a thermodynamic analysis.[17–19] However, the reaction systems modelled previously rarely take into consideration the competing side reactions and by-products. The dehydrogenation of C_3H_8 in the presence of CO_2 can be significantly affected by various competing side reactions which are deleterious to the targeted C_3H_6 yield, selectivity and indeed to the stability of the operating catalyst. Hence in this paper, the thermodynamic analysis was operated across a more comprehensive reaction system which covered all the possible by-products together with the generation of carbon – particularly important when considering catalyst operating lifetimes.

When considering the side reactions for the dehydrogenation of C_3H_8 in the presence of CO_2, the major cause of by-products is the further cracking effect on C_3H_6,[20] and the competitive adsorption from CO_2 is claimed to be an effective way to inhibit that process.[16] Our analysis reveals that the availability of active sites on the catalyst surface does indeed play a critical role in the reaction system. To observe the different performance properties of a catalyst when the reactants are saturated, to the availability of active sites, the reactions were operated over a ZrO_2-supported Cr_2O_3 catalyst with careful control of the various partial pressures of C_3H_8 but keeping a constant C_3H_8/CO_2 ratio. The reaction results exhibited high C_3H_6 selectivity by inhibiting the selectivity of the by-products, which is a promising result if these processes are to be applied in any larger scale industrial process.

2. Thermodynamics and (selected) chemical reactions of carbon dioxide

In Fig. 1 we illustrate the underlying thermodynamic considerations for the chemical utilisation of CO_2, where the Gibbs free energy of formation of CO_2 and various related substances are shown for comparison. Any attempt at utilising CO_2 as a chemical reactant must therefore take into account the relative stability

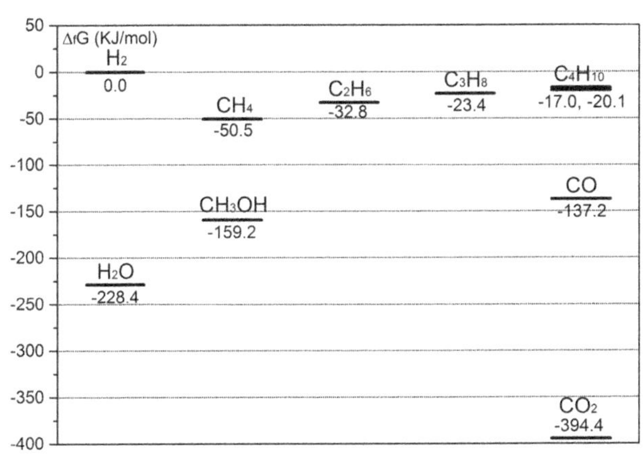

Fig. 1 $\Delta_f G_m^0$ of CO_2 and related substances at normal temperature and pressure (NTP).

(Gibbs free energy) of the CO_2 utilisation reaction products, as compared to CO_2 and the other reactant(s).

Both contributing terms (ΔH and $T\Delta S$) for the Gibbs free energy are not favourable in converting CO_2 to other molecules. The carbon–oxygen bonds are strong and substantial energy is needed for their dissociation and subsequent reduction. Similarly, the entropy term ($T\Delta S$) typically makes little or no contribution to the thermodynamic driving force for any CO_2 utilisation reaction.

The resulting Gibbs free energy of the CO_2 chemical utilisation reaction, ΔG, provides information as to the ultimate yield of the reaction products at equilibrium, through the relationship $\Delta_r G_m^\theta = -RT\ln(K^\theta)$.

From Fig. 2, however, the attractive option that the Gibbs free energy changes of CO_2 reaction becomes increasingly favourable by moving to higher members of the alkane homologous series. One notes, of course, that a ready solution for CO_2 utilisation is its conversion to the more reactive CO *via* the reverse water gas shift reaction ($H_2 + CO_2 \rightarrow H_2O + CO$), and subsequent use of syngas chemistry to yield the desired products. However, this attractive route clearly needs a ready source of H_2 and ideally a source of sustainable H_2 derived from non-fossil fuel routes. In the absence of hydrogen from low cost, low (zero) carbon sources, this process will yield a net CO_2 emission for the total wheel-to-wheel analysis.

Our research therefore looks in detail at the thermodynamics of the dehydrogenation of C_3H_8 by CO_2 as a potential prospect for CO_2 utilisation in temperature and reaction conditions where the ultimate application of solar thermal energy may be a promising technology. This is a specific example of the broader challenge facing *any* CO_2 utilisation technology; namely due to the inevitable input energy required to convert CO_2 to useful products, reducing CO_2 emissions through CO_2 utilisation will only be possible if the energy inputs are from renewable sources. We believe that Low-to-Medium Temperature Solar Thermochemical processes offer considerable opportunities in that regard and the research outlined here — setting out the complete thermodynamic analysis

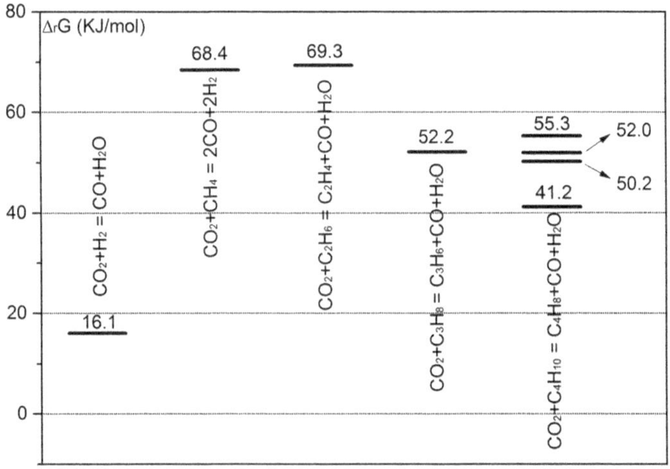

Fig. 2 $\Delta_r G_m^\theta$ of the reactions between CO_2 and other substances at 400 °C and normal pressure.

together with a catalytic chemistry study — may present a way forward for CO_2 utilisation.

3. Thermodynamic calculations and simulations

3.1 Methods

In order to fully understand the thermodynamic basis of the dehydrogenation of C_3H_8 in the presence of CO_2, in connecting also with the solar thermal technology which is normally set close to 500 °C in the form of a high pressure steam, the standard change of reaction in Gibbs free energy, $\Delta_r G_m^\theta$, of the dehydrogenation of C_2–C_4 alkanes with CO_2 utilisation ($C_nH_{2n+2} + CO_2 \rightarrow C_nH_{2n} + H_2O + CO$, $n = 2, 3, 4$) were calculated in the temperature range of 300–500 °C. The cracking of C_1–C_4 alkanes and olefins to form carbon $\left(C_nH_m \rightarrow nC + \frac{m}{2}H_2, n = 1,\ 2,\ 3,\ 4\right)$ were also included in these calculations, in this temperature range, to indicate the potential coking ability on operating catalysts of each substances.

Here, the $\Delta_r G_m^\theta$ of each single reaction in the temperature range of 300–500 °C were calculated as following:

$$\Delta_r H_m^\theta\ (298.15\ \text{K}) = \sum_j v_j\ \Delta_f H_m^\theta\ (298.15\ \text{K},\ j) \tag{1}$$

$$\Delta_r H_m^\theta (T) = \Delta_r H_m^\theta (298.15\ \text{K}) + \sum_j v_j \int_{298.15}^{T} C_{p,m} dT \tag{2}$$

$$\Delta_r S_m^\theta (298.15\ \text{K}) = \sum_B v_B\ S_m^\theta (298.15\ \text{K}) \tag{3}$$

$$\Delta_r S_m^\theta (T) = \Delta_r S_m^\theta (298.15\ \text{K}) + \sum_j v_j \int_{298.15}^{T} \frac{C_{p,m}}{T} dT \tag{4}$$

$$\Delta_r G_m^\theta = \Delta_r H_m^\theta - T \Delta_r S_m^\theta \tag{5}$$

In the functions above, $\Delta_f H_m^\theta$ is the standard molar enthalpy of formation, S_m^θ is the standard molar entropy, while $C_{p,m}$ is the parameters of molar heat capacity at constant pressure. The property parameters can be looked up in the chemical properties hand books, and these parameters for the related substances are list in Table 1 below.

3.2 Thermodynamic advantages of the dehydrogenation of C_3H_8 in the presence of CO_2

Fig. 3 exhibits the standard change of reaction in Gibbs free energy ($\Delta_r G_m^\theta$) of the reactions being compared. In the temperature range of 300–500 °C, all dehydrogenation reactions show a similar trend. The highest $\Delta_r G_m^\theta$ is found for the process to dehydrogenate C_2H_6 with CO_2, indicating that it is much harder to operate this reaction as compared to other low-n alkanes at 300–500 °C. Importantly, $\Delta_r G_m^\theta$ does not keep falling as the carbon number of alkane increases, and the dehydrogenation of C_3H_8 with CO_2 appears to be more feasible than some C_4 reactions.

Table 1 Standard thermodynamic properties of chemical substances: standard molar enthalpy of formation ($\Delta_f H_m^\theta$),[25] standard molar entropy (S_m^θ)[25] and parameters of molar heat capacity at constant pressure ($C_{p,m}$)[26]

Substance	$\Delta_f H_m^\theta$ (298.15 K) (kJ mol^{-1})	S_m^θ (298.15 K) (J K^{-1} mol^{-1})	$\frac{C_{p,m}}{R} = A + BT + CT^2 + DT^3 + ET^4$ (R = 8.3145 J K^{-1} mol^{-1}) A($\times10^0$)	B($\times10^{-3}$)	C($\times10^{-5}$)	D($\times10^{-8}$)	E($\times10^{-11}$)
H_2	0	130.684	2.883	3.681	−0.772	0.692	−0.213
CO	−110.53	197.67	3.912	−3.913	1.182	−1.302	0.515
CO_2	−393.51	213.74	3.259	1.356	1.502	−2.374	1.056
H_2O	−241.82	188.83	4.395	−4.186	1.405	−1.564	0.632
CH_4	−74.81	186.26	4.568	−8.975	3.631	−3.407	1.091
C_2H_4	52.26	219.56	4.221	−8.782	5.795	−6.729	2.511
C_2H_6	−84.68	229.6	4.178	−4.427	5.660	−6.651	2.487
C_3H_6	20.42	267.05	3.834	3.893	4.688	−6.013	2.283
C_3H_8	−103.85	269.91	3.847	5.131	6.011	−7.893	3.079
1-C_4H_8	−0.13	305.71	4.389	7.984	6.143	−8.197	3.165
cis-2-C_4H_8	−6.99	300.94	3.689	19.184	2.230	−3.426	1.256
trans-2-C_4H_8	−11.17	296.59	5.584	−4.890	9.133	−10.975	4.085
i-C_4H_8	−17.10	295.29	3.231	20.949	2.313	−3.949	1.566
n-C_4H_{10}	−126.15	310.23	5.547	5.536	8.057	−10.571	4.134
i-C_4H_{10}	−134.73	291.82	3.351	17.883	5.477	−8.099	3.243
C (graphite)[a]	0	5.74	−0.977	9.458	−1.118	0.739	−0.207

[a] The parameters of $C_{p,m}$ are calculated by regressing the molar heat capacity at various temperature in handbook.[27]

The $\Delta_r G_m^\theta$ of alkane and olefin cracking to generate carbon and H_2 are shown in Fig. 4, The C_4 alkanes have multiple curves because of their constituent isomers. In general, a lower value of $\Delta_r G_m^\theta$ indicates a higher possibility of

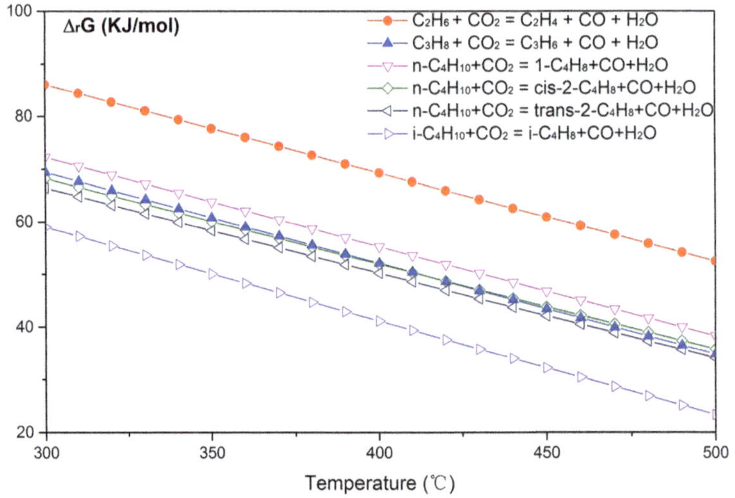

Fig. 3 $\Delta_r G_m^\theta$ of the dehydrogenation of light alkane in the presence of CO_2 at 300–500 °C.

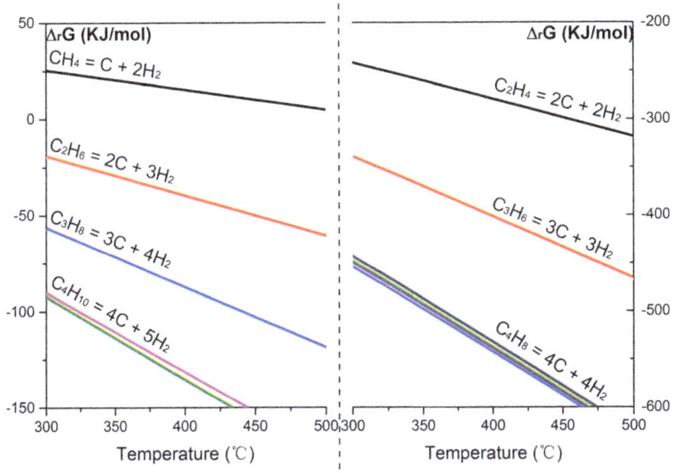

Fig. 4 $\Delta_r G_m^\theta$ of the alkanes (left) and olefins (right) cracking to generate carbon and H_2 at 300–500 °C.

cracking. The olefins therefore show a much higher cracking possibility than any alkanes, which implies that the selectivity of the target products will be greatly affected by further cracking of the olefins. For instance, C_3H_6 will be further cracked to lower level hydrocarbons or even carbon, so the selectivity will be reduced. It is also obvious that CH_4 has the highest resistance to coking, and as a sequence of stepwise mechanisms of hydrocarbon decomposition, it will be a major by-product in these reaction systems when looking at olefins only. Furthermore, it is obvious that the carbon number is the most important indicator of the cracking possibility. With a similar $\Delta_r G_m^\theta$ of dehydrogenation of alkane with CO_2, C_3H_6 has a much higher resistance to cracking than any isomers of butene. Although C_2H_4 has even higher resistance to subsequent cracking, it is even more difficult for C_2H_6 to be activated with CO_2 when the temperature is kept below 500 °C. From this analysis, the dehydrogenation of C_3H_8 in the presence of CO_2 is recognisably the best alkane to be targeted for a chemical utilisation process for CO_2 operating in this temperature range, a range specifically chosen for the ready-availability of solar heating technology for these conditions.

3.3 Modelled calculations of the dehydrogenation of C_3H_8 in the presence of CO_2

In reality, of course, the dehydrogenation of C_3H_8 in the presence of CO_2 is not a simple single reaction but a multiple reaction system, as shown in Table 2, mainly coupling the traditional dehydrogenation of C_3H_8 (reaction (1)) and reverse water gas shift (reaction (2)).

In this system, (1) and (2) are the main reactions considered; (3) and (4) are the further cracking of C_3 species where lighter by-products are formed; (5) is the reaction to gasify the carbon deposition.

The $\Delta_r G_m^\theta$ of reaction (1)–(3) and (5) are shown in Fig. 5, while the $\Delta_r G_m^\theta$ of hydrocarbons cracking (reaction (4)) are already displayed in Fig. 4. From 300–500 °C, only the cracking of C_2 and C_3, including reaction (3), can reach a negative value of

Table 2 Modelled reaction system for the dehydrogenation of C_3H_8 in the presence of CO_2

Reaction number	Reaction formula
(1)	$C_3H_8 \rightarrow C_3H_6 + H_2$
(2)	$H_2 + CO_2 \rightarrow H_2O + CO$
(3)	$C_3H_8 \rightarrow C_2H_4 + CH_4$
(4)	$C_nH_m \rightarrow nC + \frac{m}{2}H_2$
(5)	$C + CO_2 \rightarrow 2CO$

$\Delta_rG^\theta_m$, which means these reactions are thermodynamically favoured. On the contrary, the decomposition of CH_4 is not favoured in this temperature range. Considering C_3H_8 is the feedstock in this system, and the stepwise mechanism of the hydrocarbons to be decomposed, the generation of CH_4 can be an important indicator to measure the degree and extent of side reactions during the reaction process. CO_2, as a mild oxidant, is not favoured to gasify the formed carbon deposition. Hence, CO_2 is always applied in order to shift the equilibrium without increasing the temperature. It is obvious that all other reactions in this system depend strongly on the conversion of C_3H_8 in reaction (1). Thus, when considering only reaction (1), this is a decomposition reaction which is favoured under a lower partial pressure of C_3H_8. With the equilibrium constants calculated from $\Delta_rG^\theta_m$ at the temperature range of 300–500 °C, the conversions of C_3H_8 under a different partial pressure are calculated as shown in Fig. 6.

$$K^\theta = \exp\left[\frac{-\Delta_r G^\theta_m}{RT}\right]$$

The C_3H_8 conversion is observed to be significantly affected by the partial pressure of C_3H_8 within the temperature range of 300–500 °C, it is for this reason

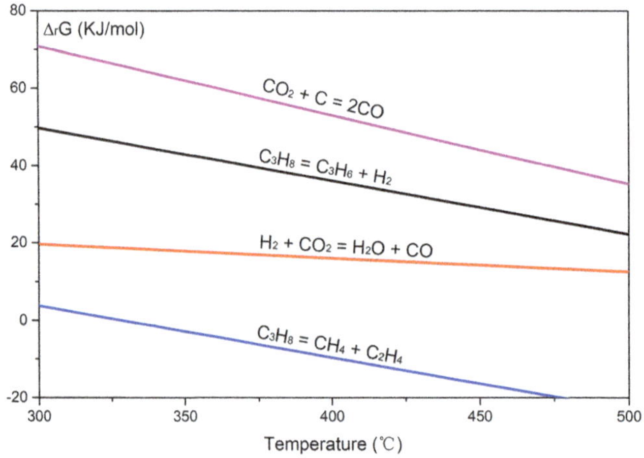

Fig. 5 $\Delta_rG^\theta_m$ of reactions (1)–(3) and (5) in the modelled system, at 300–500 °C.

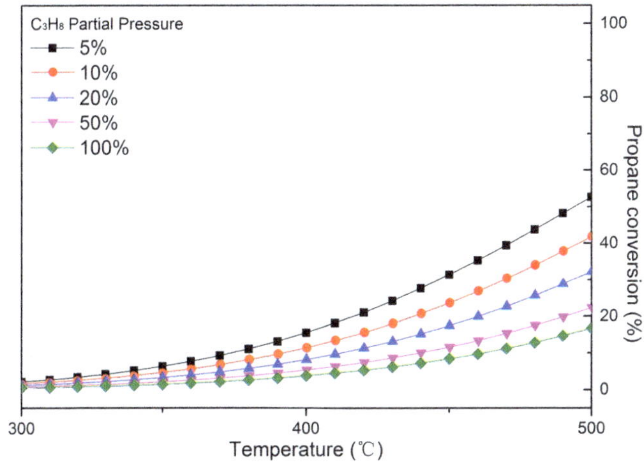

Fig. 6 Calculated C_3H_8 conversions (mol%) at equilibrium state, considering reaction (1) only; various C_3H_8 partial pressure; 300–500 °C.

that the dehydrogenation of C_3H_8 in the presence of CO_2 should be operated with a very low partial pressure of C_3H_8 when designing experiments.

Apart from the improvement of catalysts, higher C_3H_6 selectivity can also be achieved by changing the conditions of reaction. A general way to inhibit the reaction extent is to limit the contact time of reactants on a catalyst surface by increasing the space velocity of gas flow.[21] As a comparison we model this and also increase the partial pressure of reactants operating under these conditions.

4. Experimental

4.1 Catalyst preparation

To prepare the 5 wt% of the Cr_2O_3/ZrO_2 catalyst by a precipitation method, ZrO_2 (Alfa-Aesar, 90 m^2 g^{-1}, 99%) was pre-heated at 600 °C for 6 hour before being ground and sieved to <125 μm in particle size. Subsequently, the $Cr(NO_3)_3 \cdot 9H_2O$ precursor (99%, Sigma-Aldrich) was dissolved in distilled water, and this solution was mixed with the ground ZrO_2 support and stirred at room temperature for 24 hours. To achieve Cr_2O_3-ZrO_2, the as prepared suspension was dried to obtain a slurry or paste, which was finally calcined at 600 °C for 6 hours in a muffle furnace using a heating ramp of 10 °C min^{-1}. The solid sample was ground to obtain fine particles <125 μm.

4.2 Catalyst characterisation

All the samples, including the ZrO_2 support, were characterized with high resolution X-ray Diffraction (XRD) using a PANalytical X'Pert PRO diffractometer with CuKα radiation (45 kV, 40 mA). When being scanned, the samples were flat loaded in the custom-built sample holders and scanned from 20° to 30° 2θ with a step size of 0.0084° and a scanning speed at 0.017778° s^{-1}.

The amount of carbon deposition on the spent catalyst was measured *via* thermal gravimetric analysis (TGA). The instrument employed was a TA

Instrument, SDT Q-600, using flowing air at 100 ml min^{-1} from 50 °C to 1000 °C with a ramp rate of 10 °C min^{-1}. The TGA curves were also derived as D-TGA to show the rate of weight loss of the samples.

4.3 Catalyst testing

The stability test of C_3H_8 dehydrogenation in the presence of CO_2 was operated in an M-R-10A micro-reactor (KUNLUN YONGTAI Company, China) over the Cr-based catalyst prepared as above. All tests were operated under atmospheric pressure and 500 °C for 5 hours, and the C_3H_8/CO_2 mixture, whose mole ratio was kept at $C_3H_8/CO_2 = 1:2$. As shown in Table 3, the conditions of tests differed from each other by two variables, the mole fractions of C_3H_8 (and CO_2) and overall gas hourly space velocity (GHSV). N_2 balance was applied to dilute the gas mixture. Considering that the reactions were operated under atmospheric pressure, the partial pressure of C_3H_8 in "R-4800 × 2" and "R-4800 × 3" was increased, and the GHSV of C_3H_8 was changed to the same level as "R-9600" and "R-14 400" respectively.

The composition of outlet gas was tested by online Gas Chromatography (Perkin-Elmer, Clarus 580 GC), and the conversions of C_3H_8 and CO_2, the selectivity of products, and the carbon balance in the gaseous products can be determined using the following equations:

$$\text{Conversion }(\%) = C_\text{i} = 1 - \frac{X_\text{i}^\text{outlet} \times X_{\text{N}_2}^\text{inlet}}{X_\text{i}^\text{inlet} \times X_{\text{N}_2}^\text{outlet}} \times 100 \quad (\text{i} = \text{C}_3\text{H}_8 \text{ or } \text{CO}_2)$$

$$\text{Flow of gas in product}\left(\text{ml min}^{-1}\right) = F_\text{j}^\text{outlet} = \frac{X_\text{j}^\text{outlet} \times \text{flow of N}_2}{X_{\text{N}_2}^\text{outlet}}$$

$$(\text{j} = \text{H}_2, \text{CO}, \text{CO}_2, \text{CH}_4, \text{C}_2\text{H}_4, \text{C}_2\text{H}_6, \text{C}_3\text{H}_6, \text{C}_3\text{H}_8)$$

$$\text{Selectivity of Hydrocarbon }(\%) = S_{\text{C}_n\text{H}_m} = \frac{n \times F_{\text{C}_n\text{H}_m}^\text{outlet}}{3 \times F_{\text{C}_3\text{H}_8}^\text{inlet} \times C_{\text{C}_3\text{H}_8}} \times 100$$

$$\text{Yield of CO }(\%) = Y_\text{CO} = \frac{F_\text{CO}^\text{outlet}}{2 \times F_{\text{CO}_2}^\text{inlet}} \times 100$$

Table 3 Mole fractions of C_3H_8 ($X_{C_3H_8}$), overall GHSV and C_3H_8 GHSV of the coded reactions

Reaction code	$X_{C_3H_8}$ (mol%)	GHSV (ml h^{-1} g$_{cat}^{-1}$)	GHSV of C_3H_8 (ml h^{-1} g$_{cat}^{-1}$)
R-4800	5	4800	240
R-4800 × 2	10	4800	480
R-4800 × 3	15	4800	720
R-9600	5	9600	480
R-14 400	5	14 400	720

Carbon Balance (%) =

$$B_C = \frac{F_{CH_4}^{outlet} + 2 \times F_{C_2H_4}^{outlet} + 2 \times F_{C_2H_6}^{outlet} + 3 \times F_{C_3H_6}^{outlet} + 3 \times F_{C_3H_8}^{outlet} + F_{CO}^{outlet} + F_{CO_2}^{outlet}}{3 \times F_{C_3H_8}^{inlet} + F_{CO_2}^{inlet}} \times 100$$

$$\text{Oxygen Balance (\%)} = B_O = \frac{2 \times F_{CO_2}^{outlet} + F_{CO}^{outlet}}{2 \times F_{CO_2}^{inlet}} \times 100$$

5. Results and discussion

5.1 Catalytic test results

The C_3H_8 conversion and C_3H_6 selectivity of the reaction under different C_3H_8 partial pressures and overall GHSV are shown in Fig. 7. The results reveal effective C_3H_8 conversions at each specified condition while the C_3H_6 selectivity drops at the beginning of reactions. The results with different mole fractions of C_3H_8 were displayed with solid symbols and the conclusion drawn from the data in these figures is an inhibition of C_3H_8 conversion, but also an improvement in C_3H_6 selectivity. However, with the same increase in C_3H_8 fractional GHSV (by increasing the overall GHSV), the C_3H_8 conversion was inhibited to a similar extent while the C_3H_6 selectivity also decreased. These comparative results indicate that even though the C_3H_8 conversion was similarly inhibited *via* the two routes to increase the C_3H_8 fractional GHSV, the underpinning mechanisms were fundamentally different.

Fig. 8 exhibits the production of CH_4, the main by-product. The initial selectivity of CH_4 was decreased from 6% to 3% as an increase of C_3H_8 fractional GHSV. The CH_4 selectivity can roughly indicate the cracking capability of the catalyst on higher level hydrocarbons, which indirectly exhibits the extent of further cracking effects on C_3H_6 during the reaction.

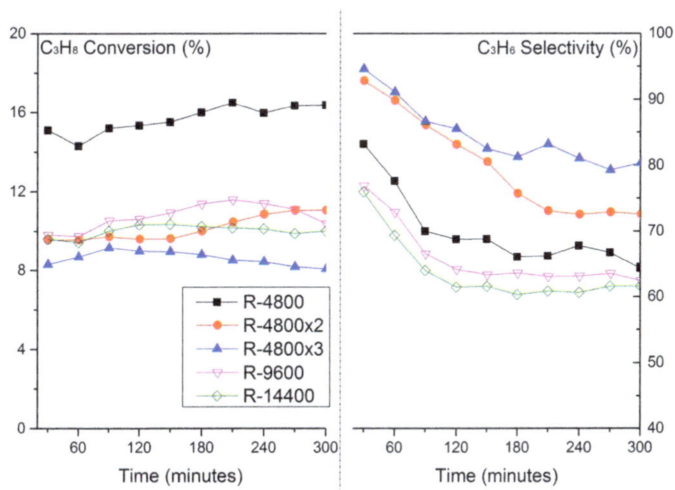

Fig. 7 C_3H_8 conversion and C_3H_6 selectivity over 5 wt%-Cr_2O_3/ZrO_2 upon operation times at various C_3H_8 partial pressures and overall GHSV; 500 °C and C_3H_8/CO_2 = 0.5.

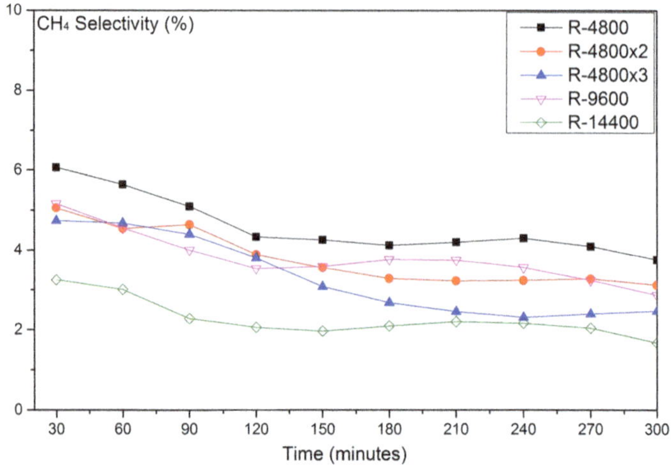

Fig. 8 Dependence of the CH_4 selectivity with time on stream over 5 wt%-Cr_2O_3/ZrO_2 at various C_3H_8 partial pressures and overall GHSV; 500 °C and $C_3H_8/CO_2 = 0.5$.

5.2 Discussion on the changes of C_3H_6 selectivity

The conversion of reactants and the mass balances of carbon and oxygen elements are shown in Table 4. In contrast to the high stability of the C_3H_8 conversion, the CO_2 conversions dropped drastically over the first 5 hours. This may be due to the various and different sites for C_3H_8 and CO_2 to be adsorbed on the catalyst surface respectively; C_3H_8 is usually attached to Cr species while CO_2 can be adsorbed at the interface between the Cr_2O_3 dopant and ZrO_2 support.[22] The high level of C-balance indicates the low generation of solid and liquid C-containing products; this confirms the high stability of C_3H_8 conversions in each test. The O-balance, which was higher than 100% at the end of the test, indicates additional, extraneous oxygen must have entered into the gaseous phase during the reaction, and the oxygen source can be attributed to the reduction of high valance state chromium species (Cr^{6+}/Cr^{5+}) on the catalyst surface.[23] Weckhuysen *et al.*[24] claimed that Cr^{6+} plays as a precursor for the Cr dehydrogenation centres, and the reduction of Cr species is assumed to be one of the primary deactivation mechanisms.

Table 4 Conversion of reactants ($C_{C_3H_8}$, C_{CO_2}) and balance of elements (B_C, B_O, B_H) in gaseous products over 5 wt%-Cr_2O_3/ZrO_2 at various C_3H_8 partial pressures and overall GHSV; 500 °C and $C_3H_8/CO_2 = 0.5$

C_3H_8 partial pressure (atm)	$C_{C_{3H8}}$ (%)		C_{CO_2} (%)		B_C (%)		B_O (%)	
	0 h	5 h	0 h	5 h	0 h	5 h	0 h	5 h
R-4800	15.09	16.37	6.95	1.80	99.27	99.17	96.64	103.80
R-4800 × 2	9.52	11.03	4.62	2.79	99.90	99.30	97.80	103.16
R-4800 × 3	8.28	8.05	3.29	1.78	99.99	99.91	99.04	101.45
R-9600	9.77	10.35	5.15	1.51	98.65	98.87	96.93	102.42
R-14 400	9.57	10.00	4.50	1.19	98.55	98.80	97.30	101.78

Table 5 Yield of CO (Y_{CO}) and selectivity of hydrocarbons (S_i) over 5 wt%-Cr_2O_3/ZrO_2 at various C_3H_8 partial pressures and overall GHSV; 500 °C and $C_3H_8/CO_2 = 0.5$

C_3H_8 partial pressure (atm)	Y_{CO} (%)		S_{CH_4} (%)		$S_{C_{2Hx}}$ (%)		$S_{C_{3H6}}$ (%)	
	0 h	5 h	0 h	5 h	0 h	5 h	0 h	5 h
R-4800	3.59	3.54	6.06	3.76	1.32	1.00	83.23	64.30
R-4800 × 2	2.43	2.42	5.07	3.13	1.52	1.36	92.75	72.47
R-4800 × 3	1.94	1.89	4.74	2.48	1.66	1.88	94.56	80.30
R-9600	2.08	1.80	5.17	2.88	1.50	1.33	76.81	62.31
R-14 400	1.80	1.57	3.26	1.68	1.46	1.35	75.91	61.53

Table 5 is a compilation of the selectivity of gaseous products. The reduced level of C_2H_x production as compared to that of the selectivity of CH_4 was observed. This experimental trend matched the prediction (highlighted in Fig. 4) that it is much more difficult to thermally crack CH_4 at 500 °C than corresponding C_2 molecules. The decrease in CO yield was not as much as the trend of CO_2 conversion; we believe that this is due to the carbon gasification by reducing the chromium species (Cr^{6+}/Cr^{5+}) on the catalyst surface, which matches the observed increasing O-balance.

The XRD patterns of 5 wt%-Cr_2O_3/ZrO_2 before and after the catalytic tests are displayed in Fig. 9, and the pattern of the ZrO_2 support is also shown here as a reference. The peak at $2\theta = 36.18°$ corresponds to the 1-1-0 phase of Cr_2O_3 with rhombohedral crystal symmetry. No peak shift or new peaks were observed from the post-reaction catalyst, which indicated that no phase change occurred during the catalytic test. Importantly, this also matched the high stability of C_3H_8 conversions, as shown in Fig. 7.

Thermal gravimetric analysis (TGA) results and the corresponding derivative weight-to-temperature (D-TGA) plots are displayed in Fig. 10. We attribute the

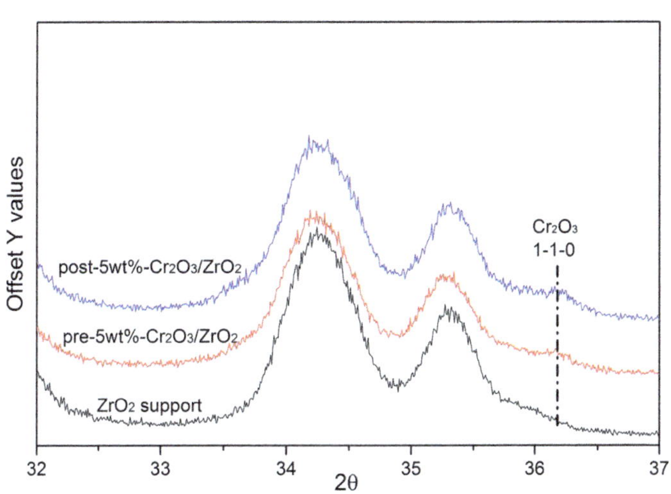

Fig. 9 X-ray diffraction (XRD) results over the catalysts both before (pre-5 wt%-Cr_2O_3/ZrO_2) and after (post-5 wt%-Cr_2O_3/ZrO_2) the catalytic tests, with ZrO_2 support as the reference.

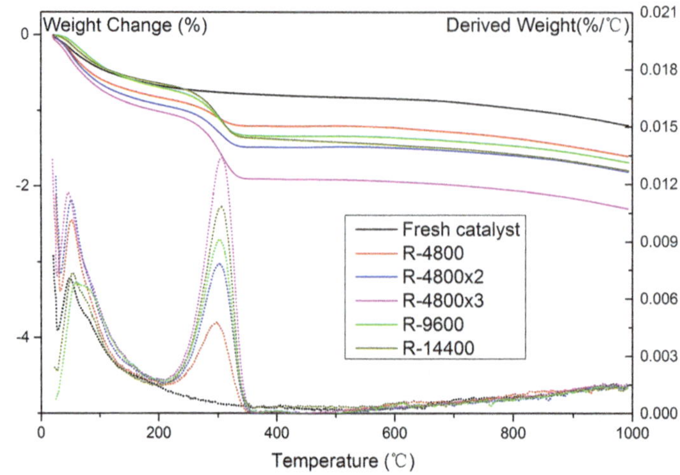

Fig. 10 TGA (solid lines) and D-TGA (dot lines) results over 5 wt%-Cr_2O_3/ZrO_2 at various C_3H_8 partial pressure and overall GHSV; 500 °C and C_3H_8/CO_2 = 0.5.

weight loss, starting from 200 °C, to the combustion of amorphous carbon in air. The integrated area of the derivative weight equals the weight loss, which displayed more coke formed with an increase of the C_3H_8 fractional GHSV.

In general terms, the mechanism of C_3H_8 dehydrogenation in the presence of CO_2 involves the primary activation of C_3H_8 on the catalyst surface, while CO_2 provides a contribution to the reaction equilibrium shift by oxidising H_2 generated from the C_3H_8 dehydrogenation (*via* the reverse water gas shift reaction, $H_2 + CO_2 \rightarrow H_2O + CO$). From the macro perspective, the average contact time for each C_3H_8 molecule is shortened with the higher C_3H_8 fractional GHSV, and this is the cause of a decreased C_3H_8 conversion by optimising both methods to modify the C_3H_8 fractional GHSV. However, the observed changes of C_3H_6 selectivity indicate that only increasing the partial pressure of reactants can effectively inhibit the further cracking of C_3H_6. This arises because the competitive adsorption between C_3H_6 and other substances are strengthened from the micro perspective.

6. Conclusions

With the aim of attempting to (ultimately) correlate the dehydrogenation of C_3H_8 in the presence of CO_2 with the energy source from solar heating technology, thermodynamic calculations were operated for chemical processes operating in the temperature range 300–500 °C. An important outcome is the great potential of C_3H_8 as compared to its lighter and heavier homologues (C_2H_6 and C_4H_{10}, respectively) from both the perspective of CO_2 activation and also coking resistance across this temperature range. The $\Delta_r G_m^\theta$ of each reaction in our thermodynamic modelling of the dehydrogenation of C_3H_8, in the presence of CO_2, also indicated that neither the reverse water gas shift nor the coke gasification with CO_2 are favoured at 300–500 °C.

A series of catalytic tests were carried out over 5 wt%-Cr_2O_3/ZrO_2 at various C_3H_8 partial pressures and overall GHSV. It was shown that the increase in C_3H_8

partial pressure is highly beneficial for the enhancement of C_3H_6 selectivity. Meanwhile, the C_3H_8 conversion was sacrificed due to the competitive adsorption between the produced C_3H_6 and inactivated C_3H_8 molecules, which was the major reason for the inhibition of further cracking of C_3H_6. As a comparison, the increase of the overall GHSV can also decrease the C_3H_8 conversion to a similar extent. However, the further cracking of C_3H_6 cannot be limited as only the contact time of C_3H_8 on the catalyst surface was shortened from this micro perspective.

With respect to the ultimate application of this particular CO_2 utilisation process, we believe that the enhancement of selectivity of the (target) C_3H_6 product, whilst sacrificing some of the reactants' ultimate conversion, is acceptable. The efficiency can be improved by a cycle system involving the reuse of the unconverted feedstock by separating it from the outlet mixture and subsequently cycling it back to the inlet mixture. It is recognised that this kind of cycle system is beneficial to the reactions with low conversion but very high selectivity. The high thermo-stability of the present catalyst makes it attractive to scale-up to a moving bed or even a fluid bed reactor for catalyst regeneration.

We believe that the results presented here are promising in terms of the underpinning catalyst science for establishing the potential industrialisation of the process of CO_2 utilisation through C_3H_8 dehydrogenation. This type of CO_2 utilisation, operating in the relatively low temperature range of 300–500 °C, and easily accessible by solar thermochemical routes, could help shift the focus of CO_2 interest from the disposal of an inconvenient by-product – typified by the process of Carbon Capture and Storage (CCS) – towards the production and use of CO_2 as a commodity chemical in Carbon Capture and Utilisation (CCU). However, one must stress again that in order for any proposed CCU process to be realistic for emission reduction potential, it can only be beneficial if any necessary energy input is from renewable sources. As noted here, our view is that solar thermal chemical processes for CO_2 utilisation, accessible across these temperature ranges, offer considerable potential in this regard.

Acknowledgements

We thank the China Scholarship Council (CSC) for financial support to X. Du and KACST and EPSRC for their continued support.

References

1 C. Song, *Catal. Today*, 2006, **115**, 2–32.
2 F. Schierbaum, *Carbon Dioxide as Chemical Feedstock*, ed. Michele Aresta, 2010.
3 Z. Jiang, T. Xiao, V. L. Kuznetsov and P. P. Edwards, *Philos. Trans. R. Soc., A*, 2010, **368**, 3343–3364.
4 F. T. Zangeneh, S. Sahebdelfar and M. T. Ravanchi, *J. Nat. Gas Chem.*, 2011, **20**, 219–231.
5 S. J. Bennett, D. J. Schroeder and S. T. McCoy, *Energy Procedia*, 2014, **63**, 7976–7992.
6 Presented in part at the ADP technical Expert meetings, Bonn, Germany, 2014.
7 C. Ampelli, S. Perathoner and G. Centi, *Philos. Trans. R. Soc., A*, 2015, **373**, 1–35.

8 P. Phelan, T. Otanicar, R. Taylor and H. Tyagi, *J. Therm. Sci. Eng. Appl.*, 2013, **5**, 021003.

9 G. Maag, G. Zanganeh and A. Steinfeld, *Int. J. Hydrogen Energy*, 2009, **34**, 7676–7685.

10 D. B. Fox, E. H. Lee and M.-H. Rei, *Ind. Eng. Chem. Prod. Res. Dev.*, 1972, **11**, 444–446.

11 P. Michorczyk and J. Ogonowski, *React. Kinet. Catal. Lett.*, 2003, **78**, 41–47.

12 K. Takehira, Y. Ohishi, T. Shishido, T. Kawabata, K. Takaki, Q. Zhang and Y. Wang, *J. Catal.*, 2004, **224**, 404–416.

13 A. L. Lapidus, N. A. Gaidai, Y. A. Agafonov, N. V. Nekrasov, D. V. Trushin, A. V. Makashov, M. A. Botavina, G. Martra and S. Coluccia, *DGMK Tagungsber.*, 2008, **3**, 275–282.

14 M. Chen, J.-L. Wu, Y.-M. Liu, Y. Cao, L. Guo, H.-Y. He and K.-N. Fan, *Appl. Catal., A*, 2011, **407**, 20–28.

15 P. Michorczyk, P. Kustrowski, A. Kolak and M. Zimowska, *Catal. Commun.*, 2013, **35**, 95–100.

16 T. Shishido, K. Shimamura, K. Teramura and T. Tanaka, *Catal. Today*, 2012, **185**, 151–156.

17 L. Liu, H. Liu, C.-y. Li and S.-f. Ji, *Beijing Huagong Daxue Xuebao, Ziran Kexueban*, 2005, **32**, 9–13.

18 R. Shangguan, X. Ge, J. Wang and J. Shen, *Shiyou Yu Tianranqi Huagong*, 2002, **31**, 5–7.

19 K. Mueller, A. Baumgaertner, L. Mokrushina and W. Arlt, *Chem. Eng. Technol.*, 2014, **37**, 1261–1264.

20 F. Solymosi and P. Tolmacsov, *Catal. Lett.*, 2002, **83**, 183–186.

21 H. Liu, Z. Zhang, H. Li and Q. Huang, *J. Nat. Gas Chem.*, 2011, **20**, 311–317.

22 S. Naito, M. Tsuji and T. Miyao, *Catal. Today*, 2002, **77**, 161–165.

23 S. Deng, S. Li, H. Li and Y. Zhang, *Ind. Eng. Chem. Res.*, 2009, **48**, 7561–7566.

24 B. M. Weckhuysen and R. A. Schoonheydt, *Catal. Today*, 1999, **51**, 223–232.

25 P. W. Atkins and J. de Paula, *Atkins' Physical Chemistry*, 7th edn, 2002.

26 J. Kunesh, *The Properties of Gases and Liquids*, ed. B. Poling, J. Prausnitz and J. O'Connell, 5th edn, 2002.

27 W. M. Haynes, *CRC Handbook of Chemistry and Physics on DVD, Version 2012*, 2011.

Faraday Discussions

PAPER

Hydrothermal conversion of carbon dioxide into formate with the aid of zerovalent iron: the potential of a two-step approach

K. Michiels,[ab] B. Peeraer,[a] W. Van Dun,[a] J. Spooren*[a] and V. Meynen[b]

Received 9th June 2015, Accepted 16th June 2015
DOI: 10.1039/c5fd00104h

Our research focuses on the hydrothermal conversion of carbon dioxide into formate with the aid of zerovalent iron. Conventionally, a one-step approach is applied wherein both (i) the production of hydrogen gas, through the oxidation of zerovalent iron in an aqueous medium and (ii) the conversion of carbon dioxide with this hydrogen gas into formate/formic acid, are performed under the same reaction conditions at a temperature of approximately 300 °C. Until now, the yields of formate/formic acid mentioned in the literature are, in the absence of a catalytic substance, low (13.5%). Recently, we developed a hydrothermal hydrogen gas production method based on the oxidation of zerovalent iron and performed under mild conditions (temperature of 160 °C). This synthesis method produces hydrogen gas with a high purity (>99 mol%) and a significant yield (approximately 80 mol%). These experimental results suggested that the optimal hydrothermal reaction conditions for the production of hydrogen gas and the conversion of carbon dioxide, are strongly different in case of applying zerovalent iron as the reducing agent. Therefore, this paper studies the potential of a two-step approach to enhance the carbon conversion yields. The first step is the production of hydrogen gas *via* the developed method at 160 °C. The second step is the conversion of carbon dioxide at higher temperatures (250–350 °C). This study reveals that the solubility of hydrogen gas into the aqueous solution is a key parameter in order to achieve a high amount of carbon conversion. Therefore, a high temperature, the degree of filling and the initial hydrogen gas amount are necessary to successfully perform the carbon dioxide conversion step with high carbon conversion yields. Applying these insights have led to the experimental observation that *via* a two-step approach the conversion of potassium hydrogen carbonate into potassium formate can be successfully performed with higher carbon conversion yields, up to 77.9 wt%, and a selectivity of at least 81% when applying a reaction temperature of 280 °C for 24 hours, a degree of filling with water of 50 vol% and an initial amount of hydrogen gas of 100 mmol.

[a]*Waste Recycling Technologies, Sustainable Materials Management, Flemish Institute for Technological Research, VITO N.V., Boeretang 200, 2400 Mol, Belgium. E-mail: jeroen.spooren@vito.be; Tel: +32 14335633*
[b]*Laboratory of Adsorption and Catalysis, Department of Chemistry, University of Antwerp, UAntwerp, Universiteitsplein 1, 2610 Wilrijk, Belgium*

Introduction

Natural hydrothermal systems occur at different places on our planet.[1] Bevan M. French[2,3] was the first scientist to recognise these natural hydrothermal systems as plausible sites for abiotic synthesis of organic compounds such as methane, CH_4.[4] Abiotic synthesis can be defined as the formation of compounds through purely chemical processes, without the participation of biological organisms.[5] Following the work by French, scientists have been focusing on lithospheric hydrothermal environments as potential sites for abiotic synthesis of organic molecules.[4] This interest arose due to the proposition that the origin of life on Earth occurred in submarine hydrothermal systems where abiotic synthesis may have supplied the prebiotic organic compounds from which life emerged. That is, as CH_4 and other organic compounds in hydrothermal fluids may be able to provide sources of metabolic energy and fixed carbon for biological communities at the sea floor and in the overlying water column.[4–6] Generally, the abiotic formation of hydrocarbons and other organic compounds (carboxylic acids, amino acids, …) in natural hydrothermal systems takes place *via* the conversion of carbon dioxide CO_2, or other inorganic carbon sources such as carbon monoxide CO and hydrogen carbonate HCO_3^-, with the aid of hydrogen H_2.[5] Such conversions occur *via* mechanisms which may include Fischer–Tropsch type reactions.[5,7] These reactions can be defined as surface-catalysed reduction and polymerisation reactions of oxidised single carbon compounds.[5,7] High H_2 concentrations can result from fluid–rock interactions, represented by reaction (1), which is a general reaction.[5]

$$2(FeO)_{rock} + H_2O \rightarrow (Fe_2O_3)_{rock} + H_2 \quad (1)$$

Herein, water reacts with iron(II)-containing minerals $(FeO)_{rock}$, such as olivine, pyroxene and pyrrhotite, whereby Fe(II) oxidises to Fe(III), which precipitates as magnetite Fe_3O_4 or in general Fe(III)-containing minerals $(Fe_2O_3)_{rock}$, and H_2O reduces to H_2.[5]

In 2003, Yamasaki[8] proposed how information obtained from the study of fundamental Earth principles can inspire the development of new technologies in order to establish a sustainable society. One of the proposals described the possibility to reduce CO_2 to organic compounds under hydrothermal conditions in the presence of H_2 that could be produced *via* redox reactions between zerovalent iron or/and iron(II)-containing compounds and water.[8] Since then scientific research papers on the technological possibilities of the conversion of carbon dioxide under hydrothermal conditions are being published on a regular basis.[9]

The earliest reports on laboratory scale hydrothermal experiments date back to 1845.[1] Hydrothermal synthesis has evolved from being used in geological studies to materials chemistry, where it enabled controlled synthesis of materials with practical applications, such as zeolites, transition metal oxides and even nitrides or sulphides.[10] The definition of laboratory scale hydrothermal reactions is, according to Rabenau,[11] '*reactions which take place in a sealed reaction container, such as an autoclave, if water is used as a solvent and the temperature is raised above 100 °C*'. These conditions generate an autogenous pressure (*i.e.* self-developing and not externally applied pressure).[10] This pressure increases with temperature, but also depends on other factors, such as the percentage fill of the vessel.[10,12,13]

Furthermore, the properties of water change by applying hydrothermal conditions. For example, density, dielectric constant, dynamic viscosity, heat capacity and ionic product are significantly different in comparison with standard ambient temperature and pressure conditions.[14] One can state that a completely 'new solvent' is found at high temperature and elevated pressure,[10] which may be beneficial for the involved reaction system, as the following examples prove.

- A low dielectric constant influences the solubility of different products. For example, the solubility of hydrophobic organic compounds increases. The solubility of certain salts, on the other hand, such as Na_2SO_4, decreases.[14]
- A low viscosity has a positive impact on the mobility of dissolved ions and molecules.[10] Therefore, in combination with a high diffusion rate, interphase mass transfer resistances are substantially reduced or even eliminated and efficient heat transfer is guaranteed.[15,16] These factors will make reactions fast, especially in heterogeneous catalytic reaction systems.[16]
- The ionic product of water is relatively high at certain conditions. These high levels of H^+ and OH^- ions may accelerate many acid- or base-catalysed reactions.[14]

Therefore, the properties of water under hydrothermal conditions may play a defining role in the conversion of a very stable molecule such as CO_2. Based on the used H_2 source, hydrothermal CO_2 conversion methods can be divided in two main categories: metal based methods and non-metal based methods. Metal based methods make use of water as a H_2 source, *via* redox reactions between a zerovalent metal M and water, as generally described in reaction (2).

$$xM + yH_2O \rightarrow M_xO_y + yH_2 \tag{2}$$

Largely available and cheap metals with a low redox potential are therefore preferred. Zerovalent iron,[17–27] zinc,[25,28–34] aluminium,[25,30,35–37] manganese[38] and magnesium[25] have been already studied as potential reductants. Instead, non-metal based methods make use of different H_2 sources, as the following examples prove. Sulphur containing compounds such as hydrogen sulphide H_2S, which may be *in situ* released (desulphurisation) *via* the hydrothermal cracking of certain types of polymer waste such as ethylene propylene diene monomer EPDM, can reduce water in H_2.[39] Furthermore, the conversion of alcohols such as isopropanol C_3H_7OH and glycerine $C_3H_8O_3$ into acetone C_3H_6O and lactic acid $C_3H_6O_3$, respectively, can occur together with the hydrogenation of CO_2 and its derivatives (transfer hydrogenation).[40,41] H_2 can also be generated by the *in situ* decomposition of hydrazine N_2H_4.[42] However, since the work presented in this paper is related to metal based methods, non-metal based methods will not be discussed further.

So far, metal based methods are reported as one-step processes: where the production of H_2 and the conversion of CO_2 take place under the same reaction conditions in the same reactor. However, it is unclear if this one-step approach is (always) synergic. Experimental observations indicate that a one-step method may be beneficial. On the one hand, the oxidised metal phases, including possible intermediates, may enhance the conversion of CO_2, *e.g.* by a catalytic activation.[33,34,38] However, often the addition of catalytic substances, such as metallic nickel or metallic copper, is necessary in order to obtain efficient kinetics.[31,32,37] We mention 'catalytic substances' because in most cases significant, non-catalytic amounts of the additives are required to enhance the kinetics: *e.g.* the mole ratio M : Ni or M : Cu is frequently at least 1 : 1. On the other hand, the consumption of produced H_2, by reacting with CO_2, may shift the equilibria to the product side of both the total H_2 production reaction and the conversion reactions of CO_2, depending on the involved kinetics of the reactions that take place (*cf.* Le Chatelier's principle).[33,38] Concretely, in most cases it is unknown whether the production of H_2 and the conversion of CO_2 take place simultaneously, or take place simultaneously for a certain time period, or take place as subsequent reactions.

Metal based methods experimentally show the possibility to convert CO_2 into a range of products. Depending on the applied method and reaction parameters, different products such as formic acid HCOOH, methanol CH_3OH or methane CH_4 have been obtained with high selectivity coinciding in many cases with a high conversion yield. The total CO_2 conversion reactions are shown in reactions (3), (4) and (5). For example, a highly selective (98%) conversion with a high percentage yield (76%) into sodium formate, the sodium salt of HCOOH, was observed under the following conditions: 8 mmol Mn, mole ratio Mn : $NaHCO_3$ = 8 : 1, degree of filling 55 vol%, a reaction temperature of 325 °C, a reaction time of 1 hour and in the absence of a catalytic substance.[38]

$$CO_2 + H_2 \rightarrow HCOOH \tag{3}$$

$$CO_2 + 3H_2 \rightarrow CH_3OH + H_2O \tag{4}$$

$$CO_2 + 4H_2 \rightarrow CH_4 + 2H_2O \tag{5}$$

However, without the addition of a catalytic substance, the conversion yields of formate are strongly different when comparing different metals, as shown in Table 1.[21,25,33,38] Upon applying Fe, the percentage yield of formate is found to be significantly low compared to the other metals. However, no (possible) explanation can be found in the literature. In our opinion, the oxidation degrees of the applied metals and, more specifically, the corresponding amounts of produced H_2 may explain the observed differences. The hydrogen yield, which is sometimes estimated based on the formate yield, the amount of H_2 in the gas phase after reaction, and the maximal producible amount of H_2, seemed to be significantly lower in the case of Fe compared to the corresponding amounts obtained with other metals. As a matter of fact, Fe is the most noble metal of the four mentioned metals. Hence, in the case of Fe, we suggest that the low formation of H_2 led to a low percentage yield of formate.

In the case of Fe, the addition of a potential catalytic substance, such as Ni or Cu, leads to an increase of the percentage yields of formate, as shown in Table 2.[26] However, the extents of increase of the percentage yields of formate are dissimilar. In the case of applying Ni, the highest reported percentage yield of formate is 20.7% while under similar conditions with Cu, the obtained percentage yield of formate is 48%. The highest percentage yield of formate, 76.7%, is in the case of adding Cu. The corresponding conditions in order to achieve this yield, such as the high amount of Fe and the degree of filling, are remarkable compared to the other described experiments. Unfortunately, and to the best of our knowledge,

Table 1 Highest reported percentage yields of formate, with corresponding selectivities and hydrogen yield, as a function of different metals in the case of applying $NaHCO_3$ as the CO_2 source and in the absence of an added catalytic substance[21,25,33,38]

Metal (74 μm)	Metal amount (mmol)	Mole ratio metal : $NaHCO_3$	Temperature (°C)	Degree of filling (vol%)	Reaction time (hours)	Percentage yield of formate (%)	Selectivity to formate (%)	(Indication of) H_2 yield (%)
Fe	6	6 : 1	325	35	2	13.5	98	25[a]
Zn	10	10 : 1	325	35	1.5	78.0	±100	±100
Al	6	6 : 1	300	35	2	62.8	±100	99.4
Mn	8	8 : 1	325	55	1	76.0	98	77.5[b]

[a] After a hydrothermal treatment of 1 mmol $NaHCO_3$ in the presence of 6 mmol Fe at 300 °C for 2 hours (degree of filling, 35 vol%), a percentage yield of formate of 10.5% was observed. The amount of H_2, which was present in the gas phase after reaction, was recovered and quantified as 48 mL (20 °C, 1 atm). Based on the formate yield (0.105 mmol H_2), the amount of H_2 in the gas phase after reaction (2 mmol) and the maximal producible amount of H_2 (8 mmol), we estimate the hydrogen yield on 25% (dissolved amount of H_2 at 20 °C not included). [b] After a hydrothermal treatment of 1 mmol $NaHCO_3$ in the presence of 8 mmol Mn at 300 °C for 2 hours (degree of filling of 35 vol%), a percentage yield of formate of 43% was observed. The amount of H_2 which was present in the gas phase after reaction was recovered and quantified as 140 mL (20 °C, 1 atm). Based on the formate yield (0.43 mmol H_2), the amount of H_2 in the gas phase after reaction (5.8 mmol) and the maximal producible amount of H_2 (8 mmol), we estimate the hydrogen yield on 77.5% (dissolved amount of H_2 at 20 °C not included).

Table 2 Highest reported percentage yields of formate in the case of applying Fe, $NaHCO_3$ as the CO_2 source, and the addition of a catalytic substance[26]

Fe amount (mmol)	Mole ratio Fe : $NaHCO_3$	Mole ratio Fe : additive	Temperature (°C)	Degree of filling (vol%)	Reaction time (hours)	Percentage yield of formate (%)
12	6 : 1	1 : 1 (Ni)	300	35	2	20.7
12	6 : 1	1 : 1 (Cu)	300	35	2	48.0
13.2	6 : 1	1 : 1 (Cu)	300	35	2	63.6
108	6 : 1	1 : 1 (Cu)	300	55	2	76.7

these results obtained *via* the addition of potential catalytic substances, are only mentioned in a review paper[26] since the herein referred article was never published. Hence, no discussion is available on the influence of the addition of Ni or Cu, or the influence of reaction parameters such as the applied degree of filling. In our opinion, Ni and especially Cu may enhance the formation of H_2, and may even change the mechanism of the conversion of CO_2, which could lead to a higher formate yield.

Recently, we have shown that the hydrothermal production of H_2, from (zerovalent) Fe and water, is optimal under milder conditions of temperature in comparison to the conditions applied in a conventional one-step CO_2 conversion method (generally approximately 300 °C).[43] In the presence of a specific amount of carbonate ions CO_3^{2-} (0.0375 M), a highly pure (purity > 99 mol%) and significant amount (yield of approximately 80 mol%) of H_2 was experimentally obtained after a reaction time of 16 hours, and at a reaction temperature of 160 °C. It was proven that in the studied reaction system the presence of a specific amount of carbonate ions (0.0375 M) was crucial since the formation of H_2 is carbonate assisted. More specifically, the oxidation of Fe into Fe_3O_4 is suggested to take place *via* the formation of iron(II) carbonate $FeCO_3$ as an intermediate. The catalytic effect of carbonate ions decreases significantly, and eventually fades out, as higher temperatures or/and other amounts of carbonate ions are applied. At lower temperatures (>100 °C and <160 °C), the kinetics of the occurring reactions is slow. Hence, our experimental research suggests that the optimal reaction conditions for the two occurring reactions, *i.e.* production of H_2 and conversion of CO_2, are strongly different in the case of the hydrothermal conversion of CO_2 in the presence of Fe. This hypothesis of an imbalance of reaction conditions could explain why an efficient one-step hydrothermal conversion of CO_2 in the case of Fe, and in the absence of a catalyst, is difficult at elevated temperatures such as 300 °C. Hence, the goal of this paper is to study the potential of a two-step approach for the hydrothermal conversion of CO_2 into formate with Fe: the first step is the production of H_2 *via* the developed method at 160 °C, the second step is the conversion of CO_2 at a higher temperature (250–350 °C). More specifically, the influence of different reaction parameters on the second step will be discussed, with a specific focus on the degree of filling and the (initial) amount of H_2. In conclusion, with the herein described work, we aim at elucidating the chemical feasibility of a two-step approach for the hydrothermal conversion of CO_2 into formate with Fe.

Materials and methods

Materials

Metallic iron powder (min. 99.5 wt%) with an average grain size of 10 μm and magnetite powder (min. 97 wt%) with a grain size smaller than 44 μm were both obtained from Merck Millipore. Potassium hydroxide, KOH (min. 85 wt%), potassium hydrogen carbonate, $KHCO_3$ (min. 99.5 wt%), and anhydrous potassium carbonate, K_2CO_3 (min. 99.9 wt%), were purchased from Alfa Aesar, UCB and JT Baker respectively. All hydrothermal experiments made use of degassed Milli-Q water. The applied high purity gases, namely carbon dioxide (99.5 vol%), nitrogen (99.998 vol%) and hydrogen (99.999 vol%), were supplied by Air Products.

Experimental

The hydrothermal experiments were performed in closed 100 mL stainless steel (1.4435) stirred premex reactor ag autoclaves. For safety reasons, the maximal reaction temperature and the maximal degree of filling of the autoclaves were 280 °C and 50 vol%, respectively. A digital control unit and a logging unit allowed for the temperature, pressure and stirring speed to be adjusted and logged online. The heating time, defined as the time to increase the temperature to the set point, is dependent on the desired reaction temperature. Inherent to the settings of the applied heating system (PID controlled), a short-time temperature overshoot of maximum 5 °C was observed. However, in all cases the set point was reached in less than 30 minutes. Based on the applied H_2 source, two variants of hydrothermal CO_2 conversion reactions were performed and studied in this work of which the procedures are described hereafter.

CO_2 conversion reactions which apply hydrogen gas produced *via* a hydrothermal method. The optimisation of the production of H_2 by a mild hydrothermal method is described elsewhere.[43] In brief, this synthesis procedure of 15 bar absolute H_2 (15 bar, 20 °C, 70 mL volume intake: approximately 42 mmol H_2) is described here. Firstly, 40 mmol Fe was loaded into an autoclave. Secondly, 40 mL of a 1 M KOH solution was added. Subsequently, the autoclave was flushed with N_2 to create an oxygen free atmosphere. Thereafter, the autoclave was filled with 6 bar absolute of CO_2 (6 bar, 20 °C, 60 mL volume intake: approximately 15 mmol CO_2). Finally, the temperature was raised to 160 °C and stirring was applied at 250 rpm. The reaction was carried out for 16 hours, which includes the heating time and excludes the time to cool down to 35 °C. During cooling down, stirring was still performed at 250 rpm to prevent aggregation of the solid phase.

Subsequently, CO_2 conversion reactions were carried out in the same autoclave. Concretely, when the autoclave was cooled down after the production of H_2, the temperature was increased to 280 °C. This temperature was held constant for 24 hours. Stirring was still continuously performed at 250 rpm. Note that some of the experiments are performed under different conditions. In such cases, the differences in conditions are explicitly mentioned.

CO_2 conversion reactions which apply commercial hydrogen gas. When applying commercially produced and pressurised H_2, the following general method was applied. Firstly, 9 mmol $KHCO_3$ was loaded in an autoclave, serving as the CO_2 source. Secondly, a specific amount of degassed Milli-Q water (30, 40 or

50 mL) was added. Subsequently, after stirring manually to dissolve $KHCO_3$ in a homogeneous way and measuring pH, the autoclave was flushed with N_2 to create an oxygen free atmosphere. Finally, before increasing the temperature to 280 °C, and stirring at 250 rpm, the autoclave was filled at approximately 25 °C with a specific amount of H_2 (42, 84 or 100 mmol) with the aid of a pressure regulator. In general, each reaction was performed for 24 hours, which includes the heating time and excludes the time to cool down to room temperature (<35 °C). Note that some of the experiments were performed under different conditions. In such cases, the differences in conditions are explicitly mentioned.

Analysis. At the end of each hydrothermal CO_2 conversion reaction, the autoclave was cooled down to room temperature (<35 °C) with continuous stirring at 250 rpm. At room temperature, the gas phase constituents were identified and quantified by coupling the autoclave to a Bruker 450-GC system. The TCD channel (helium as reference) was calibrated for H_2, CO_2, N_2, CH_4 and CO with the aid of calibration gas mixtures (defined in mol%) delivered by Air Products. The FID channel was used to check if other hydrocarbons than CH_4 were present in the gas phase.

If applicable, after opening the autoclave, the solid phase and liquid phase were rapidly separated over a Whatman mixed cellulose ester membrane filter ME25 (0.45 μm pore size). The liquid phase was collected in a closed polyethylene bottle and stored in a refrigerator at approximately 2 °C.

Different measurements were performed on the liquid phase. Firstly, the pH was measured using a METTLER TOLEDO SevenMulti system, which was calibrated on a daily basis. Subsequently, the amount of non-purgeable organic carbon NPOC was measured to have an estimation of the converted amount of carbon (dioxide). In our opinion, NPOC measurements provide valid estimations since the formation of purgeable organic carbon compounds, such as benzene and derivatives and halogenated hydrocarbons, is expected to be low to non-existent. Measurements were performed using a Analytik Jena AG multi N/C 3100 system. The applied procedure is as follows. Firstly, acidified diluted liquid samples were purged for 10 minutes with oxygen. In this way, inorganic carbon, for example derived from hydrogen carbonate, is removed through the formation of CO_2. Next, the liquid sample is catalytically combusted at 850 °C. Finally, the corresponding generated CO_2 is measured *via* infrared detection. Based on a calibration curve, the amount of carbon can be determined. This procedure was successfully validated for a prepared formic acid solution of 1000 mg C per L: a recovery percentage of approximately 100 wt% was obtained. Throughout the paper, the amount of carbon conversion is mentioned. This amount is defined as the weight percentage of the absolute amount of NPOC (mg) to the absolute initial amount of carbon before reaction (mg). Furthermore, the selectivity of the conversion of CO_2 was checked by analysing the liquid samples with proton NMR spectroscopy (Bruker Avance II 400 MHz, measuring method with water suppression). The liquid samples were diluted with D_2O, TMS was used as reference. In addition, in case of high carbon conversion amounts, HPLC measurements were performed on the pure, non-acidified liquid samples in order to quantify the amount of potassium formate/formic acid. These measurements were performed with an Agilent 1200 series system equipped with an Agilent 1260 Refraction Index Detector, which operated at 55 °C. An ion-exchange ligand-exchange column, namely Agilent Hi-Plex H (7.7 × 300 mm, 8 μm particle size), was applied at 60 °C. The mobile phase, a 10 mM sulphuric acid solution, was used in an isocratic way (1 mL min^{-1}). The quantification was performed based on a calibration curve. Based on the concentration of potassium formate/formic acid (mg L^{-1}) and the amount of NPOC (mg L^{-1}), the selectivity to potassium formate/formic acid was calculated, and expressed as a percentage (%).

Results and discussion

The possibilities of a two-step approach: exploratory experiments

As described above, H_2 is optimally produced *via* the developed iron-based hydrothermal method at 160 °C in the presence of a small amount of CO_2, which completely dissolves in a 1 M KOH solution and is mainly present as CO_3^{2-} ions (approximately 15 mmol). In order to verify if any conversion of CO_2 has already taken place during the developed production method of H_2, the gas and liquid phases obtained at the end of the H_2 production method were analysed. GC analyses showed that the gas phase did not contain any detectable amounts of hydrocarbons. However, proton NMR spectroscopic analyses on the corresponding liquid phase proved the presence of a complex mixture of organic compounds. Especially between 0.5 and 2.5 ppm [δ_H (400 MHz; D_2O; Me_4Si)] different peaks were observed, which may indicate the presence of aliphatic compounds. Furthermore, the presence of formate was confirmed [δ_H (400 MHz; D_2O; Me_4Si), 8.47, 1H, s]. Therefore, reactions between CO_2 and the produced H_2 appear to have taken place under the applied reaction conditions of the developed H_2 production method although unselective. Nevertheless, the amount of the formed organic compounds was small since NPOC analyses suggested a carbon conversion amount of approximately 0.7 wt%.

The mild reaction conditions of this process compared to one-step metal based CO_2 conversion methods reported in the literature (Table 1), could be the reason for this low carbon conversion amount. Therefore, an experiment was performed in which, after performing the hydrothermal H_2 production at 160 °C for 16 hours (step 1), the temperature of the autoclave was increased to 280 °C for 24 hours (step 2). GC analyses demonstrated that the gas phase did contain traces of CO and CH_4 (much below 1 mol%). Furthermore, the presence of a complex mixture of organic compounds was confirmed *via* proton NMR spectroscopic analyses on the corresponding liquid phase. In the proton NMR spectrum, the proton peak of formate [δ_H (400 MHz; D_2O; Me_4Si), 8.47, 1H, s] had the highest intensity by far and thus formate seemed to be the main product. NPOC analyses suggested a carbon conversion amount of approximately 9.7 wt%. Hence, hydrothermal reactions between H_2 and CO_2 appear to take place better at 280 °C, as expected. However, the observed carbon conversion was still low. Therefore, in order to enhance the conversion of CO_2 *via* the proposed two-step method, significant modifications are necessary. Concretely, a systematic study of all reaction parameters of the second step should be performed in order to improve the hydrothermal conversion of CO_2 with H_2.

In the study of the influence of different reaction parameters on this second step, we chose to switch from mainly CO_3^{2-} ions, *in situ* produced through the dissolution of 15 mmol CO_2 gas in 40 mL of a 1 M KOH solution, to mainly HCO_3^- ions as CO_2 source, in order to be able to compare to the experimental results obtained with similar one-step hydrothermal methods described in literature. A hydrogen carbonate salt, such as $NaHCO_3$, is often dissolved in water and used as a CO_2 source (natural pH approximately 8.6) in the literature.[21] In the above described experiments, on the other hand, the CO_2 source is mainly CO_3^{2-} ions (pH approximately 11), obtained by dissolving CO_2 in a 1 M KOH solution. In order to obtain mainly HCO_3^- ions, the addition of extra CO_2 is necessary after the first step of hydrothermal H_2 production. However, performing an exactly quantified addition of CO_2 without losing the produced amount of hydrogen gas is complicated, as we experimentally encountered. Therefore, we conducted our study on the influence of different reaction parameters on the hydrothermal CO_2 conversion reaction in a separate autoclave with commercial $KHCO_3$ and commercial H_2. We chose to use 9 mmol of $KHCO_3$ for each reaction throughout this study. This amount is a midpoint setting, based on the two extremities mentioned in literature considering the conversion of CO_2 *via* one-step iron based hydrothermal methods: the minimum amount of applied $NaHCO_3$ is 1 mmol,[21] the corresponding maximum amount is 18 mmol.[26]

The possibilities of a two-step approach: parameter study of the CO_2 conversion reaction

In literature, parameter studies describe the impact of different reaction parameters, such as the amount of Fe, mole ratio Fe : $NaHCO_3$, reaction temperature, reaction time and alkalinity, on the hydrothermal conversion of CO_2 in the presence of Fe as the reductant, and in the absence of a catalyst.[21] However, no study investigated the effect or the optimisation of (i) the degree of filling and (ii) the (initial) amount of H_2. Hence, we designed a series of experiments in order to understand the role of these parameters (Table 3). The minimal amounts of both parameters, namely 30 vol% and 42 mmol H_2, were selected based on the obtained parameters *via* the developed H_2 production method.

The maximal amounts of both parameters, namely 50 vol% and 100 mmol H_2, were chosen based on the safety sheets of the autoclaves. In particular, both the filling degree and the introduction of H_2 will augment the pressure during reaction, which is limited to 200 bar for the used autoclaves.

Table 3 Design of a series of experiments, with a constant amount of 9 mmol $KHCO_3$, to investigate the role of the degree of filling with water and the presence of a significant amount of H_2

Experiment	Degree of filling with water (vol%)	Initial amount of H_2 (mmol) (corresponding absolute pressure in bar at 25 °C)	Mole ratio H_2 : $KHCO_3$
1	30	42 (15)	4.67
2	30	84 (30)	9.33
3	30	100 (36)	11.11
4	40	42 (17)	4.67
5	40	84 (35)	9.33
6	40	100 (42)	11.11
7	50	42 (21)	4.67
8	50	84 (42)	9.33
9	50	100 (51)	11.11

Table 4 Amount of carbon conversion and selectivity to formate/formic acid in case of conducting Experiment 5 for 2, 4 and 24 hours at 280 °C

Time in hours	Amount of carbon conversion (wt%)	Selectivity to formate/ formic acid (%)
2	14.7	97
4	22.6	±100
24	46.2	±100

Furthermore, to perform this series of experiments, other reaction parameters such as reaction temperature, reaction time and the presence/absence of magnetite had to be defined. Based on literature, the temperature has to be approximately 300 °C up to 325 °C.[21] However, the experiments were conducted at 280 °C, which is the maximal operating temperature of the applied autoclaves in this study. Since this reaction temperature is lower, a longer reaction time was expected than those reported in literature from 1 up to 2 hours.[21] Hence, we chose to empirically determine the required time to convert a sufficient amount of CO_2. Therefore, we chose to conduct Experiment 5, the midpoint of the experimental design with a filling degree of 40 vol% and an initial hydrogen content of 84 mmol, for 2, 4 and 24 hours at a reaction temperature of 280 °C. The obtained amounts of carbon conversion are shown in Table 4.

GC analyses showed that the gas phases did not contain any detectable amounts of hydrocarbons after a reaction time of 2 and 4 hours. After a reaction time of 24 hours, traces of CO and CH_4 (much below 1 mol%) were detected in the gas phase. However, in all cases, the selectivity to potassium formate/formic acid was approximately 100%. The highest amount of carbon conversion, namely 46.2 wt%, was found after 24 hours, showing that the kinetics of the occurring reaction(s) is slow.

The evolution of the reactor pressure as a function of time of the corresponding experiment is shown in Fig. 1. When the set reaction temperature of 280 °C is reached and kept constant, the reactor pressure decreases as a function

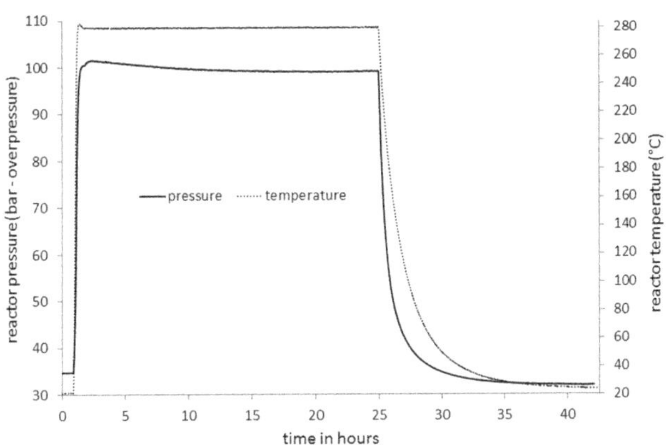

Fig. 1 Evolution of the reactor pressure and temperature as a function of time when conducting Experiment 5 for 24 hours at 280 °C.

of time. However, the rate of decrease is variable within this time range of a constant temperature: in the beginning the rate is significantly higher compared with at the end. We suggest that this decrease of reactor pressure is coupled to the consumption of dissolved H_2 due to the conversion of CO_2. Because of the consumption of dissolved H_2, the concentration of H_2 in the liquid phase decreases. On the one hand, this will drive more H_2 in the gas phase to dissolve according to Le Chatelier's principle. On the other hand, the decrease in H_2 present in the gas phase will weaken the H_2 solubility due to the loss in pressure and eventually lead to a decrease in concentration of H_2 in the liquid phase. This might mean that, after a certain time, the conversion of CO_2 becomes diffusion/reagent limited. Considering Experiment 5, a reaction time of 15 up to 16 hours could be the turning point of these changes in kinetics because the reactor pressure did not significantly decrease any more from that moment (end point of reaction). Based on the graph of the evolution of the reactor pressure as a function of time (Fig. 1), and the amounts of carbon conversion obtained after different reaction times (Table 4), in combination with the possibility that a different degree of filling with water or/and a different initial amount of H_2 might have an impact on kinetics, we defined 24 hours as the reaction time for the series of experiments.

The last parameter which had to be defined before being able to perform the series of experiments, concerns the influence of a significant amount of Fe_3O_4 being present in the reaction solution. Since Fe is oxidised into Fe_3O_4 during the hydrogen production step, Fe_3O_4 will be available in a significant amount, as a solid present in the solution after this step: approximately 11 mmol Fe_3O_4 is obtained *via* the developed hydrogen production process. Hence, we chose to investigate whether the presence of 9 mmol Fe_3O_4 (mole ratio $KHCO_3 : Fe_3O_4 = 1 : 1$)

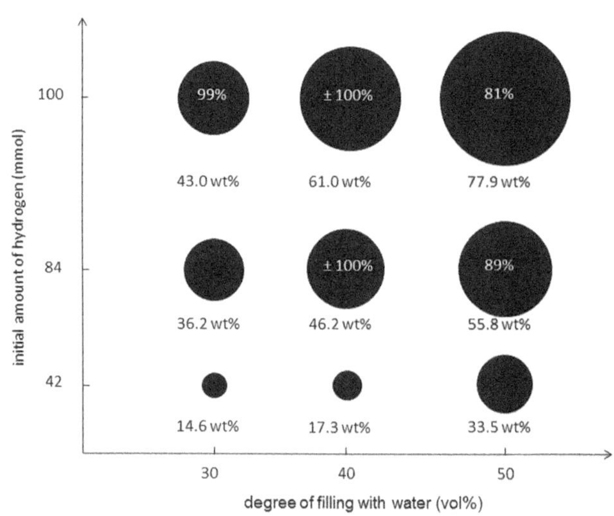

Fig. 2 Amounts of carbon conversion as a function of the applied filling degree and initial amount of H_2. When obtaining an amount of carbon conversion comparable to or higher than the amount of carbon conversion obtained *via* Experiment 5, the midpoint (46.2 wt%), the corresponding selectivity to potassium formate/formic acid was determined. These selectivities can be found in the spheres.

would influence the conversion of CO_2 when performing Experiment 5 for 24 hours at 280 °C.

We observed an identical evolution of the reactor pressure as a function of time and an identical amount of carbon conversion (46.2 wt%) in comparison with a similar experiment in which Fe_3O_4 is not present (Fig. 1 and Table 4). The selectivity to potassium formate/formic acid was approximately 100%, and in the gas phase, no detectable amounts of hydrocarbons were found. Hence, in this case, we did not observe any indication that Fe_3O_4, available in a significant amount, influences the kinetics nor the occurring reactions of the carbon dioxide conversion. Hence, in order to make handling easier and to prevent possible effects of Fe_3O_4 under other conditions, we chose to perform the series of experiments in absence of Fe_3O_4.

In summary, the selected constant reaction parameters were defined as follows: a reaction temperature of 280 °C, a reaction time of 24 hours and the absence of magnetite. Hence, the series of experiments to understand the role of the degree of filling with water and the presence of a significant initial amount of H_2 on the carbon dioxide conversion step (Table 3) could be performed. In the different gas phases, none to very low (much below 1 mol%) amounts of reduced CO_2 species (CO and CH_4) were detected. The amounts of carbon conversion obtained in the different experiments are graphically shown in Fig. 2. When an amount of carbon conversion was obtained comparable to or higher than the amount of carbon conversion obtained *via* Experiment 5, the midpoint of the design (46.2 wt%), the corresponding selectivity to potassium formate/formic acid was determined. These selectivities can also be found in Fig. 2. The observed results prove the importance of the filling degree and the initial amount of H_2: increasing the degree of filling with water and/or increasing the initial amount of H_2 results in a significant increase of the amount of carbon conversion within the selected reaction conditions.

Table 5 Data obtained from the graphs of the evolution of the reactor pressure as a function of time for the different experiments

Experiment	Operating pressure range in bar (overpressure)	(Estimated) amount of converted H_2[a] (mmol)	Endpoint of reaction[b] (hours)
1	65–70	5.2	7–8
2	90–95	7.0	14–15
3	100–105	7.7	>24
4	65–70	4.4	14–15
5	100–105	7.6	15–16
6	110–115	9.6	>24
7	75–80	6.5	22–23
8	110–115	7.9	22–23
9	125–130	10.8	23–24

[a] Value is equal to the difference of the present H_2 pressures before and after reaction at 25 °C. Since the applied pressure values are not highly accurate, these calculated values have to be considered as estimations rather than absolute values. [b] Estimated reaction time, represented as a range in hours, at which the reactor pressure did not significantly decrease any more.

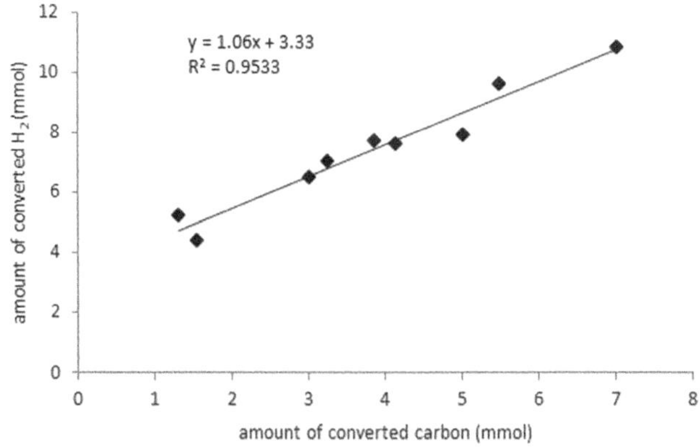

Fig. 3 Estimated, non-absolute amounts of converted H_2 plotted *versus* the amounts of converted carbon. A relative strong correlation ($R^2 = 0.9533$) can be observed.

Hence, the highest amount of carbon conversion, namely 77.9 wt%, was obtained with a degree of filling with water of 50 vol% and an initial amount of H_2 of 100 mmol. We believe that these observations can be explained by the solubility of H_2 in water at hydrothermal conditions: increasing the degree of filling with water and/or increasing the initial amount of H_2 leads to a higher absolute amount of dissolved H_2 to execute the conversion of CO_2, here mainly present as dissolved HCO_3^- ions. This postulation is enhanced by the data obtained from the graphs of the evolution of the reactor pressure as a function of time for the different experiments (Table 5). Firstly, the (estimated, non-absolute) amounts of converted H_2 are directly proportional with the amounts of carbon conversion, as shown in Fig. 3. Secondly, the estimated end point of the reaction seems to shift to longer times as the (estimated, non-absolute) amount of converted H_2 increases up to values greater than 7.5 mmol, which is the case when the reaction

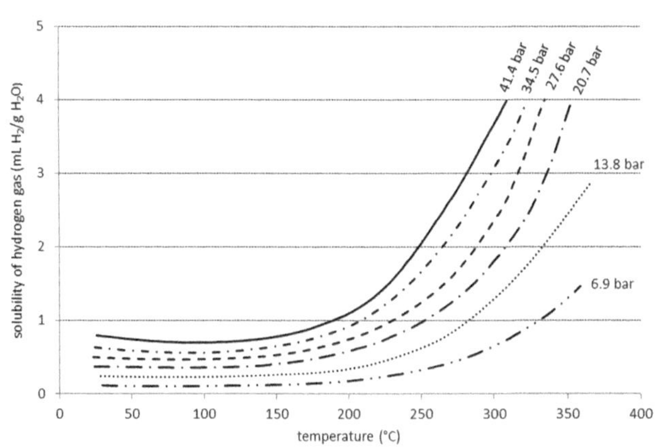

Fig. 4 Relationship between the solubility of hydrogen gas and the applied temperature and the partial pressure of hydrogen gas. Adapted with permission from H. A. Pray, C. E. Schweickert and B. H. Minnich, *Ind. Eng. Chem.*, 1952, **44**, 1146–1151. Copyright 1952 American Chemical Society.

conditions (high degree of fill and/or H_2 pressure) allow for more H_2 to dissolve. These two observations further suggest that the conversion of CO_2 might become diffusion/reagent limited after a certain reaction time due to the consumption of H_2 and the consequent lower solubility of H_2 in solution.

However, it should be mentioned that the selectivity to potassium formate/formic acid seemed to decrease as the filling degree increased to 50 vol%. As mentioned above, the corresponding gas phases contained none to very low amounts of reduced CO_2 species such as CO and CH_4. Hence, we suggest, in the case of higher degrees of filling, the occurrence of side reactions of CO_2 conversion or the further reduction of formate in the liquid phase, leading to the formation of other carbon containing products. In the case of the highest amount of carbon conversion (77.9 wt%), which was obtained with a degree of filling of 50 vol%, the selectivity to potassium formate/formic acid was 81%.

The possibilities of a two-step approach: evaluation and discussion

The experimental results above suggest that the solubility of H_2 in water is a key parameter. Since, as expected, the reduction of CO_2, mainly present as dissolved hydrogen carbonate ions, with H_2 seems to be taking place in the water phase and not in the gas phase. The absolute amount of H_2 that dissolves into an aqueous solution is dependent on the operating temperature, the initial hydrogen pressure and the operating pressure (of the system), as well as the volume of the aqueous solution which is correlated with the degree of filling. The relationship between the solubility of H_2 and the applied temperature and partial pressure of H_2 is shown in Fig. 4.[44] The solubility of H_2 in an aqueous solution decreases as a function of temperature up to a certain temperature (mostly between 100–150 °C).[44] Above this temperature, the solubility of H_2 increases rapidly as a function of temperature.[44] Besides, pressure has an influence on the solubility of a gas in an aqueous solution: a pressure increase leads to a higher solubility (Henry's Law). As described above, the autogenous pressure depends on factors such as temperature and the percentage fill of the vessel with water.[12,13] However, a percentage fill lower or equal to 50 vol% has an especially large influence on the reactor pressure above the supercritical point of water.[11] At 280 °C and a degree of filling lower than 80 vol%, a subcritical regime is present and the autogenous pressure is in accordance with the vapour–pressure curve, namely approximately 65 bar.[11]

Table 6 Overview of the observed and suggested influences of reaction temperature, degree of filling with water and initial H_2 pressure

	Temperature increase	Increase of the degree of filling with water[a]	Initial H_2 pressure increase
Absolute amount of dissolved H_2	↑	↑	↑
Total pressure	↑	↑	↑
Chemical kinetics	↑	↑	↑

[a] In the case of a constant initial hydrogen amount and a reaction temperature of 280 °C.

Therefore, the differences in operating pressure are due to a variation of the initial amount of H_2. However, the degree of filling with water influences the (absolute) dissolved amount of H_2. In the case of a constant initial hydrogen pressure, a higher degree of filling with water leads to a higher absolute dissolved amount of H_2 (only a volume effect occurs, concentration of dissolved H_2 is in this case independent of the degree of filling). In the case of a constant initial hydrogen amount, a higher degree of filling with water requires a higher initial hydrogen pressure, which leads to a higher dissolved amount of H_2 due to (i) an increase of the absolute dissolving of H_2 (volume effect) and (ii) an increase of the concentration of dissolved H_2 due to the higher initial hydrogen pressure (concentration of H_2 is in this case dependent on the degree of filling). We conducted our experiments with a constant initial hydrogen amount.

In the case of the performed experiments, the highest amount of carbon conversion, namely 77.9 wt%, was observed at the conditions at which the solubility of H_2 is the highest, namely a degree of filling with water of 50 vol% and an initial amount of H_2 of 100 mmol at a temperature of 280 °C. We expect an even higher amount of carbon conversion can be obtained by further optimising these three parameters. Unfortunately, due to restrictions of the operating conditions of our autoclaves, we are not able to experimentally prove this expectation. We already applied a relatively high initial pressure of H_2 (100 mmol H_2: 51 bar absolute at 25 °C, taking in a volume of 50 mL) and degree of filling with water (50 vol%) in our best case. Therefore, we expect that the most room for improvement can be achieved by varying the temperature. Increasing the temperature will have a significant effect on the solubility of hydrogen.[44] A temperature of 300 °C already leads to a significant difference in the solubility of hydrogen compared with a temperature of 280 °C.[44] Nevertheless, the selectivity to potassium formate might decrease due to a possible lower stability of this product at higher temperatures or the occurrence of side reactions of carbon dioxide conversion. An overview of the observed and suggested influences of the three parameters on the amount of carbon conversion can be found below in Table 6.

Conclusions

This study investigated the potential benefits or drawbacks of applying a one-step approach for the hydrothermal conversion of carbon dioxide to formate/formic acid with the aid of *in situ* produced hydrogen gas, obtained through the oxidation of zerovalent iron in an aqueous medium. However, we have shown that the optimal reaction conditions for the two occurring reactions, *i.e.* production of hydrogen gas and conversion of carbon dioxide, are strongly different due to thermodynamic and kinetic limitations, respectively. The imbalance of reaction conditions could explain why an efficient one-step hydrothermal conversion of carbon dioxide in the case of zerovalent iron, and in the absence of a catalyst, is difficult at elevated temperatures around 300 °C. Hence, the experimental study further focused on a two-step approach for the hydrothermal system: the first step is the production of hydrogen gas *via* a recently developed hydrothermal method which operates at 160 °C, and the second step is the conversion of carbon dioxide, present as dissolved potassium hydrogen carbonate, at higher temperatures (250–350 °C). More specifically, the influence of the process parameters, on this second step were investigated and discussed, with a specific focus on the degree of filling and the (initial) amount of hydrogen.

This study revealed that a high solubility of hydrogen gas into the aqueous solution during reaction is a key parameter in order to achieve a good carbon conversion. Thus, the presence of a higher amount of hydrogen gas in solution, where the conversion reaction between hydrogen gas and potassium hydrogen carbonate into potassium formate takes place, shifts the thermodynamic equilibrium to the product side. As a result, a high temperature, degree of filling with water and initial hydrogen pressure are necessary to successfully perform the second step of the approach. Applying these insights have led to the experimental observation that *via* a two-step approach, separating hydrogen production conditions from carbon dioxide conversion conditions, the conversion of potassium hydrogen carbonate into potassium formate can be successfully performed with high carbon conversion yields, up to 77.9 wt%, and a selectivity of at least 81%. This result was obtained by applying a reaction temperature of 280 °C, a degree of filling with water of 50 vol% and an initial amount of hydrogen of 100 mmol for 24 hours. However, it should be mentioned that the selectivity to potassium formate/formic acid seemed to decrease as the filling degree increased to 50 vol%. We suggest, in case of higher degrees of filling, the occurrence of side reactions of carbon dioxide conversion or the further reduction of formate in the liquid phase, leading to the formation of other carbon containing products. In our opinion, an even higher amount of carbon conversion can be obtained by further optimising the reaction conditions of the second step reaction, in particular the reaction temperature. Unfortunately, we were not able to perform these experiments due to restrictions of the operating conditions of the autoclave systems we used. We expect the most room for improvement would be by increasing the reaction temperature: a rise to 300 °C could lead to both a higher hydrogen gas solubility and enhanced carbon dioxide conversion reaction kinetics. Nevertheless, the selectivity to potassium formate might decrease due to a possible lower stability of this product at higher temperatures or the occurrence of side reactions of carbon dioxide conversion. Finally, this study did not take into account the addition of moieties with catalytic activity to further improve the kinetics and selectivity of the second reaction step.

Acknowledgements

The authors would like to gratefully acknowledge the trust and financial support of the Agency for Innovation by Science and Technology in Flanders IWT (Ph.D. grant No. 111324). In addition, the authors greatly appreciate the expertise provided by Ellen Poelmans (VITO N.V.), Miranda Maesen (VITO N.V.) and Heidi Seykens (University of Antwerp) concerning the NPOC, HPLC and proton NMR measurements, respectively.

Notes and references

1 K. Byrappa and M. Yoshimura, in *Handbook of Hydrothermal Technology*, ed. W. Andrew, Elsevier, Oxford, 2nd edn, 2013.

2 B. M. French, (1) Synthesis and Stability of Siderite, $FeCO_3$, (2) Progressive Contact Metamorphosis of the Biwabik Iron Formation on the Mesabi

Range, Ph.D. thesis, The Johns Hopkins University, Baltimore, Minnesota, 1964.

3 B. M. French, Stability Relations of Siderite, FeCO3, Determined in Controlled fO2 Atmospheres, Document X-644-70-102, Goddard Space Flight Center, Planetology Branch, NASA, Greenbelt, MD, 1970.

4 N. G. Holm and E. Andersson, *Astrobiology*, 2005, **5**, 444–460.

5 T. M. McCollom and J. S. Seewald, *Chem. Rev.*, 2007, **107**, 382–401.

6 M. Yoshimura and K. Byrappa, *J. Mater. Sci.*, 2008, **43**, 2085–2103.

7 T. M. McCollom, *Rev. Mineral. Geochem.*, 2013, **75**, 467–494.

8 N. Yamasaki, *J. Ceram. Soc. Jpn.*, 2003, **111**, 709–715.

9 G. Demazeau, *Res. Chem. Intermed.*, 2011, **37**, 107–123.

10 R. I. Walton, *Chem. Soc. Rev.*, 2002, **31**, 230–238.

11 A. Rabenau, *Angew. Chem., Int. Ed. Engl.*, 1985, **24**, 1026–1040.

12 G. C. Kennedy, *Econ. Geol.*, 1950, **45**, 623–629.

13 R. A. Laudise, *Chem. Eng. News*, 1987, **65**, 30–43.

14 S. S. Toor, L. Rosendahl and A. Rudolf, *Energy*, 2011, **36**, 2328–2342.

15 A. A. Peterson, F. Vogel, R. P. Lachance, M. Fröling, M. J. Antal Jr and J. W. Tester, *Energy Environ. Sci.*, 2008, **1**, 32–65.

16 Y. Long and Z. Fang, *Biofuels, Bioprod. Biorefin.*, 2012, **6**, 686–702.

17 Y. Yashiro, H. Takahashi, N. Yamasaki and T. Kori, *Proceedings of the Seventh International Symposium on Hydrothermal Reactions*, Changchun, 2003.

18 H. Takahashi, L. H. Liu, Y. Yashiro, K. Ioku, G. Bignall, N. Yamasaki and T. Kori, *J. Mater. Sci.*, 2006, **41**, 1585–1589.

19 H. Takahashi, T. Kori, T. Onoki, K. Tohji and N. Yamasaki, *J. Mater. Sci.*, 2008, **43**, 2487–2491.

20 G. Tian, H. Yuan, Y. Mu, C. He and S. Feng, *Org. Lett.*, 2007, **9**, 2019–2021.

21 B. Wu, Y. Gao, F. Jin, J. Cao, Y. Du and Y. Zhang, *Catal. Today*, 2009, **148**, 405–410.

22 G. Tian, C. He, Y. Chen, H. Yuan, Z. Liu, Z. Shi and S. Feng, *ChemSusChem*, 2010, **3**, 323–324.

23 C. He, G. Tian, Z. Liu and S. Feng, *Org. Lett.*, 2010, **12**, 649–651.

24 H. Zhong, F. Jin, B. Wu, H. Chen and G. Yao, *AIP Conf. Proc.*, 2010, **1251**, 213–216, 2nd International Symposium on Aqua Science, Water Resource and Low Carbon Energy, Sanya Hainan, 2009.

25 F. Jin, Y. Gao, Y. Jin, Y. Zhang, J. Cao, Z. Wei and R. L. Smith Jr, *Energy Environ. Sci.*, 2011, **4**, 881–884.

26 F. Jin, X. Zeng, Z. Jing and H. Enomoto, *Ind. Eng. Chem. Res.*, 2012, **51**, 9921–9937.

27 Z. Liu, G. Tian, S. Zhu, C. He, H. Yue and S. Feng, *ACS Sustainable Chem. Eng.*, 2013, **1**, 313–315.

28 P. Yan, F. Jin, J. Cao, B. Wu and G. Zhang, *AIP Conf. Proc.*, 2010, **1251**, 242–245, 2nd International Symposium on Aqua Science, Water Resource and Low Carbon Energy, Sanya Hainan, 2009.

29 M. Hu, F. Jin, C. Ma and H. Chen, *AIP Conf. Proc.*, 2010, **1251**, 238–241, 2nd International Symposium on Aqua Science, Water Resource and Low Carbon Energy, Sanya Hainan, 2009.

30 M. Cheng, B. Wu, F. Jin, X. Duan and Y. Wang, *AIP Conf. Proc.*, 2010, **1251**, 234–237, 2nd International Symposium on Aqua Science, Water Resource and Low Carbon Energy, Sanya Hainan, 2009.

31 J. Liu, X. Zeng, M. Cheng, J. Yun, Q. Li, Z. Jing and F. Jin, *Bioresour. Technol.*, 2012, **114**, 658–662.

32 Z. Huo, M. Hu, X. Zeng, J. Yun and F. Jin, *Catal. Today*, 2012, **194**, 25–29.

33 F. Jin, X. Zeng, J. Liu, Y. Jin, L. Wang, H. Zhong, G. Yao and Z. Huo, *Sci. Rep.*, 2014, **4**, 4503.

34 X. Zeng, M. Hatakeyama, K. Ogata, J. Liu, Y. Wang, Q. Gao, K. Fujii, M. Fujihira, F. Jin and S. Nakamura, *Phys. Chem. Chem. Phys.*, 2014, **16**, 19836–19840.

35 H. Yao, X. Zeng, M. Cheng, J. Yun, Z. Jing and F. Jin, *Bioresearch*, 2012, **7**(1), 972–983.

36 X. Zeng, F. Jin, H. Yao and M. Cheng, *Adv. Mater. Res.*, 2012, **347–353**, 3677–3680.

37 L. Lyu, F. Jin, H. Zhong, H. Chen and G. Yao, *RSC Adv.*, 2015, **5**, 31450–31453.

38 L. Lyu, X. Zeng, J. Yun, F. Wei and F. Jin, *Environ. Sci. Technol.*, 2014, **48**, 6003–6009.

39 X. Zeng, F. Jin, Z. Huo, T. Mogi, A. Kishita and H. Enomoto, *Energy Fuels*, 2011, **25**, 2749–2752.

40 Z. Shen, Y. Zhang and F. Jin, *Green Chem.*, 2011, **13**, 820–823.

41 Z. Shen, Y. Zhang and F. Jin, *RSC Adv.*, 2012, **2**, 797–801.

42 F. Chen, G. Yao, Z. Huo and F. Jin, *RSC Adv.*, 2015, **5**, 11257–11260.

43 K. Michiels, J. Spooren and V. Meynen, *Fuel*, 2015, **160**, 205–216.

44 H. A. Pray, C. E. Schweickert and B. H. Minnich, *Ind. Eng. Chem.*, 1952, **44**, 1146–1151.

Faraday Discussions

PAPER

Novel process and catalytic materials for converting CO_2 and H_2 containing mixtures to liquid fuels and chemicals

Nora Meiri, Yakov Dinburg, Meital Amoyal, Viatcheslav Koukouliev, Roxana Vidruk Nehemya, Miron V. Landau and Moti Herskowitz*

Received 21st April 2015, Accepted 24th June 2015
DOI: 10.1039/c5fd00039d

Carbon dioxide and water are renewable and the most abundant feedstocks for the production of chemicals and fungible fuels. However, the current technologies for production of hydrogen from water are not competitive. Therefore, reacting carbon dioxide with hydrogen is not economically viable in the near future. Other alternatives include natural gas, biogas or biomass for the production of carbon dioxide, hydrogen and carbon monoxide mixtures that react to yield chemicals and fungible fuels. The latter process requires a high performance catalyst that enhances the reverse water-gas-shift (RWGS) reaction and Fischer–Tropsch synthesis (FTS) to higher hydrocarbons combined with an optimal reactor system. Important aspects of a novel catalyst, based on a Fe spinel and three-reactor system developed for this purpose published in our recent paper and patent, were investigated in this study. Potassium was found to be a key promoter that improves the reaction rates of the RWGS and FTS and increases the selectivity of higher hydrocarbons while producing mostly olefins. It changed the texture of the catalyst, stabilized the Fe–Al–O spinel, thus preventing decomposition into Fe_3O_4 and Al_2O_3. Potassium also increased the content of Fe_5C_2 while shifting Fe in the oxide and carbide phases to a more reduced state. In addition, it increased the relative exposure of carbide iron on the catalysts surface, the CO_2 adsorption and the adsorption strength. A detailed kinetic model of the RWGS, FTS and methanation reactions was developed for the Fe spinel catalyst based on extensive experimental data measured over a range of operating conditions. Significant oligomerization activity of the catalyst was found. Testing the pelletized catalyst with CO_2, CO and H_2 mixtures over a range of operating conditions demonstrated its high productivity to higher hydrocarbons. The composition of the liquid (C_{5+}) was found to be a function of the potassium content and the composition of the feedstock.

Chemical Engineering Department, Blechner Center for Industrial Catalysis and Process Development, Ben-Gurion University of the Negev, Beer-Sheva, 84105, Israel. E-mail: herskow@bgu.ac.il; Fax: +972 8 6479427; Tel: +972 8 6461482

1. Introduction

Crude oil is the dominant feedstock for the production of liquid fuels, as clearly stressed in a recent International Energy Agency (IEA) report.[1] Most of the foreseen output rise over the next three decades will be produced in the Middle East. The biofuels share is currently only 8%, and is expected to remain at this level until 2040, supported by increasing subsidies. The IEA report sends a very strong message regarding the CO_2 emissions, which are expected to increase dramatically, unless low-carbon investments will increase by a factor of four beyond current values. All those projections indicate that the search for alternative, renewable and sustainable sources for the production of fungible and competitive fuels needs to be accelerated. Biomass has been the main source for renewable fuels, mainly ethanol and biodiesel. A fierce debate related to future use of biomass has been conducted over the past decade. A recent report[2] presents a less optimistic view of the future of bioenergy in Germany concluding that with the exception of the use of biogenic waste, the larger scale use of biomass as an energy source is limited. A proper assessment of the impact of biofuels is critical.[3]

Liquid fuels are mixtures of hydrocarbons that meet specific standards updated from time to time. Carbon dioxide, an environmentally damaging greenhouse gas, and water are the most abundant and low-cost sources of carbon and hydrogen needed for the production of hydrocarbons. The current scientific and technological challenge is to develop and implement environmentally-friendly processes that convert CO_2 and H_2O into commercially-viable, fungible and compatible fuels, using green energy. A viable route is the reaction of captured carbon dioxide from flue gases with hydrogen produced from water. Methods for both CO_2 capture and water splitting have been recently reviewed.[4]

Solar water splitting using photoelectrochemical cells has been extensively studied and is aimed at developing efficient, robust and scalable processes for the production of low-cost, commercially-competitive hydrogen.[5–8] In spite of the extensive scientific effort to provide innovative solutions, the question raised in a recent publication[8] "Will Solar-Driven Water-Splitting Devices See the Light of Day?" remains open. The solar thermochemical cycle method is another promising route being pursued, although no large scale facilities have been developed.[9] Actually the only commercially-proven technology and operated at a large scale for water splitting is electrolysis.[10] The cost of hydrogen produced by this method depends mainly on the cost of electricity, estimated to be ~\$3 kg^{-1} at an electricity cost of <\$0.055 kW h^{-1}. This is by far higher than the cost of hydrogen[11] (~\$1 kg^{-1}) produced using the dominant commercial technology, steam reforming of natural gas. The projected cost of hydrogen[11] produced by other water splitting methods is also significantly higher (\$3.5–10 kg^{-1}).

An economic analysis[12] of the production of liquid solar fuels from carbon dioxide and water was reviewed recently. Employing a modified Fischer–Tropsch process for reacting captured carbon dioxide from flue gas with hydrogen produced by electrolysis for water (using electricity from wind energy) yielded an estimated cost of \$5.25 per gal of gasoline. The relative high cost of the fuel resulted from the high cost of hydrogen. Another study[13] reviewed sustainable hydrocarbon fuel production by recycling CO_2 and H_2O with renewable or nuclear energy and concluded that \$3 per gal gasoline could be achieved at an electricity cost of \$0.04–0.05 kW h^{-1}. The significantly different production cost of the fuels estimated by the two studies results from the different assumptions made in estimating the cost of the electrolysis and the Fischer–Tropsch processes.

A recent paper[14] reviews the catalytic CO_2 hydrogenation to hydrocarbons and describes a novel Fe spinel catalyst coupled with a system of three packed-bed reactors in series with interim removal of water and condensed hydrocarbons that reached a CO_2 conversion of 89% and C_{5+} hydrocarbons productivity of >0.5 kg kg_{cat}^{-1} h^{-1}. The pure-spinel catalyst displayed a significantly higher activity and selectivity than the other Fe catalysts published in the literature. This process produces hydrocarbon mixtures that can be readily converted to liquid fuels.

Although the CO_2 hydrogenation process is a very promising route to renewable and fungible fuels, its implementation is limited by the availability of low-cost hydrogen produced from water. Only specific locations with access to extremely low-cost electricity (<\$0.03 kW h^{-1}) will operate such competitive processes unless a significantly higher-cost fuel is acceptable for specific purposes like military use. Therefore, other sources of hydrogen should be sought like fossil natural gas, renewable biogas or biomass. Natural gas has been applied commercially[15] as feedstock for the production of fuels through conversion to syngas followed by the Fischer–Tropsch synthesis (FTS). The syngas is generated by steam reforming, partial oxidation and autothermal reforming which produce a mixture of hydrogen and carbon monoxide at a molar ratio close to 2. Carbon dioxide generated in these processes is normally separated because it could be detrimental to selectivity or at best is a diluent. Biomass is a potential renewable feedstock for gasification to syngas, containing CO, CO_2 and H_2, which is fed to the FTS process. Comprehensive reviews[16,17] describe the potential catalysts, effect of impurities, the economics and the challenges of such processes. A comparison of several routes for the production of fuels from biomass, carbon dioxide and electricity[18] based on economical evaluation indicated that the thermochemical route (gasification and conversion to fuels) is by far the preferred route. Biogas is another renewable feedstock that can be readily converted to syngas. This syngas contains a significant amount of CO_2, as pointed out in a recent review.[19] Therefore, the development of suitable FTS catalysts that handle mixtures of CO_2, CO and H_2 and are the basis for environmentally-acceptable and competitive processes is a major challenge.

Iron-based catalysts[20,21] that display significant reverse water gas shift (RWGS) activity are employed in processes for converting CO, CO_2 and H_2 mixtures to fuels and chemicals. Specifically, a recent paper[20] discusses K/Fe–Cu–Al catalyst performance along with carbon deposition by variation of the CO_2/CO ratio. The authors report that in experiments operated for 40 h at 300 °C, 20 bar, $CO_2 : CO : H_2$ molar ratio of 1 : 1 : 7 and weight hourly space velocity (WHSV) of the carbon oxides of about 0.83 h^{-1}, the conversion of CO was 96% while the conversion of CO_2 was only 12%. The selectivity to methane was about 10% while the selectivity to C_{5+} was about 60%. The significant conclusion of this study was that operating at a high CO_2/(CO + CO_2) ratio and high pressure and temperature prevents carbon deposition on the Fe-based catalyst. A recent patent[22] extends the information on the novel catalytic system[14] from mixtures of CO_2 and H_2 to CO_2, CO and H_2 mixtures.

Potassium was identified as an important component of the Fe-based catalysts for the FTS. In a recent review,[23] the authors concluded that the presence of potassium reduces the formation of CH_4 while increasing the selectivity to higher hydrocarbons. A similar effect was reported in an early study[24] of CO_2 hydrogenation. This effect was related to the enhanced carburization of the Fe catalyst in the presence of potassium. A recent study[25] reported the effect of potassium on CO_2 in the hydrogenation of mixtures of CO and CO_2. Apparently, increasing the K content did not increase the catalyst activity while the average product molecular weight increased.

It is generally accepted that K acts as electronic promoter when adsorbed at the surface of $Fe^0/FeO_x/FeC_x$ phase particles lowering the iron work function due to interactions between the alkali metal valence state and conduction band states in catalytic phases.[26,27] Donating electron density to the vacant d-orbital of iron enhances the dissociative adsorption of CO while lowering the H_2 adsorption ability.[27] Depressing hydrogenation activity is associated with lower rates of CH_4 formation, higher selectivity to olefins and long-chain hydrocarbons. The oxidative action of CO_2 and excessive water on Fe-carbide phases[28] during CO_2 hydrogenation creates more oxidized iron on the catalysts that may modify the action of the K promoter.

The hydrogenation of mixtures of CO and CO_2 is a complex process that requires optimization through modeling and simulations. A recent study[29] describes a simulation that includes steam reforming of natural gas with addition of CO_2. The FTS was simulated using experimental results previously published[20] by the authors. While numerous kinetic models have been proposed for Fe-based FT catalysts, limited information has been published on the kinetic behavior of catalysts for the hydrogenation of CO_2. A kinetic study[30] of CO_2 hydrogenation proposed two rate expressions, one for reverse WGS and the other for the FTS. The experiments were conducted at 10 bar and a stoichiometric H_2/CO_2 ratio of 3. The highest measured CO_2 conversion was 45% at 300 °C to 60% at 400 °C due to the equilibrium limitation of the reverse WGS. Another kinetic study[31] was conducted at high H_2/CO_2 molar ratios of 4 and 8 that produced a light product ($\alpha = 0.2$–0.3). A similar kinetic model that included two reactions (reverse WGS and FTS) was used. The highest CO_2 conversion was 44% at a H_2/CO_2 molar ratio of 8. Increasing the temperature and residence time did not increase the conversion.

A more detailed kinetic study of the FTS on K/Fe–Cu–Mn–Al_2O_3 catalysts has been published.[32] It includes 10 reactions of CO and H_2, nine of them to light paraffins (C_1–C_4) and olefins (C_2–C_4). The last reaction represents the production of C_{5+} hydrocarbons lumped in one component C_6H_{14}. The selection of hexane means that the liquid product was very light and mostly paraffinic. No oxygenates were mentioned in this study.

Potential applications of catalytic processes for converting mixtures of CO_2, CO and H_2 to hydrocarbons and oxygenates to be readily converted to liquid fuels and chemicals requires sustained research aimed at improving the performance of the catalysts (selectivity to higher hydrocarbons and olefins, activity and stability) and optimization of the processes. One of the most important components of Fe-based catalysts is potassium. Developing a better understanding of the effects of potassium as promoter is a key factor in improving performance. Optimization of the process requires a reliable kinetic model to be implemented in modeling and simulations. The scope of this study is to determine some of the fundamental effects of potassium in the Fe spinel catalyst, to develop a detailed kinetic model of the CO_2 hydrogenation on the Fe spinel catalyst and present effects of the feed composition and operating conditions on the performance of the pelletized Fe spinel catalyst in a fixed bed reactor.

2. Experimental

2.1. Catalyst preparation

The K-promoted, pure Fe–Al–O spinel $Fe(Fe_{0.5}Al_{0.5})_2O_4$ catalyst used in this study was prepared and characterized according to procedures described elsewhere.[14] Potassium was added to the dried Fe–Al–O spinel precursor by incipient wetness impregnation with an aqueous solution of K_2CO_3, followed by overnight drying at 110 °C and calcination in air at 450 °C for 6 h. Three Fe-based spinel catalysts with different K content were prepared: unpromoted 100 spinel/0 wt% K (Cat. Fe–0K), 100 spinel/2 wt% K (Cat. Fe–2K) and 100 spinel/4 wt% K (Cat. Fe–4K).

2.2. Catalyst characterization

The spent catalysts were characterized after He treating and passivation. After testing, the catalysts were treated under a He flow at 300 °C for 3 h for removal of residual adsorbed organics and passivated under a CO_2 flow at 25 °C for 1 h. The textural properties of the catalysts, surface area, pore volume and pore size, were calculated from N_2 adsorption–desorption isotherms using conventional BET and BJH methods. The isotherms were recorded using a NOVA 3200e (Quantachrome) at the temperature of liquid nitrogen after degassing under vacuum at 100 °C. The phase composition of the catalysts was determined based on their X-ray diffraction patterns. XRD measurements were conducted using a Phillips 1050/70 powder diffractometer fitted with graphite monochromator, at 40 kV and 28 mA using software developed by Crystal logic. The phase identification was performed by using a SBDE ZDS computer search/match program coupled with the ICDD. The relative content of iron oxide and carbide phases was obtained using Rietveld refinement of the XRD profile by using the DBWS-9807 program. The elemental composition of the catalysts was measured using the EDS method (Quanta-200, SEM-EDAX, FEI Co instrument). The CO_2-TPD measurements were performed using a Chemisorption Analyzer Autochem II 2920 instrument, Micrometrics Co. The samples were saturated with CO_2 at 40 °C before recording the desorption spectra with a heating rate of 5 °C min^{-1} in 5% CO_2/He flow of 25 ml min^{-1}. XPS data were collected using X-ray photoelectron spectrometer ESCALAB 250 ultrahigh vacuum (1×10^{-9} bar) apparatus with an Al Kα X-ray source and a monochromator. The spectral components of C and Fe signals were found by fitting a sum of single component lines to the experimental data by means of a non-linear least-square curve fitting. The EX05 argon gun system performed controlled removal of surface layers allowing depth profile analyses of elements.

2.3. Catalysts performance testing

Measurements of the catalysts activities and selectivities including data for the kinetic model were carried out in two units:

1. A computer-controlled bench rig equipped with three continuous-flow fixed bed reactors connected in series including devices for removing water and organic liquid between reactors[14] for the kinetic runs (packed with 50–200 μm

powder) and CO_2, CO and H_2 mixtures (packed with 1.4–1.7 mm pellets) runs. The pellets were prepared by pressing the powder catalyst. The catalyst was mixed with silica powder diluent and packed in the reactors.

2. One fixed-bed reactor packed with diluted (silica) powder catalyst followed by a cooler and a gas–liquid separator for studying the effect of potassium.

The hydrocarbon liquid mixture and oxygenates from aqueous products were separated and analyzed using GC-MS (Agilent Technologies 6890N network GC system equipped with 5973 Network mass-selective detector). CO_2, H_2, CO and C_1–C_6 hydrocarbons were analyzed on line using a GC instrument Agilent Technologies, model 7890A equipped with 5 valves/7 columns (PLOT and packed)/2 TCD/FID detectors. The organic content in the aqueous product was measured by TOC analysis (instrument TOC-V_{CPN} Shimadzu Co.).

3. Results and discussion

3.1. The effect of potassium as a promoter of the Fe-spinel catalyst

The catalysts were activated *in situ* by H_2/CO carburization with a stream of 0.03 : 0.03 : 0.14 N mL min^{-1} $g_{cat}{}^{-1}$ of CO, H_2 and He, respectively, at 300 °C for 3 h. The catalysts were tested at constant conditions of 20 bar, 320 °C and a H_2/CO_2 molar ratio of 3. The WHSV of CO_2 was adjusted to reach about 30% CO_2 conversion so as to achieve differential operation of the reactor and calculate directly the reaction rates. The catalysts reached steady state yielding constant CO_2 conversion/products selectivity after 120 h on stream. After the end of the run, the catalysts were treated with He and then passivated with CO_2 before being discharged from the reactor.

The effect of potassium on the performance of the Fe–Al–O spinel catalyst in CO_2 hydrogenation. The performance measured with the three catalysts, Cat. Fe–0K, Cat. Fe–2K and Cat. Fe–4K, listed in Table 1 demonstrates a significant effect of potassium on the catalyst activity and selectivity. The RWGS rate of reaction was calculated from the differential measurements of the CO_2 conversion (30%) and WHSV. The rate of methanation was calculated from the molar flow rate of methane and the catalyst weight. Similarly, the rate of the FTS reaction was calculated from the molar flow rate of all hydrocarbons (without methane) and oxygenates and the catalyst weight.

Adding 2 wt% potassium increased dramatically both the RWGS and FTS rates of reaction and decreased the rate of methanation. Further increasing the

Table 1 Effect of potassium on the performance of Fe–Al–O spinel catalyst in CO_2 hydrogenation. Testing conditions: 20 bar, 320°C, molar H_2/CO_2 = 3, 120 h on stream

	Reaction rate, mmol $g_{cat}{}^{-1}$ h^{-1}				Product selectivity, wt%					
							C_2–C_4			
Catalyst	RWGS	FTS	CH_4	FTS/RWGS	CO	CH_4	Olefins	Paraffins	Oxygenates	C_{5+}
Cat. Fe–0K	125	62	46	0.50	15	40	15	15	8	7
Cat. Fe–2K	179	112	34	0.63	22	23	24	11	4	16
Cat. Fe–4K	123	95	6	0.77	22	6	24	5	10	33

potassium content decreased the reaction rate of RWGS. The methanation rate displayed an extreme drop thus increasing the selectivity of the desired products. The FTS reaction rate decreased mildly. Actually, the ratio of the FTS/RWGS reaction rates increased gradually as the potassium content increased which has a very important impact on the process. The effect of potassium on selectivity was very impressive and dominant. Beyond the drop in methane selectivity, the significant increase in the selectivity to higher hydrocarbons (C_{5+}) is a substantial factor. These findings are in agreement with the results presented in other studies of Fe-catalysts in CO and CO_2 hydrogenation.[23,33]

The effect of potassium on the texture, chemical/phase composition and surface chemistry of the Fe–Al–O spinel. The texture measurements of the spent catalysts after operation of 120 h listed in Table 2 indicate a significant decrease of the surface area, pore volume and diameter as the potassium content increased. This may be attributed to increasing crystal size or aggregation of nanoparticles of the Fe-oxide and carbide phases produced from the Fe–Al–O spinel during activation and self-organization in the presence of K as was observed in ref. 34 for other Fe-catalysts. It may also be attributed to the blocking of pores with coke or wax which was increasingly deposited as the potassium content increased.

The chemical composition of the spent catalysts expressed as the weight ratio related to Fe is shown in Table 3. The catalyst containing 2 wt% K accumulated about the same amount of total carbon as the catalyst with no K. Increasing the amount of potassium to 4 wt% increased the total carbon content from 0.23 to 0.42. In all cases, the atomic ratio C/Fe was greater than unity, much higher than in any iron carbide phases FeC_x stable under the testing conditions (*i.e.* Fe_5C_2, Fe_7C_3). This means that the spent catalysts contained a significant amount of coke or wax carbon deposited during self-organization.

The quantitative XRD phase analysis revealed a strong effect of potassium promoter on the phase composition of the spent catalysts as illustrated by the results listed in Table 4. The calculations of the phase content and their average crystal size were conducted based on their X-ray diffraction patterns shown in Fig. 1. All catalysts contain two Fe-carbide phases: χ-Fe_5C_2 and Fe_7C_3 formed as a result of spinel activation and self-organization when part of the iron was removed from the spinel phase. This decreases the x value in the formula of the spinel phase $Fe^{II}[(Fe_x^{III}Al_{1-x})_2]O_4$ from 0.5, in fresh catalyst, to 0.2–0.3 in Fe-poor spinel residua measured according to the positions of characteristic reflections in the X-ray diffractograms.

A very significant result is the removal of Fe from the Fe–Al–O spinel in catalyst Cat. Fe–0K that was partially decomposed into Fe-poor spinel (27 wt%),

Table 2 Effect of potassium on the texture parameters of Fe–Al–O spinel spent catalysts. Testing conditions: 20 bar, 320 °C, molar $H_2/CO_2 = 3$, 120 h on stream

Catalyst	Surface area, $m^2\ g^{-1}$	Pore volume, $cm^3\ g^{-1}$	Average pore diameter, nm
Cat. Fe–0K	78	0.23	6.0
Cat. Fe–2K	55	0.16	5.6
Cat. Fe–4K	42	0.10	4.6

Table 3 Chemical composition of the Fe–Al–O spinel spent catalysts. Testing conditions: 20 bar, 320°C, molar $H_2/CO_2 = 3$, 120 h on stream

	Chemical composition							
	wt ratio				Atomic ratio			
Catalyst	Fe	Al	C	K	Fe	Al	C	K
Cat. Fe–0K	1	0.21	0.25	0.0	1	0.44	1.2	0.0
Cat. Fe–2K	1	0.22	0.23	0.05	1	0.45	1.1	0.08
Cat. Fe–4K	1	0.20	0.42	0.10	1	0.42	1.9	0.13

amorphous alumina whose reflections cannot be observed in the diffractogram and magnetite (18 wt%). Adding potassium prevents the formation of a magnetite phase while the content of the Fe-poor spinel increased to 41–42 wt%. Potassium also enhances carburization of the Fe–Al–O spinel precursor to the active χ-Fe_5C_2 phase.[35] Interestingly, the content of the χ-Fe_5C_2 phase decreased as the potassium content increased from 2 wt% to 4 wt%. A relatively little change in the crystal size of the two iron carbide phases in two potassium promoted catalysts was observed.

The XPS spectra of the spent unpromoted and K-promoted Fe–Al–O spinel catalysts, as-received and after Ar-etching, are shown in Fig. 2 and 3 respectively. The BE (binding energy) for detected peaks and surface concentrations of corresponding atoms are given in Tables 5 and 6. Two main peaks were detected in the XPS spectra of iron. The first is attributed to the iron carbides at lower binding energy around 707.1–707.5 eV corresponding to the near-metallic state of iron and the second belongs to iron ions in oxide phases with higher binding energies in the range of 710.4.5–711.5 eV. The latter peaks represent superposition of signals contributed by Fe^{2+} with a BE of 709.5 and of Fe^{3+} with a BE of 711.2 eV.[36,37] Insertion of 2 wt% potassium shifted the BE of the carbide and oxide iron atoms to lower values: from 707.5 (707.4 after etching) to 707.3 eV and from 711.5 (710.8 after etching) to 710.8 (710.4 after etching) eV. This corresponds to formation of more reduced states of iron in carbides and in Fe-poor spinel due to electron-donation of the K-promoter.[26,27]

Table 4 Phase composition of the Fe–Al–O spinel spent catalysts. Testing conditions: 20 bar, 320°C, molar $H_2/CO_2 = 3$, 120 h on stream

Catalyst	Spinel wt%	Crystal size, nm	Magnetite wt%	Crystal size, nm	χ-Fe_5C_2 wt%	Crystal size, nm	Fe_7C_3 wt%	Crystal size, nm	χ-Fe_5C_2/(χ-Fe_5C_2 + Fe_7C_3)
Cat. Fe-0K	27	8	18	35	34	25	21	35	62
Cat. Fe-2K	41	5	0	—	47	30	12	40	80
Cat. Fe-4K	42	6	0	—	42	35	16	40	72

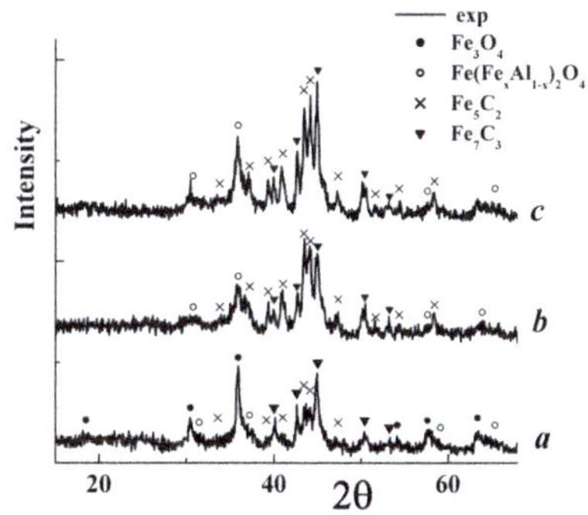

Fig. 1 XRD patterns of the Fe–Al–O spinel catalyst after reaching steady state in the catalytic testing run. (a) "Cat. Fe–0K", (b) "Cat. Fe–2K", (c) "Cat. Fe–4K". Testing conditions: $P = 20$ bar, $T = 320$ °C, $H_2/CO_2 = 3$, 120 h on stream.

Insertion of K increased the surface concentration or relative exposure of carbide iron. Deconvolution of XPS spectra recorded for the C 1s core shows the coexistence of three groups of signals: peaks with BE 283.6–284.3 eV corresponding to Fe-carbide carbon, 284.8–285.1 eV belonging to C–C sp^2 and sp^3 in carbon and hydrocarbon (–CH_2=CH) deposits and 285.4–288.6 eV characteristic for adsorbed carbon oxygenate moieties containing C=O, C–O, CH_2–O, O–C=O and other oxygen groups.[36,38,39] Insertion of potassium increases the relative exposure of carbide carbon at the spent catalysts surface (from 11 to 19%). This is in agreement with growing of the exposure of carbide iron. Potassium also significantly increases the relative exposure of carbon belonging to graphite and hydrocarbon deposits from 31% (36% after etching) in the unpromoted catalyst to

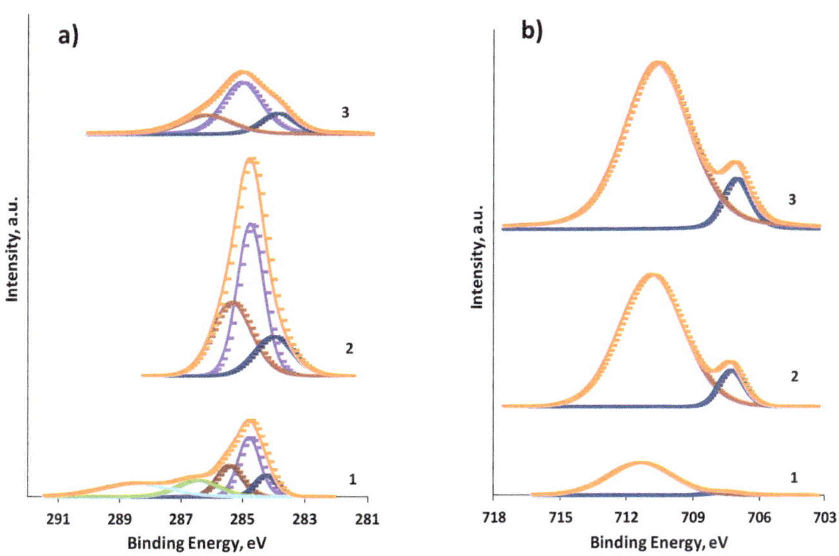

Fig. 2 XPS spectra recorded with as-received spent Fe–Al–O spinel catalysts: (a) – C 1s core; (b) – $Fe_{3/2}$ core: 1 – unpromoted catalyst; 2 – 2 wt% K; 3 – 4 wt% K.

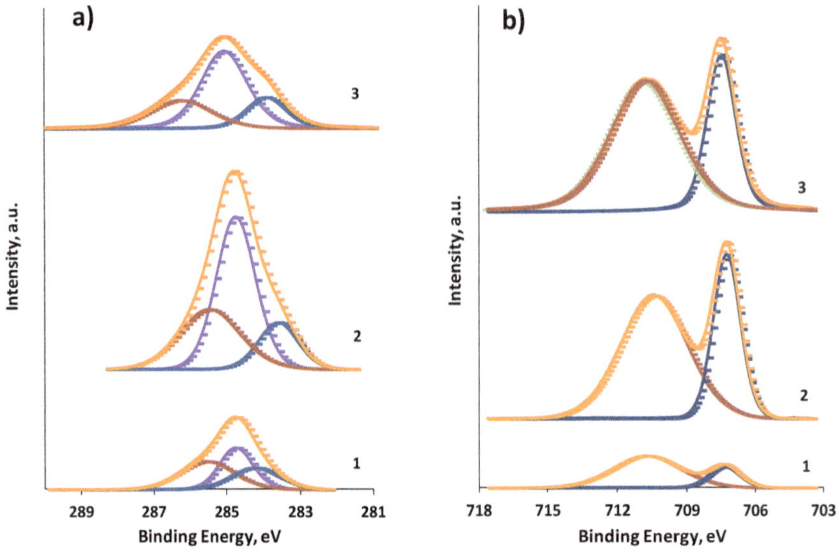

Fig. 3 XPS spectra recorded with spent Fe–Al–O spinel catalysts after Ar-etching: (a) – C 1s core; (b) – $Fe_{3/2}$ core: 1 – unpromoted catalyst; 2 – 2 wt% K; 3 – 4 wt% K.

48–55% in the catalysts containing 2–4 wt% K. At the same time insertion of potassium removes the oxygenated moieties with a high C 1s BE from 58% (39% after etching) in the unpromoted catalyst to 26–34% in the K-promoted catalysts.

Potassium is expected to increase the basicity of the Fe–Al–O spinel thus increasing the surface coverage with acidic CO_2 reagent molecules.[40] The CO_2-TPD spectra illustrated in Fig. 4 indicate that the 2 wt% K catalyst displayed one CO_2 desorption peak centered at ~150 °C which represents relatively weak surface basic sites. Increasing the K content to 4 wt% further increased the weak basicity by a factor of 2.5. Furthermore, at the high K content of 4 wt% there

Table 5 Binding energies and concentrations of surface atoms of the Fe–Al–O spinel spent catalyst. Testing conditions: 20 bar, 320°C, molar H_2/CO_2 = 3, 120 h on stream

	C 1s core		Fe $2p_{3/2}$ core		
Catalyst name	BE, eV	% atomic	BE, eV	% atomic	Fe_{carb}/Fe_{ox}
Cat. Fe–0K	284.3	11	707.5	4	0.04
	284.8	31			
	285.5	19	711.5	96	
	286.5	17			
	288.6	22			
Cat. Fe–2K	284.1	18	707.3	10	0.11
	284.8	48	710.8	90	
	285.4	34			
Cat. Fe–4K	284.0	19	707.1	11	0.12
	285.1	55	710.7	89	
	286.3	26			

Table 6 Binding energies and concentrations of surface atoms of the Ar-etched Fe–Al–O spinel spent catalyst. Testing conditions: 20 bar, 320°C, molar H_2/CO_2 = 3, 120 h on stream

	C 1s core		Fe $2p_{3/2}$ core		
Catalyst	BE, eV	% atomic	BE, eV	% atomic	Fe_{carb}/Fe_{ox}
Cat. Fe–0K	284.3	26	707.4	23	0.30
	284.8	36	710.8	77	
	285.6	39			
Cat. Fe–2K	283.6	16	707.3	36	0.56
	284.8	53	710.4	64	
	285.5	31			
Cat. Fe–4K	284.0	19	707.5	34	0.52
	285.1	55	710.8	66	
	286.3	26			

appeared two additional CO_2 desorption peaks centered at 550 and 850 °C, corresponding to very strong basic sites. These results may be explained assuming strong interaction between Fe–Al–O spinel and potassium. At low concentrations of K, it mainly stabilizes the Fe–Al–O spinel phase, as detected by XRD. However, at thwe high concentration of 4 wt%, potassium may exist in a free form of adsorbed K_2O, displaying strong surface basicity.

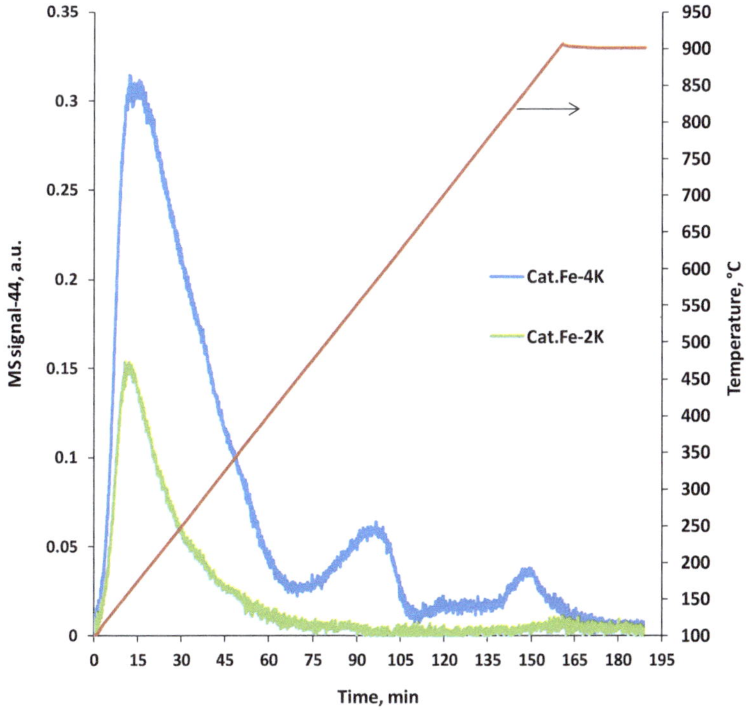

Fig. 4 CO_2-TPD spectra recorded with fresh K/Fe–Al–O spinel catalysts.

3.2. Kinetic model

Kinetic data were measured for the 4 wt% spinel Fe catalyst in a three-reactors-in-series system over a range of operating conditions (300–340 °C, 20–28 bar, molar $H_2/CO_2 = 2.9$, WHSV = 1–6 h^{-1}). In essence two processes took place in series: RWGS and FTS. Oligomerization of light olefins was also found to be important. In contrast to the recent kinetic model[32] that employed power law expressions for all 10 reaction rates of FTS, our model consists of LHHW (Langmuir–Hinshelwood–Hougen–Watson) expressions that include an adsorption term in the denominator. This term contains two components: H_2O and CO_2. The reactions were selected so as to reflect the main findings of the kinetic study: the RWGS reaction, methanation reaction, 7 FTS reactions and 4 oligomerization reactions. The 7 FT reactions (4)–(10) produce light paraffins (C_2–C_5), four light olefins (C_2, C_3, C_4 and C_5), C_{6+} represented by decene and oxygenates represented by butanoic acid. The oligomerization reactions (11)–(14) convert the light olefins to C_{6+} products represented by decene. The reactions are listed in Table 7. The mass balances expressed by eqn (1) assume plug-flow with no internal or external mass and heat transfer resistances:

$$\frac{d\omega_i}{d\left(\frac{1}{\text{WHSV}}\right)} = R_i \omega_{CO_2,\,\text{in}} M_{w_i} \tag{1}$$

where ω_i is the mass fraction of component i, WHSV is the weight hourly space velocity of CO_2, R_i is the rate of change in the reaction where component i is a reactant/product, $\omega_{CO_2,\text{in}}$ is the mass fraction of CO_2 at inlet and M_{w_i} is the molecular weight of component i.

The reactions and rate expressions of kinetic model are as follows:

$$CO_2 + H_2 \leftrightarrow CO + H_2O \quad r_1 = k_1 \frac{p_1 p_2 - \dfrac{p_3 p_4}{K_{eq}}}{1 + K_4 p_4 + K_1 p_1} \tag{2}$$

$$CO + 3H_2 \rightarrow CH_4 + H_2O \quad r_2 = k_2 \frac{p_3 {p_2}^{0.2}}{1 + K_4 p_4 + K_1 p_1} \tag{3}$$

$$CO + 2H_2 \rightarrow \frac{1}{2} C_2H_4 + H_2O \quad r_3 = k_3 \frac{p_3 {p_2}^{0.2}}{1 + K_4 p_4 + K_1 p_1} \tag{4}$$

Table 7 Key components of the kinetic model

Comp.	Name	Formula	Comp.	Name	Formula
1	Carbon dioxide	CO_2	7	Propylene	C_3H_6
2	Hydrogen	H_2	8	Propane[a]	C_3H_8
3	Carbon monoxide	CO	9	1-Butene	C_4H_8
4	Water	H_2O	10	Pentene	C_5H_{10}
5	Methane	CH_4	11	Decene[b]	$C_{10}H_{20}$
6	Ethylene	C_2H_4	12	Butanoic acid[c]	$C_4H_8O_2$

[a] C_2–C_5 paraffins. [b] C_{6+}. [c] Oxygenates.

$$CO + 2H_2 \rightarrow \frac{1}{3}C_3H_6 + H_2O \quad r_4 = k_4\frac{p_3 {p_2}^{0.2}}{1 + K_4p_4 + K_1p_1} \tag{5}$$

$$CO + \frac{7}{3}H_2 \rightarrow \frac{1}{3}C_3H_8 + H_2O \quad r_5 = k_5\frac{p_3 {p_2}^{0.2}}{1 + K_4p_4 + K_1p_1} \tag{6}$$

$$CO + 2H_2 \rightarrow \frac{1}{4}C_4H_8 + H_2O \quad r_6 = k_6\frac{p_3 {p_2}^{0.2}}{1 + K_4p_4 + K_1p_1} \tag{7}$$

$$CO + 2H_2 \rightarrow \frac{1}{5}C_5H_{10} + H_2O \quad r_7 = k_7\frac{p_3 {p_2}^{0.2}}{1 + K_4p_4 + K_1p_1} \tag{8}$$

$$CO + 2H_2 \rightarrow \frac{1}{10}C_{10}H_{20} + H_2O \quad r_8 = k_8\frac{p_3 {p_2}^{0.2}}{1 + K_4p_4 + K_1p_1} \tag{9}$$

$$CO + \frac{6}{4}H_2 \rightarrow \frac{1}{4}C_4H_8O_2 + \frac{2}{4}H_2O \quad r_9 = k_9\frac{p_3 {p_2}^{0.2}}{1 + K_4p_4 + K_1p_1} \tag{10}$$

$$C_2H_4 \rightarrow \frac{1}{5}C_{10}H_{20} \quad r_{10} = k_{10}\frac{{p_6}^{1.3}}{1 + K_4p_4 + K_1p_1} \tag{11}$$

$$C_3H_6 \rightarrow \frac{1}{3.3}C_{10}H_{20} \quad r_{11} = k_{11}\frac{{p_7}^{1.8}}{1 + K_4p_4 + K_1p_1} \tag{12}$$

$$C_4H_8 \rightarrow \frac{1}{2.5}C_{10}H_{20} \quad r_{12} = k_{12}\frac{{p_9}^{1.6}}{1 + K_4p_4 + K_1p_1} \tag{13}$$

$$C_5H_{10} \rightarrow \frac{1}{2}C_{10}H_{20} \quad r_{13} = k_{13}\frac{{p_{10}}^{1.3}}{1 + K_4p_4 + K_1p_1} \tag{14}$$

Kinetic modeling of the experimental data was conducted employing MATLAB, using 216 data points measured at the outlet of the first, second and third reactor. The MATLAB function "ode113" solved the differential equations system. Kinetic models were fitted to the data using the "Fmincon" function used to find the minimum sum of squared error (SSE) value. The kinetic constants, including pre-exponent coefficient and activation energy, are listed in Table 8. The adsorption constants, including the adsorption pre-exponential factor and the heat of adsorption, are listed in Table 9. A parity plot of ω_i of all components at the outlet of each reactor that reflects the good agreement of the experimental data with values calculated with the model (R^2 calculated using regression statistics is 0.99) is depicted in Fig. 5. The predicted and experimental weight fractions of CO_2 and CO at the outlet of each one of the three reactors at 320 °C and 20 and 28 bar, depicted in Fig. 6, are in good agreement. They illustrate the significant contribution of each reactor and the relatively weak effect of total pressure.

The kinetic developed for the catalyst containing 4 wt% K will be further extended to account for the significant effect of potassium. Furthermore, it will be employed to evaluate the effectiveness factor of the reactions in the model for pelletized catalyst. CO in the process feed will also be examined. This will

Table 8 Kinetic constants of the rate expressions

k_j	k_{0j}	Units	E_j, kJ mol^{-1}	k_j	k_{0j}	Units	E_j, kJ mol^{-1}
k_1	$5.5 \times 10^{+6}$	mol g^{-1} h^{-1} MPa^{-2}	72.2	k_8	$4.0 \times 10^{+3}$	mol g^{-1} h^{-1} $MPa^{-1.2}$	33.4
k_2	$1.0 \times 10^{+9}$	mol g^{-1} h^{-1} $MPa^{-1.2}$	98.0	k_9	$3.6 \times 10^{+3}$	mol g^{-1} h^{-1} $MPa^{-1.2}$	31.5
k_3	$1.2 \times 10^{+4}$	mol g^{-1} h^{-1} $MPa^{-1.2}$	45.0	k_{10}	$2.7 \times 10^{+3}$	mol g^{-1} h^{-1} $MPa^{-1.3}$	25.8
k_4	$1.2 \times 10^{+5}$	mol g^{-1} h^{-1} $MPa^{-1.2}$	53.6	k_{11}	$1.4 \times 10^{+4}$	mol g^{-1} h^{-1} $MPa^{-1.8}$	35.5
k_5	$1.8 \times 10^{+4}$	mol g^{-1} h^{-1} $MPa^{-1.2}$	46.4	k_{12}	$3.7 \times 10^{+4}$	mol g^{-1} h^{-1} $MPa^{-1.6}$	40.7
k_6	$1.1 \times 10^{+3}$	mol g^{-1} h^{-1} $MPa^{-1.2}$	31.5	k_{13}	$6.0 \times 10^{+7}$	mol g^{-1} h^{-1} $MPa^{-1.3}$	88.8
k_7	$9.0 \times 10^{+2}$	mol g^{-1} h^{-1} $MPa^{-1.2}$	31.5				

facilitate the application of the kinetic model to simulate the performance of the process at operating conditions that are relevant to commercial applications.

3.3. Conversion of CO_2, CO and H_2 mixtures to higher hydrocarbons

The process described in this study was conducted with a feed containing CO_2 and H_2 and a mixture that contained CO_2, CO and H_2. The data presented here were measured with pelletized (1.4–1.7 mm) catalysts containing 3 wt% and 4 wt% K over a range of feed compositions and operating conditions to examine the potential of the process under conditions relevant for commercial applications. Furthermore, the data will be employed in the simulations of the process.

Experiments were carried out in the three-reactor system at 20 and 30 bar, 320 and 330 °C, over a range of operating conditions, including feed composition and WHSV. The results of 5 runs, 4 of them at 20 bar and run P-63-300 at 30 bar, are listed in Table 10. The oxygenates listed in Table 10 are dissolved in water. The composition of the organic liquid collected in runs P-63-164 and P-65-69 is given in Table 11. WHSV is calculated as the mass flow rate of CO_2 and CO divided by the mass of catalyst. The other parameters are defined in Table 10. sCO_x is defined so that a molar $H_2/sCO_x = 1$ is stoichiometric for all feed compositions. The mass balance in all 4 runs was >94%.

Run P-63-164 was carried out with catalyst containing 4 wt% K at 20 bar with stoichiometric feed. The conversion was relatively high and the selectivity to C_{5+} was >50 wt%. P-63-300 was conducted with a sub-stoichiometric ratio. Thus the CO_2 conversion was lower than that of H_2. WHSV and the pressure were increased from 1.1 h^{-1} and 20 bar to 1.3 h^{-1} and 30 bar, respectively. Increasing the WHSV should lower the conversion while increasing the pressure should increase the conversion. Apparently, the two changes offset each other thus the conversion changed little. The selectivity in the runs was rather similar, although the

Table 9 Adsorption constants of CO_2 and H_2O

K_i	k_{0i}, MPa^{-1}	ΔH_{ai}, kJ mol^{-1}
K_1	2.1×10^{-2}	33.4
K_4	2.4×10^{-2}	31.5

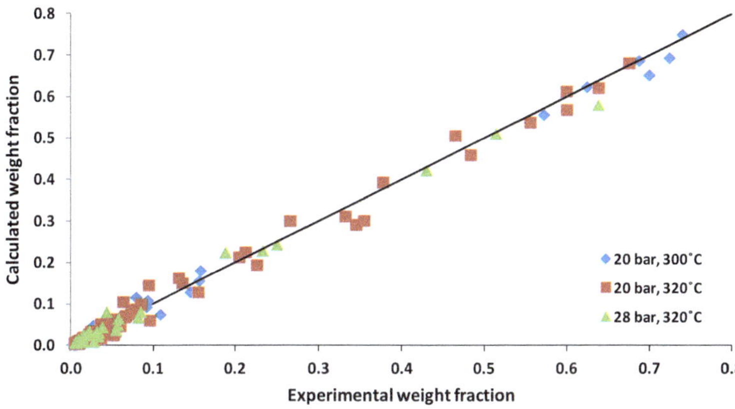

Fig. 5 Parity plot of the calculated and experimental data.

methane selectivity in run P-63-300 was slightly lower and the C_{5+} selectivity slightly higher.

Introducing CO to the feed in the system packed with catalyst containing 4 wt% K led to a gradual deterioration of the catalyst performance as shown in run P-65 depicted in Table 10. This was rectified in Run P-66 by replacing the catalyst in the first reactor with a 3 wt% K catalyst that displayed stable performance. The performance of two reactors in series, P-66-396 R1 (after one reactor) and R2 (after two reactors) listed in Table 10, indicates that the conversion of CO is much higher than the CO_2 conversion in both reactors. The selectivity in both reactors is similar.

The composition of the organic phase in runs P-63-164 and P-66-396 is listed in Table 11. As expected, the liquid contained much more olefins (>65 wt%) than paraffins (<25 wt%). A comparison of the product distribution between naphtha and distillates in the two experiments and the data measured with the 3 wt% K catalyst published elsewhere[14] is depicted in Fig. 7. Surprisingly, the first reactor packed with 3 wt% K catalyst in run P-66-396 yielded a slightly heavier product

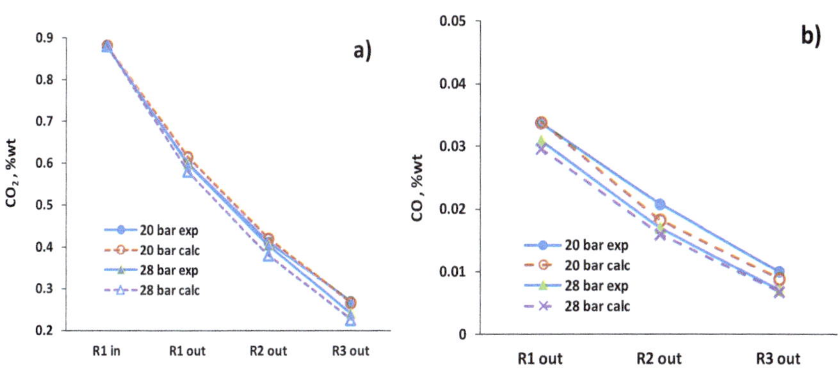

Fig. 6 Calculated *vs.* experimental weight fraction of CO_2 (a) and CO (b) at the outlet of the three reactors system at molar $H_2/CO_2 = 2.9$, 320 °C, WHSV = 3 h^{-1}.

Table 10 Performance of CO_2 and CO hydrogenation process on pelletized catalyst[a]

Run	WHSV h^{-1}	Temp. °C	H_2/ sCO_x molar	CO_2/ CO_x molar	x_1 %	x_2 %	x_3 %	S_4 %	S_5 %	S_6 %	S_7 %	S_8 %	S_9 %	S_{10} %
P-63-164	1.0	320	1.0	1.0	82	81.7	—	10.5	6.2	18.5	2.0	2.7	51.9	8.2
P-63-300	1.3	320	0.9	1.0	77.6	87.9	—	9.6	5.5	15.6	2.1	2.6	54.1	10.5
P-65-45	0.9	330	1.1	0.7	63.8	70.9	92.0	11.6	7.0	19.7	1.8	2.5	49.1	7.9
P-65-69	0.9	330	1.1	0.7	61.6	68.7	90.3	11.3	6.6	17.9	1.8	2.5	51.8	8.1
P-65-93	0.9	330	1.1	0.7	58.6	68.6	90.0	12.0	6.7	18.2	2.0	3.1	50.4	7.8
P-65-141	1.0	330	1.1	0.7	48.3	63.0	83.2	13.8	7.5	12.7	2.4	3.5	50.7	9.8
P-66-396 R1	1.0	320	0.7	0.8	20.6	33.9	48.7	12.9	7.7	22.4	1.9	2.9	45.0	7.2
P-66-396 R2	1.0	320	0.7	0.8	39.0	66.8	86.7	11.1	6.9	21.0	2.3	2.9	47.8	7.9

[a] The reactants are defined as: $CO_2 \equiv A_1$; $H_2 \equiv A_2$; $CO \equiv A_3$ while the products are listed as: $CH_4 \equiv A_4$; $C_2H_4 \equiv A_5$; $C_3H_6 + C_4H_8 \equiv A_6$; $C_2H_6 \equiv A_7$; $C_3H_8 + C_4H_{10} \equiv A_8$; $C_{5+} \equiv A_9$; oxygenates $\equiv A_{10}$; $H_2O \equiv A_{11}$; $sCO_x = 3CO_2 + 2CO$; $CO_x = CO_2 + CO$; x_j is the conversion of reactant j; S_j is the selectivity of product j expressed as the mass fraction of the carbon converted to that product.

compared with the product in the second reactor of the same run and in the three combined reactors in run P-63-164, packed with 4 wt% K catalyst. This is probably due to the fact that the feed to the first reactor contained a significant

Table 11 Composition of the organic liquid in runs P-63-164 and P-66-396

		P-66-396		
		Reactor 1	Reactor 2	P-63-164
C_5–C_{10}	Non α-olefins	10.8	9.2	10.1
	α-Olefins	16	19.9	21.4
	Non *n*-paraffins	5.1	8.4	6.1
	n-Paraffins	5	5.9	6.0
	Aromatics	2.2	2	2.6
	Oxygenates	1.6	1.7	3.3
C_{11}–C_{22}	Non α-olefins	10.5	12.4	12.7
	α-Olefins	20.7	18	17.1
	Non *n*-paraffins	9.5	5.9	5.8
	n-Paraffins	7.8	5.8	5.4
	Aromatics	2.3	3	4.8
	Oxygenates	1.3	3.2	3.0
>C_{22}	Non α-olefins	1	0.6	0.4
	α-Olefins	5.5	3.4	1.3
	Paraffins	0.7	0.6	0.0

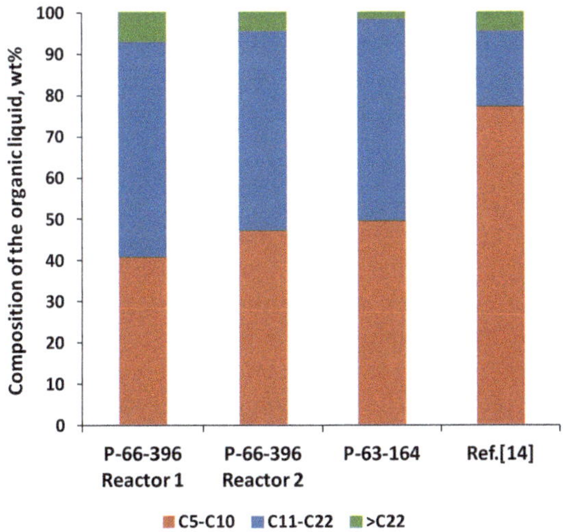

Fig. 7 A comparison of the liquid organic product between the first two reactors in run P-66-396, P-63-164 and ref. 14. First reactor in P-66-396 is loaded with 3 wt% K, second reactor in P-66-396 is loaded with 4 wt% K. P-63-164 is loaded with 4 wt% K.

concentration of CO. Furthermore, the product obtained with a stoichiometric mixture of H_2 and CO_2 in a three-reactor unit packed with a 3 wt% K catalyst[14] yielded a much lighter product than produced with a 4 wt% K catalyst.

4. Conclusions

Potassium was found to be a key promoter that enhances significantly the reaction rates of RWGS and FTS and increases the selectivity to higher hydrocarbons while producing mostly olefins. The methanation rate was decreased extensively on the promoted catalysts. These effects are a result of the lower hydrogenation activity, consistent with the observation of more reduced states of iron atoms at the spent steady state Fe–Al–O catalysts surface after addition of potassium. They could be attributed to the donation of electron density to vacant d-orbitals of iron in both Fe-poor spinel and carbide phases thus enhancing the dissociative adsorption of CO while lowering the H_2 adsorption ability. The stabilization of the Fe-poor spinel phase that prevented decomposition to the magnetite phase contributed to the higher activity and selectivity of the catalyst. Furthermore, potassium enhanced the catalysts surface coverage with CO_2 as shown by CO_2-TPD data. The FTS reactions were accelerated by the enrichment of the catalyst surface with carbide iron, especially the more active Fe_5C_2 phase as established by the XRD and XPS data.

A detailed kinetic model was developed based on extensive experimental data. It includes the RWGS, methanation, 7 FTS, 4 oligomerization and oxygenates formation reactions. The fit of the kinetic model to the experimental data was very good. Specifically, the significant oligomerization activity of the catalyst that converts light olefins to higher hydrocarbons should be mentioned. Running mixtures of CO_2, CO and H_2 in a three reactor system packed with pelletized catalyst yielded a high productivity of higher hydrocarbons. The composition of the C_{5+} product was a function of the potassium content and the composition of the feed. A heavier product was measured with a catalyst containing 4 wt% potassium compared with the 3 wt% potassium catalyst and with a feed containing CO_2, CO and H_2 compared with a feed containing only CO_2 and H_2.

Acknowledgements

This work was supported by the Israel – Strategic Alternative Energy Foundation (I-SAEF), the I-CORE Program of the Planning and Budgeting Committee and The Israel Science Foundation (grant No. 152/11) and grant No. 582/13 of the Israel Science Foundation. The authors are grateful to Dr A. Erenburg for XRD characterization of catalysts and to Dr N. Froumin for characterization of catalysts by XPS.

References

1 World Energy Outlook, IEA, 12 November, 2014.
2 B. Friedrich, B. Schink and R. K. Thauer, *Bioenergy – Chances and Limits*, German National Academy of Sciences Leopoldina, 2012, pp. 1–118, http://www.leopoldina.org.
3 F. Creutzig, C. von Stechow, D. Klein, C. Hunsberger, N. Bauer, A. Popp and O. Edenhoferb, *Economics of Energy & Environmental Policy*, 2012, vol. 1, pp. 65–82.
4 J. A. Herron, J. Kim, A. A. Upadhye, G. W. Huber and C. T. Maravelias, *Energy Environ. Sci.*, 2015, **8**, 126–215.
5 G. Centi and S. Perathoner, *ChemSusChem*, 2010, **3**, 195–208.
6 K. S. Joya, Y. F. Joya, K. Ocakoglu and R. van de Krol, *Angew. Chem., Int. Ed.*, 2013, **52**, 10426–10437.
7 B. A. Pinaud, J. D. Benck, L. C. Seitz, A. J. Forman, Z. Chen, T. G. Deutsch, B. D. James, K. N. Baum, G. N. Baum, S. Ardo, H. Wang, E. Millere and T. F. Jaramillo, *Energy Environ. Sci.*, 2013, **6**, 1983–2002.
8 J. R. McKone, N. S. Lewis and H. B. Gray, *Chem. Mater.*, 2014, **26**, 407–414.
9 C. Agrafiotis, M. Roeb and C. Sattler, *Renewable Sustainable Energy Rev.*, 2015, **42**, 254–285.
10 S. P. S. Badwal, S. Giddey and C. Munnings, *WIREs Energy Environ.*, 2013, **2**, 473–487.
11 J. A. Trainham, J. Newman, C. A. Bonino, P. G. Hoertz and N. Akunuri, *Curr. Opin. Chem. Eng.*, 2012, **1**, 204–210.
12 J. Newman, P. G. Hoertz, C. A. Bonino and J. A. Trainham, *J. Electrochem. Soc.*, 2012, **159**(10), A1722–A1729.
13 C. Graves, S. D. Ebbesen, M. Mogensen and K. S. Lackner, *Renewable Sustainable Energy Rev.*, 2011, **15**, 1–23.
14 M. V. Landau, R. Vidruk and M. Herskowitz, *ChemSusChem*, 2014, **7**, 785–794.
15 D. A. Wood, C. Nwaoha and B. F. Towler, *J. Nat. Gas Sci. Eng.*, 2012, **9**, 196–208.
16 A. H. Lillebø, A. Holmen, B. C. Enger and E. A. Blekkan, *WIREs Energy Environ.*, 2013, **2**, 507–524.
17 R. Rauch, J. Hrbek and H. Hofbauer, *WIREs Energy Environ.*, 2014, **3**, 343–362.
18 I. Hannula, *Biomass Bioenergy*, 2015, **74**, 26–46.

19 L. Yang, X. Ge, C. Wan, F. Yu and Y. Li, *Renewable Sustainable Energy Rev.*, 2014, **40**, 1133–1152.
20 S. C. Kang, K.-W. Jun and Y.-J. Lee, *Energy Fuels*, 2013, **27**, 6377–6387.
21 Y. Yao, X. Liu, D. Hildebrandt and D. Glasser, *Ind. Eng. Chem. Res.*, 2011, **50**, 11002–11012.
22 M. V. Landau, R. Vidruk and M. Herskowitz, WO Patent 2014111919, 2014.
23 J. Yang, W. Ma, D. Chen, A. Holmen and B. H. Davis, *Appl. Catal., A*, 2014, **470**, 250–260.
24 M.-D. Lee, J.-F. Lee and C.-S. Chang, *Bull. Chem. Soc. Jpn.*, 1989, **62**, 2756–2758.
25 M. Martinelli, C. G. Visconti, L. Lietti, P. Forzatti, C. Bassano and P. Deiana, *Catal. Today*, 2014, **228**, 77–88.
26 H. P. Bonzel, *Surf. Sci. Rep.*, 1987, **8**, 43–125.
27 M. E. Dry, T. Shingles, L. J. Boshoff and G. J. Oostzuizen, *J. Catal.*, 1969, **15**, 190–199.
28 E. de Smit and B. M. Weckhuysen, *Chem. Soc. Rev.*, 2008, **37**, 2758–2781.
29 C. Zhang, K.-W. Jun, K.-S. Ha, Y.-J. Lee and S. C. Kang, *Environ. Sci. Technol.*, 2014, **48**, 8251–8257.
30 T. Riedel, G. Schaub, K.-W. Jun and K.-W. Lee, *Ind. Eng. Chem. Res.*, 2001, **40**, 1355–1363.
31 M. Iglesias, C. de Vries, M. Claeys and G. Schaub, *Catal. Today*, 2015, **242**, 184–192.
32 N. Park, J.-R. Kim, Y. Yoo, J. Lee and M.-J. Park, *Fuel*, 2014, **122**, 229–235.
33 M. E. Dry, *Catal. Today*, 1990, **6**, 183–206.
34 H. Wan, B. Wu, C. Zhang, H. Xiang and Y. Li, *J. Mol. Catal. A: Chem.*, 2008, **283**, 33–42.
35 B. H. Davis, *Catal. Today*, 2009, **141**, 25–33.
36 J. B. Butt, *Catal. Lett.*, 1990, **7**, 61–82.
37 K. B. Sunil, A. V. Anupama, B. Bikramjit and S. Balaram, *J. Phys. Chem. C*, 2015, **119**, 6539–6555.
38 http://xpssimplified.com/elements/carbon.php.
39 A. Furlan, U. Jansson, J. Lu, L. Hultman and M. Magnuson, *J. Phys.: Condens. Matter*, 2015, **27**, 045002.
40 X. An, B. Wu, H.-J. Wan, T.-Z. Li, Z.-C. Tao, H.-W. Xiang and Y.-W. Li, *Catal. Commun.*, 2007, **8**, 1957–1962.

Faraday Discussions

PAPER

Plasma-based conversion of CO_2: current status and future challenges

Annemie Bogaerts,* Tomas Kozák, Koen van Laer and Ramses Snoeckx

Received 3rd May 2015, Accepted 29th June 2015
DOI: 10.1039/c5fd00053j

This paper discusses our recent results on plasma-based CO_2 conversion, obtained by a combination of experiments and modeling, for a dielectric barrier discharge (DBD), a microwave plasma and a packed bed DBD reactor. The results illustrate that plasma technology is quite promising for CO_2 conversion, but more research is needed to better understand the underlying mechanisms and to further improve the capabilities.

1 Introduction

The conversion of greenhouse gases (mainly CO_2 and CH_4) into value-added chemicals and liquid fuels is considered as one of the great challenges for the 21st Century.[1] A lot of research has been carried out to develop energy-efficient technologies.[2–4] One of these technologies gaining increasing interest is plasma technology.

Plasma is a partially ionized gas, consisting of a large number of neutral species (different types of molecules, radicals and excited species), as well as electrons and various types of ions. These species can all interact with each other, making plasma a highly reactive chemical cocktail and of interest for many applications.[5,6]

The great potential of plasma technology for CO_2 conversion is due to the presence of energetic electrons. Indeed, plasma is created by applying electric power to a gas, causing the gas to breakdown, *i.e.*, the formation of electrons and positive ions (in addition to other reactive species). As the electrons are much lighter than the other plasma species, they gain the most energy from the electric field, and they do not lose their energy so efficiently by collisions with the other plasma species, explaining their higher energy. These energetic electrons can activate the (inert) gas by electron impact ionization, excitation and dissociation, and the reactive species (*i.e.*, ions, excited species and radicals, respectively) created in this way, will easily undergo other reactions, yielding the formation of new molecules. Thus, the gas itself (*e.g.*, CO_2) does not have to be heated as a whole, but can remain near room temperature. In this way, even strongly

Research Group PLASMANT, University of Antwerp, Department of Chemistry, Universiteitsplein 1, Antwerp, Belgium. E-mail: annemie.bogaerts@uantwerpen.be; Fax: +32-3-265-23-43; Tel: +32-3-265-23-77

endothermic reactions, like CO_2 splitting and dry reforming of methane (DRM), can occur with reasonable energy consumption under mild reaction conditions. However, as the plasma is created by applying electrical power, the energy efficiency of this process is still an important issue.

The most common types of plasma used for CO_2 conversion are dielectric barrier discharges (DBDs),[7–31] microwave (MW) plasmas[32–37] and gliding arc (GA) discharges.[38–45] The highest energy efficiency was reported for a MW plasma, *i.e.*, up to 90%,[32] but this was under very specific conditions, *i.e.*, supersonic gas flow and reduced pressure (~100–200 Torr), and a pressure increase to atmospheric pressure, which would be desirable for industrial applications, yields a dramatic drop in energy efficiency. Indeed, at normal flow conditions and atmospheric pressure, an energy efficiency up to 40% was reported.[6] A GA plasma also exhibits a rather high energy efficiency, even at atmospheric pressure, *i.e.*, around 43% for a conversion of 18% in the case of CO_2 splitting,[45] and even around 60% for a conversion of 8–16%, for DRM.[38] The energy efficiency of a DBD is more limited, *i.e.*, in the order of 2–10%,[8–17] but as demonstrated already for other applications,[46] it should be possible to improve this energy efficiency by inserting a (dielectric) packing into the reactor, *i.e.*, a so-called packed bed DBD reactor. Moreover, it also operates at atmospheric pressure, and has a very simple design, which is beneficial for upscaling, as has been demonstrated already for the large scale production of ozone,[47] and therefore it also has high potential for industrial applications. Finally, when combined with catalytic packing, it should enable the selective production of targeted compounds.[18–23]

Within our research group PLASMANT, we investigate both pure CO_2 conversion[15,48–50] and DRM,[16,17] by means of experiments and computer modeling. We have also investigated the effect of adding H_2 or CH_4, as a means to better separate the product gases,[24] the effect of adding He or Ar,[31] or N_2 [51] which is mostly present in industrial gas flows. Our research up to now was focused on DBD and MW plasma. By means of computer simulations, validated by experiments, we try to elucidate the underlying reaction chemistry, and to investigate how the process can be optimized in terms of conversion and energy efficiency. In this paper, we give an overview of some characteristic, recent results obtained within our group to illustrate the state-of-the-art of plasma-based CO_2 conversion, and we will also try to identify the most important challenges for the future.

2 Experimental setup

The experiments were all carried out in a DBD reactor, of which a schematic drawing is presented in Fig. 1. It consists of a central grounded electrode with variable diameter (between 8 and 13 mm), surrounded by a coaxial dielectric tube, with inner and outer diameter of 17 and 22 mm, respectively. The dielectric tube is covered by a stainless steel mesh, which serves as the outer electrode and is powered by a high voltage power source. The length of this outer electrode is 9 cm, and this defines the length of the plasma zone.

The DBD reactor is coupled to a gas chromatograph to analyse the gas flowing out of the reactor, and to calculate the CO_2 conversion, the yields and selectivities of the formed products, and the energy efficiency, as calculated from the conversion, the power input in the plasma and the gas flow rate (see *e.g.*, ref. 15 for more details).

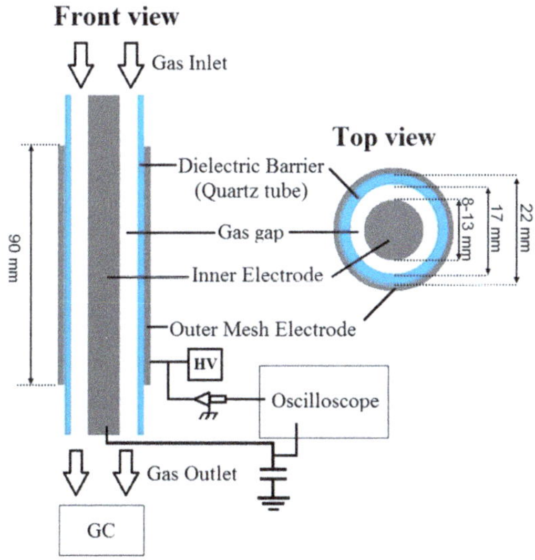

Fig. 1 Schematic diagram of the experimental setup, in front view and top view.

We performed experiments in a normal (*i.e.*, empty) DBD reactor, as well as in a DBD reactor filled with dielectric packing, *i.e.*, a packed bed DBD reactor. For this purpose, we used the smallest rod diameter, yielding a discharge gap (*i.e.*, distance between the central electrode and dielectric tube) of approximately 4.5 mm, and a discharge volume of about 15 cm^3. We introduced ZrO_2 beads (SiLi-Beads), with five different bead size ranges, *i.e.*, 0.90–1.00, 1.00–1.18, 1.25–1.40, 1.60–1.80 and 2.00–2.24 mm diameter.

3 Computational model

We used a zero-dimensional (0D) chemical reaction kinetics model to describe the underlying plasma chemistry of CO_2 conversion. It consists of solving balance equations for the species densities, based on production and loss rates, as determined by the chemical reactions (see *e.g.* ref. 16 and 48 for details).

One balance equation is solved for each species included in the model, *i.e.*, different types of molecules, radicals, ions, excited species, as well as electrons (see below). These balance equations yield the time-evolution of the species densities, averaged over the plasma reactor volume. Indeed, because it is a 0D model, it only accounts for time-variations, while spatial variations, due to transport in the plasma, are not considered. However, based on the gas flow rate, we can translate the time-variation into a spatial variation, *i.e.*, as a function of distance travelled through the plasma reactor. Besides the species densities, the average electron energy is also calculated, based on an energy balance equation, again with the energy source and loss terms as defined by the chemical reactions.

We developed a 0D chemical kinetics model for different gas mixtures relevant for CO_2 conversion, *i.e.*, pure CO_2,[15,48–50] CO_2/CH_4 [16,17] and CO_2/N_2,[51] but here we only show results for pure CO_2. Table 1 gives an overview of the species included in the pure CO_2 model.

Table 1 Overview of the species included in the CO_2 model

Molecules	Charged species	Radicals	Excited species
CO_2, CO	CO_2^+, CO_4^+, CO^+, $C_2O_2^+$, $C_2O_3^+$, $C_2O_4^+$, C_2^+, C^+, CO_3^-, CO_4^-	C_2O, C, C_2	CO_2(Va, Vb, Vc, Vd), CO_2(V1–V21), CO_2(E1, E2), CO(V1–V10), CO(E1–E4)
O_2, O_3,	O^+, O_2^+, O_4^+, O^-, O_2^-, O_3^-, O_4^- Electrons	O	O_2(V1–V4), O_2(E1–E2)

The vibrational levels of CO_2 can play an important role in CO_2 conversion, depending on the type of plasma to be studied. Indeed, while they are of minor importance in a DBD,[48] they are crucial for CO_2 splitting in a MW plasma.[49,50] This will be illustrated in Section 4.2 below. For this reason, we have developed an extensive chemical kinetics model, taking into account the CO_2, CO and O_2 vibrational levels.[49,50] Hence, the symbols "V" and "E" between brackets for CO_2, CO and O_2 represent the vibrationally and electronically excited levels of these species respectively. Details about these notations can be found in.[49,50]

These species will all chemically react with each other. Hence, a large number of chemical reactions are incorporated in these models, including electron impact reactions, electron–ion recombinations, ion–ion, ion–neutral and neutral–neutral reactions. All details about these chemical reaction sets, as well as the corresponding rate coefficients, can be found in ref. 15, 48–50.

4 Results and discussion

We present here the results obtained from our experiments and computer simulations, first for CO_2 splitting in a DBD reactor, followed by a comparison of a DBD reactor and MW plasma, and finally, we will show how the CO_2 conversion and energy efficiency can be improved in a packed bed DBD reactor.

4.1 CO_2 splitting in a DBD reactor

Fig. 2 illustrates the CO_2 conversion and the corresponding energy efficiency as a function of the specific energy input (SEI), measured in a DBD reactor. The SEI is the ratio of plasma power over gas flow rate. Therefore, different combinations of power and gas flow rate can give rise to the same SEI. The SEI is typically considered as the major determining factor for conversion and energy efficiency.

It is clear that the CO_2 conversion rises with SEI, which is logical as more energy is put into the system, either by applying more power for the same amount of gas, or by applying a lower gas flow rate (which corresponds to a longer residence time) at the same power. However, above a certain SEI, the measured conversion seems to saturate, and we did not obtain higher conversion values than 35% in our experiments. On the other hand, the energy efficiency drops upon increasing SEI, which is also logical, because the energy efficiency is proportional to the conversion but inversely proportional to the SE (see formula in ref. 15). Indeed, it is obvious from this formula that when the conversion does not rise to the same extent as the SEI, the energy efficiency will drop.

The highest energy efficiency obtained in this case is 8%, but this corresponds to a very low conversion of only a few %. On the other hand, the highest

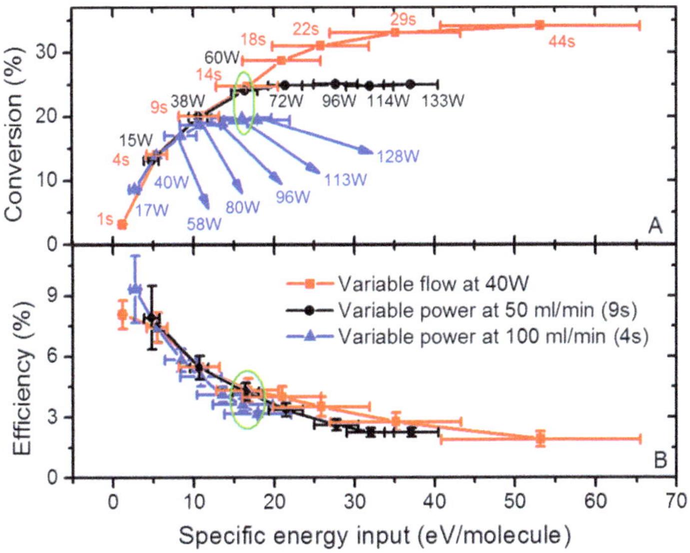

Fig. 2 Measured CO_2 conversion (A) and energy efficiency (B) in a DBD reactor, as a function of the specific energy input (SEI), using alumina dielectrics. The corresponding values of plasma power, resulting in certain SEI values at fixed gas flow rates of 50 and 100 ml min^{-1} (black and blue curves), as well as the corresponding values of the residence time, resulting in certain SEI values at a fixed plasma power of 40 W (red curve), are also shown in (A). The calculation of the error bars is based on the uncertainties of the power, the flow rate and the GC measurements. For the sake of clarity, the error bars are only presented for the energy efficiency. Adopted from ref. 15 with kind permission of Wiley-VCH Verlag.

conversion of 35% corresponds to a very low energy efficiency of only 2%. Thus, there is clearly a trade-off between conversion and energy efficiency as a function of SEI.

The values obtained for conversion and energy efficiency are comparable to the data reported in the literature for similar conditions (*e.g.*, ref. 7 and 8). We can conclude that the obtained CO_2 conversion is reasonable in a DBD reactor, but the energy efficiency is clearly too low for industrial implementation. Indeed, Spencer *et al.* estimated that if all the electrical energy for CO_2 splitting originated from fossil fuels, an energy efficiency of 52% would be needed to ensure that more CO_2 can be split in the plasma than the amount of CO_2 created by fossil fuel combustion in the electricity production needed for sustaining the plasma.[52] When using renewable electricity, this criterion might be somewhat less severe, but still, the energy efficiency is an important issue.

In addition to the experiments, we have also developed a detailed model for CO_2 conversion in a DBD plasma, to investigate in detail the role of the various processes contributing to CO_2 splitting.[48] The calculations predict that electron impact dissociation of ground-state CO_2 is the dominant process for CO_2 conversion in a DBD, and the role of the CO_2 vibrational levels is limited in this case[48] (see more details in Section 4.2 below).

Subsequently, the model was extended to calculate the CO_2 conversion in real time.[15] The obtained CO_2 conversion and energy efficiency were in very good agreement with measured data.[15] Therefore, the model can be used to elucidate the underlying reaction paths of CO_2 conversion.

A simplified reaction scheme of CO_2 splitting, as obtained from the model, is illustrated in Fig. 3. It is clear that the actual CO_2 splitting is quite straightforward. The most important reactions are electron impact dissociation into CO and O (reaction r1), electron impact ionization into CO_2^+ (r2), which recombines with electrons or O_2^- ions into CO and O and/or O_2 (r3, r4), and electron dissociative attachment into CO and O^- (r5). The created CO molecules are relatively stable, but at a long enough residence time, they can recombine with O^- ions or O atoms, to form CO_2 again (r6, r7). This explains, among others, why the CO_2 conversion tends to saturate at high SEI values (corresponding to low gas flow rates or long residence times). At shorter residence times, the O atoms will, however, almost immediately recombine into O_2 or O_3. Moreover, there are several other reactions possible between O, O_2 and O_3, sometimes also involving O^- and O_2^- ions. The details of these reactions are not indicated in Fig. 3, but can be found in ref. 15. These reactions will affect the balance between the formation of O_2 and O_3 as stable products. Our model indeed predicts that the selectivity towards CO formation is always close to 50%, but the selectivity towards O_2 formation varies between 45 and 50%, depending on O_3 production.

4.2 Comparison of CO_2 splitting in DBD and MW plasma

The fact that the energy efficiency for CO_2 splitting is quite limited in a DBD plasma, as mentioned above, is because the reduced electric field is quite high (typically above 200 Td or 200×10^{-21} V m^2), yielding an average electron energy of 2–3 eV,[48] which is somewhat too high for efficient population of the CO_2

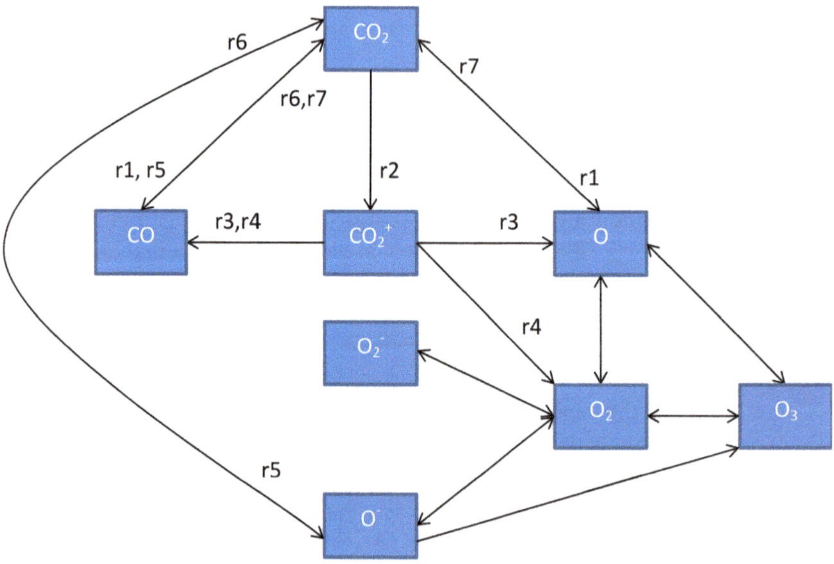

Fig. 3 Chemical reaction scheme, illustrating the chemistry of CO_2 splitting and further reactions between O, O_2 and O_3, as predicted by the model. The labels of the arrows are explained in the text.

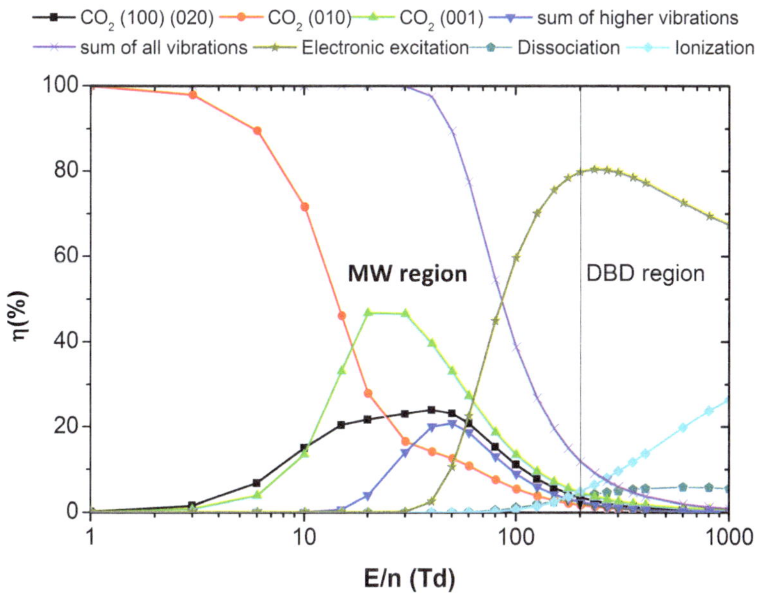

Fig. 4 The fraction of electron energy transferred to different channels of excitation as well as ionization and dissociation of CO_2, as a function of the reduced electric field (E/n), as calculated from the corresponding cross sections of the electron impact reactions. The E/n region characteristic for MW plasma and DBD plasma are indicated.

vibrational levels.[49] To illustrate this, Fig. 4 depicts the fraction of the electron energy transferred to different channels of excitation, ionization and dissociation of CO_2, as a function of the reduced electric field (E/n), as calculated from the cross sections of the corresponding electron impact reactions. It is clear that in the region above 200 Td, indicated as "DBD region", 70–80% of the electron energy goes into electronic excitation, while the remaining 20–30% is used for ionization (increasing with rising E/n) and about 5% goes into dissociation. Note that this electron impact dissociation is also induced by electronic excitation, and thus requires a lot of energy.[48] The fraction of the electron energy going into vibrational excitation is 12% at $E/n = 200$ Td, but drops quickly upon increasing E/n. Thus, vibrational excitation is of minor importance in the "DBD region".

In a MW plasma, the reduced electric field is typically around 50 Td, which is most appropriate for vibrational excitation of CO_2; see Fig. 4. The green curve is especially important, as this represents the first vibrational level of the asymmetric stretching mode, which is known to provide the most important channel for dissociation.[6,49] Thus, we can deduce already from Fig. 4 that a MW plasma will give rise to a high population of CO_2 vibrational levels, which contribute to energy-efficient CO_2 splitting.

Fig. 5 shows a comparison of the CO_2 conversion and energy efficiency in a MW plasma and DBD reactor as a function of SEI, as predicted by our model taking into account the CO_2 vibrational levels.[49] Note that the results of the MW plasma are obtained for a reduced pressure of 2660 Pa (20 Torr), as used in the experiments of ref. 33 and 34 while the DBD results are for atmospheric pressure. Therefore, to compare both discharges, we need to show them at the same SEI in

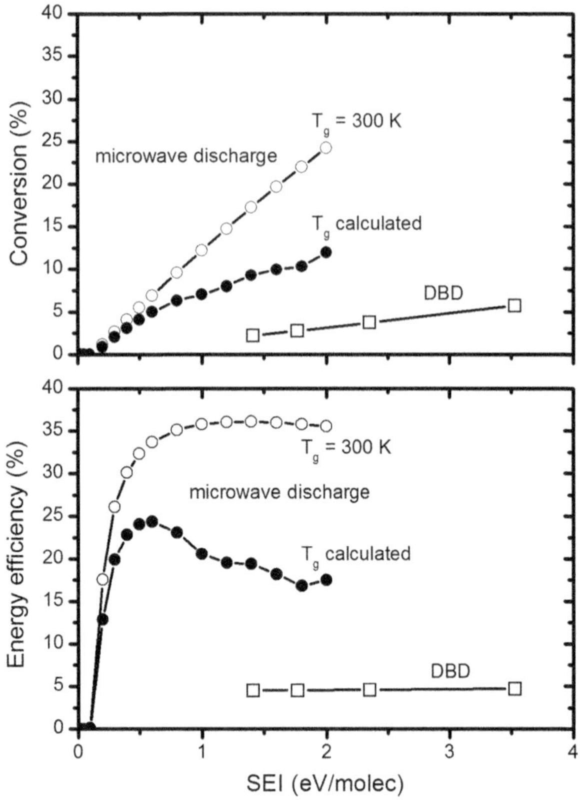

Fig. 5 Calculated CO_2 conversion (top) and energy efficiency (bottom) in a moderate pressure MW discharge and atmospheric pressure DBD reactor, as a function of SEI. The pressure in the MW plasma is 2660 Pa (20 Torr). Calculations are performed for two different gas temperatures, *i.e.*, 300 K, like assumed in the DBD reactor, and a more realistic self-consistently calculated gas temperature as a function of time, reaching values up to 1000 K.

eV per molec, because this is the most fundamental parameter for comparison. Furthermore, the DBD results are obtained at a gas temperature of 300 K, while the MW results are shown both for a fixed gas temperature of 300 K (to allow a more direct comparison with the DBD results, as the rate coefficients of most chemical reactions are a function of gas temperature), as well as for a more realistic gas temperature, self-consistently calculated in the model as a function of time (or distance in the reactor).[50] In this case, values up to 1000 K are reached; see temperature profile in ref. 50.

It is clear that both the CO_2 conversion and energy efficiency are calculated to be much higher in the MW plasma than in the DBD reactor. The conversion rises as a function of SEI in both cases, which is logical (see also previous section). However, in the DBD reactor, the conversion reaches only about 5%, at an SEI of 3.5 eV per molecule, while in the MW plasma with a realistic (calculated) gas temperature, the CO_2 conversion is already 12% at an SEI of 2 eV per molecule. Note that the calculated conversion in the MW plasma at a fixed gas temperature of 300 K is much higher (and thus overestimated), compared to the result at the higher (more realistic) gas temperature. The reason is that the higher gas temperature gives rise to more vibrational-translational (VT) relaxation collisions of the CO_2 vibrational levels, which is the most important loss mechanism for the vibrational population.[49,50] This explains the lower conversion in the case with a higher gas temperature. Nevertheless, the CO_2 conversion is still significantly higher than in the DBD reactor operating at 300 K.

The latter is certainly also true for the energy efficiency, which is calculated to be around 5% in the DBD reactor (more or less independent from the SEI, as the conversion rises proportionally with the SEI), and it reaches values above 35% (when assuming a constant gas temperature of 300 K; thus overestimated) and around 25% (in the case of the self-consistently calculated gas temperature). In the latter case, the energy efficiency reaches its maximum at an SEI around 0.6 eV per molecule, which is in good agreement with the theoretical and experimental results presented in ref. 6, although in that case, energy efficiencies up to 80–90% were reported. Below, we will discuss the major effects that limit the maximum energy efficiency in our case.

As mentioned above, the reason for the higher CO_2 conversion and energy efficiency in the MW plasma is attributed to the higher population of the CO_2 vibrational levels. This is indeed apparent from Fig. 6, where the calculated vibrational distribution function of the CO_2 asymmetric mode levels (*i.e.*, the mode which is most important for CO_2 splitting; *cf.* above) is plotted, for both the DBD and MW plasma, for an SEI of 0.6 eV per molecule. In the DBD reactor, the population of the vibrational levels drops over several orders of magnitude compared to the ground state density, even for the lowest levels. The corresponding vibrational temperature is calculated to be 961 K. In the MW plasma, the vibrational distribution drops much more smoothly, yielding a vibrational temperature of 4115 K. Although the population of the highest vibrational levels is much lower than the ground state density, they still play an important role in the CO_2 splitting process, which explains the higher CO_2 conversion and energy

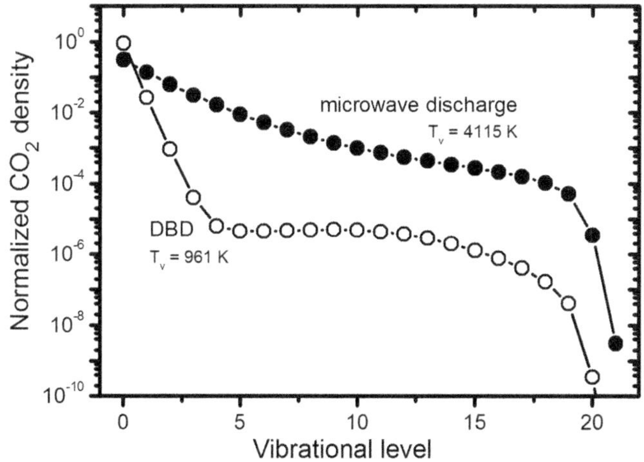

Fig. 6 Normalized vibrational distribution function of the asymmetric mode levels of CO_2 in a moderate pressure MW discharge and atmospheric pressure DBD plasma, at an SEI of 0.6 eV per molecule, both taken at the time of maximum vibrational temperature.

efficiency. Indeed, while in the DBD reactor, electron impact excitation–dissociation from the CO_2 ground state is mainly responsible for CO_2 splitting (*cf.* previous section), in the MW plasma, the CO_2 splitting predominantly proceeds by electron impact vibrational excitation of the lowest vibrational levels, followed by vibrational–vibrational (VV) collisions, gradually populating the higher vibrational levels, which then lead to dissociation of the CO_2 molecule. This stepwise vibrational excitation process, or so-called "ladder-climbing" process, is schematically illustrated in Fig. 7, and is indeed responsible for the much higher energy efficiency in a MW plasma.

Our model also allows us to identify the discharge conditions that favour the highest energy efficiency for CO_2 conversion. The highest value reached in our calculations, in the case of the realistic gas temperature, was around 32%. This value was obtained at an SEI in the range of 0.4–1.0 eV per molecule and a reduced electric field in the range of 50–100 Td.[50] Moreover, our calculations predict that a shorter residence time favours a higher energy efficiency, because in that case the time for VT relaxation, which depopulates the vibrational levels, is longer than the residence time of the gas within the plasma. This corresponds well with the fact that the highest energy efficiencies were reported at supersonic flow conditions.[32,35]

The best energy efficiencies obtained experimentally in a MW plasma at moderate pressure were 80% in subsonic flow, and up to 90% in supersonic flow conditions.[6,32] These results were obtained in 1983, and to our knowledge, nobody has been able to reproduce them since. Recently however, Goede *et al.* were able to reach energy efficiencies as high as 55% with a MW plasma at moderate pressure

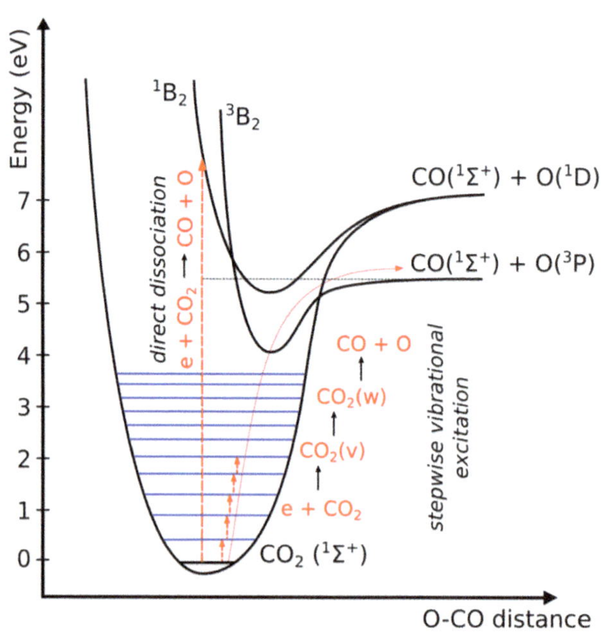

Fig. 7 Schematic diagram of some CO_2 electronic and vibrational levels, illustrating that much more energy is needed for direct electronic excitation–dissociation than for stepwise vibrational excitation, *i.e.*, the so-called ladder climbing process.

and again under supersonic flow conditions,[35] which is still higher than the values obtained by our model. Thus, to better understand the limitations in the energy efficiency, we have analysed how the vibrational energy of CO_2 is consumed by individual reactions. Our model predicts that up to 60% of the energy available in the CO_2 vibrational levels can be used for CO_2 dissociation, at least at high enough electron density (order of 10^{20} m^{-3} at a pressure of 100 Torr). The remaining fraction of the energy is largely lost by VT relaxation, which gives rise to the gas heating. *Vice versa*, because a higher gas temperature gives rise to higher VT relaxation rates, it is desirable to keep the gas temperature as low as possible, to minimize VT relaxation losses in the vibrational population. This is also one of the reasons why the energy efficiency drops upon increasing gas pressure, because of the increasing V-T relaxation processes. One way to reduce this effect is by using a fast gas flow, as mentioned above.

4.3 CO_2 splitting in a packed bed DBD reactor

Finally, we have also investigated whether we can improve the energy efficiency in a DBD plasma, by adding a dielectric packing in the reactor. More specifically, we have inserted ZrO_2 beads, with five different bead size ranges, *i.e.*, 0.90–1.00, 1.00–1.18, 1.25–1.40, 1.60–1.80 and 2.00–2.24 mm diameter, in a DBD reactor with a gap size of 4.5 mm.

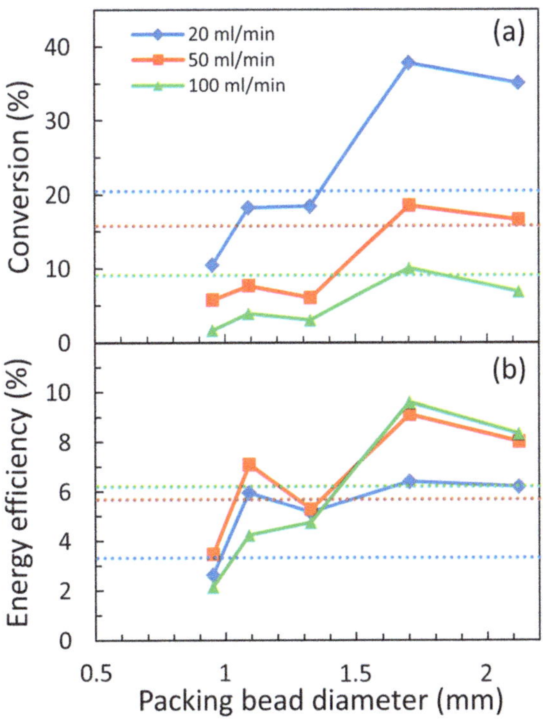

Fig. 8 Measured CO_2 conversion (a) and corresponding energy efficiency (b), as a function of packing bead diameter for ZrO_2, at a power of 60 W and three different gas flow rates. The corresponding results without packing are indicated with dashed horizontal lines.

Fig. 8 shows the measured CO_2 conversion (a) and corresponding energy efficiency (b), as a function of bead diameter, for three different gas flow rates, *i.e.* 20, 50 and 100 ml min^{-1} at an applied power of 60 W. The results are also compared with experiments without packing (dashed horizontal lines), which serve as a benchmark to define the improvement in conversion and energy efficiency.

It is clear that a packed bed reactor can result in a better conversion and energy efficiency than without packing, but only for bead diameters above 1.4 mm. Indeed, the results for lower bead sizes are even worse than without packing. This can be explained because the residence time in the reactor filled with smaller beads is probably too low to benefit from the enhancement effect due to the presence of the packing material. The best results, in terms of both conversion and energy efficiency, are obtained for a flow rate of 20 ml min^{-1} and a bead diameter of 1.6–1.8 mm. In this case, the conversion reaches 38%, which is almost a factor 2 higher than without packing, while the energy efficiency is 6.4%, which is also nearly a factor 2 higher than without packing. In particular, the fact that both conversion and energy efficiency are improved simultaneously is quite promising. The combination of maximum conversion and energy efficiency obtained here, is comparable to or slightly better than the results reported in the literature for a packed bed DBD reactor with various types of dielectric materials (silica gel, quartz, α-Al_2O_3, γ-Al_2O_3, $CaTiO_3$ and $BaTiO_3$) under similar conditions.[8,25]

To explain why the packed bed reactor yields a better conversion and energy efficiency, we have developed a 2D fluid model within the COMSOL Multiphysics software, for a simplified axisymmetric geometry of a packed bed reactor, consisting of only two beads with a diameter of 2.25 mm. This simplified geometry in 2D is needed to keep the simulation time reasonable. Indeed, to resolve the plasma behaviour near the contact points of the beads, a very narrow mesh size is needed (*i.e.*, typically around 50 μm in the bulk, but up to 0.1 μm near the contact points). Thus, the simulation domain contains in the order of 100 000 mesh points, leading to calculation times of a few weeks for a few periods of the applied voltage, even for this simple geometry. Therefore, this model is, in the first instance, developed for a helium plasma instead of a CO_2 plasma, because this yields a more simple plasma chemistry, and it is sufficient to explain the behaviour of the packed bed effect.

Fig. 9 illustrates the calculated electric displacement field distribution within the beads and in the plasma and the resulting electron temperature profile in the plasma in this simplified geometry. It is clear that the electric displacement field is much higher near the contact points. This is attributed to polarization of the dielectric material as a result of the applied potential. At the contact points there will thus be local charges of opposite sign close together, which leads to a higher electric displacement field, as well as a locally enhanced electric field in the plasma. The latter gives rise to an enhanced electron temperature for the same applied power. Indeed, the electron temperature is up to 8 eV near the contact points, while it is only 2–3 eV in an empty DBD reactor (or far away from the contact points; see Fig. 9). This means that the applied electric power is used more efficiently for heating the electrons, which can then transfer their energy to CO_2 splitting, by electron impact ionization and excitation–dissociation, and this explains the higher CO_2 conversion and energy efficiency, as shown in Fig. 8.

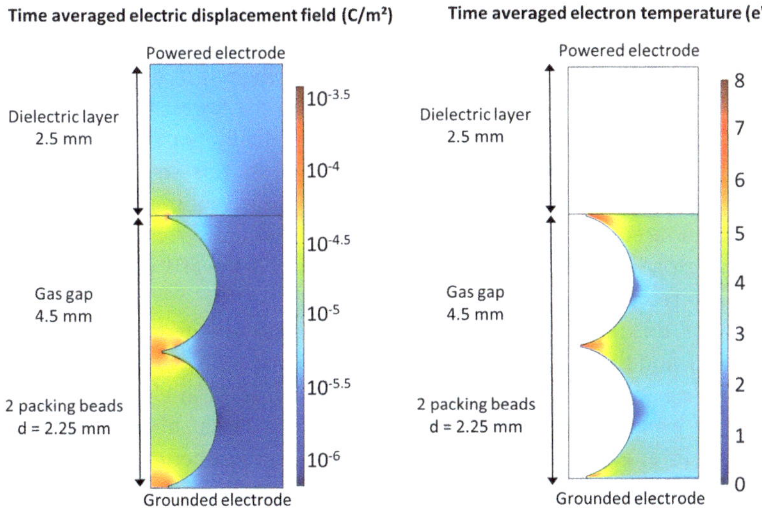

Fig. 9 Calculated time averaged electric displacement field and electron temperature over one period of the applied potential.

5 Conclusion and outlook for future work

We have presented our recent results on CO_2 conversion, obtained by means of combined experimental and computer modeling efforts, for a DBD reactor and a MW plasma. A DBD plasma provides a reasonable conversion, in the order of 30%. However, the corresponding energy efficiency is only in the order of 10%, and this is probably too low for industrial implementation. Indeed, when all the electrical energy to sustain the plasma originates from fossil fuels, it was estimated that an energy efficiency of 52% would be needed for CO_2 conversion, to compensate for the CO_2 production by the fossil fuel combustion. On the other hand, a DBD plasma is very flexible, as it can easily be switched on and off. Therefore, it has great potential to be combined with renewable energy sources (wind turbines or solar panels), *i.e.*, for the storage of peak renewable energy into chemicals or fuels.

Furthermore, there is still room for improvement in the conversion and energy efficiency of a DBD plasma, by inserting dielectric beads, *i.e.*, in a packed bed DBD reactor, as demonstrated in this paper for ZrO_2. The reason for this higher conversion and energy efficiency is the enhanced electric field near the contact points of the beads, yielding a higher electron temperature, which facilitates electron impact dissociation of CO_2, as could be explained by our model. Moreover, when inserting catalytic packing in a DBD reactor, the selective production of specific products can be targeted. This was already demonstrated many times in the literature for air pollution control (*e.g.*, ref. 46 and 53–55), but it has also great potential for CO_2 conversion.[18–23] Nevertheless, we believe that a lot of research will still be needed to find out which catalyst materials are most promising for the selective production of specific compounds.

When comparing a DBD plasma and the MW plasma, it is clear that MW plasma exhibits a much better CO_2 conversion and energy efficiency. This is attributed to the important role of the CO_2 vibrational levels. Indeed, our model calculations have elucidated that the CO_2 conversion proceeds by direct electron impact excitation–dissociation in a DBD plasma, whereas in a MW plasma, the dominant process is electron impact excitation to the lowest vibrational levels, followed by vibrational–vibrational collisions, gradually populating the higher vibrational levels, which give rise to dissociation. As this stepwise vibrational excitation process, or so-called ladder climbing process, requires significantly less energy than direct electron impact excitation–dissociation from the CO_2 ground state, this explains the much better energy efficiency in the MW plasma compared to the DBD reactor.

However, this good energy efficiency is obtained at a moderate pressure (order of 3000 Pa), and this was also the case for the experimental data published in the literature (*e.g.*, ref. 32 and 35). This is not so practical for high-throughput processing of exhaust gases. Increasing the pressure leads, however, to a clear reduction in energy efficiency,[6,36,37] although at atmospheric pressure, a CO_2 conversion of 45% with an energy efficiency of 20% were recently reported,[37] which is still better than the results obtained with a DBD plasma.

We also illustrated that computer modelling can contribute to a better insight in the underlying plasma chemistry of CO_2 conversion, and this will be useful for further improving the performance of plasma technology for this application.

To conclude, we believe that plasma technology is very promising for CO_2 conversion into value-added chemicals and new fuels, but still a lot of research will be needed to further improve the capabilities.

Acknowledgements

We thank R. Aerts and W. van Gaens for setting up the experimental systems and for the interesting results obtained during their PhD study in our group. We also acknowledge financial support from the IAP/7 (Inter-university Attraction Pole) program 'PSI-Physical Chemistry of Plasma-Surface Interactions' by the Belgian Federal Office for Science Policy (BELSPO), the Fund for Scientific Research Flanders (FWO) and the EU-FP7-ITN network "RAPID".

References

1 *Climate Change 2007: The Physical Science Basis*, Contribution of Working Group I to the Fourth Assesment Report of the Intergovernmental Panel on Climate Change, S. Solomon, D. Qin, M. Manning, M. Marquis, K. Averyt, M. M. B. Tignor, H. L. Miller and Z. Chen, Cambridge University Press, New York, 2007.

2 G. Centi, E. A. Quadrelli and S. Perathoner, *Energy Environ. Sci.*, 2013, **6**, 1711.

3 M. Aresta, A. Dibenedetto and A. Angelini, *Chem. Rev.*, 2014, **114**, 1709.

4 P. Styring, E. A. Quadrelli and K. Armstrong, *Carbon Dioxide Utilization: Closing the Carbon Cycle*, Elsevier, Amsterdam, 2015.

5 A. Bogaerts, E. Neyts, R. Gijbels and J. J. A. M. van der Mullen, *Spectrochim. Acta, Part B*, 2002, **57**, 609.

6 A. Fridman, *Plasma Chemistry*, Cambridge University Press, Cambridge, 2008.

7 S. Paulussen, B. Verheyde, X. Tu, C. De Bie, T. Martens, D. Petrovic, A. Bogaerts and B. Sels, *Plasma Sources Sci. Technol.*, 2010, **19**, 034015.

8 Q. Yu, M. Kong, T. Liu, J. Fei and X. Zheng, *Plasma Chem. Plasma Process.*, 2012, **32**, 153.
9 Q. Wang, B.-H. Yan, Y. Jin and Y. Cheng, *Plasma Chem. Plasma Process.*, 2009, **29**, 217.
10 V. Goujard, J.-M. Tatibouët and C. Batiot-Dupeyrat, *Appl. Catal., A*, 2009, **353**, 228.
11 X. Zhang and M. S. Cha, *J. Phys. D: Appl. Phys.*, 2013, **46**, 415205.
12 X. Tu and J. C. Whitehead, *Appl. Catal., B*, 2012, **125**, 439.
13 Y. Zhang, Y. Li, Y. Wang, C. Liu and B. Eliasson, *Fuel Process. Technol.*, 2003, **83**, 101.
14 A. Ozkan, T. Dufour, G. Arnoult, P. De Keyzer, A. Bogaerts and F. Reniers, *J. CO2 Util.*, 2015, **9**, 74.
15 R. Aerts, W. Somers and A. Bogaerts, *ChemSusChem*, 2015, **8**, 702.
16 R. Snoeckx, R. Aerts, X. Tu and A. Bogaerts, *J. Phys. Chem. C*, 2013, **117**, 4957.
17 R. Snoeckx, Y. X. Zeng, X. Tu and A. Bogaerts, *RSC Adv.*, 2015, **5**, 29799.
18 A.-J. Zhang, A.-M. Zhu, J. Guo, Y. Xu and C. Shi, *Chem. Eng. J.*, 2010, **156**, 601.
19 A. Gómez-Ramírez, V. J. Rico, J. Cotrino, A. R. González-Elipe and R. M. Lambert, *ACS Catal.*, 2014, **4**, 402.
20 M. Scapinello, L. M. Martini and P. Tosi, *Plasma Processes Polym.*, 2014, **11**, 624.
21 Q. Wang, Y. Cheng and Y. Jin, *Catal. Today*, 2009, **148**, 275.
22 Q. Wang, B.-H. Yan, Y. Jin and Y. Cheng, *Energy Fuels*, 2009, **23**, 4196.
23 B. Eliasson, U. Kogelschatz, B. Xue and L.-M. Zhou, *Ind. Eng. Chem. Res.*, 1998, **37**, 3350.
24 R. Aerts, R. Snoeckx and A. Bogaerts, *Plasma Processes Polym.*, 2014, **11**, 985.
25 D. Mei, X. Zhu, Y. He, J. D. Yan and X. Tu, *Plasma Sources Sci. Technol.*, 2015, **24**, 015011.
26 J. Wang, G. Xia, A. Huang, S. L. Suib, Y. Hayashi and H. Matsumoto, *J. Catal.*, 1999, **185**, 152.
27 R. Li, Q. Tang, S. Yin and T. Sato, *J. Phys. D: Appl. Phys.*, 2007, **40**, 5187.
28 S. Wang, Y. Zhang, X. Liu and X. Wang, *Plasma Chem. Plasma Process.*, 2012, **32**, 979.
29 C. De Bie, T. Martens, J. van Dijk, S. Paulussen, B. Verheyde and A. Bogaerts, *Plasma Sources Sci. Technol.*, 2011, **20**, 024008.
30 N. R. Pinhao, A. Janeco and J. B. Branco, *Plasma Chem. Plasma Process.*, 2011, **31**, 427.
31 M. Ramakers, I. Michielsen, R. Aerts, V. Meynen and A. Bogaerts, *Plasma Processes Polym.*, 2015, **12**, 755–763.
32 R. I. Asisov, V. K. Givotov, E. G. Krasheninnikov, B. V. Potapkin, V. D. Rusanov and A. Fridman, *Soviet Physics - Doklady*, 1983, **271**, 94.
33 N. Britun, T. Godfroid and R. Snyders, *Plasma Sources Sci. Technol.*, 2012, **21**, 035007.
34 T. Silva, N. Britun, T. Godfroid and R. Snyders, *Plasma Sources Sci. Technol.*, 2014, **23**, 025009.
35 A. P. H. Goede, W. A. Bongers, M. F. Graswinckel, M. C. M. van de Sanden, M. Leins, J. Kopecki, A. Schulz and M. Walker, *3rd European Energy Conference*, Hungary, Budapest, 2013.
36 A. Vesel, M. Mozetic, A. Drenik and M. Balat-Pichelin, *Chem. Phys.*, 2011, **382**, 127.

37 L. F. Spencer and A. D. Gallimore, *Plasma Sources Sci. Technol.*, 2013, **22**, 015019.

38 X. Tu and J. C. Whitehead, *Int. J. Hydrogen Energy*, 2014, **39**, 9658.

39 A. Indarto, D. R. Yang, J.-W. Choi, H. Lee and H. K. Song, *J. Hazard. Mater.*, 2007, **146**, 309.

40 A. Indarto, J. Choi, H. Lee and H. K. Song, *Environ. Eng. Sci.*, 2006, **23**, 1033.

41 Z. Bo, J. Yan, X. Li, Y. Chi and K. Cen, *Int. J. Hydrogen Energy*, 2008, **33**, 5545.

42 Y.-C. Yang, B.-J. Lee and Y.-N. Chun, *Energy*, 2009, **34**, 172.

43 N. Rueangjitt, T. Sreethawong, S. Chavadej and H. Sekiguchi, *Chem. Eng. J.*, 2009, **155**, 874.

44 C. S. Kalra, Y. I. Cho, A. Gutsol, A. Fridman and T. S. Rufael, *Rev. Sci. Instrum.*, 2005, **76**, 025110.

45 T. Nunnally, K. Gutsol, A. Rabinovich, A. Fridman, A. Gutsol and A. Kemoun, *J. Phys. D: Appl. Phys.*, 2011, **44**, 274009.

46 H. L. Chen, H. M. Lee, S. H. Chen and M. B. Chang, *Ind. Eng. Chem. Res.*, 2008, **47**, 2122.

47 U. Kogelschatz, *Plasma Chem. Plasma Process.*, 2003, **23**, 1.

48 R. Aerts, T. Martens and A. Bogaerts, *J. Phys. Chem. C*, 2012, **116**, 23257.

49 T. Kozák and A. Bogaerts, *Plasma Sources Sci. Technol.*, 2014, **23**, 045004.

50 T. Kozák and A. Bogaerts, *Plasma Sources Sci. Technol.*, 2015, **24**, 015024.

51 S. Heijkers, R. Snoeckx, T. Kozák, T. Silva, T. Godfroid, N. Britun, R. Snyders and A. Bogaerts, *J. Phys. Chem. C*, **119**, 12815–12828.

52 L. F. Spencer and A. D. Gallimore, *Plasma Chem. Plasma Process.*, 2011, **31**, 79.

53 H. L. Chen, H. M. Lee, S. H. Chen, M. B. Chang, S. J. Yu and S. N. Li, *Environ. Sci. Technol.*, 2009, **43**, 2216.

54 J. van Durme, J. Dewulf, C. Leys and H. van Langenhove, *Appl. Catal., B*, 2008, **78**, 324.

55 H.-H. Kim and A. Ogata, *Eur. Phys. J.: Appl. Phys.*, 2011, **55**, 13806.

Faraday Discussions

PAPER

Taming microwave plasma to beat thermodynamics in CO_2 dissociation

G. J. van Rooij,*[a] D. C. M. van den Bekerom,[a] N. den Harder,[a] T. Minea,[a] G. Berden,[b] W. A. Bongers,[a] R. Engeln,[c] M. F. Graswinckel,[a] E. Zoethout[a] and M. C. M. van de Sanden[ac]

Received 29th April 2015, Accepted 10th June 2015

DOI: 10.1039/c5fd00045a

The strong non-equilibrium conditions provided by the plasma phase offer the opportunity to beat traditional thermal process energy efficiencies *via* preferential excitation of molecular vibrations. Simple molecular physics considerations are presented to explain potential dissociation pathways in plasma and their effect on energy efficiency. A common microwave reactor approach is evaluated experimentally with Rayleigh scattering and Fourier transform infrared spectroscopy to assess gas temperatures (exceeding 10^4 K) and conversion degrees (up to 30%), respectively. The results are interpreted on a basis of estimates of the plasma dynamics obtained with electron energy distribution functions calculated with a Boltzmann solver. It indicates that the intrinsic electron energies are higher than is favorable for preferential vibrational excitation due to dissociative excitation, which causes thermodynamic equilibrium chemistry to dominate. The highest observed energy efficiencies of 45% indicate that non-equilibrium dynamics had been at play. A novel approach involving additives of low ionization potential to tailor the electron energies to the vibrational excitation regime is proposed.

1 Introduction

The emission of carbon dioxide into the atmosphere is widely regarded as a severe environmental issue due to concern over its effect on climate change. The consequent increasingly stronger limits on CO_2 exhausts are driving the transition to sustainable energy sources. The intermittent character of, particularly, solar photovoltaics and wind imposes the need for energy storage. At the same time, these sustainable energy sources produce electricity, whereas, globally, less

[a]*Dutch Institute for Fundamental Energy Research, P.O. box 6336, 5600 HH Eindhoven, The Netherlands. E-mail: g.j.vanrooij@differ.nl; Tel: +31 40 333 49 99*

[b]*Radboud University, Institute for Molecules and Materials, FELIX Laboratory, Toernooiveld 7c, 6525 ED Nijmegen, The Netherlands*

[c]*Department of Applied Physics, Plasma and Materials Processing Group, Eindhoven University of Technology, P.O. Box 513, 5600 MB Eindhoven, The Netherlands*

than 20% of the energy is consumed in that form. Converting temporary electrical energy surpluses into chemical fuels would, thus, be advantageous in view of energy density as well as to address CO_2 emissions in other areas of energy consumption (*e.g.* transportation). A promising approach for this conversion involves the non-equilibrium chemistry in the plasma phase. The present paper deals with the particular case of plasma chemical CO_2 reduction as the first step in the production of carbon-based fuels.

Electrical discharges are usually sustained by electric fields through which energy is, in the first instance, transferred to the free electrons. Subsequently, this energy is transferred by collisions to the (neutral) gas phase heavy particles (*e.g.* CO_2 molecules). Due to the large mass difference between collision partners, the elastic energy transfer is inefficient and hence there may be large differences between their temperatures. It is under such far from thermodynamic equilibrium conditions that it is possible to intensify traditional chemical processes and to achieve the highest energy efficiencies.[1]

If the energy of the electrons is high enough, the heavy gas particles may be excited into higher electronic states or even be ionized. Ionization is required for sustaining the plasma discharge. However, in the present context, the subsequent dissociative recombination should not become the dominant pathway to dissociation as it is energetically an inefficient way of initiating chemical reactions. After all, ionization of CO_2 requires ~14 eV per molecule, whereas its "net" dissociation energy is ~3 eV (considering the "net" reaction $CO_2 \rightarrow CO + \frac{1}{2}O_2$). This consideration implies a 20% maximum energy efficiency for dissociation *via* ionization.

On the other hand, low energy electron collisions can excite vibrational modes in the molecule. Such vibrationally excited molecules will further interact with each other, collisionally exchanging vibrational energy up along the energy scale until the dissociation limit is reached. In this scheme, the electrons, that were energetically "expensive" to create, are used many times to deliver energy specifically to the bonds that are to be broken up to the point where dissociation of the molecule is achieved. It is this qualitative mechanism that has been put forward to explain the up to 90% energy efficiencies that were demonstrated in the 70s and 80s.[2–7]

Having established that it is energetically attractive, we note that the plasma chemical approach for reduction of CO_2 is also inherently appropriate in the context of alleviating intermittency of sustainable energy sources. The main reason for this is the low "inertia" of a plasma reactor, *i.e.* the plasma can be turned off and on and the input power can be regulated quickly (sub-second time scale, no heating of *e.g.* catalytic surfaces is required) and thus can follow the availability of energy surpluses. Secondly, the reactor is compact and power (*e.g.* microwave) supplies are cheap (0.1 euro per W), thus the investment costs can be low enough to economically allow intermittent use. Both aspects are not self-evident for, *e.g.*, electrolysis installations.

Although the plasma chemical approach for CO_2 reduction has been proven to be promising in terms of energy efficiency, it is presently not clear how this holds under industrially relevant throughputs. Two aspects play a role here. Firstly, the highest (90%) energy efficiencies were achieved at only a moderate (20%) conversion efficiency. Conversely, the highest (90%) conversion efficiencies were achieved at a moderate (20%) energy efficiency.[1] Secondly, these experiments were performed at a strongly sub-atmospheric pressure. At elevated pressures, the efficiency remained very low level, typically <10%.

A microwave plasma approach for CO_2 dissociation has been recognized as the most favourable in terms of energy efficiency.[6,8] It is generally assumed that this is due to a low reduced electric field and hence preferential vibrational excitation. In recent years, a number of attempts have been undertaken with microwave discharges to reproduce these record values, experimentally as well as numerically.[8–12]

This paper also focuses on a microwave approach for CO_2 dissociation. Simple molecular physics considerations are discussed to illustrate the different CO_2 dissociation pathways. Conversion and energy efficiency are determined (from Fourier transform infrared spectroscopy, FT-IR, measurements) in scans of power, gas flow and reactor pressure. The level of non-equilibrium is assessed on the basis of gas temperatures that are determined from Rayleigh scattering density measurements combined with reactor pressure readings. These experimental results are interpreted in a comparison with estimated reaction rates on the basis of electron energy distribution calculations with a Boltzmann solver. Finally, a novel methodology to enhance the non-equilibrium is proposed.

2 Vibrational *versus* electronic excitation – elementary molecular physics

We discern three reaction pathways that may be relevant under common low temperature plasma conditions as expected in microwave CO_2 plasma. Each has its own intrinsic energy efficiency. These pathways are schematically drawn in Fig. 1 and the (in-)efficiency of each pathway is indicated by the amount of energy that is intrinsically converted to heat.

Firstly, dissociation may occur due to step-by-step vibrational excitation (see Fig. 1a). The reaction products may be formed without a significant amount of kinetic energy and thus it is the desirable channel for CO_2 dissociation. As

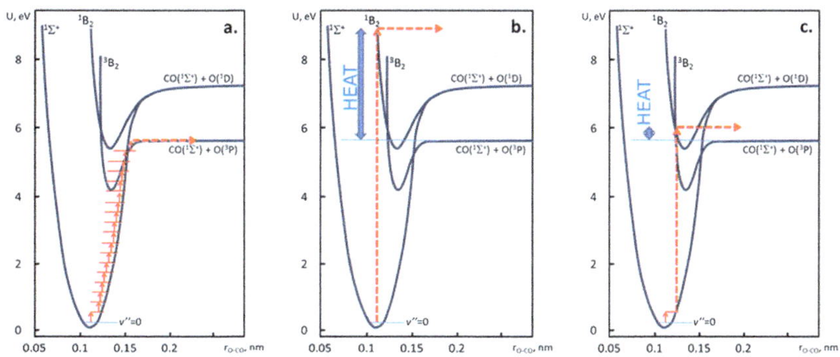

Fig. 1 Potential energy diagrams (as a function of one O–CO bond length) that illustrate the molecular physics behind three variants of CO_2 dissociation. Diagram (a) represents stepwise vibrational excitation that, after term crossing, causes ideally efficient dissociation. Contrary is the dissociative excitation in diagram (b), which involves a significant amount of heat to be released. A mix of these two cases is shown in diagram (c), vibrationally activated dissociative excitation, which is shown here as highly efficient as well.

mentioned before, this mechanism was put forward to explain the high energy efficiencies achieved. Typical electron temperatures of ~1 eV ensure that a major portion of the discharge energy is transferred from the plasma electrons to the lower CO_2 vibrational levels. At the same time, conversion of vibrational energy into translational energy can be kept to a minimum by ensuring a low gas temperature. Instead, vibrational–vibrational exchange may lead to overpopulation of the higher vibrational levels. The non-adiabatic transition $^1\Sigma^+ \rightarrow {}^3B_2$ opens the most effective dissociation pathway $CO_2(^1\Sigma^+) \rightarrow CO(^1\Sigma^+) + O(^3P)$.

Secondly, dissociation may be induced by dissociative excitation (Fig. 1b). It is evident that this pathway involves the highest dissociation barrier and will be the least efficient. Electronic excitation becomes increasingly effective if the plasma temperature rises above ~1 eV. This occurs for example when the input power exceeds the power loss to vibrational excitation (which leads to more electronic excitation upon an increase of the electron temperature).

Finally, vibrational stimulated electronic excitation (Fig. 1c) may occur as a mixture of the previous two. In the example sketched, a single vibrational excitation in the electronic ground state is sufficient to significantly decrease the excitation energy to the $^1B_2$ upper state. Again, the non-adiabatic term crossing presents a dissociation pathway with 90% energy efficiency.

The plasma temperature (or more precisely, the electron energy distribution function) will in the end determine the relative importance of these reaction pathways as it determines the balance between vibrational excitations and electronic excitations.

3 Experimental

3.1 Microwave flow reactor

A 1 kW (continuous wave) microwave source is used to produce 2.45 GHz microwaves that are transferred through WR340 waveguides to the flow reactor (Fig. 2). A 3-stub tuner is placed in the microwave circuit to match the load impedance and adjust the power transfer. A sliding short is positioned a quarter wavelength away from the center of the quartz tube. A quartz tube, which is placed perpendicularly through the broad wall of the waveguide, serves as flow reactor. A 27 mm inner diameter tube was used during the Rayleigh scattering measurements while during the FTIR measurements a 18 mm diameter tube was used.

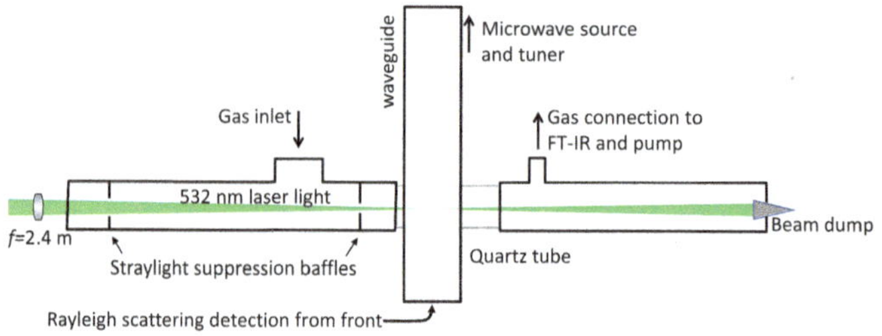

Fig. 2 Schematic of the microwave plasma source used for CO_2 dissociation. Indicated is the laser beam path that is used for the Rayleigh scattering experiment.

Gas flows of up to 15 slm of pure CO_2 were supplied. The operation pressure in the reactor was independently controlled in the range of 10–250 mbar by varying the effective pumping speed.

3.2 Rayleigh scattering

A Nd:YAG laser light was used for the Rayleigh scattering measurements. The laser delivered 0.5 J per pulse at 532 nm with a repetition frequency of 10 Hz. An anti-reflection coated UV fused silica window was used to couple the laser into the reactor. A 2.4 m lens was used to focus the laser beam at the detection volume in the center of the waveguide. Since the divergence of the laser beam was 0.45 mrad, this resulted in a 1.1 mm diameter beam at the focus. A 20 mm diameter hole in the 14 mm thick sliding short enabled the observation of laser scattering with minimal microwave leakage.

To suppress parasitic scattering of the laser light, *e.g.* from imperfections in the entrance window, baffles were installed in the vacuum components. A critical aperture of 6 mm diameter was installed 505 mm from the scattering volume, and a subcritical aperture with 12 mm diameter was placed at 100 mm distance. The beam dump was placed 2 meters after the scattering volume, and was installed in a vacuum.

The magnification of the collection optics resulted in an axial detection range of approximately 20 mm. A fiber packet was used to transport the scattered light from the collection optics to the spectrometer, a single-pass spectrometer in Littrow arrangement with a 0.3 m focal distance Littrow lens. The spectrometer was equipped with a third generation image intensifier of type EPM102G-04-22S and a Manta-G145B camera.

3.3 FT-IR effluent measurements

FT-IR spectra were taken with a Varian 670 FTIR spectrometer at a resolution of 0.09 cm^{-1}, so that individual ro-vibrational CO-peaks were resolved. A least square fit of the spectrum was used to obtain a species concentration c for different parameter settings using Beer–Lambert's law:

$$\ln\left(\frac{I}{I_0}\right) = -\varepsilon(p, T)cl. \tag{1}$$

The values for the molecule specific absorptivity ε were obtained from the HITEMP database,[13] where pressure and Doppler-broadening, as well as instrument-broadening are taken into account. The temperature was assumed to be constant at room temperature. Detailed fits showed this to be accurate within 15 K. At higher pressures and CO-concentrations, the most intense CO-peaks were saturated and thus disregarded in the fitting. The CO fraction f in the effluent was determined from the concentration and pressure, and subsequently converted into the conversion factor α using the relation:

$$f = \frac{\alpha}{(1 + \alpha/2)}. \tag{2}$$

For conversions lower than 2%, the fitting routine didn't converge and a 'best-fit' was estimated by inspection. Knowing the conversion efficiency, the energy efficiency can be calculated using:

$$\eta = \frac{q\alpha E_{\mathrm{d,\,CO_2}}}{P_{\mathrm{in}}}, \tag{3}$$

where q is the gas flow rate in molecules per second, $E_{\mathrm{d,CO_2}}$ is the effective dissociation energy per CO_2 molecule (2.9 eV) and P_{in} is the input power.

4 Experimental results

4.1 Discharge modes – spontaneous light emission

Depending on the operational conditions, particularly on the ratio between pressure and electric field, different modes of plasma operation are known to occur in microwave plasma discharges: diffuse, contracted, and a combination thereof.[1] The contracted plasma region forms at high electron densities. The high density prevents electromagnetic waves penetrating into the core of the plasma and, instead, the microwave power is absorbed in a skin layer at the boundary of the high density region. Generally, the plasma becomes also thermal in this mode, *i.e.* the neutral gas temperature approaches the plasma temperature. It is believed that a combined operation, in which the small volume of thermal plasma is surrounded by a large volume of non-equilibrium plasma, is best to achieve high energy efficiency conversion.

These different modes of operation and transitions in between occurred in the experiments that are discussed in this work as well. To illustrate the effect, images of the spontaneous light emission of the plasma inside the waveguide are shown in Fig. 3. The two cases differed mainly in reactor pressure and correspond to the extremes of the parameter space that was covered: 10 and 250 mbar (input power was 650 and 550 W, respectively; gas flow rate was 13 slm in both cases).

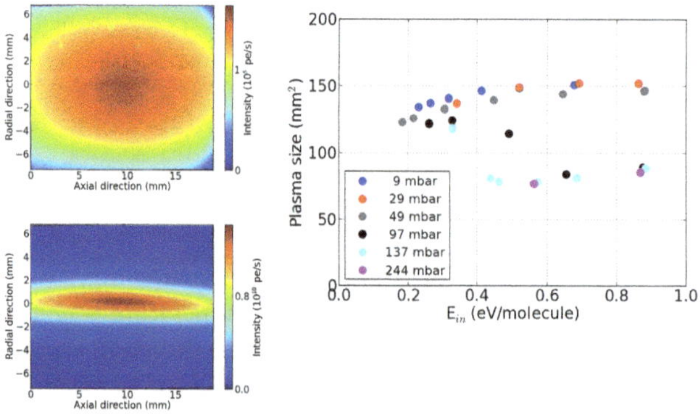

Fig. 3 Spontaneous light emission from the plasma inside the waveguide for (left top) 10 mbar (650 W, 13 slm) and (left bottom) 250 mbar (550 W, 13 slm), illustrating the two modes of operation: diffuse and contracted.[1] It is noted that vignetting at the sides of the images was not corrected for. The plot on the right shows the plasma cross sectional area determined from such emission images in dependence of pressure and specific energy. Plasma cross sectional areas of 120–150 mm^2 and 80 mm^2 correspond to diffuse and contracted operations, respectively.

The emission images were used to estimate the (cross sectional) plasma size in order to qualify the mode of operation throughout the parameter space covered by the present experiments. The results are presented in dependence of the specific energy input in the plot in Fig. 3. In essence, full width half maxima values were used. The plot clearly resolves the diffuse mode with typical plasma cross sectional areas of 120–150 mm^2 and the contracted mode with a size of typically 80 mm^2. Reducing the specific energy input evidently causes back transition of the high pressure contracted mode to a high pressure diffuse mode.

4.2 Conversion degree and energy efficiency

The performance of microwave plasma dissociation of CO_2 was characterized firstly with respect to conversion degree (α). The conversion degrees were inferred from FT-IR measurements of the effluent. The results are plotted in Fig. 4 for scans of specific energy (input power, range 300–1000 W, divided by gas flow rate) and pressure (25–200 mbar) at gas flow rates of 5 and 15 slm. The following observations are made. (i) The pressure series that yield the highest conversions line up along a common linear dependence on the specific energy input. (ii) An optimum exists in operation pressure. This is most clearly seen for the 5 slm data, in which the 100 and 150 mbar series are at maximum conversion whereas the 37 and 200 mbar series show decreased performance. The 15 slm data step up in conversion as pressure goes up and reach a maximum for pressures of 150 and 200 mbar. The latter also implies that (iii) the optimum operation pressure shifts up with gas flow rate. Furthermore, returning to the previous discussion on the different modes of operation, (iv) it was observed that the highest conversion series all correspond to the plasma operating in the contracted mode. Finally, (iv) the high flow data peak at 0.8 eV per molecule energy input. We presently believe that this relates to a transition to the combined diffuse-contracted mode.

The same data is plotted in Fig. 5 in terms of energy efficiency for CO formation, this time with a logarithmic specific energy input axis. As to be expected, the trends are similar as before. Efficiencies range from 7% up to ~45%. In other words, the record high energy efficiencies of 90% are not reproduced in

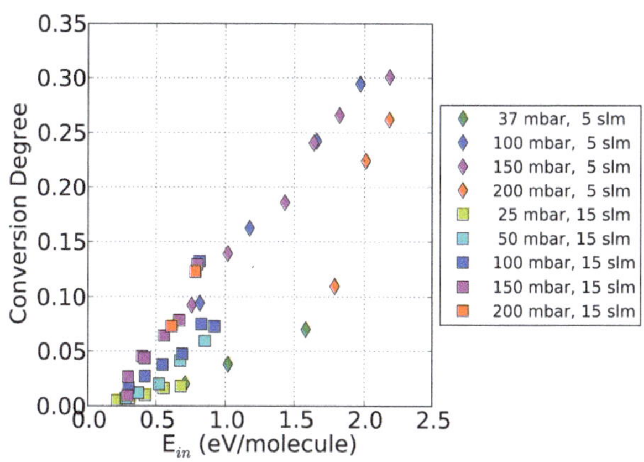

Fig. 4 Conversion degree measured in pure microwave CO_2 plasma as a function of the specific energy input, at 5 and 15 slm gas flow rate, in scans of the reactor pressure.

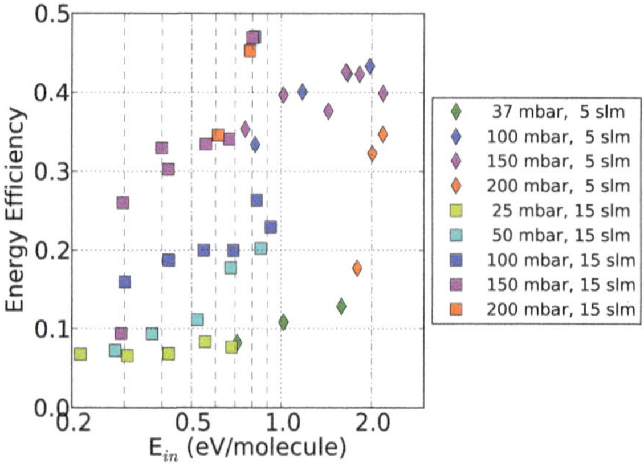

Fig. 5 Energy efficiency measured in pure microwave CO_2 plasma as a function of the specific energy input, at 5 and 15 slm gas flow rates, in scans of the reactor pressure. The underlying experimental data are identical to those of Fig. 4.

the present experiments, although similar conditions in terms of pressure and specific energy were covered.[2]

4.3 Gas temperature

A signature of reaching the desired non-equilibrium conditions is a low gas temperature that is maintained whilst vibrational temperatures (up to 8000 K [12,14]) and plasma temperatures (1–5 eV) are high. Usually, rotational distribution temperatures are determined by molecular spectroscopy as a measure for the gas temperature. This assumes equilibration between rotational and translational degrees of freedom, which is not always necessarily the case.[15] In this work, we rely on direct, *in situ* and spatially resolved, density measurements with laser Rayleigh scattering, from which temperatures are inferred using reactor pressure readings and the ideal gas law. Taking the parameter space covered by the conversion measurements as reference, we selected a high and low pressure condition (20 and 135 mbar) at a high flow rate (13 slm) to perform scans of the specific energy. The temperature results are presented in Fig. 6.

In the analysis, pure CO_2 was assumed as the scattering species. However, in the case of significant dissociation, considerable amounts of CO and O_2 would also have been present. This would disturb the temperature analysis as the total Rayleigh scattering caused by CO and $\frac{1}{2}O_2$ is a factor 1.5 less than the scattering from CO_2. So if locally all CO_2 was dissociated, the temperature would be overestimated by the same factor of 1.5. A perhaps more realistic estimate of the maximum error is to assume at the highest specific energy input an energy efficiency of 50%, *i.e.* a maximum conversion degree of 15%. This would imply an overestimation of the neutral gas temperature of only 5%.

It is seen that gas temperatures are around 2000 K for the low pressure conditions but rise steeply to values that are usually assumed for the temperature of the plasma electrons for the high pressure conditions. Putting these observations in the context of the earlier conversion measurements: the highest conversions are obtained as gas temperatures are also maximal.

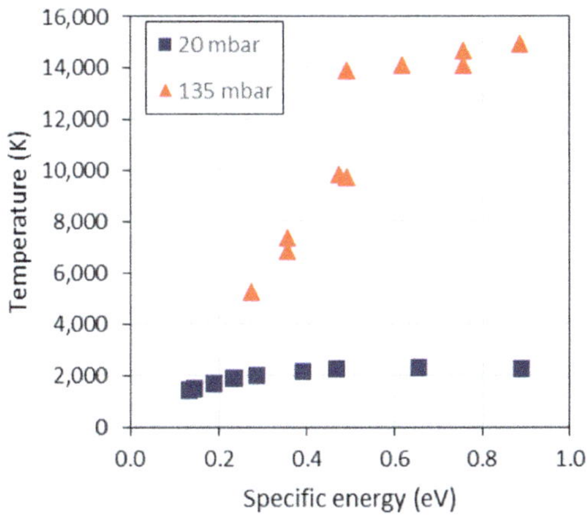

Fig. 6 Gas temperatures determined from neutral gas density measurements with Rayleigh scattering and reactor pressure readings in a scan of specific energy input. The gas flow was at 13 slm similar to the high flow conditions in Fig. 4 and 5, the pressure was tuned to 20 and 135 mbar, respectively.

The temperature data exhibit a sudden rise at a specific energy of about 0.5 eV. A closer inspection of the 137 mbar data series in the plot of the discharge cross sectional areas (Fig. 3) shows that this is exactly where the discharge undergoes a transition from diffuse to contracted. Apparently, contracted discharges have a central neutral gas temperature that is approximately constant at 1.4×10^4 K.

5 Electron energy distribution evaluation

The temperature of the electrons, or equivalently the mean energy, in a discharge is set by the requirement that on average an electron should be able to replace itself by an ionization event before it escapes to the wall or it recombines with a positively charged heavy particle in the volume. In other words: the production by ionization must be equal to the losses, or, the electron temperature is determined by the particle balance. For a discharge as studied here (simplified in the following by a uniform electron density n_e, electron temperature T_e, and gas density n_0 in a reactor of radius r), we write the particle balance as:

$$n_e n_0 k_{ion}(T_e) = n_e^2 k_{DR}(T_e) + \frac{n_e c_s(T_e)}{2r}. \tag{4}$$

The left hand side (LHS) of the equation is the production of ions (with $k_{ion}(T_e)$ the effective ionization rate per unit of volume). The first term on the RHS (right hand side) is volume recombination due to dissociative recombination, with the dissociative recombination rate $k_{DR} = 4.2 \times 10^{-13}(T_e/300\ K)^{-0.75}\ m^3\ s^{-1}$.[16] The second term indicates wall recombination due to Bohm flux to the wall, *i.e.* ions reach sound speed c_s at the sheath in front of the wall. Note that plasma flow is disregarded as typical flow velocities of 100–500 m s^{-1} are negligible compared to the sound speed.

The electron density is determined by the requirement that the number of collisions between electrons and heavy particles is sufficient to transfer the power injected into the discharge, P_{in}, to the heavy particles. In other words, the electron density is determined by the power balance, which is written as:

$$\frac{P_{\text{in}}}{V} = n_e n_0 \sum_i (k_i U_i) \tag{5}$$

(with V the reactor volume and k_i and U_i the rate and energy loss of an energy transfer process). In the ideal case, vibrational excitation would dominate and would be the only term on the RHS.

In case of Maxwellian velocity distributions, these equations are straightforward to solve. However, the free electrons in microwave discharges are recognized to be more likely Druyvesteyn distributed.[1] This means that the high energy tail is decreased, which drives the mean electron energy up to still supply sufficient ionization. In such cases, a Boltzmann solver can be used to model the electron energy distribution function (EEDF), in which case effective rates of the relevant processes are calculated on a basis of the cross sections. The publicly available Bolsig$^+$ solver,[17] was used, which has been described by Hagelaar and Pitchford.[18]

The results of solving the balanced equations assuming Maxwellian distributions and EEDF calculations with Bolsig$^+$ are compared in Fig. 7. In the case of a Maxwellian distribution, the mean electron energy is taken as $\frac{3}{2}k_B T_e$. It is seen that the mean electron energy is indeed increased due to the non-Maxwellian nature of the discharge. Moreover, with increasing input power, the mean electron energy increases, thus opening direct electron excitation channels in addition to the preferred vibrational excitation.

At this point it is of interest to note that the mean electron energies calculated here are different from values that are calculated or estimated in other work,

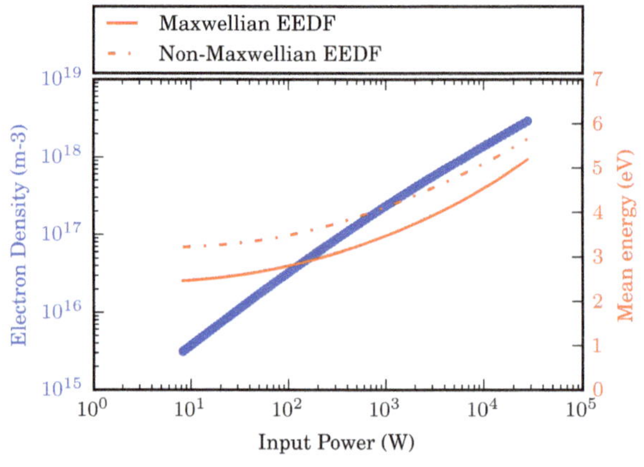

Fig. 7 Electron density (left axis) and mean energy (right axis) calculated as a function of input power for a CO_2 microwave discharge. Compared are solutions of the balanced equations assuming a Maxwellian EEDF and solutions which involved detailed Bolsig$^+$ calculations of the EEDF. The non-Maxwellian nature leads to increased mean electron energies. Dissociative recombination becomes more important as density increases, which causes the increase in mean electron energy with input power. Assumed was a neutral gas density of 10^{23} m^{-3}, which is similar to measured values.

e.g. by Kozák *et al.*[8] or Silva *et al.*[12] Usually, microwave discharges in CO_2 are assumed to operate at electron temperatures of 1–2 eV, *i.e.* mean energies of 1.5–3.5 eV. At modest power inputs of 10 W there is quantitative agreement with these literature values, but with increasing power, the mean electron energy increases significantly. One may be suspicious about the simplified wall losses in the particle balance that are used here (eqn (4)). After all, the electron density will in reality be lower near the walls compared to the center, especially in the contracted mode of operation, thus Bohm fluxes calculated on the basis that central densities overestimate the losses significantly. In fact, a radial ambipolar diffusion model (as used by *e.g.* Vijvers *et al.* for thermal arcs[19]) would have been more appropriate. This was not implemented since it is not significant compared to dissociative recombination. The main characteristic of dissociative recombination is its dependence on the square of the electron density (eqn 4). As a result, it causes the mean electron energy to increase with electron density, which is seen in Fig. 7. Already around an input power of 100 W, the electron energy is seen to increase, indicating that dissociative recombination is at play in the particle balance.

Knowing the EEDF enables calculation of the effective rates on the basis of the cross sections. These rates can be normalized to yield electron energy loss fractions giving more insight into the loss processes of the electrons. In Fig. 8 these loss fractions are shown for various processes. A major insight is obtained from this plot: the Bolsig$^+$ evaluations predict that at reactor relevant values of input power, the vibrational excitation is no longer dominant in the power balance. This is indirect, *via* increasing of the mean energy due to dissociative recombination.

Important in the present context is that dissociative recombination not only shortens the ion lifetime and therewith increases the electron energy, but also converts ionization energy into additional heating of the neutral gas. Since the vibration–translation exchange increases with gas temperature,[20] the extra gas heating might be sufficient to quench the vibrational excitation.

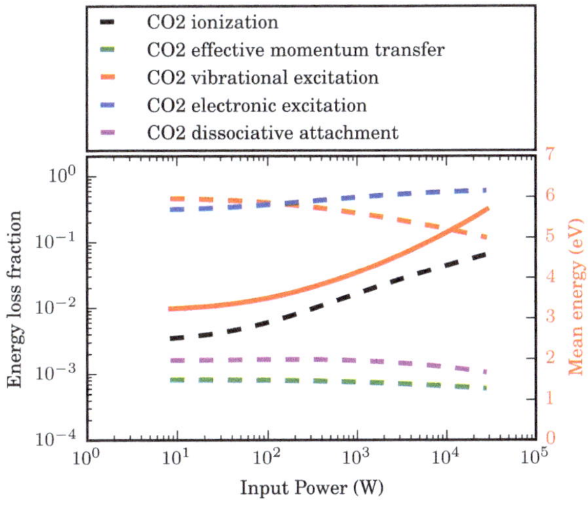

Fig. 8 Energy loss fractions for the various processes at play in a CO_2 microwave discharge. It shows that vibrational excitation is not the dominant energy loss channel for the electrons in conditions as investigated in the present work.

6 Discussion

Having explored CO_2 dissociation in microwave plasma both experimentally in terms of dissociation degree, energy efficiency, and neutral gas temperature, as well as numerically in terms of EEDF and energy loss fractions for the main reaction processes at play in the plasma, we summarize the main findings as follows:

• the energy efficiency for CO formation never exceeded the thermodynamic equilibrium value of ~45% in the investigated pressure range;

• the conversion degree for CO formation never approached the thermodynamic maximum of ~55%;

• the best performance was observed at pressures of around 150 mbar and also when the plasma operated in the contracted mode (in which gas temperatures in the center are too high to expect strong non-equilibrium conditions in view of vibration-to-translation relaxation);

• the electron energy estimates indicate that vibrational excitation is not dominant in the present conditions.

In other words, it is questionable whether vibrational excitation has been at play in the present experiments. Here, we will put these findings in the context of thermal equilibrium dissociation efficiencies, hypothesize on the mechanism of dissociation at play and how this relates to the observed scalings of dissociation degree and energy efficiency with input power.

As reference, thermodynamic equilibrium conversion data were calculated with the CEA package,[21] which are plotted in Fig. 9. Also included in the plot are energy efficiencies calculated assuming instantaneous cooling.[22]

Let's first put the *in situ* temperature measurements (Fig. 6) in the context of the chemical equilibrium data. Starting with the low pressure series (20 mbar, in a diffuse discharge mode), which were rather constant at around 2000 K, these

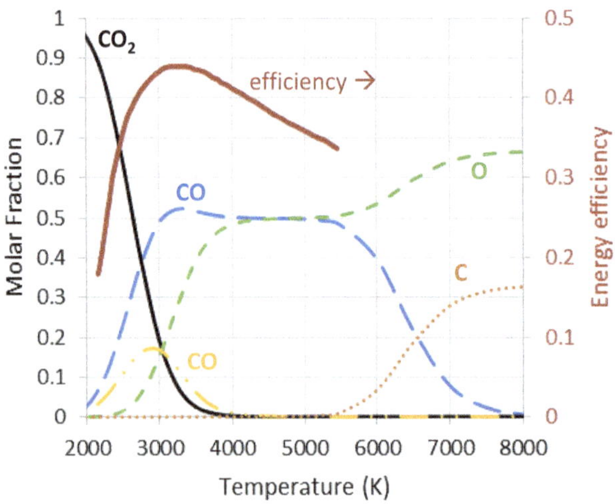

Fig. 9 Chemical equilibrium calculated molar fraction of the products of CO_2 dissociation at a fixed pressure of 100 mbar. Calculated energy efficiencies assuming instantaneous quenching after Polak *et al.* are plotted on the right axis.

would produce chemical equilibrium conversion degrees of about 5%. The measured data at a similar pressure (25 mbar, Fig. 4) increased with power from close to zero to up to ~2%. These measured conversion degrees thus make perfect sense if one realizes that the radial temperature must have peaked. As a result of the peaking, only the hotter centers would have contributed to dissociation. At increasing input power the radius of the dissociation zone would have increased thus producing more dissociation and hence the measured linear behavior.

Next is the high pressure series (135 mbar). The measured temperatures in the center of these contracted discharges started at 5000 K and increased to over 15 000 K. In terms of chemical equilibrium conversion, this corresponds to starting at a maximum conversion of 50% that subsequently drops down to 0% CO due to full dissociation. The measured conversions, however, go up from a few % to nearly 10%. The same temperature peaking argument holds again. However, this time the CO formation region moves out of the center as the temperature increases beyond the CO formation region. This extra hot center becomes even smaller as the discharge contracts. In this situation, the plasma acts as a central heat source that provides the outer regions with thermal heat and induces the dissociation. The amount of gas in the extra hot center is relatively small. Not only does geometry play a role here (the central cross section being smaller than the boundary area), but also the strong temperature peaking in the center in radially constant pressure implies rarefaction of the center.

An aspect that has not been considered in the present work is the quenching of the reaction products. In fact, no measures had been in place to optimize the quenching rate. Instead, the effluent had been flowing through a constant radius tube, only being cooled by the colder tube walls. Roughly estimating, this gives a quenching rate of only 10^6 K s^{-1} (assuming 1000 K cooling over 10 cm with a flow velocity of 100 m s^{-1}). As a consequence, back reactions might have deteriorated energy efficiencies in the effluent by a factor of more than 2 compared to inside the reactor.

These equilibrium features of the observed dissociation degrees are in line with the mean electron energy evaluations and the consequence that vibrational excitation is not the main electron energy loss process. However, both aspects are especially valid for the central regions of the reactor, *i.e.* the hotter plasma region and its direct hot neutral gas surroundings. So the non-equilibrium dynamics may still be at play further out, even closer to the reactor walls where the gas is colder and still an appreciable density of low energy electrons is available for vibrational excitation. Indeed, closer inspection of the energy efficiencies that are presented in Fig. 5 do show a sign thereof. Especially the outliers at about 0.8 eV specific energy, which were already indicated to correspond to the combined diffuse-contracted mode, point to partly non-equilibrium CO production. The achieved energy efficiency of just over 45% approaches the maximum of the calculated efficiency for instantaneous quenching (see Fig. 9), and it is thus improbable that this could have been achieved by thermal equilibrium chemistry alone in the present reactor geometry.

Having established that thermodynamic equilibrium chemistry dominates in the present work, the question arises how to move in operational space to strong non-equilibrium dominated conditions. One aspect that should help in this respect is lowering the pressure even further, to the 1 mbar range, *e.g.* by installing larger diameter flow tubes. This is based on the one hand on observing lower neutral gas temperatures at lower pressure, and on the other hand on dissociative recombination strongly falling off with electron density. In the next Section, we propose an alternative, entirely novel, approach that relies on tailoring the EEDF by adding low ionization potential species to the reactor input.

7 Adding traces of low ionization potential alkali metals

Having established that the mean electron energy has been too high for preferential vibrational excitation leading to a strong non-equilibrium, the relevant question to ask is how to mitigate this effect induced by dissociative recombination. In other words, how to tame the CO_2 microwave discharge? We postulate that this can possibly be achieved by adding traces of low ionization potential species, *e.g.* alkali metals such as lithium, sodium or potassium. The effect of this action will be twofold. Firstly, the lower ionization potential will cause a lower plasma temperature *via* the particle balance. Secondly, the positive charge is no longer carried by molecular ions, so that volume recombination *via* dissociative recombination is no longer at play. Atomic ions can only recombine in the volume *via* three-body recombination, a process which has exceedingly small reaction rates. In the absence of dissociative recombination, the previously observed electron density dependence on the mean electron energy, as shown in Fig. 7, will vanish.

To estimate the effect of seeding sodium to the discharge, the addition of alkali impurities was modeled using Bolsig$^+$. Firstly, the calculated electron density and mean energy are presented as a function of input power in Fig. 10. It is seen that the mean electron energy is indeed significantly decreased, to 1 eV (*i.e.* in the case of a near Maxwellian distribution, this would mean 0.7 eV electron temperature) and its previous dependence on electron density (or input power) is no longer present.

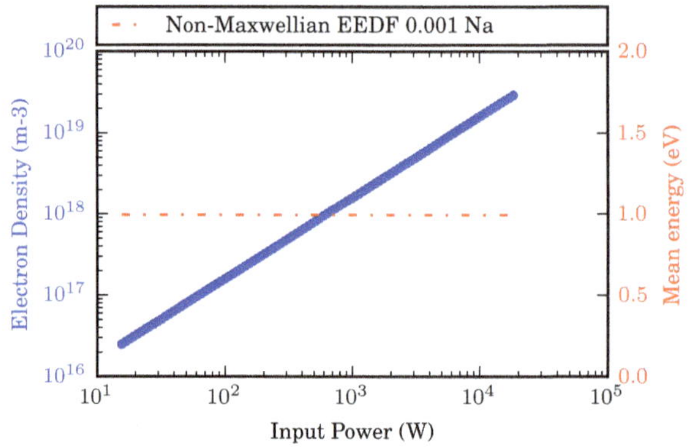

Fig. 10 The effect of sodium being added to a CO_2 microwave discharge on the mean electron energy as calculated with the Bolsig$^+$ solver. Assumed was 0.01% sodium added to a neutral gas density of 10^{23} m^{-3}. The mean electron energy is significantly decreased and lost any dependence on input power (compared to Fig. 7).

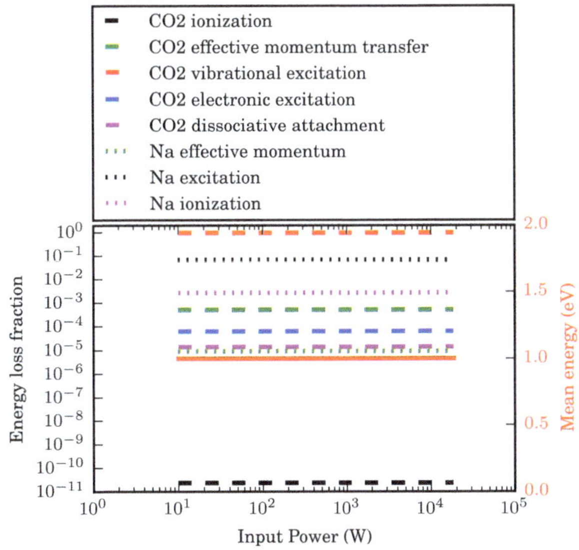

Fig. 11 The effect of 0.01% sodium added to a CO_2 microwave discharge on the relative energy loss fractions for the main processes at play. It shows that vibrational excitation has become the dominant energy loss channel for the electrons, which does not change within the conditions investigated in the present work.

As the mean electron energy has become independent of input power, so have the energy loss fractions, as can be seen in Fig. 11. More importantly, vibrational excitation has become the dominant energy loss channel for the electrons. In other words, these evaluations support the beneficial effect of adding low ionization species for reaching a strong non-equilibrium in a CO_2 microwave discharge.

An important result from analyzing the energy loss fractions is that adding sodium at amounts of 0.01% (which allows equal levels of ionization degree) does not add a significant amount of radiative losses to the power balance. This becomes important at levels of 0.1–1% and thus sets the limit on the amount of sodium that can be added. It is noted that this balance may be different for other alkali metals.

8 Conclusions

The present experiments and EEDF evaluations show that retaining a low electron temperature (mean energy) is required for realizing strongly non-equilibrium conditions that yield high energy efficiency CO_2 conversion. One possible route to follow is enhancing the volume of the non-equilibrium regions in the reactor compared to the high density chemical equilibrium plasma part. However, this will always mean compromising between high power input (*i.e.* reaching application relevant throughput) and achieving high efficiency. A successful development of the here proposed alkali additives would offer an alternative in which the plasma dynamics can be tailored to the non-equilibrium in the entire discharge volume. In effect, it would overcome the compromise, the trade-off between conversion degree and energy efficiency. It would tame the microwave plasma to beat thermodynamic equilibrium performance in CO_2 dissociation.

References

1 A. A. Fridman, *Plasma Chemistry*, Cambridge University Press, New York, 2008.
2 R. I. Azizov, A. K. Vakar, V. K. Zhivotov, M. F. Krotov, O. A. Zinov'ev, B. V. Potapkin, V. D. Rusanov, A. A. Rusanov and A. A. Fridman, *Sov. Phys. Dokl.*, 1983, **28**, 567.
3 I. A. Semiokhin, Y. P. Andreev and G. M. Panchenkov, *Russ. J. Phys. Chem.*, 1964, **38**, 1126.
4 A. N. Maltsev, E. N. Eremin and L. V. Ivanter, *Russ. J. Phys. Chem.*, 1967, **41**, 633.
5 Y. P. Andreev, I. A. Semiokhin, Y. M. Voronkov, V. A. Sirotkina and V. A. Kaigorodov, *Russ. J. Phys. Chem.*, 1971, **45**, 1587.
6 V. D. Rusanov, A. A. Fridman and G. V. Sholin, *Sov. Phys.-Usp.*, 1981, **24**, 447.
7 R. I. Asisov, A. A. Fridman, V. K. Givotov, E. G. Krasheninnikov, B. I. Petrushev, B. V. Potapkin, V. D. Rusanov, M. F. Krotov and I. V. Kurchatov, *5th Int. Symp. on Plasma Chemistry*, Scotland, Edinburgh, 1981, p. 774.
8 T. Kozák and A. Bogaerts, *Plasma Sources Sci. Technol.*, 2015, **24**, 015024.
9 L. F. Spencer and A. D. Gallimore, *Plasma Chem. Plasma Process.*, 2011, **31**, 79.
10 L. F. Spencer and A. D. Gallimore, *Plasma Sources Sci. Technol.*, 2013, **22**, 015019.
11 A. Goede, W. A. Bongers, M. F. Graswinckel, M. C. M. van de Sanden, M. Leins, J. Kopecki, A. Schulz and M. Walker, *3rd European Energy Conf.*, Hungary, Budapest, 2013.
12 T. Silva, N. Britun, T. Godfroid and R. Snyders, *Plasma Sources Sci. Technol.*, 2014, **23**, 025009.
13 L. S. Rothman, I. E. Gordon, R. J. Barber, H. Dothe, R. R. Gamache, A. Goldman, V. I. Perevalov, S. A. Tashkun and J. Tennyson, *J. Quant. Spectrosc. Radiat. Transfer*, 2010, **111**, 2139–2150.
14 A. K. Vakar, V. K. Zhivotov, F. F. Karimova, E. G. Krashenninikov, V. D. Rusanov and A. A. Fridman, *Sov. Tech. Phys. Lett.*, 1981, **7**, 429.
15 N. den Harder, D. C. Schram, W. J. Goedheer, H. J. de Blank, M. C. M. van de Sanden and G. J. van Rooij, *Plasma Sources Sci. Technol.*, 2015, **24**, 025020.
16 A. A. Viggiano, A. Ehlerding, F. Hellberg, R. D. Thomas, V. Zhaunerchyk, W. D. Geppert, H. Montaigne, M. Larsson, M. Kaminska and F. Österdahl, *J. Chem. Phys.*, 2005, **122**, 226101.
17 http://www.bolsig.laplace.univ-tlse.fr/.
18 G. J. M. Hagelaar and L. C. Pitchford, *Plasma Sources Sci. Technol.*, 2005, **14**, 722.
19 W. A. J. Vijvers, D. C. Schram, A. E. Shumack, N. J. L. Cardozo, J. Rapp and G. J. van Rooij, *Plasma Sources Sci. Technol.*, 2010, **19**, 15.
20 E. Treanor, J. W. Rich and R. G. Rehm, *J. Chem. Phys.*, 1968, **48**, 1798–1807.
21 S. Gordon and B. J. McBride, *Computer Program for Calculation of Complex Chemical Equilibrium Compositions and Applications II: Users Manual and Program Description*, NASA Reference Publication, 1996, vol. 1311.
22 A. A. Fridman, in *Plasma Chemistry*, Cambridge University Press, New York, 2008, p. 263, Fig. 5–4.

Faraday Discussions

PAPER

Novel windows for "solar commodities": a device for CO_2 reduction using plasmonic catalyst activation

Alexander Navarrete,*[a] Sergio Muñoz,[a] Luis M. Sanz-Moral,[a] Juergen J. Brandner,[b] Peter Pfeifer,[b] Ángel Martín,[a] Roland Dittmeyer[b] and María J. Cocero[a]

Received 12th June 2015, Accepted 2nd July 2015
DOI: 10.1039/c5fd00109a

A novel plasmonic reactor concept is proposed and tested to work as a visible energy harvesting device while allowing reactions to transform CO_2 to be carried out. Particularly the reverse water gas shift (RWGS) reaction has been tested as a means to introduce renewable energy into the economy. The development of the new reactor concept involved the synthesis of a new composite capable of plasmonic activation with light, the development of an impregnation method to create a single catalyst reactor entity, and finally the assembly of a reaction system to test the reaction. The composite developed was based on a Cu/ZnO catalyst dispersed into transparent aerogels. This allows efficient light transmission and a high surface area for the catalyst. An effective yet simple impregnation method was developed that allowed introduction of the composites into glass microchannels. The activation of the reaction was made using LEDs that covered all the sides of the reactor allowing a high power delivery. The results of the reaction show a stable process capable of low temperature transformations.

Introduction

CO_2 as a renewable energy vector

One of the most industrially promising heterogeneous catalytic processes is carbon dioxide hydrogenation. In the process of catalytic hydrogenation, hydrogen obtained from carbon neutral energy sources (*e.g.* wind or solar) is reacted with CO_2 to obtain products such as fuels.[1] Thus, this process serves a double purpose: firstly, as chemical storage of the surplus energy generated by fluctuating renewable energies; and secondly, to reduce emissions of CO_2.

[a]*University of Valladolid, Department of Chemical Engineering and Environmental, Technology, High Pressure Processes Group, Paseo Prado de la Magdalena s/n, 47005 Valladolid, Spain. E-mail: anm@iq.uva.es*
[b]*Institute for Micro Process Engineering, Karlsruhe Institute for Technology, Eggenstein-Leopoldshafen, Germany*

The successful introduction of CO_2-use technologies requires plausible and profitable processes that use efficiently renewable energy. The most abundant and evenly distributed of such energies is the one provided by the Sun. If stored in the form of "solar commodities" such as methanol or olefins new opportunities for the use of CO_2 could be considered.[2] The production of methanol from CO_2 and hydrogen involves the following reactions:

$$CO_2 + H_2 \Leftrightarrow CO + H_2O;\ \Delta H^0 = +41.19\ \text{kJ mol}^{-1} \tag{1}$$

$$CO + 2H_2 \Leftrightarrow CH_3OH;\ \Delta H^0 = -90.70\ \text{kJ mol}^{-1} \tag{2}$$

$$CO_2 + 3H_2 \Leftrightarrow CH_3OH + H_2O;\ \Delta H^0 = -49.51\ \text{kJ mol}^{-1} \tag{3}$$

In order to capture that energy in chemical bonds there are several useful methods. From a chemical point of view, they can be grouped into photocatalytic, thermal and electrical routes. In electrical routes, the solar energy is first converted into electricity and then, the resultant electrical energy is used in the chemical transformation of CO_2.[3–7] The thermal routes concentrate solar radiation and convey that energy directly into the reactor.[8–10] Photocatalytic CO_2 conversion involves either water splitting connected to a CO_2 reduction reaction, or a process combining both in one "single pot".[11–13]

Selective use of visual light with plasmon catalysts

The surface plasmon resonance (SPR) phenomenon is commonly found in metallic (or carbon) nanostructures and allows the range of the solar spectrum used on a given photoinduced process to be increased.[14] This effect is the result of the response of the conduction electrons to the oscillations of the electric field of the light radiation. Increased energy absorption by the electrons is possible at selected wavelengths under the proper particle size and shape of the nanoparticles for a given surrounding media (fluid or catalyst). This phenomenon produces a high light concentration up to the point that a reduction in the amount of semiconductor of three orders of magnitude for the same amount of light has been possible.[15] On the other hand, increased light capture with SPR for photothermal conversions is leading to breakthroughs in energy systems such as solar collectors.[16]

This work explores the reduction of CO_2 to CO as a first step in a solar-based process to produce methanol. Thus, it is based on the reverse water gas shift (RWGS) reaction as described by eqn (1). Recent approaches have used gold and semiconductor composites for plasmonic enhancement of the reduction.[17,18]

This process is commonly activated in industry using Cu/ZnO based catalysts.[19] Here, we have developed plasmonic catalytic composites in mesoporous silica structures (aerogels). For this, we have used the plasmon-tuneable Cu/ZnO catalyst reported by Tan *et al.* (2013).[20]

A plasmonic microreactor as a light harvesting device

The efficiency of the chemical reactions is not ruled only by the catalytic material but also by the reactor configuration and their mass and energy transport characteristics. On many occasions, promising catalytic materials have failed to reach industrial success due to the disconnection between the catalytic structure and the reactor-level phenomena.[21,22]

Here we propose a novel concept for visual energy harvesting: a plasmonic microreactor device. It integrates a plasmon catalyst and reactor as one entity with a sole response to light (Fig. 1).

Microreactors allow efficient energy and mass transport while are easily scalable (numbering up). Thus, the combination of efficient microstructured devices and direct plasmonic absorption of solar energy by the catalyst would represent a major breakthrough in the CO_2-use field.

In this work, a plasmon-tuneable composite is integrated with a microchannel based reaction system under visual LED illumination for the RWGS reaction. This involved the synthesis of the new composite; the development of an impregnation method to create a catalyst-reactor entity; and finally, the assembly of a reaction system to test the reaction.

Methods

Plasmo-catalytic composite synthesis

The chemicals used during this stage are detailed: zinc acetate dihydrate (>98%), oleylamine (70%), tetramethyl orthosilicate (98%), ammonia (28–30%) and triethyleneglycol (99%) were purchased from Sigma-Aldrich. Ethyleneglycol (99.5%) (Merck) and copper acetate monohydrate (99.9%) were purchased from Alfa Aesar. Methanol (99.8%) was provided by Panreac. All chemicals were used without further purification.

Synthesis of the Cu/ZnO bimetallic catalyst

This bimetallic (Cu : ZnO, 1 : 2) catalyst was synthesised following the procedure proposed by Tan *et al.* (2013).[20]

Briefly, first ZnO nanorods were prepared. Zinc acetate (3 mmol) was added to 1.3 mmol of oleylamine in a two necked flask. The oleylamine was not of high purity (70%), requiring a step where the reactants were degassed at 80 °C for 45

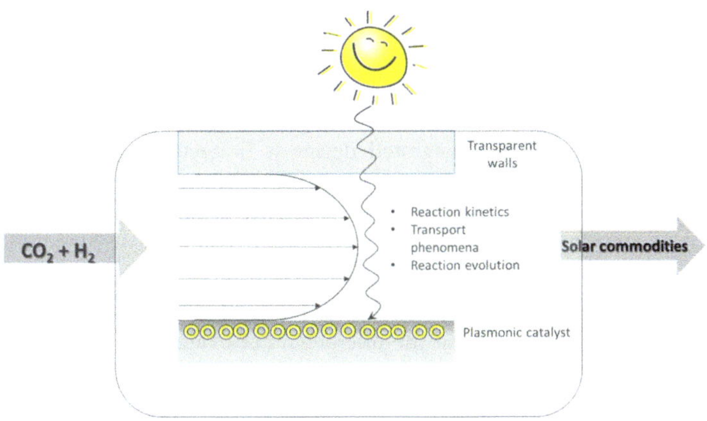

Fig. 1 Concept of the plasmonic microreactor device.

min under a vacuum atmosphere. Then, the temperature was increased to 220 °C under nitrogen purging. During the heating process, the solution turned white upon reaching 180 °C. After 15 min of heating, the mixture was cooled to atmospheric temperature, washed with ethanol and centrifuged in order to isolate the precipitate. It was washed with 6 ml of ethanol three times to ensure complete removal of the reactants or byproducts.

The ZnO nanorods thus prepared were then redispersed in 20 ml of triethyleneglycol by sonication for two hours, followed by stirring under room conditions overnight. Ethyleneglycol (2 ml) was added to the ZnO dispersion, and the mixture was degassed at room temperature for 5 min before heating to 190 °C. Simultaneously, a second solution of copper acetate monohydrate was prepared by dissolving in ethyleneglycol. This mixture required sonication in order to dissolve the copper acetate in the liquid. This mixture was added to the ZnO mixture in a dropwise manner over 10 min. After this, 5 more min heating at 190 °C was allowed before the composite was washed with isopropanol, centrifuged for 15 minutes at 4500 rpm (centrifuge Kubota 5100, Japan) and isolated from the mixture.

Synthesis of the mesoporous silica composites

Light transmission to the catalytic structures is essential while enough surface area has to be provided in order to have enough metal loading to capture light. Transparent aerogels are mesoporous materials combining high surface areas and good light transmission.[23]

Aerogels were synthesised following the sol–gel route. The precursor for the silica hydrogel selected was tetramethyl orthosilicate (TMOS). The molar ratio of TMOS : CH_3OH : H_2O : NH_4 was 1 : 2.3 : 3.84 : 0.012.

Firstly, methanol was used to disperse the nanoparticles formed in the catalyst synthesis. Sonication (15 min) was applied to ensure a good dispersion of the nanoparticles in the liquid phase. Methanol with the particles and TMOS were mixed together. While this solution was stirred, a second solution of ammonium hydroxide and water was prepared and stirred. After a few minutes of stirring, both solutions were mixed, and the gelation process began.

In this moment, the gelation process of the silica hydrogel has started, but it is still liquid for a few minutes. This time lapse, before gelation, must be used to impregnate the solution inside the microchannels of the microreactors.

Integration of the composites and microreactor

In order to have a single integrated device it is necessary to integrate light transmission and composite activation in the same structure. Here, we have developed a method to integrate transparent aerogels in glass microchannels. The method used to impregnate the sol–gel that showed the best results was the suction of the liquid with a syringe, which was previously adapted to the microreactor on its top (Fig. 2). With this method, placing the microreactor in a vertical position, it was very easy to fill the microchannel placing the tip in the liquid. After a short time, the gelation process finished and the hydrogel formed had a good adherence inside the microchannels.

It is worth mentioning that one of the most important parameters during this step is the amount of ammonia, as it acts as a catalyst for the gelation process. It is

Fig. 2 Syringe filling of the microchannels.

necessary to use an amount of ammonia low enough that allows enough time to impregnate the hydrogel inside the microreactor, because it must be still a fluid. On the other hand, the amount of ammonia cannot be very low, because if the gelation process is too slow, the nanoparticles start to precipitate and they will not remain inside the silica net. Finding an equilibrium between these two factors is the key to achieving good impregnation of the nanoparticles supported in the silica gels inside the microreactors. The microreactor consisted of a 0.5 mm ID glass capillary with an external diameter of 5 mm (Schott Duran, USA).

In order to ensure good adherence of the aerogel to the walls of the microchannels, it is necessary to perform a pretreatment to the glass microreactors (15.0 cm) to clean the walls of the microchannels. For this cleaning process, the most common option is to use a piranha solution, which reacts violently with most organic materials. The solution used was a mixture of sulfuric acid and hydrogen peroxide that can be prepared in different proportions; the most usual being 4 : 1 in concentrated sulfuric acid. For 15 to 30 minutes the material is submerged in the solution, then removed, washed with plenty of Milli-Q water and dried carefully.

The empty glass microreactors were put inside a glass pot, and the sulfuric acid was first added. Then, the hydrogen peroxide was added with extreme care because the reaction is very exothermic; the temperature is suddenly increased and some vapor can be formed. After 20 minutes, the slides were removed from the piranha solution, washed with Milli-Q water and dried carefully.

After the introduction of the nanoparticles in the sol–gel and its introduction into the microreactors, these were put in a vessel with methanol for aging. This vessel was carefully closed to avoid methanol evaporation, and it was heated to 50 °C in an oven. With this procedure the water contained in the silica net was replaced with methanol, resulting in alcogels.

After 24 hours of heating, the alcogels were dried using supercritical carbon dioxide. The microreactors were put in a high pressure vessel, and this vessel was filled completely with methanol. Carbon dioxide was introduced slowly in the vessel to allow a good diffusion into the methanol. The pressure was raised to 100 bar and the temperature to 40 °C, above the critical point of carbon dioxide. Three cycles of 45 minutes were performed, renewing the carbon dioxide between each cycle to complete the drying process.[24] After this, silica aerogels were correctly obtained, keeping the adherence to the walls of the microchannels.

Composite characterization

Scans of the bimetallic catalysts to check absorbance of visual light were carried out using a UV-vis spectrometer (UV 2550, Shimadzu). XRD analysis was carried out using a Bruker Discover D8 diffractometer. The porosity measurements were carried out using a Surface Area and Porosity Analyzer (ASAP 2020, Micromeritics).

Proof-of-concept setup

Once the Cu/ZnO based plasmonic composites were integrated into the glass microchannels, a test of these devices was made. In order to test the reactor concept, a reaction system had been built that included visual LED illumination and control of the reaction temperature while a precise control of the flow and pressure was provided. A scheme of the experimental plant is presented in Fig. 3.

Hydrogen and carbon dioxide were introduced in the system, and their flows were controlled with two different flow mass meter/controllers (EL-Flow F-200, Bronkhorst) with ranges from 0.02 to 1 ml min^{-1}.

Before the reaction was initiated, hydrogen and carbon dioxide were mixed in a 3 : 1 proportion, and sent to the vent while both flows were stabilized.

When the flows were correctly controlled, the mixture of the gases moved to the second part of the setup. In this part the gases were heated, together with the glass microreactor, in a gas chromatography oven (Agilent 7890).

A second vent was used to take out the gases while the pressure was increased to 20 bar. Pressure was controlled by a pressure meter/controller (EL-Press series, Bronkhorst). When the pressure and flow were stable at 20 bar, the valve for extraction was closed, and the oven and LEDs were turned on.

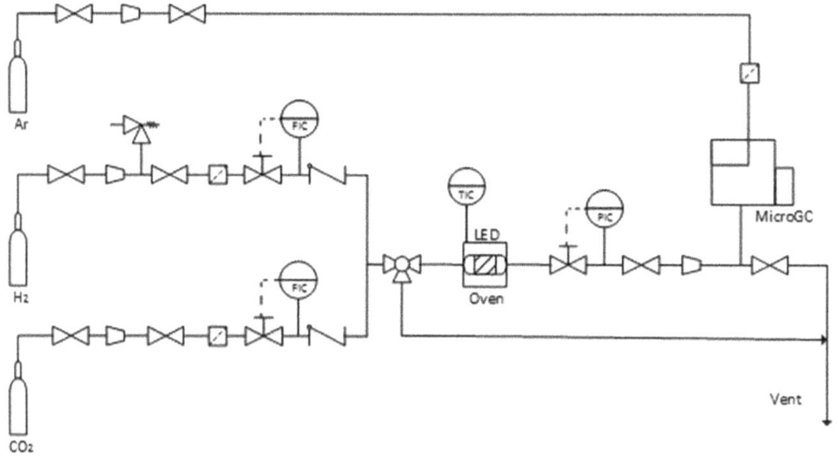

Fig. 3 Schematic flow diagram of the plant.

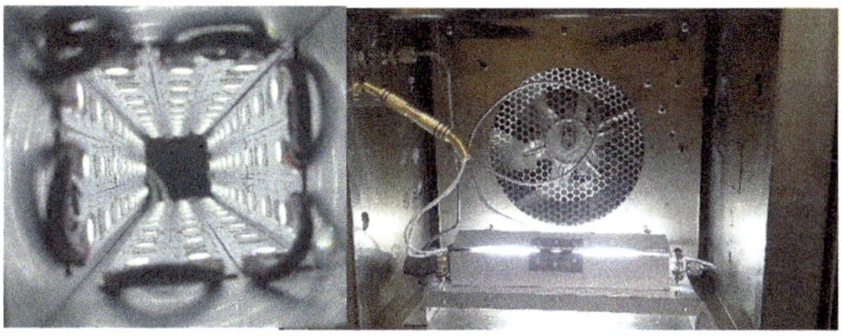

Fig. 4 LED light configuration. Left, detail of the inner LED distribution. Right, microreactor in the oven surrounded by the LEDs.

Visual light stimulation was provided by 36 LEDs (superbright, inspired LED) surrounding the microreactor as shown in Fig. 4. In total, they provided the equivalent of a nominal power of 9780 W m^{-2} of white light.

The reaction began and the products of the reaction were measured in a Micro Gas Chromatograph (CP-4900, Varian) equipped with two columns: a poraplot 10 m and a 5A molsieve. Before the micro GC, the pressure of the gas stream was reduced to less than 5 bar.

Results and discussion

Composite characterization

After the synthesis UV-vis scans were carried out in order to check the absorption of visual light from both the ZnO nanorods and the bimetallic Cu/ZnO catalyst (Fig. 5).

It can be seen that the bimetallic catalyst has an absorption peak at 498 nm. This corresponds to absorption in the range close to the green colour.[25] The transparent aerogels change and acquire colour once the composite is formed (Fig. 6).

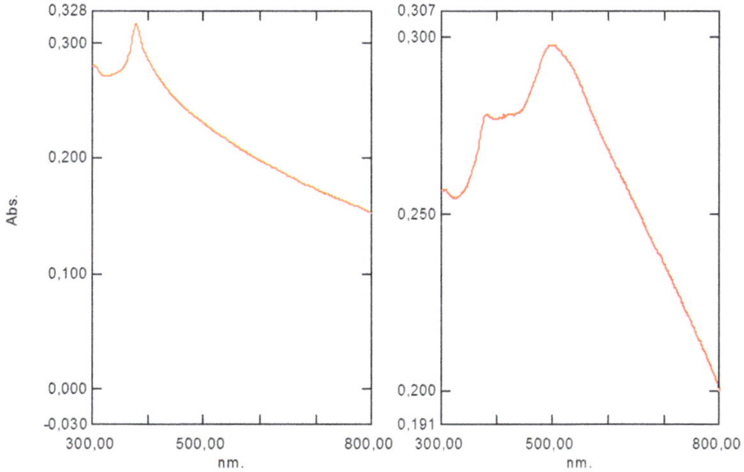

Fig. 5 Absorption spectra of ZnO (left) and Cu/ZnO (right).

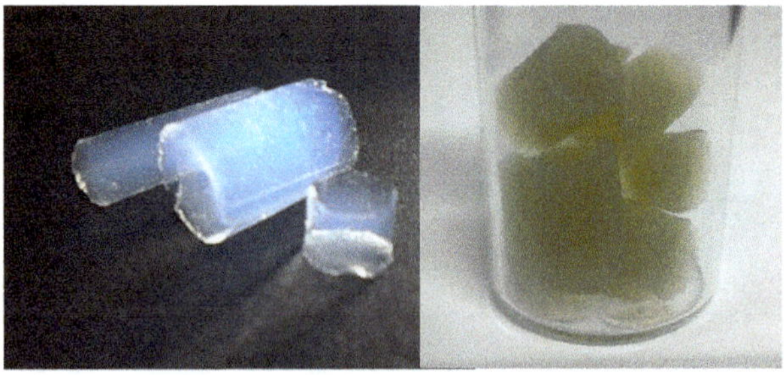

Fig. 6 Silica aerogel before (left) and after (right) catalyst impregnation.

The XRD pattern shows the presence of ZnO and metallic copper in the silica amorphous structure. The ZnO planes 100 ($2\theta = 31.7°$), 002 ($2\theta = 34.4°$), 101 ($2\theta = 36.1°$) and 110 ($2\theta = 56.4°$) can be identified in Fig. 7. Copper cannot be seen due to its lower proportion in the structure.

BET surface area and pore volume of the samples were calculated from N_2 isotherms. The adsorption–desorption curve shows a type IV isotherm curve typical for mesoporous silica aerogels[26] (Fig. 8). The BET surface area is 945.8 m^2 g^{-1} which indicates that the inclusion of the bimetallic catalyst does not have a significant influence on the textural properties of the aerogel. The BJH pore volume is equal to 2.29 cm^3 g^{-1} reinforcing that the structure is not affected.

Reaction test of the concept

The full power of the LEDs was applied and the evolution of the compounds were followed. In order to test the influence of the main variables of the process, changes in flow and temperature were made during the reaction (Fig. 9).

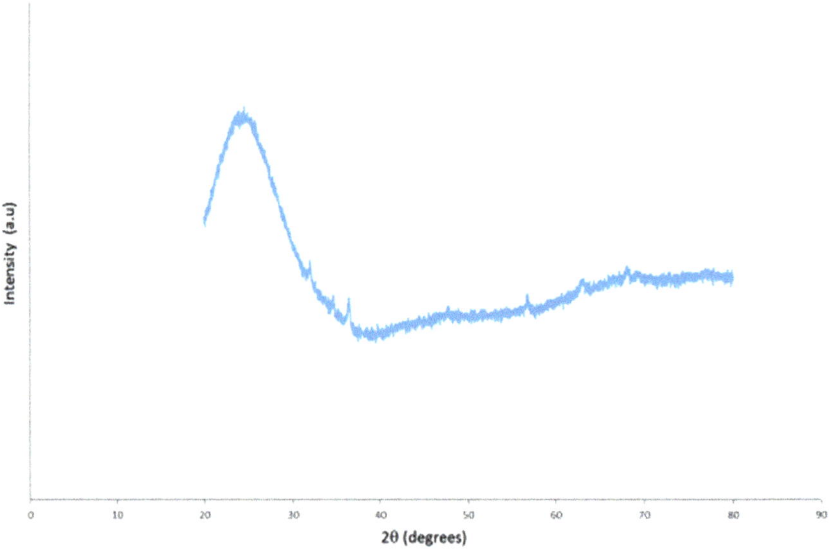

Fig. 7 XRD pattern of the plasmonic composite.

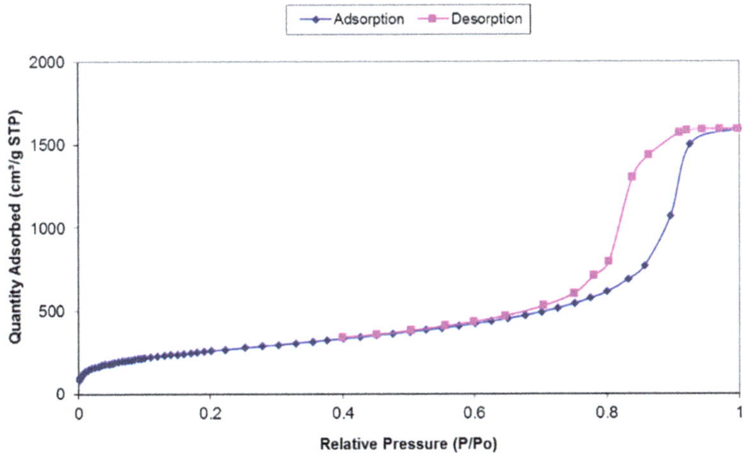

Fig. 8 Adsorption–desorption isotherms of the composites.

It could be observed that the reaction was stable at 50 °C during more than 100 minutes. Then the temperature was increased at 70 °C and tested during the same time span. Finally, the flow was reduced to half of the initial conditions. No significant changes were observed during almost 250 minutes at constant flow, which indicates the suitability of this system to test several catalyst loads under different thermodynamic and flow conditions.

It is important to note that the average conversion rate at higher flow is similar to the one obtained in other works at temperatures around 200 °C.[17] Thus, the integrated plasmonic reactor concept proposed here opens new avenues to couple low temperature solar collectors and chemistry as a mean to introduce renewable energy in the economy, particularly for the conversion of CO_2.

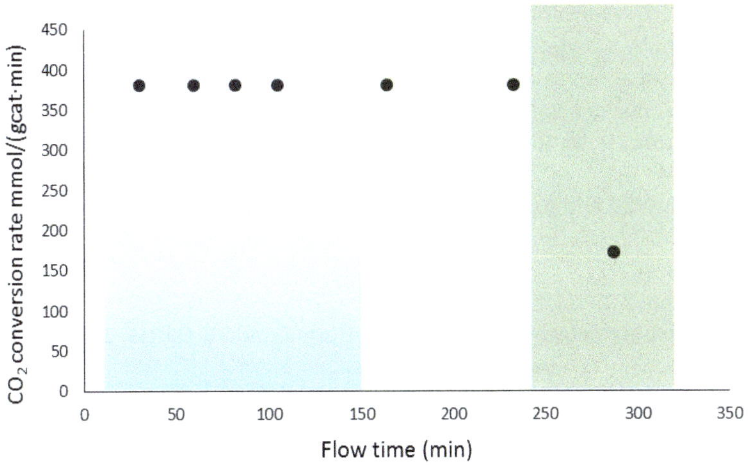

Fig. 9 Evolution of the reaction.

Conclusions

A novel plasmonic composite was developed that can absorb light from the visible spectrum. Its characterization evidenced a high surface area and proper integration with the metallic components of the catalyst.

It was possible to introduce this composite into glass microchannels in order to obtain a single entity that acts as both a light-harvesting device and as a reactor. This is possible due to the development of a simple yet effective impregnation method that allows the synthesis of the aerogels *in situ*.

The microreactor obtained was tested for RWGS in a system that allowed not only control of the reaction variables such as pressure, temperature and flow, but also, delivers visual light in an elegant way.

The CO_2 conversion rates were in accordance with other works testing plasmonic catalysts at higher temperatures. This can be due to the increased light energy delivery, the high surface area of the material and to the integrated and efficient way to deliver the radiant energy possible in this reactor. This is then a novel opportunity to gain understanding in the field of CO_2- use by applying solar energy.

Acknowledgements

The authors acknowledge the kind support of the FP7 Shyman European project (Project reference: 280983).

References

1 G. Centi, E. A. Quadrelli and S. Perathoner, *Energy Environ. Sci.*, 2013, **6**, 1711–1731.
2 G. Centi and S. Perathoner, in *Green Carbon Dioxide: Advances in CO_2 Utilization*, ed. G. Centi and S. Perathoner, John Wiley & Sons, 2014.
3 K. Ogura, H. Yano and T. Tanaka, *Catal. Today*, 2004, **98**, 515–521.
4 R. Angamuthu, P. Byers, M. Lutz, A. L. Spek and E. Bouwman, *Science*, 2010, **327**, 313–315.
5 H. Li, P. H. Opgenorth, D. G. Wernick, S. Rogers, T.-Y. Wu, W. Higashide, P. Malati, Y.-X. Huo, K. M. Cho and J. C. Liao, *Science*, 2012, **335**, 1596.
6 B. A. Rosen, A. Salehi-Khojin, M. R. Thorson, W. Zhu, D. T. Whipple, P. J. A. Kenis and R. I. Masel, *Science*, 2011, **334**, 643–644.
7 E. E. Barton, D. M. Rampulla and A. B. Bocarsly, *J. Am. Chem. Soc.*, 2008, **130**, 6342–6344.
8 G. P. Smestad and A. Steinfeld, *Ind. Eng. Chem. Res.*, 2012, **51**, 11828–11840.
9 I. Ermanoski, J. E. Miller and M. D. Allendorf, *Phys. Chem. Chem. Phys.*, 2014, **16**, 8418–8427.
10 M. Forster, *J. CO_2 Util.*, 2014, **7**, 11–18.
11 A. Dhakshinamoorthy, S. Navalon, A. Corma and H. Garcia, *Energy Environ. Sci.*, 2012, **5**, 9217–9233.
12 P. D. Tran, L. H. Wong, J. Barber and J. S. C. Loo, *Energy Environ. Sci.*, 2012, **5**, 5902–5918.
13 Ș. Neațu, J. Maciá-Agulló and H. Garcia, *Int. J. Mol. Sci.*, 2014, **15**, 5246.
14 S. Linic, P. Christopher and D. B. Ingram, *Nat. Mater.*, 2011, **10**, 911–921.

15 C. Hägglund, S. P. Apell and B. Kasemo, *Nano Lett.*, 2010, **10**, 3135–3141.
16 M. A. Garcia, *J. Phys. D: Appl. Phys.*, 2011, **44**, 283001.
17 A. A. Upadhye, I. Ro, X. Zeng, H. J. Kim, I. Tejedor, M. A. Anderson, J. A. Dumesic and G. W. Huber, *Catal. Sci. Technol.*, 2015, **5**, 2590–2601.
18 C. Wang, O. Ranasingha, S. Natesakhawat, P. R. Ohodnicki, M. Andio, J. P. Lewis and C. Matranga, *Nanoscale*, 2013, **5**, 6968–6974.
19 M. Behrens, F. Studt, I. Kasatkin, S. Kühl, M. Hävecker, F. Abild-Pedersen, S. Zander, F. Girgsdies, P. Kurr, B.-L. Kniep, M. Tovar, R. W. Fischer, J. K. Nørskov and R. Schlögl, *Science*, 2012, **336**, 893–897.
20 Z. Y. Tan, D. W. Y. Yong, Z. Zhang, H. Y. Low, L. Chen and W. S. Chin, *J. Phys. Chem. C*, 2013, **117**, 10780–10787.
21 J. Gascon, J. R. van Ommen, J. A. Moulijn and F. Kapteijn, *Catal. Sci. Technol.*, 2015, **5**, 807–817.
22 B. Louis, G. Laugel, P. Pale and M. M. Pereira, *ChemCatChem*, 2011, **3**, 1263–1272.
23 G. M. Pajonk, *J. Non-Cryst. Solids*, 1998, **225**, 307–314.
24 L. M. Sanz-Moral, M. Rueda, R. Mato and Á. Martín, *J. Supercrit. Fluids*, 2014, **92**, 24–30.
25 NASA, What Wavelength Goes With a Color?, http://science-edu.larc.nasa.gov/EDDOCS/Wavelengths_for_Colors.html.
26 R. Al-Oweini and H. El-Rassy, *Appl. Surf. Sci.*, 2010, **257**, 276–281.

Faraday Discussions

DISCUSSIONS

CO_2 reduction reactions: general discussion

Michael North, Pedro Abrantes, Eryk Remiezowicz, André Bardow, Jennifer Dodson, Troy Manning, Jonathan Albo, Daniel Reed, Derek Harris, Ian Ingram, Alexander Cowan, Matthew Moss, Gonçalo V. S. M. Carrera, Alexander Foote, Eric Fraga, Peter Styring, Annemie Bogaerts, Gabriele Centi, Alexander Navarrete, James Comerford, George Dowson, Anthony Coogan, Thomas Butterworth, Katie Lamb, Michele Aresta, Csaba Janaky, Richard Heyn, William Webb, Gerard van Rooij, Niklas Meine, Vera Meynen, Roman Michez, Dina Churina, Kusman Dossumov, Xian Du, Jeroen Spooren, Moti Herskowitz, Koen Michiels and Carmine Capacchione

DOI: 10.1039/c5fd90080h

Jonathan Albo opened the discussion of the paper by Gabriele Centi: Our group recently found the formation of methanol (as the major CO_2 electroreduction product) at Cu-based surfaces in a similar two-compartment cell configuration.[1] Actually, the production of methanol at Cu-based electrocatalysts is commonly reported in the literature.[2] How do you explain (in terms of reaction mechanisms) the formation of acetic acid and only trace amounts of methanol in your liquid phase system?

1 J. Albo, A. Saez, J. Solla-Gullon, V. Montiel, A. Irabien, *Appl. Catal. B-Environ.*, 2015, **176**, 709-717.
2 J. Albo, M. Alvarez-Guerra, P. Castaño, A. Irabien, *Green Chem.*, 2015, **17**, 2304-2324.

Gabriele Centi replied: We believe that due to the type of electrodes we use (gas diffusion electrodes, with the electrocatalyst at close contact with the Nafion membrane), methanol forms but further reacts under these experimental conditions to give acetic acid, with the tentative mechanism reported in the paper.

Jonathan Albo continued: Could the differences in pH explain the productivity and selectivity variations between the liquid and gas phase CO_2 electroreduction?

Gabriele Centi responded: No, we do not believe it is a problem of pH, but rather of effective concentration of CO_2 at the electrocatalyst surface. It is the higher concentration in the gas-phase approach (as it is not limited from the solubility and electrical double layer) which explains the formation of products > C1.

George Dowson queried: During the description of the proposed mechanism for the formation of acetic acid during your reaction, you suggest that methanol, the product of formaldehyde hydrogenation, is not observed, however the very next section describes how the methanol may essentially be being scavenged by formic acid in the generation of methyl formate. Would it not therefore be more reasonable to conclude that only some of the formaldehyde is converted into acetic acid, with the rest being converted to methanol for methyl formate production?

Additionally, in your paper, the reaction profiles in Fig. 5 suggest that the methyl formate is being consumed during the latter stages of the reaction. As this coincides with the increased formation of acetic acid in both cases, could this indicate that perhaps methyl formate is being consumed to form acetic acid? This reaction is known to occur in the literature, both *via* insertion of CO and by rearrangement on the metal surfaces. Finally, is any methyl acetate observed as a product?

Gabriele Centi replied: Yes, we believe that methanol and methyl formate further react under our experimental conditions to produce acetic acid. This is one of the possible mechanisms of C-C bond formation during the CO_2 reaction, and also may be of interest in terms of application. We are currently analyzing this mechanism in more detail, as it is currently only in a tentative stage. Methyl acetate, on the contrary, is currently only observed in minor amounts. For now, we have focused our attention on the main products.

Csaba Janaky asked: Could you comment on the amount of water transferred with H^+ ions through the Nafion membrane (electroosmotic water drag)? What is the temperature in the cathode side in the "gas-phase" reactor?

Gabriele Centi answered: We feed, in the gas phase configuration, some water vapor together with CO_2. If the electrocatalytic side joins with a photocatalyst, water vapor is also coifed on the other side (for the chemical reported in the paper, a water solution is present on the cathode side). The water feed (on the anode side, together with CO_2) is necessary to maintain a wet Nafion membrane and (likely) has a thin water film over the electrocatalyst (in contact with the Nafion membrane). This helps in enhancing the H^+ surface transport.

Csaba Janaky enquired: You mention the role of the O-functions of the carbon nanotubes on the electrocatalytic effect towards CO_2-reduction. I don't know the exact potential of your working electrode, but it is very likely that at the applied cathodic potential, these O-functions are electrochemically reduced. Can you comment on this issue?

Gabriele Centi replied: After the oxygen functionalization of CNT (with HNO_3) there is a thermal annealing to remove less stable oxygen species. The remaining oxygen functionalities are stable in the potential range studied.

Gonçalo V. S. M. Carrera commented: Unlike the gas phase cell configuration, when the reaction is carried out in aqueous electrolyte the products are mainly carboxylic acids. I wonder if this difference in behaviour is caused by the stabilization of the carboxylates by water in relative alkaline medium, as this type of stabilization is not possible in a gas phase cell configuration. Of course, other parameters influence the reaction outcome; nevertheless the results in my opinion suggest this explanation. What is your perspective on this issue?

Gabriele Centi responded: In our experiments, we use gas diffusion type electrodes, thus the electrocatalyst is at close contact with Nafion both in the liquid and gas phase operations. Under these conditions, the environment close to the catalyst is strongly acidic, which also determines the type of products we observed. However, I agree that under different pH conditions, there may be a different reaction path and type of products.

Michael North extended the discussion by addressing both Alexander Cowan and Gabriele Centi: How do the electrochemical processes described in each of your papers cope with being turned on and off repeatedly? The ability to cope well under these conditions is essential if they are to be linked with renewable energy.

Alexander Cowan answered: This is an interesting point. Our electrochemical system is stable if left in the dark, turned off, and it can be used for several cycles. Indeed most electrochemical systems for CO_2 reduction that I am aware of are able to operate intermittently. In some of the electrodes presented here we actually find that a period of in-operation can be advantageous as it provides a period for the CO_2 concentration at the electrocatalytic sites to increase - we believe that the diffusion of CO_2 through the polymer is one of the limiting factors in the experiment.

Gabriele Centi replied: The fixed costs of plants in electrochemical processes is typically over 50%, thus continuous operations are necessary to decrease amortization costs. When coupled with renewable energy sources (except for in few cases, such as hydropower), a storage of electrical energy would be necessary. Therefore continuous operations (with adequate energy storage) are requested to minimize the costs and the device should not operate with a fast on/off switch.

Ian Ingram addressed Alexander Cowan and Gabriele Centi: High Aspect Ratio Nanoparticles (HARNs), and carbon nanotubes in particular, are the subject of growing concern and regulation because of their toxicity. Konarka, an organic photovoltaics company, went out of business in 2012 partly because their process used large quantities of chlorobenzene and other toxic solvents which proved to be very expensive to handle safely in a production environment, even though they are of very little concern in the research laboratory. Could you both comment on the viability of using carbon nanotubes at scale in an industrial process, given the concerns about their safety?

Alexander Cowan responded: We note that CNTs are already in use in several industrial processes, however, this is of course not sufficient evidence for us to propose that they can be safely used on very large scales. A full assessment of the health impacts of CNT manufacture would be required if these materials, developed for fundamental studies, where to be deployed on a large scale. Of relevance to this question is the stability of the polymer/CNT films, which is relatively good. We do not see evidence for CNT loss into the electrolyte which would decrease the potential exposure to CNTs.

Gabriele Centi answered: Health and safety aspects are clearly critical and definitively we should avoid using toxic solvents or dangerous materials. However, the toxicity of CNTs has often been overestimated, being made on materials which are not those in practical use. By themselves, CNTs are not toxic, for example they are used for drug delivery, but it is the surface radical-chemistry or the adsorption of some organic compounds which may make them problematic. However, in practical materials where high conductivity is necessary, these aspects are not different from those of many other materials used to prepare electrodes.

Csaba Janaky commented: When discussing the scalability of CO_2 electrolysis, one should consider that water electrolysis (H_2-evolution) was already performed on an industrial scale in the USA in the 1970s. This shows that the technology itself can be scaled, it's just that more research is needed to find efficient and selective electrocatalysts.

Gabriele Centi responded: Yes, we agree. There are also large electrochemical processes in operation (besides Cl_2, adiponitrile, *etc.*), even if the technology is somewhat different. However, the key problem of cost should be addressed by increasing the productivity of the electrodes, *i.e.* by developing 3D-like electrodes.

Peter Styring addressed Gabriele Centi and Alexander Cowan: I agree with the good point raised by Professor Centi regarding the use of 3D electrodes. Related to scalability, we've talked about the materials but what about the process itself? Cells operate in a laboratory environment, but how do you scale out into tonnes of product?

Gabriele Centi replied: The cost of the cell is the critical element in electrocatalytic reactions, and it is therefore necessary to increase productivity per surface of the electrode in order to decrease the costs to the levels required for industrial exploitation. Developing 3D-like electrodes is therefore a priority in order to proceed in this direction. There are problems of scalability, related to homogeneous utilization in large electrodes, but these technological problems are already solved in part for other applications.

Alexander Cowan answered: This is an interesting point. I would like to start by noting that even at the laboratory scale significant technical challenges still need to be overcome. Overpotentials for many systems are still larger than many would consider practical, furthermore, obtaining selectivities towards a particular CO_2 reduction product is challenging, especially when aqueous electrolytes are used. Finally, the role of potential contaminants in CO_2 is very rarely even studied and this must be addressed if we are to utilise industrial CO_2 sources. However, work is underway to address these problems and the potential of this approach is enormous. I can think of no insurmountable reason why the sort of materials discussed could not be scaled up to these levels. As discussed, particular attention will need to be paid to the development of 3D high surface areas that are able to maintain suitably high current densities, and additional effort will be required to develop suitable reactor designs. Many very high tonnage electrochemical synthesis processes already exist which does provide confidence that such issues can be addressed.

Niklas Meine asked: In industry, safety is one of the major issues, and for this reason we are training people all around the world to guarantee very high safety standards. With respect to electrochemical cells, I wonder how you consider up scaling this process, and how the choice of electrode materials is therefore considered?

Gabriele Centi answered: I fully agree that high safety standards are important in the manufacture of electrodes. The electrode we discussed could be manufactured in quite a similar manner to those for fuel cells, by screen printing or similar methods that can also be automated to avoid direct contact of operators with the materials used. There are therefore currently no specific issues over their industrial manufacture.

Jonathan Albo remarked: The transport of protons across the membrane may be key in this process. Have you tested other ionic transport membrane materials for your system? I also wonder if you detected CO_2 reduction products in the anodic compartment (membrane crossover)?

Gabriele Centi responded: The transport of protons from the two electrodes is the element limiting productivity, and likely the transport from the photocatalyst to the membrane and from the membrane to the electrocatalyst (rather than across the membrane). While transport in the Nafion membrane is well established, the measure of the other two aspects is not easy, but it's greatly influenced by the reaction conditions (a thin liquid film is likely necessary) as well as the electrode characteristics (hydrophilicity). We feed CO_2 gas into both the anode and cathode compartments to avoid crossover.

Michael North commented: Regarding the benefits of water *versus* acetonitrile as the electrolyte, polar aprotic solvents, including acetonitrile, are on the EU REACH at risk list and could soon be banned. Furthermore, cyclic carbonates would be a suitable alternative to polar aprotic solvents, and have the advantage of being made from CO_2.

Michele Aresta replied: Organic carbonates are already used as support electrolytes, also if at a low volume. Their use is driven by the operation conditions: they cannot be used in acidic conditions as they would be converted back into alcohol and CO_2.

Peter Styring continued the general discussion of the paper by Gabriele Centi: A general comment on using acetonitrile is that it would have to be used in a closed system for safety reasons. However, there are economic concerns as it is getting harder to source acetonitrile. Are there any other solvent systems that could be used that are more sustainable and safe?

Gabriele Centi responded: It is unlikely that an organic electrolyte like acetonitrile could be practically used in industrial operations under strong acid/base conditions. For this reason, our experiments are undertaken in aqueous solutions.

Michele Aresta enquired: Cu electrodes have been reported as being able to produce ethene from CO_2. Have you observed any formation of this alkene under your experimental conditions?

Gabriele Centi answered: No, we have searched for them but they were not detected, although traces of hydrocarbons were detected. Alkenes form at a high potential, in conditions where fast deactivation occurs.

Daniel Reed asked: Can you comment on the durability/life span of the cells, and do you have any special handling requirements for them? For example, are they brittle and do they require careful handling? Furthermore, how easily can they be scaled up?

Gabriele Centi responded: The cell itself does not have any special requirements for handling and the electrodes are quite robust. One of the problems, however, is that some of the product may remain adsorbed on the electrode, causing deactivation. One of the reasons for operating above r.t. is also to increase the rate of desorption of some of the reaction products from the electrode. Productivity is still the main issue, so we have not made long-term durability tests (beyond the usual lab scale). In terms of scale up there are no main concerns, at least up to 10 x 10 cm size.

Alexander Cowan replied: Once formed, the polymer/catalyst/CNT films are quite stable in the dark. The biggest issue regarding stability arises through operation at very high current densities, when excessive gas production can cause film rupture. We are working on new electrode designs in an attempt to alleviate this issue. One of the advantages of the proposed approach is its simplicity. Film fabrication involves preparation of a polymer/catalyst/CNT ink that is either drop-cast or doctor bladed onto the carbon electrode. The ability to simply print the electrodes appears to be quite amenable to scale up.

George Dowson continued the general discussion of the paper by Alexander Cowan: Comparing the previous literature performance of catalyst 2 against the performance you see post-immobilisation, there is an extremely large drop in turnover frequency. Since the higher voltage used in the non-immobilised form was partially responsible for this discrepancy, do you think that perhaps there is an economic argument that it is worth increasing the voltage used to achieve these higher frequencies? Concordantly, would it not be worth testing your catalyst system at the same and even higher voltages as used in the previous literature?

Alexander Cowan responded: Thank you for your question. One of the reasons for the lower turnover frequency here is that the numbers are based upon the inclusion of all of the Mn centres within the film, of these we know that only a small subset are actually electroactive, so if we quoted the turnover frequency for electroactive centres it would be far higher. You are also correct in noting that we study the system at a different applied potential. We agree that ideally the potential dependence of the turnover frequency would be fully reported in both cases so that the catalysts can be fairly compared, however, in our experiments we find that at higher applied potentials the selectivity towards carbon dioxide decreases and increased hydrogen production occurs. So although higher TOF may be achievable we believe that the system is of less interest at greater applied potentials.

Richard Heyn remarked: Rather than performing the electrochemical CO_2 reduction in water, is there any benefit to be gained if one could do the reduction on aqueous carbonate or bicarbonate species?

Alexander Cowan answered: Many studies are indeed carried out using bicarbonate based electrolytes as you suggest. We chose to use a phosphate buffer in this particular study as it simplifies the analysis of the control experiments done in the absence of carbon dioxide. During experiments using bicarbonate/carbonate, the electrolyte is a potential carbon source for reduction, which in practical applications may be advantageous, but for fundamental studies such as this one, complicates the analysis.

Katie Lamb communicated: I have a question regarding the manganese complexes used for electrochemical carbon dioxide reduction. I understand that many manganese complexes are light sensitive, and therefore many of the reactions in this study were performed in the dark. I was therefore wondering whether you see this as a potential hindrance in the future for these catalysts, and if this will also hinder potential industrialisation of these catalysts?

Alexander Cowan communicated in reply: The complexes employed are all handled in air and are quite stable in air. The question is completely correct that they are however very light sensitive and all experiments are carried out in the dark. This is a potential disadvantage. If this system, which is still at an early stage of research, was to be developed into an industrial process then the reactors and electrode preparation areas would have to be darkened.

William Webb communicated: I was wondering if you observed, or had measured, the leaching of your Mn system from the electrode into solution during the course of the reaction? Given that, if any leaching is present can you comment on the lifetime of the catalyst system within the electrode and do you know of a way to improve this?

Alexander Cowan communicated in response: This is an important question. We examine the electrolyte solutions after the experiment using a range of techniques (UV/Vis, NMR and electrochemically) to check for catalyst leaching. In all of the systems tested we have not observed any catalyst leaching. To date, the longest experiments have been on the timescale of hours, and in the future if this system were to be developed further it would be important to also confirm that leaching does not occur over much longer periods (weeks to months).

Roman Michez communicated: In your paper, degradation of complex 4 and a decrease in the faradaic efficiency for the formation of CO is observed after long time electrolysis is performed in acetonitrile; apparently correlating to the low catalytic activity of this system. However, the same behavior can also be observed during experiments in which the potential of the reference electrode is not stable, due to the accumulation of the reduction products. Have you observed any indications of the stability or non-stability of the reference electrode you used?

Alexander Cowan communicated in reply: You make a good point that we did not fully consider in our original study. I agree that the CO produced may poison the Ag reference electrode that was used in the anhydrous acetonitrile experiments after prolonged electrolysis. Since our discussion at the meeting we have also carried out some studies using a Ag/$AgNO_3$ reference electrode. However, unfortunately we still find that the catalytic activity decreases during prolonged electrolysis experiments.

Jennifer Dodson addressed Gabriele Centi and Alexander Cowan: Separation of dilute concentrations of products from solutions can be a huge processing challenge. Can you comment on the process for recovery and separation of the products from your solution-phase electrocatalytic cells?

Gabriele Centi responded: The main reason for the electrolyte-less cell design we used (*i.e.* gas phase) is the possibility of easy recovery of the reaction products, which is energy intensive when it requires recovery from an electrolyte.

Alexander Cowan answered: In this study we produce CO and H_2 as the products. The relatively low solubility of CO in water means that the product can be recovered from the reactor headspace relatively easily. At the moment the headspace gas consists of CO, H_2, CO_2 and some water vapour.

Troy Manning asked: How sensitive are these electrocatalysts and electrocatalytic processes to the composition of the CO_2 feedstock? How do impurities in the CO_2 affect the process?

Alexander Cowan replied: Thank you for the question. The potential inactivation of the catalyst through impurities within the CO_2 feedstock is an important issue. For this early stage fundamental study we use a high purity CO_2 source where water is the primary impurity. Unpublished experiments from our lab indicate that our catalysts are also reasonably tolerant to the presence of oxygen. To date, we have not tested other likely impurities such as SOx and H_2S, and this will be addressed in future work.

I would like to note that the use of a molecular electrocatalyst for CO_2 reduction may have distinct advantages when considering competition between the desired substrate and impurities. If the mechanism of catalyst inactivation can be identified then it may be possible to design new synthetic analogues with improved tolerances.

Gabriele Centi responded: We do not use a rather pure CO_2, but one compatible with that which may be derived from fermentation processes or from biogas after methane recovery. We do not believe that there is a need to recover CO_2 from flue gases, in which case some of the impurities may become critical for stability.

Ian Ingram opened the general discussion of the paper by Peter Edwards†: The boiling points of propane and propene are very close together, which makes separation difficult and more expensive. As propene is a relatively low-value product, you are making it from a fossil source and your process is thermodynamically limited to low conversions. Is the significant environmental and economic cost of separation justified?

Xian Du answered: The physical separation of propylene from propane is of course challenging, as you point out, but there are indeed well-established (and indeed new) technological processes for such separations. Regarding the price, from alkane to olefin, the price can be nearly doubled as you can see from a quick check online: from the market in China, I can see that propane is currently 3500 CNY/ton while the price for propylene is 6000 CNY/ton. In industrial terms over large volume production, this is a highly significant difference indeed! Many industrial concerns are now aggressively targeting new routes to propene production and this has been a major focus for the King Abdulaziz – Oxford Centre of Excellence in Petrochemicals, which is a part-funder of this research. More robust, higher performance catalysts are now the subject of our own studies. The major aim of our research in this paper is to find a way to minimize the by-product and coke effect, and of course this method can be easily applied over other higher performance catalysts.

Eryk Remiezowicz asked: You produce carbon monoxide, what is your idea for the usage of this product of your reaction?

Xian Du responded: We haven't yet made a complete calculation, but of course both CO and hydrogen as the synthesis gas is the route to many fuels. Of course, at the very heart of our 'vision' is the use of thermal energy in so-called 'intermediate' temperature regimes (ca 3–500 °C), where such solar thermal energy is accessible by cheap, robust collectors rather than the high technologies required for say 1000 °C thermal collectors and processes.

Gabriele Centi asked: The system is essentially the combination of two reversible reactions, propane dehydrogenation and RWGS. How fair are the results with respect to equilibrium? Did you analyse the amount of H_2 formed to demonstrate that RWGS is shifting the equilibrium in propane dehydrogenation?

† Peter Edwards' paper was presented by Xian Du, University of Oxford, United Kingdom.

Xian Du replied: According to our modelled simulation results with Aspen Plus (a software which can calculate the composition of a gas mixture at equilibrium state), at 500 °C, propane will be totally converted, no propylene would be observed in the system, and only methane, CO and CO_2 are left. When comparing this result with the calculated results of traditional propane dehydrogenation (without CO_2, which should be around 30% at 500 °C), we can see that the ultimate equilibrium of the system was effectively shifted by CO_2 addition. It's not easy to confirm that H_2 production can immediately demonstrate the existence of RWGS. H_2 was generated *via* multiple side reactions (such as cracking of propylene), and therefore it'll be tricky to prove the RWGS by only calculating H_2 production. The ideal and most straightforward way, we believe, is to observe H_2O production; so far, the amount of H_2O is nearly immeasurable, and that's why we will try to scale up the reactor to seek more details on this.

Michele Aresta enquired: The use of CO_2 as the dehydrogenating agent has different properties (thermodynamics, rate, selectivity) with respect to dioxygen. Can you comment on the different behaviours of the oxidizing agents?

Xian Du answered: So far, besides CO_2, only O_2 and H_2O have been published as oxidising agents in dehydrogenation reactions. O_2 is effective in improving the propane conversion, but the selectivity is massively reduced due to the enhancement from O_2 in the complete combustion process. We are seeking a high selectivity of propylene even though the propane conversion would be sacrificed a little bit, so we think O_2 is simply too strong in this delicate system. H_2O has similar usage to CO_2, and the generated coke can be more easily removed when using H_2O rather than CO_2, thermodynamically. However, because H_2O cannot consume H_2, the equilibrium of propane dehydrogenation ($C_3H_8 = C_3H_6 + H_2$) can hardly be shifted. This is the main reason why we are using CO_2 as the mild oxidising agent, as a potential route to these solar thermal reactions, at these more accessible temperatures, rather than highly energy intensive dry reforming processes.

Kusman Dossumov communicated: Do you take into consideration the problem of catalyst carbonization during utilization of CO_2?

Xian Du communicated in reply: Actually we didn't consider it, and also we didn't see any evidence of carbonate species generated on the post-reaction catalysts.

James Comerford remarked: Why do you specifically use a 2 : 1 ratio of CO_2:propane in your paper? Have you tried varying this?

Xian Du responded: Yes we have; varying the ratio will apparently affect the conversions and selectivity due to the competitive adsorption between CO_2 and propane. However, in this paper, we maintain the ratio at 2 : 1 due to a similar ratio being published in other papers, and we would like to make the data comparable with others.

James Comerford asked: The total concentration of propane and CO_2 used in your system is low. As such, how much propene can be produced (*i.e.* grams per hour), considering that this paper is essentially focused at developing an industrial process? Additionally, does catalytic activity diminish after 300 min?

Xian Du replied: The production of propylene is indeed very low due to the experimental scale in our lab. Actually, as you can see in the paper, we were mainly focusing on how to inhibit by-product generation, and industrialisation, as in most of the schemes presented during this meeting, is about an outlook for the future. In fact, we do not think the production rate will be a big issue when we finally, successfully, scale this reaction up and proper heat and mass transfer are been carefully considered.

Michael North asked: Is CO_2 the actual oxidant for propane, or is the chromium oxide oxidising the propane and CO_2 reoxidising the chromium?

Xian Du answered: First, I don't think that CO_2 is the major oxidant for propane immediately, because nearly no conversion of propane can be observed if we don't use any effective catalyst. This result can also prove that the activation and dehydrogenation of propane can only happen on a suitable catalyst surface. Here we believe that CO_2 acts as a sacrificial agent of hydrogen, which is from the dehydrogenation of propane. CO_2 can also act as a reoxidising agent for the chromium catalyst. This needs further study.

Gabriele Centi opened a general discussion of the paper by Moti Herskowitz: Can you give a little more information about your catalyst, for example, whether there are supported iron particles segregated on the surface and which dimensions they have, as they are relevant for the FT reaction. Likely, it is not a true spinel phase, but a partially decomposed phase which then further transforms during the reaction.

Moti Herskowitz responded: The catalyst is a potassium promoted pure Fe-Al-O spinel $Fe(II)(Fe(III)_{0.5}Al(III)_{0.5})_2O_4$ in its as-prepared form. It is transformed after activation (carburization) and self-organization under the reaction conditions to iron carbide phases (mainly Fe_5C_2) and Fe-poor spinel phase with an Fe(III)/Al(III) ratio of 0.2–0.3. Their particle size is 20–50 nm. No metallic iron nanoparticles supported on these phases were detected in the steady state catalytic material (XRD, XPS, HRTEM-EELS).

Vera Meynen communicated: The experimental section, in your paper, indicates that the catalysts have only been calcined to 450 °C, while the TPD has been done up to 900 °C. Can you comment on the stability of the catalysts over the entire temperature profile of the TPD? Did you check the stability with a blank measurement with the same heating profile, with only a helium flow, or evaluate the presence of structural and/or physico-chemical changes after the TPD measurements?

Moti Herskowitz communicated in reply: The CO_2-TPD spectra were recorded within the temperature range of 100–900 °C in order to detect both weak and strong basic sites. It is known that the reflections representing strong basic sites in Fe-oxide based materials appear at >500 °C, as illustrated in Fig. 4. Heating of the K/Fe-Al-O spinel to 900 °C did not change its structure (as measured by XRD), it only increased the size of the primary nanocrystals.

Vera Meynen communicated: Can you also indicate if degassing or stripping of the sample was done before the TPD measurement and if so, under which conditions they occurred?

Moti Herskowitz communicated in response: During the TPD measurements the catalyst surface was saturated with CO_2 at 40 °C, and before recording the TPD spectra the samples were heated in a He flow at 40–100 °C for stripping.

Vera Meynen communicated: In your discussion of Fig. 4, you write that two additional CO_2 desorption peaks appeared centered around 550 and 850 °C. However, in my opinion there are 3 or even 4 peaks, as I see two additional smaller desorption peaks at 120 and 135 min (so in between the two peaks mentioned). Or are they not significant? Also, the first peak appearing in both the samples with 2 wt% and 4 wt% K content is not just one peak, as you have indicated in your paper, as it is asymmetric. Could you please correct this. Do they have any characterization towards physico-chemical information on which type of bonding sites these peaks coincide with?

Moti Herskowitz communicated in reply: Firstly, the level of CO_2 signal recorded during 120–135 min was significantly lower than that of the peaks mentioned in the article, and therefore they were indeed neglected. Secondly, the purpose of the CO_2-TPD measurements performed in this section was to study the effect of potassium on the surface basicity of the catalyst, differentiating between weak and strong basic sites. As a matter of fact, the peaks of the three samples centered at ~150 °C are asymmetric indicating that there could be more than one peak. The total area was integrated providing the relative amount of weak basic sites, while the strong basic sites are found at >500 °C. Finally, the characterization of the physico-chemical information is beyond the scope of this work.

Gabriele Centi enquired: What is the effect of K, for example how does it affect the product distribution? CO_2 has a stronger oxidizing effect with respect to CO and this typically affects the productivity. In your case, what is the productivity of the catalyst feed CO_2/H_2 rather than syngas (with a little amount of CO_2)? What is the role of Al in determining the catalyst performances, particularly the resistance to water inhibition?

Moti Herskowitz answered: The addition of potassium shifts the products distribution to higher carbon hydrocarbons. Productivity is lower for CO_2 hydrogenation compared with CO hydrogenation. Our spinel material is a good FT catalyst yielding almost complete conversion of CO. The role of aluminum in the composition of the K/Fe-Al-O spinel is to reduce the reduction-carbidization of iron at the activation step, thus controlling the amount of Fe-carbides. Excess Fe-carbides formed by activation of magnetite spinel in the absence of aluminum leads to deep carburization. As a result, a large amount of graphitic carbon is formed due to oxidation of carbides at the catalyst self-organization stage. This significantly reduces the catalyst surface area.

Dina Churina communicated: 1) What is the stability of the catalyst for utilising CO_2 as a chemical feedstock? 2) What is the cost of this catalyst?

Moti Herskowitz communicated in response: 1) The catalyst was tested for 500 h with no apparent deactivation. 2) The catalyst cost was not evaluated, however, given the low cost of the reagents and the simple preparation method, it is expected to be low.

Alexander Cowan remarked: Your paper states that hydrogen production for electrocatalysts is not competitive when compared to hydrogen from fossil derived resources. The use of biogas is instead proposed here – have you done a comprehensive analysis looking at the relative costs for biogas as well?

Moti Herskowitz replied: We are working on the economic evaluation of various sources of hydrogen, biogas among them.

Jennifer Dodson commented: To add to the discussion about biogas availability, I think this would be extremely limited as a feedstock. I wrote a recent overview on Anaerobic Digestion for the UK Parliament and we estimated that the country's biogas potential was less than 1% of the UK energy supply. This is already used directly as a fuel both for cars and electricity production.

Moti Herskowitz responded: Biogas availability is limited, but this may change in the future, certainly in specific locations where it could be an abundant feedstock.

Michael North opened the discussion of the paper by Koen Michiels: Did you see any evidence for phenol formation in your reactions, as there is some evidence in the literature for the formation of phenol under similar conditions? Furthermore, as you use large amounts of water, could corrosion of the reactor be a potential problem?

Koen Michiels replied: All obtained liquid samples were analysed with proton NMR spectroscopy. In all cases, no evidence was found for the presence of aromatics such as phenol. Large amounts of water and hydrogen gas create corrosive conditions, therefore, it is important to select the materials of which the autoclave and related components are made off. We chose stainless steel 1.4435 as the 'reactor' material. This type of stainless steel is known to be highly resistant to corrosion. Until now, we have not faced any corrosion problems.

Eryk Remiezowicz asked: How do you justify a CO_2 utilization process making iron, which generates a lot of CO_2 during production, knowing that the iron waste is being used by the steel industry and is valuable?

Jeroen Spooren answered: Iron production indeed generates CO_2. Therefore, it is our intention to utilise waste fine grained iron materials, such as waste generated during shaving, milling, and drilling at iron works. The fine fractions of these waste materials cannot be directly used in the iron or steel industry as secondary raw materials due to the creation of hotspots upon introduction in the oven, and even a danger of explosion. In fact, through industrial contacts, we are aware that currently a small fraction of fine grained iron waste is still introduced in furnaces by mixing it with other scrap materials. However, acceptance criteria with regard to the fine grained waste iron are getting more strict, so that blast furnaces and smelters are refusing to take in these waste materials. Therefore, it would be a good solution to recycle such waste in an alternative way and turn it into higher added value products.

Niklas Meine enquired: I think it is quite an interesting idea to use the fine iron powder as a raw material. Can you form only magnetite or would it be possible to oxidize iron to hematite as well? How could you do so?

Koen Michiels responded: The developed hydrothermal process to produce hydrogen gas is based on the oxidation of zero valent iron into magnetite Fe_3O_4. Experiments performed at different conditions of temperature, initial CO_2 amount, *etc.* – which were performed in the context of the optimisation study of hydrogen production – do not suggest the formation of hematite Fe_2O_3. In the literature, the commonly described hydrothermal iron oxidation reaction is the conversion of zero valent iron into magnetite.[1] Therefore, until now, we were not able to convert metallic iron directly into hematite *via* a hydrothermal process.

1 F. Jin, X. Zeng, Z. Jing and H. Enomoto, *Ind. Eng. Chem. Res.*, 2012, **51**, 9921–9937.

Niklas Meine asked: Does the particle size remain small after oxidation so that the iron oxide could be used as a pigment?

Koen Michiels replied: During the hydrothermal production of hydrogen gas particle aggregation occurs. In the case of applying zero valent iron grains of 5 μm as the starting material, aggregates of different sizes were obtained, ranging from 10 μm up to 1000 μm, which was confirmed by laser diffraction particle size analyses. The formation of aggregates probably has a negative effect on the formation of hydrogen gas due to incomplete conversion of metallic iron present in the inner core of the larger grains. We are indeed looking at ways to re-utilise the oxidised iron after reaction as the intention is to valorise the oxidised iron phase. Use as a pigment is indeed one of the routes that we are currently investigating. To our knowledge, Fe_3O_4 is applied as a black pigment in concrete and has a value of about 500-750 € per ton. But of course, other options are also possible.

Peter Styring returned to the general discussion of the paper by Moti Herskowitz: You say that the production of hydrocarbons from CO_2 using Fischer–Tropsch-type processes are un-economical. Have you come across the German company Sunfire's process? They are producing synthetic diesel at a cost claimed to be less that 1 euro per litre. Do you have any comments on this?

Michael North stated: CRI now have a commercial plant for methanol synthesis in Iceland which uses very cheap geothermal electricity to electrolyse water – process economics are very location dependent.

Moti Herskowitz responded: I definitely agree that there are opportunities to produce low-cost hydrogen by electrolysis if indeed very low cost electricity is available.

Gabriele Centi commented: To make economical methanol, you need to have a cost of H_2 at about 0.12–0.14 € Nm^3 H_2 – this low cost is possible in some areas where cheap electrical energy is available. This is different if you apply for an excess of energy, it is necessary to store the energy, because the fixed costs of a plant for electrolyzer/catalytic conversion account for about 70% of the overall cost (thus the amount of time it is being used should be high).

Jennifer Dodson continued the general discussion of the paper by Koen Michiels: You presented a scheme indicating that your process involves the direct reaction of CO_2 and H_2 to form formic acid, however, in reality you are reacting bicarbonate ions. Have you tested the direct reaction of CO_2 and H_2 over your catalyst?

Koen Michiels answered: Our experimental results suggest that the reduction of CO_2, mainly present as dissolved hydrogen carbonate ions, with H_2 takes place in the water phase and not in the gas phase. In the paper, we explained our choice to use potassium bicarbonate, $KHCO_3$, as the CO_2 source – in order to make a comparison with other similar one-step hydrothermal methods, which are described in the literature and which also make use of HCO_3^- ions. In the case of applying Milli-Q water as the reaction medium, we did not perform experiments with gaseous CO_2. We believe extra parameters, such as the solubility of CO_2, will become important in that case. However, when applying 1 M KOH as the reaction medium, we have performed experiments with gaseous CO_2. After producing 42 mmol H_2 *via* the developed hydrothermal method (step 1), in which 15 mmol of gaseous CO_2 dissolves in 40 mL of a 1 M KOH solution to obtain mainly carbonate CO_3^{2-} ions, we investigated different experimental options to convert 'CO_2' (step 2), as described in the paper. Firstly, the temperature was increased to 280 °C for 24 h. In this case, a carbon conversion amount of approximately 9.7 wt% was observed. One-step hydrothermal methods described in the literature make common use of a hydrogen carbonate salt, such as $NaHCO_3$, which is dissolved in water as the CO_2 source.[1] In the above described experiment, on the other hand, the CO_2 source is mainly CO_3^{2-} ions. In order to obtain mainly HCO_3^- ions, the addition of extra CO_2 is necessary after the first step of hydrothermal H_2 production. However, performing an exactly quantified addition of CO_2 without losing the produced amount of hydrogen gas is complicated due to the high pressure already present in the reactor, as we experimentally encountered. Therefore, we did not include the obtained results in the paper.

1 B. Wu, Y. Gao, F. Jin, J. Cao, Y. Du and Y. Zhang, *Catal. Today*, 2009, **148**, 405–410.

André Bardow continued the general discussion of the paper by Koen Michiels: Your title indicates that you are producing formic acid but you appear to actually be producing formate. I worry about the separation of the product which is crucial for the energy efficiency of the process but is not discussed in your work. Could you comment on a) your current final product and b) on the separation?

Koen Michiels responded: The obtained main product is potassium formate, dissolved in an aqueous solution, therefore, the title might be misleading. In order to tackle any possible confusion, we replaced the word 'formic acid' by 'formate' in the title and throughout the paper. However, in the case of acidifying the obtained aqueous solution, *e.g.* by adding hydrogen chloride, HCl, potassium formate will be converted into formic acid (and potassium chloride). We agree that product separation is an important aspect which should be taken into account when making the process economically feasible. We believe that the optimal process conditions should match the separation procedure. More concretely, for example, the separation of formate/formic acid and water might become more difficult in the case of low formate/formic acid concentrations. Low formate/formic acid concentrations might be obtained in the case of applying a high degree of filling with water. However, a high degree of filling has, according to our research, a positive influence on the amount of carbon conversion due to the increase of the (absolute) amount of dissolved hydrogen gas. Potassium formate and water can be separated by evaporation.[1] Mixtures of formic acid and water are, due to their similar properties, more difficult to separate. For example, the boiling point of formic acid is 100.7 °C, which implies that performing a distillation process is difficult. However, certain current commercial production processes of formic acid also have to deal with the separation of mixtures of formic acid and water, since these processes make use of large amounts of water in the last (hydrolysis) step. Therefore, separation methods of mixtures of formic acid and (large amounts of) water are already commercially available. A possible solution is described hereafter. Firstly, formic acid and water are separated based on a liquid–liquid extraction method with an organic base. Formic acid will dissolve in the organic base. Eventually, formic acid and the organic base are separated by distillation (http://www.intermediates.basf.com/chemicals/topstory/ideen_tradition).

1 Patent WO2008091186 A1, 2008.

Michael North continued the discussion of the paper by Moti Herskowitz: Potassium is introduced to the catalyst as potassium carbonate, then the solid is calcined. What is the chemical nature of the potassium in the active catalyst? Does it stay as carbonate or form potassium oxide or hydroxide, which may be relevant to the strongly basic sites on the catalyst.

Moti Herskowitz replied: Potassium carbonate is decomposed at the catalyst calcination step and does not form nanoparticles of oxide or hydroxide phases (XRD, HRTEM-EELS), rather atomic or small cluster species that are uniformly distributed and interact with the surface of Fe-oxide and Fe-carbide nanoparticles. The strong basic sites detected by CO_2-TPD at high potassium loadings are probably a result of this interaction.

Anthony Coogan asked: In your paper, you state: 'Carbon dioxide and water are the most abundant and low-cost sources of carbon and hydrogen needed for the production of hydrocarbons.' I understand that the cost of water may be expensive in certain places due to the cost of desalination. Have you considered this?

Moti Herskowitz responded: The cost of water produced by desalination of seawater is by far lower than the cost of electrolysis thus it is negligible compared with other costs associated with water splitting to produce hydrogen.

Alexander Foote opened a general discussion of the paper by Annemie Bogaerts: The asymmetric mode vibrational states of CO_2 play a very important role in the dissociation to CO. There are many vibrational states of CO also included in the model. Do the vibrational states of CO also play an important role in dissociating CO into C, and if so does this significantly reduce the yield of CO obtained?

Annemie Bogaerts answered: No, the vibrational states of CO play a less important role in dissociating CO into C and O, because of the higher energy needed for excitation.

Alexander Foote remarked: Why doesn't the plasma produce a significant amount of C from CO? Is it due to the energy required to break the C-O bond or high recombination of C and O, or both?

Annemie Bogaerts responded: It is mainly because of the energy required to break the C-O bond.

Gerard van Rooij said: Dissociation of CO *via* vibrational excitation, analogous to the mechanism that is put forward to explain the efficient CO_2 dissociation in the plasma phase, is inherently less efficient than CO_2 dissociation due to the higher dissociation energy and a larger vibrational constant of CO.

Annemie Bogaerts answered: Yes, that's right.

Thomas Butterworth commented: Your experimental data for CO_2 splitting in a packed bed DBD shows that with smaller particle sizes the conversion of CO_2 decreases. In your work this is attributed to a reduction in residence time. However, this reduced conversion with particle size may also be attributable to the applied electric field strength being insufficient to initiate an electrical breakdown in the void spaces of the particles *i.e.* the reactor is not completely discharging. This phenomena can be demonstrated through the use of Lissajous figures, and measuring the burning voltage at different particle sizes. This is a result that I have demonstrated in my own, currently unpublished, work. I have attached a graph (Fig. 1) showing the change in reactor burning voltage at different particle sizes from my thesis. Although the data shown in this graph very specifically applies to my experimental equipment, the general trends are likely to apply to all packed bed plasma reactors *i.e.* burning voltage increasing with decreasing particle size. In addition to this, it is worth noting a recent publication from Duan *et al.*,[1] which has demonstrated that CO_2 conversion increases with

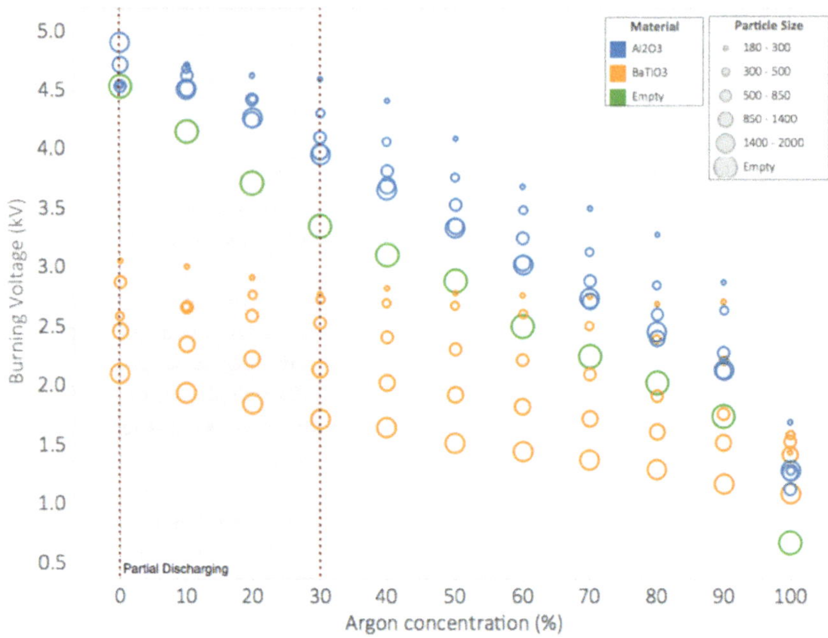

Fig. 1 A graph showing the change in reactor burning voltage at different particle sizes.

decreasing particle sizes, with the application of electric field strengths approximately 7 times greater than those used by Bogaerts *et al.*

1 X. Duan, Z. Hu, Y. Li, B. Wang, 2015, *AIChE Journal*, **61**, 898–903.

Annemie Bogaerts responded: This is a very interesting comment. To investigate this in more detail, we are analyzing the Lissajous plots for different bead sizes. The publication of Duan used a much smaller gap (0.6 mm, while we used 4.5 mm). In this smaller gap, for the same applied voltage, the electric field will be higher so the breakdown will be easier, also for smaller bead sizes. That might explain the difference with our results. This kind of DBD microplasma might be of interest for CO_2 conversion.

Matthew Moss asked: Vibrational states play an important role in the dissociation of CO_2 in a plasma, less so in a DBD but nonetheless they do play some role. As plasma modellers we aim to create accurate models yet we wish to do so within a reasonable timescale. What is a reasonable level of simplification in a typical plasma model, that still accounts for the role of vibrational levels, which reduces the computational time to a reasonable level?

Annemie Bogaerts answered: The answer depends on the type of model (0D chemical kinetics model or multi-dimensional fluid model) and on the type of plasma (DBD *vs.* microwave (MW) plasma or gliding arc (GA) discharge). In a 0D chemical kinetics model, the calculation time is not so much of an issue, as the calculations do not take too much time. In this case, it is better to include all vibrational levels of CO_2 that contribute to CO_2 splitting (*i.e.*, the levels of the asymmetric stretch mode), as we did in our model.[1,2] For the other CO_2 vibrational modes, as well as for the other molecules present in the plasma (like CO_2 and O_2), it is sufficient to include only a limited number of vibrational levels, as they play some role, but no major role, in the CO_2 splitting.

This applies especially to a MW plasma and a GA discharge, but as the calculation time is not such an issue, the same levels can also be included for a DBD (although it is not strictly necessary, as the vibrational kinetics is of minor importance in a DBD).[3] In a multi-dimensional model (1D, 2D, 3D), calculation time is much more of an issue, and an extensive reaction chemistry, as included in a 0D model, would lead to prohibitively long calculation times. My recommendation then is to reduce the vibrational kinetics scheme as much as possible for a MW and GA (*e.g.*, by combining some levels), and to neglect it completely for a DBD.

1 T. Kozák and A. Bogaerts, *Plasma Sources Sci. Technol.*, 2014, **23**, 045004.
2 T. Kozák and A. Bogaerts, *Plasma Sources Sci. Technol.*, 2015, **24**, 015024.
3 R. Aerts, T. Martens and A. Bogaerts, *J. Phys. Chem. C*, 2012, **116**, 23257–23273.

Michele Aresta commented: In the 1980′s–1990′s there was great interest in the plasma technology applied to the conversion of poor natural gas, possibly at the well mouth, instead of undergoing separation. This approach was also considered valid for biogas, rich in CO_2 (40–60%). Can you comment the state of the art in this area? What is the conversion and energy efficiency in DRM with plasma?

Annemie Bogaerts responded: There is still a lot of interest in methane conversion and specifically also dry reforming of methane (DRM) by plasma technology. The conversion and energy efficiency again depends on the type of plasma (*i.e.*, DBD, packed bed DBD, microwave (MW) plasma, or gliding arc (GA)), as for pure CO_2 splitting, as well as on the CH_4/CO_2 mixing ratio.

In a DBD, conversion up to 60%, with energy efficiencies of a few % (up to 8%) have been reported. For a packed bed DBD, a conversion of 27% with a corresponding energy efficiency of 23% was reported. For a MW plasma at atmospheric pressure, a conversion of 70%, along with an energy efficiency of 39%, was reported. Finally, for a GA, a conversion of 37% with an energy efficiency of 31% was reported, as well as a conversion of 10% with an energy efficiency of 75%.

A detailed overview of the state-of-the art for plasma-based DRM for different types of plasmas, including a table of all the data and all the corresponding references, can be found in the literature.[1]

1 R. Snoeckx, Y. X. Zeng, X. Tu and A. Bogaerts, Plasma-based dry reforming: Improving the conversion and energy efficiency in a dielectric barrier discharge, *RSC Adv.*, 2015, **5**, 29799–29808.

Michele Aresta asked: When you have a mixture of CO_2 and CH_4 in the plasma, is the CO_2 splitting and methane dehydrogenation simultaneous, or they can be separated in time?

Annemie Bogaerts replied: When we have a mixture of CH_4 and CO_2, the methane dehydrogenation and CO_2 splitting occur simultaneously. Indeed, the electrons in the plasma cause dissociation, ionization and excitation of the gas molecules (both CH_4 and CO_2), forming radicals, ions and excited species, which can recombine into new compounds. These electron impact collisions occur more or less at the same time (although CH_4 dissociation is a bit easier than CO_2 dissociation, so the CH_4 conversion will typically be a bit higher than the CO_2 conversion after a certain residence time).

Michele Aresta enquired: The DBD technology may allow the use of catalysts. Can you discuss some results of processes under plasma–catalytic conditions? Which catalysts have you used?

Annemie Bogaerts responded: DBD technology is indeed very suitable for use in plasma catalysis. We just started our work on plasma catalysis, and up to now, we mainly focused on introducing a packing in the plasma reactor, and looking at the enhancement in both CO_2 conversion and energy efficiency, which is mainly attributed to physical effects, like local electric field enhancement, and an increase in the electron energy. Up to now, we used SiO_2, ZrO_2, Al_2O_3 and $BaTiO_3$ as packing materials. We also did some first experiments for Cu and Ni catalyst coatings on the packings, but the results are still too preliminary.

In fact, the chemical effects of plasma catalysis are not yet well understood. This is partially because the physical and chemical effects are often correlated, as the catalysts are mostly introduced in a packed bed, so it is not easy to distinguish between the two effects. However, Scapinello *et al.*[1] reported an enhanced selectivity towards the production of carboxylic acids in the case of DRM, when using copper or nickel electrodes instead of stainless steel, which may be attributed to a chemical catalytic effect of these metals, as no packing was involved here.

Another reason why the chemical effects of plasma catalysis for CO_2 conversion are not yet fully understood, is that a wide variety of catalyst materials are being explored, including metals (Au, Pd, Pt, Rh, Cu),[2] zeolites (NaY, NaOH treated Y, HY, Na, NaX, Na-ZSM-5 and Linde type 5A zeolite),[3,4] La_2O_3/g-Al_2O_3, CeO_2/g-Al_2O_3,[5] n-type oxide semiconductors (ZnO, CuO), and Al_2O_3 foams with Ni, Rh or Ca catalysts.[6] Commercial Ni/Al_2O_3 catalysts, which are also used in thermal catalysis, are currently the most popular catalysts for DRM, but they are not necessarily the most suitable ones under plasma conditions. We believe that there is still a lack of insight into which catalysts should perform best in a plasma. Therefore, there is a clear need for more systematic studies on both the physical effects (of the structural packing and the support materials, *e.g.*, their dielectric constant, acid/base properties, porosity) and the chemical effects of the catalyst material (including type and coordination of the active element).

1 M. Scapinello, L. M. Martini, P. Tosi, CO_2 hydrogenation by CH_4 in a dielectric barrier discharge: catalytic effect of Ni and Cu, *Plasma Proc. Polym.*, 2014, **11**, 624.
2 S. B. Brock, M. Marquez, S. Suib, Y. Hayashi and H. Matsumoto, Plasma decomposition of CO_2 in presence of metal catalysts, *J. Catal.*, 1998, **180**, 225–233.
3 K. Krawczyk, M. Mlotek, B. Ulejczyk, K. Schmidt-Szalowski, Methane conversion with CO_2 in plasma-catalytic system, *Fuel*, 2014, **117**, 608–617.
4 C. J. Liu, R. Mallinson, L. Lobban, Methane conversion to higher hydrocarbons in a corona discharge over metal oxide catalysts with OH groups, *Appl. Cat. A: General*, 1999, **178**, 17–27.
5 X. Zhang, A. Zhu, X. Li and W. Gong, Oxidative dehydrogenation of ethane with CO_2 over catalyst under pulse corona plasma, *Catalysis Today*, 2004, **89**, 97–102.
6 J. Amouroux, S. Cavadias and A. Doubla, CO_2 reduction by non-equilibrium electrocatalysis plasma reactor, *IOP Conf Ser.: Mat. Scie. Eng.*, 2011, **19**, 012005.

Michele Aresta remarked: In the experiment made using beads, there is no apparent linear correlation between the size of the beads and both the conversion and the efficiency. Is the scattering due to an error in the measurements? Are the data replicated? Can you show the error bar for each single experiment?

Annemie Bogaerts responded: The scattering might indeed be (partially) due to an error in the measurement. These were only our first experiments. The data shown in the paper were not yet replicated, but now we are performing more experiments, with repeated packing for the same conditions, so that we will be able to add error bars in our new experimental results.

Gerard van Rooij commented: Concerning plasma production within a packed bed system; as the volume in which the plasma is to be produced becomes smaller, the ionization life time becomes lower. This increases the average electron temperature and may compromise energy efficiencies.

Annemie Bogaerts responded: This may be true. On the other hand, the polarization of the dielectric beads also gives an enhanced electric field, so that the electrons can more easily become heated, and they have a higher energy, which will facilitate the electron impact ionization and dissociation of CO_2, so that the applied power (or energy) is more efficiently used for the CO_2 splitting. Therefore, this can also increase the energy efficiency.

Alexander Foote asked: There is a lot of recombination of atomic oxygen, does this occur largely in the effluent, and how much of the atomic oxygen recombines into molecular oxygen and how much into ozone? Furthermore, what is the density of these species compared to CO?

Annemie Bogaerts answered: Most of the atomic oxygen recombines into molecular oxygen. That is also confirmed by our experiments, as we do not detect significant fractions of O_3 in our effluent (although the latter can also be attributed to the high reactivity of O_3; and other people do see some significant fractions of O_3, typically up to a few %).

In general, there is a delicate balance between O, O_2 and O_3, as explained by R. Aerts *et al.*[1] The O_2 molecule density is typically about half of the CO density (because 2 CO_2 molecules split into 2 CO and 1 O_2), while the O_3 molecule density is significantly lower.

1 R. Aerts, W. Somers and A. Bogaerts, *ChemSusChem.*, 2015, **8**, 702–716.

Alexander Foote asked: How stable is ozone in the effluent? Do you see a large drop off in the density shortly after exiting the plasma?

Annemie Bogaerts replied: In our experience, ozone is not very stable in the effluent, and that's probably one of the reasons why we don't detect ozone in our gas chromatography measurements, although other people do detect ozone, in concentrations up to a few %.

Eric Fraga opened a general discussion of the paper by Gerard van Rooij: Alexander Navarrete mentioned the issue of scale up in his paper. I would like to know if there are any issues caused by scale up in your technologies, for instance, in the uniformity of the properties of the plasmas created.

Gerard van Rooij responded: Similar to the DBD reactor, the microwave approach could be scaled up by placing larger numbers of reactors in parallel. A more attractive way (in terms of footprint) seems to be scaling up the power in a single reactor. This does involve uncertainties in maintaining favourable plasma property profiles, however, these are difficult to predict as long as these have not been solved in the present 1 kW reactors.

Annemie Bogaerts commented: In a DBD plasma, scale up is straightforward. It is realized by placing a large number of DBD reactors in parallel. This has already been successfully demonstrated many years ago for the commercial production of ozone in a DBD.[1]

1 U. Kogelschatz, *Plasma Chem. Plasma Process.*, 2003, **23**, 1.

Alexander Navarrete remarked: I assume that the penetration length of microwaves in plasmas cannot be very large (at least at atmospheric pressure), thus in principle electrons near the wall will have more energy than those in the center of the vessel containing the plasma. Therefore, I wonder if constraining the plasma to small diameters (close to or below the penetration length) could provide a more homogeneous (and perhaps efficient) absorption of the energy by the electrons throughout the plasma volume. If that assumption is true, perhaps it could be possible to design systems consisting of several parallel small cylinders, where reactions can be carried out more efficiently?

Gerard van Rooij responded: Indeed, depending on the plasma density, the penetration depth of the microwaves may become very small and power will be absorbed in the skin. There, the absorption of microwave power, as well as the electron energy, will be the highest. Electron heat conduction to the centre of the discharge will drive the discharge there, so that power transfer to the heavy particles will still occur throughout the discharge. Nevertheless, our working hypothesis, based on the Boltzmann solver results in the paper, is that mean electron energy should be decreased in order to promote vibrational excitation. Promotion of high electron energy regions, by increasing the total skin volume in a multiple discharge channel approach, has the opposite effect.

Gabriele Centi continued the general discussion of the paper by Annemie Bogaerts: Under thermal conditions in CO_2 dissociation, the back reaction has a critical role. Is this also a key issue under plasma conditions?

Annemie Bogaerts responded: This might be important, but only at high CO_2 conversions. This is evidenced by the fact that CO_2 conversion tends to saturate at high conversion values (reached at long residence times or high plasma power). A way to solve this problem could be to separate the oxygen formed.

Gabriele Centi asked: In MW plasma, the loss in energy efficiency could be reduced by having a restriction of vibration modes, which may occur under confined spaces in a catalyst (like in C-nanotubes or in zeolites)?

Annemie Bogaerts replied: Indeed, the energy efficiency could be optimized when selectively exciting the asymmetric stretch mode of CO_2, as this is the most important vibration mode for CO_2 splitting.

Csaba Janaky opened a general discussion of the paper by Alexander Navarrete: Could you comment on the mechanism of how plasmons contribute to the enhanced catalytic activity in your experiment?

Gabriele Centi asked: What is the mechanism that you have in mind for this plasmonic catalyst? Generally, in plasmonic catalysts you have small metal nanoparticles (Au, Ag) which are able, with visible light, to promote charge separation in a semiconductor. Your case seems different. Could it be possible that there is an effect of local overheating of the metal nanoparticles?

Alexander Navarrete responded: Copper based catalysts are known to be active for CO_2 hydrogenation. On the other hand, ZnO has proven to be good at providing dispersion and stability to the copper catalyst (amongst other things). Therefore, Cu/ZnO catalysts are commonly found in research dealing with methanol synthesis (200–250 °C and 20–50 bar usually). In our research, we were looking for local activation of the metallic copper nanoparticles by means of the plasmon phenomena. The related effects reported include an increase of local temperature by the generation of hot electrons, near field effects and scattering. We can therefore use this "heating" to locally activate the reaction instead of heating the complete vessel (*i.e.* it is more efficient). Thus, our work looked towards the integration of the plasmonic composite together with the reactor in order to have one single entity acting in response to visual light (not UV in our tests). Therefore, the idea was to carry out the reaction at mild temperatures (possible in a solar collector) and locally activate the copper. Now, we need to really explore the actual mechanisms taking place and know which (if any) is most prominent in order to better design these devices. I therefore propose to characterize our composite with optical and *in situ* techniques. We are presently studying the appropriate techniques to achieve this.

Alexander Cowan asked: I wonder if you could give us an insight into the activity of the material without the metal nanoparticles. ZnO is known to be photoactive under UV light. Has a control without the metal nanoparticles been carried out?

Alexander Navarrete replied: I agree we need a control. Therefore, I propose the use of different wavelengths to those where the plasmon effect is activated. We have already got blue, red and yellow LEDs in order to make those tests. Yet, we have lots of work to do! Linic *et al.* have written an interesting review dealing with the plasmon effect and its application, if you would like to read about it further.[1]

1 S. Linic, P. Christopher and D. B. Ingram, *Nat. Mater.*, **10**, 911–921.

Michele Aresta said: One of the factors affecting the net CO_2 reduction in a process based on CO_2 conversion is the emission attached to the preparation of the catalysts. Your interesting catalytic system looks quite complicated from the point of view of its preparation. Have you estimated the energy expenditure for making the catalyst? It looks like it will require high TON to compensate for the emissions attached to the catalyst preparation. Can you comment on the life of the photocatalyst?

Alexander Navarrete replied: These are initial results and we believe there is much room for optimization in terms of the composite nanostructure and composition together with the reactor operation conditions itself. Thus, we have not yet made a LCA analysis of our synthesis process. However, we are confident that the life time will be similar to the industrial Cu/ZnO based catalysts already on the market. This of course needs to be proven, and will therefore be considered among the next steps that we take.

Thomas Butterworth continued the general discussion of the paper by Gerard van Rooij: If by the addition of earth metals to the microwave plasma you limit electron energy to 1 eV specific energy, does that mean you're theoretically limiting conversion to ~30%? Am I correct?

Gerard van Rooij answered: This is not correct. The electron energy determines the excitation rates in the plasma but does not relate to the amount of power that is transferred to the feedstock. Whilst adding alkalis to the discharge will likely change the momentum loss rate of the electrons to the heavy particles, this will be compensated by a change in (integrated) electron density so that the total power transferred to the feedstock remains constant. Conversion is connected to specific input power and energy efficiency.

Thomas Butterworth asked: If your maximum conversion is only ever going to be 30% and you're operating at low pressures, isn't this a problem when using microwave plasmas in industry? Do you think that it can ever be an industrially viable technology under these conditions?

Gerard van Rooij replied: It is evident that the goal of our research is to reach energy efficient operation at higher pressures and higher conversion. In my opinion, conversion and operating pressure are not the prime indicators for success. These are energy efficiency and the total power to which the approach can be scaled. Limited conversion demands separation, and low operating pressure demands high flow speeds. Industrial viability would still heavily depend on the cost of separation and pumping.

Annemie Bogaerts enquired: It is important to improve the energy efficiency in a MW plasma to keep the gas temperature low. What is the best way to accomplish this? Keep the pressure low? Use a high gas flow rate (thus short residence time)? Pulsing the plasma? Or other possibilities?

Gerard van Rooij responded: The reasoning in our paper is that presently the effect of a high average electron energy driving the direct dissociative excitation processes is dominating. This efficiently heats the heavy particles and subsequently quenches any vibrational excitation. Gas flow rate (residence time) and pressure are connected, and were varied in the experiments reported in our paper. It strongly affected central gas temperatures but was less effective in achieving strong non-equilibrium conditions. Nevertheless, record energy efficiencies reported in the literature invoked supersonic flows, a regime far beyond the parameter space covered in the present paper. Pulsing of the plasma is surely an interesting option to investigate, although this would periodically increase electron energy even further (as has been shown in your papers on dielectric barrier discharges).

Alexander Foote asked: Is the thermal region in the plasma causing a very high degree of ionisation which, in turn, is causing dissociative recombination to be dominant over vibrational excitation, which is not desired for high energy efficiencies?

Gerard van Rooij replied: This is indeed what we believe is happening.

Alexander Foote remarked: Do you have any idea of the electron density and temperature in the thermal region of the plasma? Is there a very high electron density in the thermal region due to a high degree of ionisation?

Gerard van Rooij responded: Estimates of the electron temperature are given in the paper on the basis of Boltzmann solver calculations. Estimates of the density could be based on a simple power balance: the product of plasma volume, electron density, gas density and electron energy loss rate should match the input power. These are very sensitive to assumptions on radial profiles. Simply assuming a plasma volume of let's say $\sim 5\times10^{-6}$ m^3 for the measured central neutral densities of $\sim 10^{23}$ m^{-3} would give an electron density of $\sim 5\times10^{18}$ m^{-3}. Without gas heating, the electron density would have been 1–2 orders of magnitude lower.

James Comerford remarked: In your paper you refer to the term 'input energy', could you please define what you mean by this. My understanding is that with traditional plasma apparatus, the input energy from a generator is very different to the plasma energy. As such, the difference between the two can have a substantial effect on energy efficiency calculations.

Gerard van Rooij responded: The paper refers to the terms 'input power' and (dividing input power by gas flux) 'specific input energy'. In situations where matching circuits are used between the power generator and plasma reactor, it is often difficult to determine the input power. In our microwave experiment this is relatively straight forward. We measure the power that is launched by the microwave source as well as the power that is reflected back from the plasma reactor and absorbed in the water load. In both cases, additional power losses will decrease performance, and the latter will thus not be overestimated.

Michele Aresta asked: The formation of Group 1 cations may also cause interactions with CO_2 and, depending on the cation, a possible change of geometry (towards a bent molecule). Have you observed any effects due to such interactions?

Gerard van Rooij answered: We have not observed any such interactions. I expect that the interaction between the cations and the CO_2 will mainly have an effect on the ionization life time and therefore on the plasma properties, which is the subject of our present research. Given the low amounts of cations (well below 10^{-4}) the effect on vibrational excitation dynamics will be low.

Thomas Butterworth communicated: How do you intend to incorporate alkali metal ions in the plasma? Considering that you emphasise the importance of having 0.01% sodium ions for optimum plasma properties. How do you control the distribution of these ions in the discharge?

Gerard van Rooij communicated in reply: In our current approaches we use commercially available alkali metal dispensers. These allow controlled rates of evaporation of the metal. The exact amount and distribution of alkali atoms in the discharge is in the first instance not important. There simply should be sufficient alkali atoms present to provide the ionization. In other words, the quoted 0.01% sodium content is the minimum amount to allow an ionization degree of 10^{-4}. In this way, the plasma properties are tuned by the ionization energy of the alkali atom and not by the amount of alkali.

Richard Heyn addressed all the speakers: I understand very well that we are still at the early stages of research in the field, but I would like to be a bit provocative and ask for a frank assessment and future vision of these "alternative" techniques for CO_2 reduction: electrochemistry, photochemistry, and plasma. What are the main challenges or limitations? Which one(s) might be considered the most promising at this stage? Which one(s) have the best chance of eventually achieving some sort of commercial realization?

Alexander Cowan responded: This is a challenging question. I have experience of electrochemical and photochemical systems but it would not be appropriate for me to comment on how these compare to plasma techniques due to a lack of knowledge. Photochemical approaches are potentially very low cost, doing away with the need for separate PV panels, electrolysers and all of the costs associated with the wiring of these together, hence in my opinion the long term photochemical approaches are highly attractive. However, direct photochemical carbon dioxide reduction is currently inefficient. Typically, solar to fuel conversion efficiencies are below 1% and advances in catalyst design will be necessary in the forthcoming years. In the shorter term, electrochemical approaches offer a more viable solution. Utilising available PV panels and state of the art electrodes, other research groups have managed to convert up to 7% of the incident solar energy in the form of a carbon based fuel produced *via* carbon dioxide reduction. Further advances in the development of new low cost electrocatalysts are likely to make these systems more attractive from an economic viewpoint as well.

Gabriele Centi answered: For plasma, the real challenge is to have a different approach in the integration with the catalyst, because the catalyst should be able to work with excited gas phase species. There are still very limited attempts in this direction, but some new ideas are emerging. It is mainly a question where today's researchers have not really addressed the key problems.

Furthermore, for the photochemical approach, I believe that the problem is still not understood well. Again, it is still not clearly defined exactly what the essence of the problem is. On the other hand, electrocatalysis is somewhat more advanced, and likely within a few years (let me say 5), we will see the first commercially relevant applications, although this depends on the effective R&D efforts in this direction.

Michele Aresta continued: Plasma technology is quite old, maybe 40 years or so. Previously, in the 1970's and 1980's, discharge using CO_2 or CH_4 have already been investigated. Today, the technology is more advanced, but the fundamentals have been known for a long time. Electrochemical reduction of water to produce hydrogen for CO_2 reduction is on a commercial scale. It is a matter of cost of H_2 with respect to reforming. Photoreduction of CO_2 is in its growth phase and catalysts are designed and show interesting efficiencies of a few units %. Carboxylation is a new baby. We need to discover the band gap of semiconductors to match the potential of the reactions, depending on the substrate. Much work is still necessary, so let's wait until it grows up. The common denominator is that all technologies may use perennial primary energies – such as solar, wind or geothermal energies for producing electricity. Fossil carbon is saved anyway. I am an optimist, I was optimistic in 1975 when we discovered the first nickel complex of CO_2 and said that it opened the way to the utilization of CO_2, and now my dream has come true.

Annemie Bogaerts responded: I cannot really comment on electrochemistry and photochemistry, as I am not a specialist, but I can comment on the challenges and limitations of plasma. In my opinion, the most important challenges are to improve the energy efficiency and the selectivity towards specific chemicals.

The energy efficiency strongly depends on the type of plasma. The highest energy efficiency up to now was already reported in 1983 for a microwave (MW) plasma, *i.e.*, up to 90%, but this was at very specific conditions; *i.e.*, supersonic gas flow and reduced pressure (~100–200 Torr), and a pressure rise to atmospheric pressure, which would be desirable for industry, yields a drop in energy efficiency to 40%. Moreover, this high value of 90% has never been reproduced since then. The highest energy efficiency for a MW plasma reported recently was 55%, but again at a reduced pressure and supersonic flow. A GA plasma also exhibits a high energy efficiency, even at atmospheric pressure, *i.e.*, around 43% for a conversion of 18% in the case of CO_2 splitting, and even around 60% for a conversion of 8–16%, for DRM. The energy efficiency of a DBD is more limited, *i.e.*, up to 10% for a conversion of 30% for a DBD without packing, but by inserting a (dielectric) packing in a so-called packed bed DBD reactor, the energy efficiency can be improved. An improvement up to a factor of 12 was indeed demonstrated already, but this was for other applications (air purification) while for CO_2 conversion only improvements up to a factor of 2 were reported recently. However, the advantages of a DBD are that it also operates at atmospheric pressure, and has a very simple design, which is beneficial for up scaling, as demonstrated many years ago for the large scale production of ozone, and therefore it also has potential for industrial applications.

The other challenge, as mentioned above, is the selective production of specific chemicals. In my opinion, this is still a larger challenge. Up to now, mostly syngas is formed, and the direct production of oxygenates (methanol, formic acid, *etc.*) is very limited. The reason for this is that a variety of reactive plasma species is formed, but not in a selective way. To improve the selectivity towards specific value-added compounds, we need to combine the plasma with a catalyst, *i.e.*, so-called plasma catalysis. The latter is most easy to realize with a (packed bed) DBD reactor, as often shown already for air pollution control, and to a limited extent for CO_2 conversion. However, currently the underlying mechanisms of plasma catalysis are not yet fully understood, and it is not yet known which catalysts are most suitable for this purpose. Clearly more research is needed here.

Eric Fraga commented: The question of which technology is *best* may not be the right question. Often, the premise of the existence of a best solution is based purely on economics and often with an assessment of the local properties of the technology. However, design choices are often multi-criteria and include more than just economics. For instance, criteria may include environmental impacts, societal perception and policy directions. It is also necessary to consider the full life cycle, as much as one can, so that the impact of aspects of pretreatment or precursor preparations and also waste disposal on all the various criteria are taken into account.

Michele Aresta responded: Economics is not a factor that may be used for defining the "best" technology in the case of CO_2. We are targeting CO_2 emission reduction and therefore the quality of a process must be judged on the basis of CO_2 reduction parameters: Carbon Utilization Fraction (CUF), Carbon Footprint (CF), Waste production (E factor), Energy Utilization Fraction (EUF). Moreover, if we discuss in terms of economics the question will be "who has to pay ": consumers or producers? This is not a selective criterion in terms of environmental benefit. If the CO_2 reduction is targeted, then a more complex frame derives with implications of collective and individual responsibilities with respect to an efficient use of resources.

Gabriele Centi commented: Up to now, we have mainly worked on selectivity, putting little attention on the energy point of view – from this point of view we learn that if you just have thermal energy it's not selective – we need to find alternative solutions (like electrocatalysis, plasma catalysis, photochemistry) to supply energy selectively, but we still do not know how to create a selective catalytic path with the excited species created by these approaches. This is the challenge we face.

Pedro Abrantes commented: The solutions and technologies shown during this CO_2 meeting may come to be an industrial reality in 10–20 years time. I ask you to have in mind that in 20 years the conditions (requirements, sources of materials) will not be the same as today (see http://www.terresacree.org/

ressourcesanglais.htm); so you need to be careful and visualise how easy it will be in the future to buy some of the products required to run the process.

Keep in mind that a simple solution, which is cheap and easy to run will find its users easily. If it's more expensive and complex to operate than existing solutions, it has less chance of being industrially deployed.

Alexander Navarrete responded: I agree, as you say the technologies presented, included ours, are in the initial stages. However, thinking that there is potential in the integration of solar energy and CO_2 transformation, we have chosen to work in microreactors. This technology allows more efficient energy and mass transfer and can be scaled up by "parallelization". Furthermore, the continuous flow operation makes it adaptable and flexible enough for industrial requirements. However, we have lots of work to do yet.

Carmine Capacchione responded: The use of an iron-based catalyst to promote the cycloaddition of CO_2 to epoxides has been designed with the aim to use an abundant, cheap and biocompatible metal for this reaction. The reactions where carried out without the use of organic solvents and with low catalyst loading. Of course, there are still issues regarding energy input due to the temperature and pressure required to obtain good conversions, but we believe that we are going in the right direction to reach a financial affordable process in the near future.

Derek Harris enquired: I would like to ask whether most of the techniques we are talking about are on real situations or ones that you have recreated in the lab? I believe in real situations and that when you start with pure chemicals it doesn't work, but when you put it in real applications it works well. It may take some time before we are able to achieve good results in real situations, particularly when chemicals are prepared in distilled water.

Gerard van Rooij commented: The focus of CO_2 utilization discussions is often on emissions mitigated for each CO_2 molecule that is re-used. The plasma work aims to produce solar fuels by sustainable electricity to aid in peak shaving, by addressing intermittency issues. By enabling long term electrical energy storage and energy sector integration, it aids in reaching large scale sustainable energy deployment. Concerning realism of the plasma approach, the essential ingredients are available – particularly demonstrated by high energy efficiencies and availability of the required microwave technology. The challenge is the demonstration of it as a viable concept.

Peter Styring noted: An interesting point has been raised regarding the technological readiness of CDU technologies, and why most are presented at the fundamental stage. This is because we are looking to develop these new technologies and the level to date is appropriate for laboratory studies. You have to start from somewhere, on the fundamentals. You have to start with pure CO_2, then add impurities such as nitrogen, before going on to more simulated glue gases with more impurities. Only when this laboratory screening has been carried out can you go into the field to work on real systems. Going in to the field from the start is very difficult and expensive.

Gabriele Centi said: We are still at the stage of exploring the many possibilities which exist, but care should be taken in extrapolating the actual growth of knowledge as it is still too far from an industrial point of view. 60 years ago, the situation was similar with the transition from acetylene to olefins. The transition was extremely fast, and companies not rapidly prepared become out of business. In a PV cell, the International Energy Agency predicted in around 2000 that commercial introduction of these cells in the market would be negligible, 1–2% in 2020, however we are already above 10%. There is not a linear, but an exponential growth, when conditions (including the knowledge necessary) are ready. The same may be applied for the field of CO_2, and I would predict that within a few short years there will be quite an explosion of commercial uses, even if we are still not in the position now to indicate exactly which technology will be a priority. However, I believe that electrocatalysis is one of these technologies, which is at the foundation of the future of sustainable energy/chemical production.

Faraday Discussions

PAPER

Environmental potential of carbon dioxide utilization in the polyurethane supply chain†

Niklas von der Assen, André Sternberg, Arne Kätelhön and André Bardow*

Received 8th May 2015, Accepted 8th June 2015
DOI: 10.1039/c5fd00067j

Potential environmental benefits have been identified for the utilization of carbon dioxide (CO_2) as a feedstock for polyurethanes (PUR). CO_2 can be utilized in the PUR supply chain in a wide variety of ways ranging from direct CO_2 utilization for polyols as a PUR precursor, to indirect CO_2 utilization for basic chemicals in the PUR supply chain. In this paper, we present a systematic exploration and environmental evaluation of all direct and indirect CO_2 utilization options for flexible and rigid PUR foams. The analysis is based on an LCA-based PUR supply chain optimization model using linear programming to identify PUR production with minimal environmental impacts. The direct utilization of CO_2 for polyols allows for large specific impact reductions of up to 4 kg CO_2-eq. and 2 kg oil-eq. per kg CO_2 utilized, but the amounts of CO_2 that can be utilized are limited to 0.30 kg CO_2 per kg PUR. The amount of CO_2 utilized can be increased to up to 1.7 kg CO_2 per kg PUR by indirect CO_2 utilization in the PUR supply chain. Indirect CO_2 utilization requires hydrogen (H_2). The environmental impacts of H_2 production strongly affect the impact of indirect CO_2 utilization in PUR. To achieve optimal environmental performance under the current fossil-based H_2 generation, PUR production can only utilize much less CO_2 than theoretically possible. Thus, utilizing as much CO_2 in the PUR supply chain as possible is not always environmentally optimal. Clean H_2 production is required to exploit the full CO_2 utilization potential for environmental impact reduction in PUR production.

1. Introduction

The use of fossil fuels inherently leads to carbon dioxide (CO_2) emissions. Aiming to reduce both CO_2 emissions and fossil fuel use, CO_2 can be captured and utilized as a feedstock for fuels, materials and chemicals.[1–3] In particular, CO_2 has

Chair of Technical Thermodynamics, RWTH Aachen University, Schinkelstr. 8, 52062 Aachen, Germany. E-mail: andre.bardow@ltt.rwth-aachen.de

† Electronic supplementary information (ESI) available: ESI contains process data for polyurethane supply chain and additional figures for flexible and rigid polyurethane foams. See DOI: 10.1039/c5fd00067j

recently been successfully utilized in the production of polyurethanes (PUR), resulting in both lower CO_2 emissions and lower use of fossil fuels than conventional PUR.[4–6]

PUR production is particularly well suited for incorporation of CO_2 as PUR allows for both direct and indirect CO_2 utilization as follows: polyurethanes consist of polyols and isocyanates. In polyol synthesis, the CO_2 molecule can be directly inserted 'as such' in (poly)carbonate units, *i.e.*, without energy-intensive full cleavage of the C=O bonds.[5,6] In addition to the direct CO_2 utilization in polycarbonate units of the polyols, CO_2 can also be utilized indirectly in upstream processes of the polyol supply chain. For example, CO_2 can be converted to methanol[7] and subsequently to formaldehyde, which constitutes a potential monomer for polyols.[49]

As well as direct and indirect CO_2 utilization for polyols, CO_2 can also be used in the production of isocyanates. While the direct utilization of CO_2 for isocyanates still remains a dream in industry today,[9] conventional isocyanate production requires the feedstock carbon monoxide (CO),[10] which can be obtained by reduction of CO_2.[11]

Thus, a wide variety of options exist for the direct and indirect utilization of CO_2 in the supply chain of PUR. However, a systematic exploration and environmental evaluation of all direct and indirect CO_2 utilization options for PUR is missing. Therefore, the first goal of this article is to identify the total CO_2 utilization potential in the entire PUR supply chain. In other words, we identify the maximum amount of mass CO_2 utilized per mass PUR.

Intuitively, utilizing as much CO_2 as possible might seem environmentally most favorable. However, the energy requirements for both CO_2 capture and utilization (CCU) can lead to additional CO_2 emissions, fossil fuel use and other environmental impacts.[12] Thus, it is not always environmentally most reasonable to utilize as much CO_2 as possible. Instead, only those CO_2-based processes should be employed that allow for reductions of environmental impacts. The environmental impacts of a process can be determined by life cycle assessment (LCA). LCA is a methodology to quantify the environmental impacts of products and processes along the entire life cycle from cradle to grave. Application of LCA to CCU processes is still very limited as recently reviewed.[13] For the application of LCA to CO_2 utilization, specific guidelines have recently been developed.[12,14] Based on these guidelines, the second goal of this article is to analyze which CO_2-based processes in the PUR supply chain allow for the largest reduction of CO_2 emissions and fossil fuel use. From this analysis, we determine the minimum CO_2 emissions and minimum fossil fuel use for the PUR supply chain.

Indirect CO_2 utilization processes usually require cleavage of C=O bonds, often *via* hydrogenation.[15–18] Whether CO_2 hydrogenation is environmentally favorable compared to a fossil-based benchmark depends strongly on the provision of hydrogen (H_2).[19,20] Conventional production of H_2 *via* steam methane reforming is typically energy- and emission-intensive.[21] The impacts of H_2 production can be significantly reduced by the combination of water electrolysis with renewable electricity sources.[19] Therefore, the third goal of this article is to analyze the minimum environmental impacts of CO_2 utilization in the PUR supply chain as a function of the environmental impacts of H_2 production. Based on this analysis, we determine threshold values for the environmental impacts of

H_2 production that are tipping points for utilization of major amounts of CO_2 for PUR production.

The article is structured as follows. In Section 2, we review specific guidelines for the application of LCA for CO_2 utilization in polymers and state the goal and scope for the present LCA study of CO_2-based PUR. In Section 3, we present the considered PUR supply chain including conventional and CO_2-based processes. Furthermore, we introduce the optimization model using linear programming for the analysis of maximum amounts of CO_2 utilization and minimal environmental impacts. In Section 4, we present our findings, *i.e.*, the maximum amounts of utilized CO_2 per kg PUR, the CO_2-based processes with the largest environmental benefits, and the minimum environmental impacts of the PUR supply chain for H_2 production alternatives. Finally, in Section 5, conclusions are drawn for the utilization of CO_2 in the PUR supply chain.

2. LCA for CO_2 utilization in PUR production

Both capture and utilization of CO_2 typically require energy whose provision is often based on fossil fuels and thus causes indirect CO_2 emissions. For example, post-combustion CO_2 capture from flue gases of power plants demands the equivalent of about 20–25% of the total electricity output of the power plant.[22,23] Utilization, or more precisely, conversion of the inert CO_2 molecule usually requires direct energy input or high-energetic co-reactants such as epoxides or hydrogen (H_2).[24] Thus, the intuitively expected environmental benefits of CO_2 capture and utilization are not given by default and a detailed environmental assessment is required. Life cycle assessment (LCA) is frequently acknowledged as suitable methodology for the environmental assessment of CCU.[1,2,25–27] Recently developed guidelines[12,14] for the application of LCA to CCU have already been applied to CO_2-based polyol synthesis.[28] Relevant aspects of these guidelines are reviewed and specified for the present context of CO_2 utilization in PUR production in the following.

2.1 Functional unit and comparability

The basis for any LCA is the definition of the so-called functional unit. The functional unit is a quantitative measure for the function of the system under study.[29,30] In the case of polymer production, the functional unit could be defined as "1 kg of polymer produced". However, since polymers have very different properties and a broad range of applications, this mass-based definition may fall short of a fair comparison of different polymers.[14]

In this work, we consider the theoretical production of polyurethanes from alternative fossil- and CO_2-based monomers. In reality, the product properties of the polyurethanes will differ for alternative monomers. An integrated approach to identify environmentally optimal polyols with specified properties has recently been proposed by our group.[8] However, accurate models for prediction of PUR properties are missing. As a first step, we therefore define the functional unit of this work as "production of 1 kg of polyurethane foam" regardless of its chemical structure and resulting properties. Nevertheless, chemical limitations for the incorporation of CO_2 are taken into account (*cf.* Section 3.1). Still, our study neglects many practical constraints on PUR production to be able to explore the full design space and to provide inspiration to chemists facing the practical challenges involved. The obtained LCA results thus serve as lower bound estimates for the environmental impacts of PUR production and can guide chemists towards more sustainable PUR synthesis.

2.2 Co-product allocation

Polyurethanes are typically produced together with many co-products along the PUR supply chain. For example, most technologies for the production of propylene oxide generate co-products such as *tert*-butyl alcohol or styrol.[28] In the context of CO_2-based PUR, the 'production' of CO_2 *via* CO_2 capture from point-sources is also coupled to production of the point-sources' primary product: for example, power plants with CO_2 capture provide electricity as a primary product.[12] To account for co-products, three methods exist in LCA: *system expansion*, *allocation* and *avoided burden*.[31,32]

In *system expansion*, the scope of the study is extended to include the co-products as functions. In other words, the functional unit is defined as a basket of products: the original product, here PUR, and all co-products. As system expansion can lead to very large baskets of products, interpretation and communication can be difficult. Therefore, it is often desirable to compute product-specific impacts for PUR.

Product-specific impacts can be obtained by methods of *allocation* and *avoided burden*. For *allocation*, environmental impacts are allocated to the individual products based on criteria such as mass content, energy content or price share. However, the choice of an allocation criterion is ambiguous.[33] For *avoided burden*, the co-products are assigned with an environmental credit since co-production avoids an alternative production of the co-product and the related environmental burdens. Thus, avoided burden implies a comparison to an alternative production. Avoided burden is therefore useful for a comparison with today's production technologies.

In this work, we are interested in the reduction of impacts compared to today's situation rather than in the exact value of absolute environmental impacts. For such a comparison, a change-oriented viewpoint is recommended[34] and thus, we employ the avoided burden method to obtain environmental impact reductions compared to the conventional PUR production today. Precisely, avoided burdens are credited for the production of excess hydrogen (H_2), heat and electricity from the CO_2 supplying power plant. For some of the feedstocks, we use data from LCA databases where allocation has already been applied.[35,36]

2.3 Environmental impact categories

LCA intends to cover a broad range of environmental impacts to avoid problem shifting between impact categories. The most prominent impact category 'global warming' (also named 'climate change') aggregates CO_2 and other greenhouse gas emissions according to their global warming potential in CO_2-equivalents.[37] The impact category 'fossil fuel depletion' quantifies the use of the limited fossil resources based on their energy content in kg oil-equivalents.[38] CO_2 Capture and Utilization (CCU) aims at reducing CO_2 emissions and establishing an alternative carbon source, thus also reducing the use of fossil fuels. However, from Carbon Capture and Storage (CCS), a tradeoff is known between CO_2 emission reduction and fossil fuel use.[39–41] Therefore, CCU processes should be evaluated at least regarding impacts on global warming and fossil fuel use.[14] Of course, it is desirable to perform a more complete LCA study with a broad range of environmental impacts.[42,43]

In this work, we assess the environmental potential of CO_2 utilization in PUR production with respect to global warming impacts and fossil fuel use.

2.4 CO_2 sources and CO_2 capture

CO_2 capture from diluted CO_2 sources requires energy for the separation of CO_2 from other gases. Moreover, CO_2 capture requires operating materials such as capture solvents, and process technologies such as absorption and desorption columns. All of these efforts for CO_2 capture are typically associated with fossil fuel use and thus CO_2 emissions. Therefore, the CO_2 emission reduction of CO_2 capture is lower than 1 kg CO_2-eq. per kg $CO_{2,feed}$. Here, CO_2-eq. refers to CO_2 emissions and other greenhouse gas emissions, and $CO_{2,feed}$ refers to the captured and subsequently utilized CO_2.

In this work, we consider a coal-fired power plant as a standard CO_2 source. For the considered coal-fired power plant, CO_2 capture can reduce CO_2 emissions by 0.84 kg CO_2-eq. and increases fossil fuel use by 0.05 kg oil-eq. per kg $CO_{2,feed}$ compared to a power plant without CO_2 capture.[44] As the worst-case scenario, we also consider CO_2 capture from ambient air with CO_2 emission reductions of 0.51 kg CO_2-eq. and fossil fuel use of 0.18 kg oil-eq. per kg $CO_{2,feed}$.[44] As the best-case scenario, we consider a hypothetical, ideal CO_2 source with CO_2 emission reductions of 1 kg CO_2-eq. per kg CO_2 captured and no increase in fossil fuel use. The best-case scenario corresponds to a CO_2 source which can be readily used in the CO_2 conversion but is vented to the atmosphere today.

2.5 Temporary carbon storage in PUR

Polyurethanes have a lifetime of several decades. Incorporating CO_2 into PUR can thus be considered as temporary carbon storage during the PUR lifetime. Temporary carbon storage generally has a positive effect on climate mitigation.[45,46] The absolute effect of temporary carbon storage is, however, argued to be small for the following reasons.[12]

If conventional and CO_2-based PUR syntheses yield PURs with identical properties, then use, lifetime and end-of-life (EOL) treatment of both PURs will also be identical. Differences only occur during PUR syntheses and thus, it is sufficient to limit an LCA-based comparison of conventional and CO_2-based PUR syntheses to a so-called cradle-to-gate scope. In this case, a climate benefit can only be achieved if CO_2-based PUR synthesis causes fewer emissions than conventional synthesis. A climate benefit from temporary carbon storage cannot be expected for CO_2-based PUR synthesis compared to conventional PUR synthesis.

If, however, CO_2 utilization for PUR synthesis alters PUR properties, changes in use, lifetime and EOL treatment can occur. An increased PUR lifetime and a reduction of EOL CO_2 emissions constitute relevant changes for assessing the effect of temporary carbon storage in LCA: a longer lifetime can shift EOL CO_2 emissions into the future. As a rule of thumb, this shift of CO_2 emissions can reduce the global warming impact of EOL emissions by about 1% for each year of lifetime extension.‡ Total EOL CO_2 emissions account for about one third of the total CO_2-eq. emissions in the life cycle of conventional PUR without credits for thermal energy recovery.[35] Thus, even for a 10 year longer lifetime, the temporary carbon storage effect reduces overall PUR global warming impact by only 3%. However, this simple analysis does not consider the general environmental benefit of longer product use.[47]

For PUR, a tradeoff is expected regarding the effect of CO_2 incorporation on EOL emissions: on the one hand, CO_2 utilization in PUR typically lowers the C content in PUR and thus lowers EOL CO_2 emissions. On the other hand, the lower C content also reduces the heating value leading to reduced thermal energy recovery in EOL. The reduced thermal energy recovery usually has to be compensated by fossil-based heating leading to additional CO_2 emissions.

For the above mentioned reasons, the climate mitigation effect of temporary carbon storage of CO_2 utilization for PUR is expected to be small. Nevertheless, we suggest quantifying the exact climate mitigation effect of temporary carbon storage on a case-by-case basis using available time-dependent global warming metrics.[12]

Due to difficulties in predicting the PUR lifetime from the chemical structure and due to the expectedly small climate effect of a PUR lifetime extension, temporary carbon storage is not considered in this work.

3. The PUR supply chain optimization model

3.1 Overview of the PUR supply chain

The considered PUR supply chain is illustrated in Fig. 1. In the main article, we focus on the production of flexible PUR foams. Results for rigid PUR foams are presented in the ESI.† In the following, the considered production steps are briefly described.

3.1.1 Flexible PUR foam production. Flexible PUR foams are produced from the feedstocks polyol and toluol-2,4-diisocyanat (TDI).[36] We assume water as an indirect foam blowing agent that reacts with TDI to CO_2 as the actual blowing agent. This CO_2 is directly released to the atmosphere. We consider a fixed mass ratio of TDI and polyol of $m_{\text{TDI}}/m_{\text{polyol}} = 0.4$.[36]

3.1.2 Rigid PUR foam production. Rigid PUR foams are produced from the feedstocks polyol and methylene diphenyl diisocyanate (MDI).[36] We assume pentane as the foam blowing agent although CO_2 can be used as an alternative or co-blowing agent.[48] We consider a fixed mass ratio of MDI and polyol of $m_{\text{MDI}}/m_{\text{polyol}} = 1.6$.[36]

3.1.3 Polyol production. Polyols are synthesized from a starter (here assumed as glycerol) and from the alternative monomer building blocks polyether, polycarbonate and poly oxymethylene. Conventional polyether polyols are made mainly from propylene oxide (PO).[10] For production of PO, environmental impacts are considered according to the technology mix.[28] PO can be partly substituted by

‡ The rule of thumb of 1% global warming impact reductions is only valid for considering the absorbed radiation over a fixed time horizon of 100 years.[12] In this context, it should be mentioned that the choice of an adequate time horizon plays a key role for the assessment of temporary carbon storage.[57]

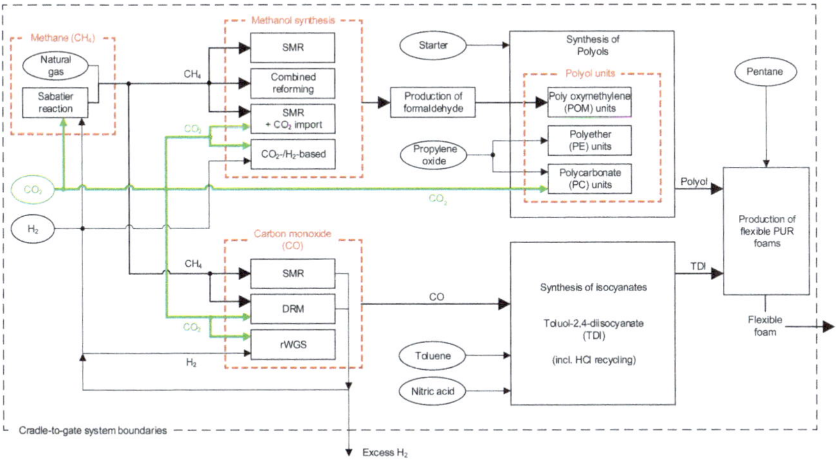

Fig. 1 The PUR supply chain for flexible foams. For simplicity, only material flows are shown; heat and electricity are also considered in the supply chain optimization model. (Green) very thick arrows indicate CO_2 feedstock flows. Large arrowheads indicate feedstock flows with carbon that potentially stems from carbon dioxide. Smaller arrow heads indicate feedstock flows from non-CO_2 sources. The (red) dashed boxes indicate choices between (i) methane sources, (ii) methanol synthesis, (iii) CO sources, and (iv) polyol units. Color online.

CO_2 which co-polymerizes with PO to polycarbonate units.[5] Furthermore, poly oxymethylene (POM) diols have been tested for polyurethane production.[49] POM is polymerized from formaldehyde, which is exclusively produced from methanol.[50]

3.1.4 Methanol production. Conventionally, methanol is synthesized from syngas produced by steam methane reforming (SMR).[51] Since syngas from SMR usually does not have the optimal composition for methanol synthesis, CO_2 can be added (SMR + CO_2 import) to obtain the desired syngas composition. A fossil-based alternative for methanol production is the combined reforming using SMR and partial oxidation of methane.[52] This process requires O_2 as an input. An exclusively CO_2-based alternative for methanol production is the direct reforming of CO_2 and H_2 to methanol.[7]

3.1.5 Methylene diphenyl diisocyanate (MDI) production. MDI is produced from phosgene and methylenedianiline (MDA). MDA is produced by the reaction of formaldehyde and aniline. Aniline results from the hydrogenation of nitrobenzene. Nitrobenzene is produced by the nitration of benzene with nitric acid. Phosgene is produced by the reaction of carbon monoxide (CO) and chlorine gas. A by-product of the MDI production is hydrochloric acid (HCl). HCl is separated by HCl electrolysis to provide chlorine as feedstock for the phosgenation. HCl electrolysis also provides H_2 which is assumed to be used internally in MDI production.[10]

For our analysis, the production of MDI described above is modelled as a single process based on stoichiometric inputs (formaldehyde, benzene, nitric acid, H_2 and CO) and outputs (MDI). Furthermore, the electricity demand for HCl electrolysis is included.

3.1.6 Toluol-2,4-diisocyanate (TDI) production. TDI is produced from phosgene and diaminotoluene (TDA). TDA results from the hydrogenation of dinitrotoluene (DNT). DNT is produced by the nitration of toluene with nitric acid.[10] Phosgene production and HCl recycling *via* electrolysis are equivalent to the processes in MDI production (see Section 3.1.5).

For our analysis, the production of TDI described above is modelled as a single process based on stoichiometric inputs (toluene, nitric acid and CO) and the main output (TDI). Furthermore, electricity demand for HCl electrolysis is included.

3.1.7 CO and hydrogen (H_2) supply. CO and hydrogen (H_2) are feedstocks for MDI and TDI production. If H_2 from internal HCl electrolysis is used (assumed here, see Sections 3.1.5 and 3.1.6), MDI production requires another 2 mol H_2 per mol CO, whereas TDI does not require additional H_2.[10] We consider the following processes for CO and H_2 supply for MDI and TDI production.

Conventionally, both CO and H_2 are supplied by SMR.[53] Part of the methane can be substituted by CO_2 through the dry reforming of methane (DRM).[11] However, for the same amount of CO, DRM produces less hydrogen compared to SMR. The complete substitution of methane is enabled by the reverse water gas shift (rWGS) reaction. For rWGS, hydrogen and CO_2 are required as feedstocks. The rWGS reaction produces only CO (and water as a by-product).[54]

3.1.8 Methane supply. Conventionally, the feedstock methane (CH_4) is supplied by natural gas. Methane can also be produced from H_2 and CO_2 through the Sabatier reaction.[55]

An overview of the employed LCA datasets for all processes is given in the ESI.†

3.2 Optimization model

To rigorously study the environmental potential of utilizing CO_2 in the PUR supply chain, a superstructure-based optimization model is used. The superstructure comprises all processes of the PUR supply chain (*cf.* Section 3.1). The optimization model is used to identify the maximum CO_2 utilization potential (Section 3.2.1) and the minimal environmental impacts (Section 3.2.2) for the PUR supply chain.

3.2.1 Maximum CO_2 amount in the PUR supply chain. As the first step, optimization is carried out maximizing the total amount of CO_2 utilized ($m_{CO_2,feed}$) to identify the maximum CO_2 utilization potential in the entire PUR supply chain. The amount of PUR produced is set to $m_{PUR} = 1$ kg in the functional unit vector f (*cf.* Section 2.1). While the total CO_2 feed ($m_{CO_2,feed}$) is to be maximized for 1 kg PUR, mass and energy balances must be fulfilled across the entire PUR supply chain. This optimization problem can be formulated as a so-called linear program (LP):

$$\begin{aligned} \max_{s} \quad & m_{CO_2,feed} = d^T s\ , \\ \text{s.t.} \quad & A\ s = f \qquad \text{(mass and energy balances in supply chain).} \end{aligned} \tag{1}$$

The so-called scaling vector s describes which processes are employed and to what extent. The vector d describes how much CO_2 is directly utilized in each process. Thus, the overall amount of CO_2 utilized is obtained by $m_{CO_2,feed} = d^T s$, to be maximized here. The matrix A contains all inputs and outputs of the individual processes in the PUR supply chain. Matrix A and vectors d and f are given explicitly in the ESI.†

3.2.2 Minimal environmental impact for PUR supply chain: effect of CO_2 utilized. To identify minimal environmental impacts for the PUR supply chain, optimization is carried out minimizing the total environmental impacts of all processes required for PUR production. In this paper, we consider the environmental impact categories 'global warming' and 'fossil fuel use' (*cf.* Section 2.3). The amount of PUR produced is again set to $m_{PUR} = 1$ kg, and mass and energy balances must be fulfilled across the PUR supply chain. The corresponding optimization problem can be formulated as follows:

$$\begin{aligned} \min_{s} \quad & g = B\,s, \\ \text{s.t.} \quad & A\,s = f \quad \text{(mass and energy balances in supply chain)}. \end{aligned} \tag{2}$$

The matrix B contains the direct environmental impacts of the individual processes in the PUR supply chain. The cradle-to-gate environmental impacts z for production of 1 kg PUR are obtained by $z = Bx$. Matrix B is given in the ESI.†

For the environmental impacts of CO_2 supply, the three cases presented in Section 2.4 are analyzed: CO_2 capture from a coal-fired power plant (standard case), CO_2 capture from ambient air (worst case) and an ideal CO_2 source (best case).

Effect of CO_2 utilized. To determine the effect of the amount of CO_2 utilized, the minimization of environmental impacts is repeated for fixed amounts of CO_2 utilized. For this purpose, the amount of CO_2 utilized is varied between zero and $m_{CO_2,\text{feed,max}}$. For this analysis, we consider three cases of hydrogen production: (i) conventional steam methane reforming (SMR), (ii) water electrolysis, and (iii) ideal hydrogen production with no environmental impacts at all. The corresponding environmental impacts for H_2 are given in Table 1.

Effect of H_2 production alternatives. To analyze the effect of hydrogen production alternatives more rigorously, the environmental impact of hydrogen supply is also varied continuously. For this purpose, the optimization is repeated for the full range of environmental impacts of the considered hydrogen production alternatives, *i.e.*, for global warming impacts from zero to 10 kg CO_2-eq. per kg H_2, and for fossil depletion impacts from zero to 5 kg oil-eq. per kg H_2.

Table 1 Considered environmental impacts for hydrogen production alternatives

	Global warming impact (kg CO_2-eq. per kg H_2)	Fossil depletion impact (kg oil-eq. per kg H_2)
Steam methane reforming[35]	10	5
Water electrolysis[a]	5	2.5
Ideal H_2 production	0	0

[a] The environmental impacts for H_2 from water electrolysis depend largely on the electricity source for electrolysis. For production of 1 kg H_2, the electricity demand of electrolysis is about 50 kW h.[54] With this electricity demand, the presented environmental impacts of water electrolysis correspond to environmental impacts of electricity generation of 100 kg CO_2-eq. and 50 kg oil-eq. per MWh (similar to the grid mix characteristics of Sweden).[35]

4. Results

In this section, we present the maximum CO_2 utilization potential and minimal environmental impacts for PUR production. In the first scenario, we consider only direct utilization of CO_2 for polycarbonate (PC) units in polyol production and indirect CO_2 utilization in the isocyanate supply chain. The utilization of poly oxymethylene (POM) units as polyol building blocks is not permitted, regardless of whether POM is produced from fossil or CO_2-based feedstocks. This first scenario ('without POM') includes technically feasible CO_2 utilization options. The utilization of POM for polyols and PUR is a promising approach; however, utilization of POM in polyols is still in the research phase.[49] As a second scenario, and for future outlook, we consider all CO_2 utilization options in the PUR supply chain including fossil- and CO_2-based POM units for polyols (scenario 'with POM').

4.1 Maximum CO_2 utilization amount in the PUR supply chain

The maximum CO_2 utilization potential refers to the maximum amount of CO_2 utilized in the entire PUR supply chain. The amount of CO_2 utilized can be greater than 1 kg CO_2 for production of 1 kg PUR since it is simply the total amount of CO_2 utilized in the PUR supply chain; it does not refer to the amount of CO_2 incorporated or the CO_2 content in the final PUR. The maximum CO_2 amount in the PUR supply chain is presented in Fig. 2 for both flexible and rigid PUR foams.

The maximum potential for direct CO_2 utilization in PC units in polyols is about twice as large in flexible PUR foam compared to rigid PUR foam: in flexible foam, up to 0.30 kg CO_2 per kg PUR can be utilized directly in polyols, while 0.16 kg CO_2 per kg PUR can be utilized for rigid foam. The larger potential for flexible foam is due to the typically higher mass content of polyols in flexible foams compared to rigid foams (*cf.* Section 3.1.1 and 3.1.2).

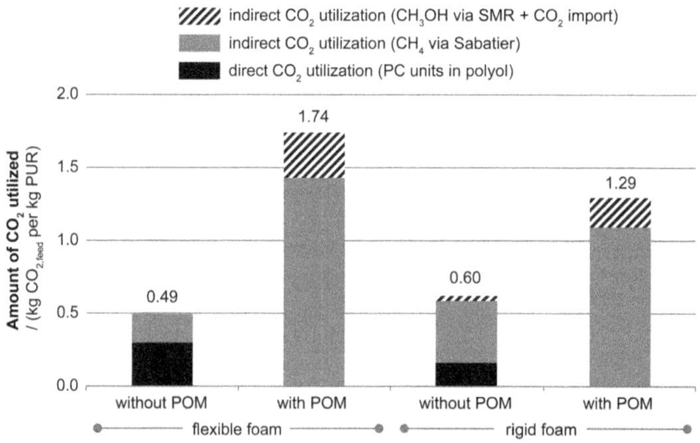

Fig. 2 CO_2 utilization potential in the PUR supply chain as the amount of CO_2 utilized per kg PUR. Scenario 'without POM' does not permit utilization of poly oxymethylene (POM) units as polyol building blocks; all other CO_2 utilization options in the PUR supply chain are possible. Scenario 'with POM' allows all CO_2 utilization options in the PUR supply chain including fossil- and CO_2-based POM units for polyols.

If POM units are not allowed for polyol production, the maximum CO_2 utilization potential can still be increased by indirect CO_2 utilization in the isocyanate supply chain: the indirect CO_2 utilization potential is 0.20 kg CO_2 and 0.46 kg CO_2 per kg PUR for flexible and rigid foams, respectively. For flexible foams, the indirectly utilized CO_2 is completely converted to methane *via* the Sabatier reaction; methane is then converted *via* steam methane reforming (SMR) to CO and H_2 for isocyanate production. For rigid foams, 93% of the indirectly utilized CO_2 is converted to methane, of which 69% is converted *via* SMR to CO and H_2 and 31% is converted *via* SMR (with CO_2 import) to methanol for subsequent formaldehyde and MDI production. Methanol production *via* SMR (with CO_2 import) utilizes the remaining 7% of the indirectly utilized CO_2.

The global maximum CO_2 utilization potential can be achieved if POM units are allowed in polyol synthesis. For flexible foams, up to 1.74 kg CO_2 can be utilized, exclusively through indirect CO_2 utilization. 82% of the indirectly utilized CO_2 are converted *via* the Sabatier reaction to methane, of which 14% are converted *via* SMR to CO and H_2 and 86% are converted *via* SMR (with CO_2 import) to methanol for subsequent formaldehyde and POM production. Methanol production *via* SMR (with CO_2 import) utilizes the remaining 18% of the indirectly utilized CO_2. For rigid foams, up to 1.29 kg CO_2 can be utilized, again exclusively through indirect CO_2 utilization. 84% of the indirectly utilized CO_2 is converted *via* the Sabatier reaction to methane, of which 27% is converted *via* SMR to CO and H_2 and 73% is converted *via* SMR (with CO_2 import) to methanol. Methanol production *via* SMR (with CO_2 import) also utilizes the remaining 16% of the indirectly utilized CO_2. The produced methanol is converted to formaldehyde for subsequent POM (84%) and MDI production (16%).

In the following part of this paper, we focus on CO_2 utilization for flexible PUR foams. The corresponding results for rigid PUR foams are presented in the ESI.†

4.2 Minimal environmental impact for PUR supply chain: effect of CO_2 utilization amount

In the previous section, maximum CO_2 utilization amounts have been identified. Since it might not be environmentally favorable to utilize as much CO_2 as possible, we now identify the minimal environmental impacts for PUR production for variable amounts of CO_2 utilized as described in Section 3.2.2.

Fig. 3 shows minimal global warming impacts for flexible PUR foams with and without POM units. For foams without POM units, increasing the amount of CO_2 utilized generally leads to a reduction of CO_2 emissions compared to conventional foams from fossil-based polyether (PE) polyols and TDI. In particular, the direct utilization of CO_2 in polycarbonate (PC) units of polyols allows for a reduction of 3.7–4.1 kg CO_2-eq. per kg CO_2 utilized. The CO_2 reductions stem from CO_2 capture (0–0.84 kg CO_2-eq. per kg CO_2 utilized, *cf.* Section 2.4) and from substitution of emission-intensive epoxides (3.1 kg CO_2-eq. per kg CO_2 utilized); *cf.* von der Assen *et al.*, 2014.[14] The potential to further reduce CO_2 emissions through indirect CO_2 utilization depends on the emissions from hydrogen production. Nevertheless, for all hydrogen production alternatives, the CO_2 reduction potential for indirect CO_2 utilization is very small in flexible PUR foams without POM units.

If POM units can be incorporated into polyols for flexible PUR foams, the potential to reduce CO_2 emissions is much larger for two reasons: first, even

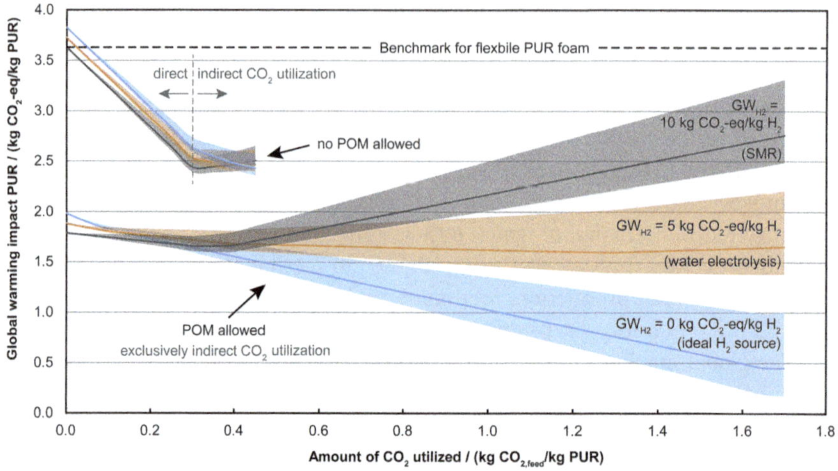

Fig. 3 Minimum global warming impact for flexible PUR foams for variable amounts of CO_2 utilized. The transparent areas indicate the range for alternative CO_2 sources: lower bounds correspond to an ideal source, upper bounds correspond to CO_2 capture from ambient air, and the solid lines correspond to CO_2 capture from a coal-fired power plant.

fossil-based production of POM units causes much lower CO_2 emissions than production of conventional PE units (*cf.* Fig. 3 for $m_{CO_2,feed} = 0$). Second, the CO_2 utilization potential is much higher for polyols with POM units (*cf.* Section 4.1). More CO_2 utilization reduces CO_2 emissions at the CO_2 source by CO_2 capture. However, for indirect CO_2 utilization, POM units require provision of hydrogen as a feedstock for methanol synthesis. Whether indirect CO_2 utilization for PUR with POM actually reduces CO_2 emissions therefore depends largely on the emissions from hydrogen production: for ideal hydrogen production with no CO_2 emissions, the computed minimal CO_2 emissions for PUR production strictly decrease with increasing amounts of CO_2 utilized. For hydrogen from water electrolysis, the computed minimal CO_2 emissions are almost constant for variable amounts of CO_2 utilized. For today's conventional hydrogen production *via* SMR, the computed minimal CO_2 emissions decrease up to a CO_2 utilization amount of 0.35 kg CO_2. Further increasing the CO_2 utilization amount increases CO_2 emissions of PUR production. Thus, increasing the amount of CO_2 utilized then leads to additional CO_2 emissions.

Fig. 3 also shows that utilization of 1 kg CO_2 does not lead to a reduction of 1 kg CO_2 emissions.[3,56] In some cases, utilization of CO_2 even increases CO_2 emissions. For this reason, the overall minimum CO_2 emissions do not necessarily occur for the maximum amount of CO_2 utilized. In most cases, CO_2 utilization reduces CO_2 emissions. Here, some processes reduce more CO_2 emissions per CO_2 utilized than others. For example, the direct utilization of CO_2 for PC units in polyols allows for the largest CO_2 reduction per amount of CO_2 utilized. However, since CO_2 is not a restricted resource, CO_2 should be utilized not only in processes with the largest CO_2 reductions but instead in such amounts that the overall minimum of CO_2 emissions is reached. The overall minimum for flexible PUR foams with POM units depends largely on the hydrogen production alternative.

In addition to global warming impacts, we identified the minimal fossil fuel use in flexible PUR foam for variable CO_2 utilization amounts. The qualitative behavior is very similar for global warming impacts and fossil fuel use, *cf.* Fig. 3 and 4. Thus, we focus on global warming impacts in the following. The corresponding results for fossil fuel use are given in the ESI.†

4.3 Minimal environmental impact for the PUR supply chain: effect of H_2 production alternatives

In the previous section, three distinct cases for hydrogen production have been analyzed in the context of minimal environmental impacts for CO_2 utilization in PUR production. Environmentally favorable hydrogen production has been identified as an important factor to increase the amount of CO_2 utilized for environmentally favorable PUR production. In contrast to the three discrete cases, this section investigates the effects of the environmental impacts of hydrogen production in more detail. In the main article, impacts on global warming are shown; the fossil fuel use is presented in the ESI.†

Fig. 5 shows that the global warming impact of flexible PUR foam can be reduced from 1.68 to 0.43 kg CO_2-eq. per kg PUR (for CO_2 captured from a coal-fired power plant, solid line) if the global warming impact of H_2 production decreases from 10 to 0 kg CO_2-eq. per kg H_2. For this decrease in global warming impact of H_2 production, the amount of CO_2 utilized increases from 0.42 to 1.68 kg CO_2 per kg PUR (right *y*-axis in Fig. 5). In particular, the amount of CO_2 utilized increases sharply if the global warming impact of H_2 production drops below 5.6 kg CO_2-eq. per kg H_2, and even further for a drop below 4.1 kg CO_2-eq. per kg H_2. The first increase in CO_2 utilization is mainly due to a switch in methanol production from SMR + CO_2 import to entirely CO_2-based methanol production. Before the first increase (above 5.6 kg CO_2-eq. per kg H_2), no hydrogen

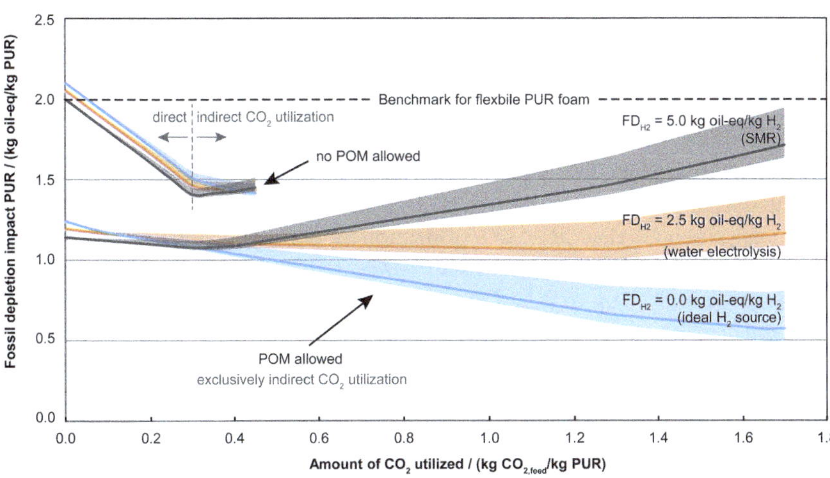

Fig. 4 Minimum fossil depletion for flexible PUR foams for variable amounts of CO_2 utilized. The transparent areas indicate the range for alternative CO_2 sources: lower bounds correspond to an ideal source, upper bounds correspond to CO_2 capture from ambient air, and the solid lines correspond to CO_2 capture from a coal-fired power plant.

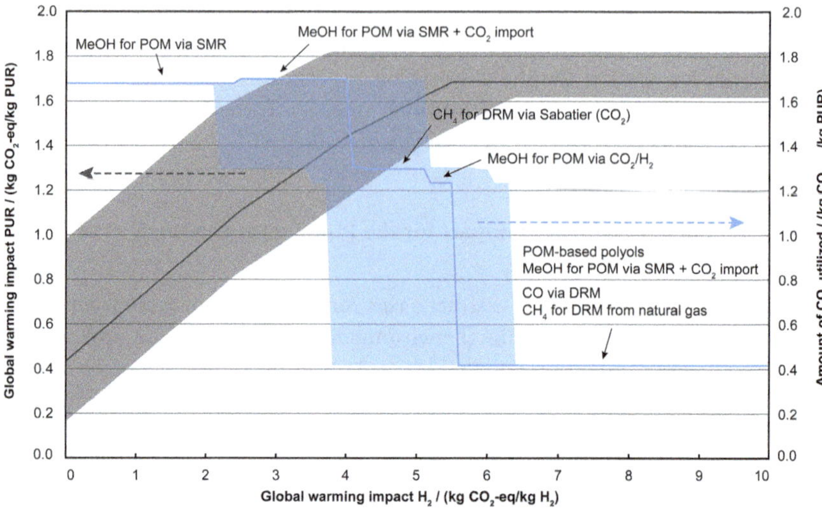

Fig. 5 Minimum global warming impacts for flexible PUR foam for a variable global warming impact of hydrogen production. The solid lines refer to CO_2 captured from a coal-fired power plant. The lower bound of the global warming impact of PUR and the upper bound of the CO_2 utilization amount refer to an ideal CO_2 source (best case). The upper bound of the global warming impact of PUR and the lower bound of the CO_2 utilization amount refer to CO_2 capture from ambient air (worst case).

is utilized for PUR production and thus, the global warming impact of PUR is independent from the global warming impact of H_2 production. The second increase is mainly due to a switch from methane from natural gas, to methane from CO_2 *via* the Sabatier reaction.

The analysis in this section highlights that hydrogen production with low environmental impacts is important for the indirect utilization of CO_2 in the PUR supply chain.

5. Conclusions

Many options exist for the utilization of CO_2 in the polyurethane (PUR) supply chain. In this paper, we present a systematic exploration and environmental evaluation of all direct and indirect CO_2 utilization options for PUR production. Our analysis shows that direct CO_2 utilization for polycarbonate units in polyols is limited in the amount of CO_2 utilized; however, direct CO_2 utilization allows for large reductions of up to 4 kg CO_2-eq. per kg CO_2 utilized. The CO_2 utilization amount can be increased by indirect CO_2 utilization for reduction to carbon monoxide in isocyanate production. However, the environmental potential of indirect CO_2 utilization in the isocyanate supply chain is rather small. Both the CO_2 utilization amount and the reduction of environmental impacts can be largely increased through indirect CO_2 utilization if poly oxymethylene (POM) units can be incorporated into polyols. In this case, large environmental impact reductions are already possible for fossil-based POM production. Additional environmental benefits from CO_2-based POM production depend largely on the required hydrogen (H_2) source. Current H_2 production *via* steam methane reforming (SMR) does not allow for additional reductions of environmental impacts. Even worse, for H_2 from SMR, increasing the amount of CO_2 utilized can even lead to additional CO_2 emissions. Thus, utilizing as much CO_2 in the PUR supply chain as possible is not always environmentally optimal. Instead, minimal environmental impacts are achieved for CO_2 utilization amounts below the maximum possible utilization amount. To still exploit the full CO_2 utilization potential for environmental impact reduction in PUR production with POM units, environmentally friendly H_2 production with CO_2 emissions below 4 kg CO_2-eq. per kg H_2 is required. The present study has neglected many practical challenges for researchers to be able to explore the full theoretical design space for environmentally optimal polyurethane production. Our work aims at inspiring future research on sustainable CO_2 utilization for polyurethanes.

Acknowledgements

Niklas von der Assen acknowledges financial support by the German Federal Ministry of Education and Research (BMBF) (Dream Polymers – Sustainable pathways to new polymers, Grant 033RC1104B). Niklas von der Assen, André Sternberg and Arne Kätelhön acknowledge financial support by the European Institute of Innovation & Technology, Climate Knowledge and Innovation Community (EIT Climate-KIC).

References

1 E. A. Quadrelli, G. Centi, J.-L. Duplan and S. Perathoner, *ChemSusChem*, 2011, **4**, 1194–1215.

2 M. Peters, B. Köhler, W. Kuckshinrichs, W. Leitner, P. Markewitz and T. E. Müller, *ChemSusChem*, 2011, **4**, 1216–1240.

3 M. Aresta, A. Dibenedetto and A. Angelini, *Chem. Rev.*, 2014, **114**, 1709–1742.

4 D. J. Darensbourg and S. J. Wilson, *Green Chem.*, 2012, **14**, 2665–2671.

5 J. Langanke, A. Wolf, J. Hofmann, K. Böhm, M. A. Subhani, T. E. Müller, W. Leitner and C. Gürtler, *Green Chem.*, 2014, **16**, 1865–1870.

6 A. M. Chapman, C. Keyworth, M. R. Kember, A. J. J. Lennox and C. K. Williams, *ACS Catal.*, 2015, **5**, 1581–1588.

7 L. K. Rihko-Struckmann, A. Peschel, R. Hanke-Rauschenbach and K. Sundmacher, *Ind. Eng. Chem. Res.*, 2010, **49**, 11073–11078.

8 N. von der Assen, M. Lampe, L. Müller and A. Bardow, in *12th International Symposium on Process Systems Engineering and 25th European Symposium on Computer Aided Process Engineering*, ed. K. V. Gernaey, J. K. Huusom and R. Gani, Elsevier, 2015, vol. 37, pp. 1235–1240.

9 T. E. Waldman and W. D. McGhee, *J. Chem. Soc., Chem. Commun.*, 1994, 957–958.

10 PlasticsEurope, Eco-profiles, 2014, http://www.plasticseurope.org/plasticssustainability/eco-profiles.aspx, http://www.plasticseurope.org/plasticssustainability/eco-profiles.aspx, accessed 17 Dec 2014.

11 E. Schwab, A. Milanov, S. A. Schunk, A. Behrens and N. Schödel, *Chem. Ing. Tech.*, 2015, **87**, 347–353.

12 N. von der Assen, J. Jung and A. Bardow, *Energy Environ. Sci.*, 2013, **6**, 2721–2734.

13 R. M. Cuéllar-Franca and A. Azapagic, *J. CO2 Util.*, 2015, **9**, 82–102.

14 N. von der Assen, P. Voll, M. Peters and A. Bardow, *Chem. Soc. Rev.*, 2014, **43**, 7982–7994.

15 P. G. Jessop, T. Ikariya and R. Noyori, *Chem. Rev.*, 1995, **95**, 259–272.

16 W. Wang, S. Wang, X. Ma and J. Gong, *Chem. Soc. Rev.*, 2011, **40**, 3703–3727.

17 S. Wesselbaum, U. Hintermair and W. Leitner, *Angew. Chem., Int. Ed.*, 2012, **51**, 8585–8588.

18 S. Wesselbaum, T. vom Stein, J. Klankermayer and W. Leitner, *Angew. Chem.*, 2012, **124**, 7617–7620.

19 G. Reiter and J. Lindorfer, *Int. J. Life Cycle Assess.*, 2015, **20**, 477–489.

20 C. van der Giesen, R. Kleijn and G. J. Kramer, *Environ. Sci. Technol.*, 2014, **48**, 7111–7121.

21 E. Cetinkaya, I. Dincer and G. Naterer, *Int. J. Hydrogen Energy*, 2012, **37**, 2071–2080.

22 International Energy Agency, *Cost and Performance of Carbon Dioxide Capture From Power Generation*, 2011, https://www.iea.org/publications/freepublications/publication/costperf_ccs_powergen.pdf, accessed 20 April 2015.

23 D. Y. Leung, G. Caramanna and M. M. Maroto-Valer, *Renewable Sustainable Energy Rev.*, 2014, **39**, 426–443.

24 T. Sakakura, J.-C. Choi and H. Yasuda, *Chem. Rev.*, 2007, **107**, 2365–2387.

25 M. Aresta and A. Dibenedetto, *Dalton Trans.*, 2007, **28**, 2975–2992.

26 G. Centi and S. Perathoner, *Greenhouse Gases: Sci. Technol.*, 2011, **1**, 21–35.

27 P. Markewitz, W. Kuckshinrichs, W. Leitner, J. Linssen, P. Zapp, R. Bongartz, A. Schreiber and T. E. Müller, *Energy Environ. Sci.*, 2012, **5**, 7281–7305.

28 N. von der Assen and A. Bardow, *Green Chem.*, 2014, **16**, 3272–3280.

29 ISO 14040:2009, *Environmental management – Life cycle assessment – Principles and framework*, European Committee for Standardisation, Brussels, 2009.

30 ISO 14044:2006, *Environmental management – Life cycle assessment – Requirements and guidelines*, European Committee for Standardisation, Brussels, 2006.

31 *Life Cycle Assessment Handbook: A Guide for Environmentally Sustainable Products*, ed. M. A. Curran, WILEY-VCH, Weinheim, 2012.

32 R. Heijungs and R. Frischknecht, *Int. J. Life Cycle Assess.*, 1998, **3**, 321–332.

33 J. Jung, N. von der Assen and A. Bardow, *Int. J. Life Cycle Assess.*, 2014, **19**, 661–676.

34 N. Pelletier, F. Ardente, M. Brandão, C. De Camillis and D. Pennington, *Int. J. Life Cycle Assess.*, 2015, **20**, 74–86.

35 PE International, *GaBi LCA Software and LCA Databases*, 2012, http://www.gabi-software.com/databases/gabi-databases, accessed nov 13, 2014.

36 Swiss Centre for Life Cycle Inventories, *Ecoinvent Data V2.2*, 2010, http://www.ecoinvent.org/, accessed Nov 13, 2014.

37 P. Forster, V. Ramaswamy, P. Artaxo, T. Berntsen, R. Betts, D. Fahey, J. Haywood, J. Lean, D. Lowe, G. Myhre, J. Nganga, R. Prinn, G. Raga, M. Schulz and R. V. Dorland, Climate Change, in *The Physical Science Basis. Contribution of Working Group I to the Fourth Assessment Report of the Intergovernmental Panel on Climate Change*, ed. S. Solomon, D. Qin, M. Manning, Z. Chen, M. Marquis, K. Averyt, M. Tignor and H. Miller,

Cambridge University Press, Cambridge, United Kingdom and New York, 2007, ch. 2, pp. 129–234.

38 M. Goedkoop, R. Heijungs, M. Huijbregts, A. D. Schryver, J. Struijs and R. V. Zelm, *ReCiPe 2008, A life cycle impact assessment method which comprises harmonised category indicators at the midpoint and the endpoint level*, 1st edn, (version 1.08) Report I: Characterisation, 2013, http://www.lcia-recipe.net, accessed 17 Dec 2014.

39 R. Sathre, M. Chester, J. Cain and E. Masanet, *Energy*, 2012, **37**, 540–548.

40 J. Marx, A. Schreiber, P. Zapp, M. Haines, J.-F. Hake and J. Gale, *Energy Procedia*, 2011, **4**, 2448–2456.

41 A. Schreiber, P. Zapp and J. Marx, *J. Ind. Ecol.*, 2012, **16**, S155–S168.

42 M. Z. Hauschild, M. Goedkoop, J. Guinée, R. Heijungs, M. Huijbregts, O. Jolliet, M. Margni, A. De Schryver, S. Humbert, A. Laurent, S. Sala and R. Pant, *Int. J. Life Cycle Assess.*, 2013, **18**, 683–697.

43 ICCA, International Council of Chemical Associations, *How to Know If and When it's Time to Commission a Life Cycle Assessment*, 2013.

44 N. von der Assen, L. Müller, A. Steingrube, P. Voll and A. Bardow, *Selecting CO_2 sources for CO_2 utilization by environmental-merit-order curves*, 2015, submitted.

45 S. V. Jørgensen, M. Z. Hauschild and P. H. Nielsen, *Int. J. Life Cycle Assess.*, 2015, **20**, 451–462.

46 M. Brandão, A. Levasseur, M. Kirschbaum, B. Weidema, A. Cowie, S. Jørgensen, M. Hauschild, D. Pennington and K. Chomkhamsri, *Int. J. Life Cycle Assess.*, 2012, **18**, 230–240.

47 M. Mequignon, L. Adolphe, F. Thellier and H. A. Haddou, *Build. Environ.*, 2013, **59**, 654–661.

48 U.S. Environmental Protection Agency, Transitioning to low-GWP foam blowing agents, 2011,http://www.epa.gov/ozone/downloads/EPA_HFC_ConstFoam.pdf, accessed 6 May 2015.

49 J. Langanke, J. Hofmann, C. Gürtler and A. Wolf, *J. Polym. Sci., Part A: Polym. Chem.*, 2015, **53**, 2071–2074.

50 G. Reuss, W. Disteldorf, A. O. Gamer and A. Hilt, in *Ullmann's Encyclopedia of Industrial Chemistry*, Wiley-VCH, Weinheim, 2012, vol. 15, ch. Formaldehyde, pp. 735–768.

51 S. G. Jadhav, P. D. Vaidya, B. M. Bhanage and J. B. Joshi, *Chem. Eng. Res. Des.*, 2014, **92**, 2557–2567.

52 N. Dave and G. A. Foulds, *Ind. Eng. Chem. Res.*, 1995, **34**, 1037–1043.

53 L. Chen, Y. Lu, Q. Hong, J. Lin and F. Dautzenberg, *Appl. Catal., A*, 2005, **292**, 295–304.

54 A. Sternberg and A. Bardow, *Energy Environ. Sci.*, 2015, **8**, 389–400.

55 E. V. Kondratenko, G. Mul, J. Baltrusaitis, G. O. Larrazabal and J. Perez-Ramirez, *Energy Environ. Sci.*, 2013, **6**, 3112–3135.

56 M. Aresta and M. Galatola, *J. Cleaner Prod.*, 1999, **7**, 181–193.

57 S. V. Jørgensen and M. Z. Hauschild, *Int. J. Life Cycle Assess.*, 2013, **18**, 747–754.

Faraday Discussions

PAPER

A framework for the analysis of the security of supply of utilising carbon dioxide as a chemical feedstock

Eric S. Fraga* and Melvin Ng

Received 16th April 2015, Accepted 1st June 2015
DOI: 10.1039/c5fd00038f

Recent developments in catalysts have enhanced the potential for the utilisation of carbon dioxide as a chemical feedstock. Using the appropriate energy efficient catalyst enables a range of chemical pathways leading to desirable products. In doing so, CO_2 provides an economically and environmentally beneficial source of C_1 feedstock, while improving the issues relating to security of supply that are associated with fossil-based feedstocks. However, the dependence on catalysts brings other supply chains into consideration, supply chains that may also have security of supply issues. The choice of chemical pathways for specific products will therefore entail an assessment not only of economic factors but also the security of supply issues for the catalysts. This is a multi-criteria decision making problem. In this paper, we present a modified 4A framework based on the framework suggested by the Asian Pacific Energy Research centre for macro-economic applications. The 4A methodology is named after the criteria used to compare alternatives: availability, acceptability, applicability and affordability. We have adapted this framework for the consideration of alternative chemical reaction processes using a micro-economic outlook. Data from a number of sources were collected and used to quantify each of the 4A criteria. A graphical representation of the assessments is used to support the decision maker in comparing alternatives. The framework not only allows for the comparison of processes but also highlights current limitations in the CCU processes. The framework presented can be used by a variety of stakeholders, including regulators, investors, and process industries, with the aim of identifying promising routes within a broader multi-criteria decision making process.

1 Introduction

Carbon dioxide (CO_2) is a non-toxic and abundant C_1 (1 carbon atom) feedstock. The release of CO_2 has contributed to global warming and the greenhouse effect making emissions an increasing concern for society and policy makers. In recent years carbon capture and sequestration (CCS) has been mooted as the solution to

Centre for Process Systems Engineering, Department of Chemical Engineering, University College London (UCL), UK. E-mail: e.fraga@ucl.ac.uk

controlling emissions but has raised the question of what to do with the increasing large amounts of CO_2. A potential solution is to use CO_2 as a chemical feedstock and a raw product in the production of chemicals. This is known as carbon capture and utilisation (CCU).

There are several benefits that arise from the utilisation of carbon dioxide as a chemical feedstock.[1] Firstly, like CCS, CCU is able to mitigate emissions of carbon dioxide through capture. CCU, unlike CCS, may allow for a closed loop recycling system. Secondly, an economic benefit may arise from the generation of economically valuable products. Finally, it may help address an individual country's security of supply for energy.

As C_1 feedstock is currently predominantly sourced from oil derived hydrocarbons, the security of supply for oil may be a concern. Most nations depend on foreign oil imports, often from areas of geo-political instability. As a result, market volatility and uncertainty disrupts national sustainability and forward planning. Through a carbon economy, reliance on volatile sources of oil could be mitigated or even eliminated through the use of secure domestic carbon sources.

Currently, there are a few chemical pathways that utilise CO_2 on a commercial scale: the production of urea, salicylic acid and sodium carbonate. However, CO_2 utilisation has a key limitation. CO_2 is a kinetically and thermodynamically stable molecule. This results in a high activation energy and a large quantity of energy may be required to react carbon dioxide. Unfortunately, energy generation currently emits more CO_2 than would be consumed in generating valuable products, leading to an overall net increase in emissions. Although these emissions could be reduced or eliminated through the use of renewable energy generation, any energy requirements could also be ameliorated through the development of suitable catalysts, as can the potentially high activation energy. This has led to an increasing level of research and development in the field of catalysis. This research is leading to new potential products that can be derived from CO_2 as a feedstock.

However, the use of catalysts brings up the issue of security of supply yet again. Many catalysts are made from rare materials and these materials may also be sourced from geo-politically unstable regions. Recycling of many of the catalysts may be difficult.[2] This paper presents a methodology for the assessment of alternative CO_2 utilisation routes, incorporating a number of factors that may characterise the security of supply. The methodology provides an attractive graphical visualisation of the characteristics of an individual route which enables the comparison of alternatives. This may help decision makers identify those routes which best achieve the potential benefits of CCU with improved security of supply.

2 Carbon dioxide as a feedstock

The pathways currently utilising CO_2 on an industrial scale are few. Examples include the Solvay process to produce sodium carbonate,[3] the production of urea *via* the Bosch–Meiser process,[4] and the production of salicylic acid through the Kolbe–Schmitt process.[5] These only account for a fraction of the number of pathways theoretically possible using CO_2 as a C_1 feedstock. Alternative pathways are being considered and a selection of these are summarised in Table 1, grouped according to the compounds used to react with CO_2.

Table 1 Summary of CO_2 pathways

Group	Substrate	Product
Oxygen containing compounds	Epoxides	Cyclic carbonate
		Alternating polycarbonates
		Aromatic polycarbonates
	Alcohols	Acyclic carbonates
Nitrogen containing compounds	Ammonia, amines	Urea
		Carbamic acid esters
		Polyurethanes
Carbon–carbon unsaturated compounds	Aromatic compounds	Carboxylic acids
	Alkynes	Carboxylic acid esters
	Alkenes	Lactones
Other	Hydrogen	Formic acid
		Methanol

Many of the routes in Table 1 are currently commercially viable. Table 2 illustrates the level of production for some of these processes. The most mature CCU process is the production of urea. The rates for the other CCU routes are orders of magnitude smaller. The level of production of each of these products correlates well with the level of development in terms of technology.[6] A summary of the utilisation levels is presented in Table 3 with polymers separated out into individual products.

Given the level of technological development, current market saturation and potential demand, we can shortlist promising CCU product targets for immediate consideration:

(1) Methanol.

(2) Urea.

(3) Formic acid.

(4) Polymers, mainly polyalkylene carbonates.

(5) Cyclic carbonates.

The promise in technologies for the production of these targets derives predominantly from catalyst development. However, in order to determine commercial viability, the security of supply for each overall process must be considered. The next section describes a methodology for assessing the security of supply, and presents this in the context of other factors that decision makers will wish to consider.

Table 2 Current levels of production of products, showing levels for both CCU and traditional (non CO_2) routes

Chemical	CO_2 utilising production (kT)	Global production (kT)	CO_2 (%)
Cyclic carbonates[7]	80	200	40
Formic acid[8]	0	300	0
Methanol[9]	4	100 000	≈0
Polypropylene carbonate[10]	76	≈0	0
Polycarbonate[11]	605	3700	16
Salicylic acid[6]	90	90	100
Urea[12]	157 000	157 000	100

Table 3 Current state of carbon capture & utilisation technologies. A double ✓ indicates significant levels of activity and a single ✓ an area that is showing increasing levels of acitivity[6]

Technology	Research	Demonstration	Feasibility	Mature market
Chemical production				
Acyclic carbonates	✓✓			
Alternating polycarbonates	✓✓	✓		
Aromatic polycarbonates	✓✓	✓		
Carboxylic acids	✓✓			
Carboxylic acid esters	✓✓			
Cyclic carbonates	✓✓			
Lactones	✓✓			
Polyurethanes	✓✓	✓		
Sodium carbonate				✓✓
Urea				✓✓
CO_2 to fuels				
Methanol		✓✓	✓	
Formic acid	✓✓	✓		

3 Methodology

The target products described above illustrate the potential for effective CO_2 utilisation. However, beyond purely economic and technical issues, other aspects may affect the adoption of any particular route to a product. One such aspect is the *security of supply*. This term is most often used in the context of energy and is defined as follows:

"the ability of an economy to guarantee the availability of energy resource supply in a sustainable and timely manner with the energy price being at a level that will not adversely affect the economic performance of the economy."[13]

In the context of energy, security of supply is dependent on 5 main factors: availability of fuels domestically and externally, the ability to acquire supply to meet demand, level of an economy's diversification, accessibility to fuel resources through sufficient infrastructure, and geo-political challenges in sourcing energy. From these factors, the Asian Pacific Energy Research Centre (APERC) proposed the categorisation of these factors into *availability*, *accessibility*, *acceptability* and *applicability*, collectively called *the 4 As of Energy Security*.[13]

We adapt this 4As approach to the security of supply to the specific case of CO_2 utilisation. The framework proposed by APERC was intended for a macro-economic analysis of the energy system. Each A value was quantified on the basis of macro-energy system characteristics.[13] In the context of CO_2 utilisation, it makes sense to also consider smaller scale analyses, *e.g.* at the process route level, while still including macro-economic aspects. The key is the suitable re-definition of the 4A categories for their application to CO_2 utilisation. In what follows, we will be defining the parameters that quantify each A with a value between 0 (bad) to 10 (good).

3.1 Availability

The availability of the supply side of the product will be defined by the catalyst. The catalyst is often the limiting factor in the production rate. Further, few of the catalysts used for the products noted above are replaceable because many are tailored for the specific reaction. The availability of the catalyst will be estimated using a combination of parameters:

Crustal abundance. Crustal abundance is a measure of the scarcity of a metal on a macro scale with the abundance measured in parts per million.[14] For example, the abundance of ruthenium is 0.00057 ppm, which indicates a high risk catalyst, whereas aluminium, with an abundance of 84 149 ppm, would be low risk. Each metal within each of the catalysts is given a score based on its abundance, as quantified in Table 4.

Production concentration. Production concentration is an indicator of the distribution of regions in which the metal is produced, and data are provided by the British Geological Survey.[15] A commodity with a limited distribution of production is at a higher risk than one produced in many places around the world. For instance, 91% of iridium is produced in South Africa, and the top 3 countries producing iridium control 98% of the global production. In contrast, the cumulative production of the top 3 copper producers accounts for less than half of the global production. A score is allocated to the geographic concentration of metals production concentration as shown in Table 5.

Reserves concentration. Reserves concentration is similar to the production concentration but relates to the distribution of the reserves. A score is allocated to the metals reserve concentration as shown in Table 6.

Political corruption and stability. Political corruption has become an increasing concern as black-market dealings and market inconsistencies reduce the transparency of what should be a free market. Societal pressure is also increasingly becoming a factor. Whilst corruption has an effect on stability, both factors are included separately. This is because political instability will describe the current situation, whilst political corruption could lead to increased instability due to pressures on the system. The measures for corruption and stability are based on data from the World Bank.[16] These data provide an index rating of political corruption and stability in [1,100], where 1 is the most corrupt/unstable and 100 the least. We have mapped the index rating to scores for our framework as shown in Table 7. This factor contributes 4 values to the overall availability measure: a value of stability and of corruption separately for each of two countries; the top producer and the country that has the most reserves.

The maximum score for availability based on the above is 28 (4×7). A perfect score of 28 means that a metal has high levels of availability where both the production and reserves are not highly concentrated in single regions, and that the main producers and owners of the reserves present low levels of corruption and high levels of stability. A scaled value in the desired range was achieved by dividing the sum of all of the above by 2.8.

Table 4 Crustal abundance scoring

Abundance (ppm)	Score
≥100	4 (low risk)
50–100	3 (medium/low risk)
1–50	2 (medium/high risk)
<1	1 (high risk)

Table 5 Production concentration scoring

Concentration (%)	Score
0–25	4 points
25–50	3 points
50–75	2 points
75–100	1 point

Table 6 Reserves concentration scoring

Concentration (%)	Score
0–25	4 points
25–50	3 points
50–75	2 points
75–100	1 point

In calculating the availability of the catalyst material, we have firstly assumed that there is an equal weighting between each of the parameters in terms of importance. For example, that the extent of corruption in the country where the reserves are held, is as important as the abundance. Secondly, in measuring the abundance of a mineral, the abundance measure does not take into account the spatial distribution within a region. For example, in one region there may be an abundance of zinc although it may be spread uniformly over the whole region. This means that the crustal abundance may not necessarily be a realistic measure of the ability to extract the material through mining. Finally, this index is based on results at one point in time and does not include future projections and historical context. These assumptions can, of course, be addressed subsequently and the scorings adapted, if desired.

3.2 Affordability

The second A is economic, based on the revenue generated per unit of catalyst:

$$\text{Catalyst to revenue ratio} = \frac{\text{catalyst price}}{\text{product price} \times \text{product amount}} \quad (1)$$

This assumes an equal turnover frequency for all catalysts which is not typically the case. For instance, cheaper catalysts may have a shorter life than a more expensive or robust catalyst. However, the measure should be appropriate as an indication of affordability. Other measures of affordability can be used, of course. In any case, the mapping of ratio to score is shown in Table 8.

It is worth noting that the process cost is not considered, only the cost of any catalyst required. This is based on the assumption that the general plant cost is comparatively less volatile and hence affects security of supply negligibly. It is also assumed that other raw materials will have lower costs than the catalyst, which is generally the case.

Table 7 Political corruption & stability scoring

Measure	Score
75–100	4 points
50–75	3 points
25–50	2 points
1–25	1 point

3.3 Applicability

Applicability has been defined according to the technological readiness level[17] (TRL) of the process. The TRL rating is a systematic approach that assesses the level of maturity for a given technology, allowing for consistency in comparison. The TRL method classifies a technology into one of 9 levels, from level 1 indicating that the basic principles are understood, through to level 9 indicating that commercial operations exist. The TRL is incorporated by scaling by 10/9.

3.4 Acceptability

The final A value is dependent on two main parameters: a life cycle assessment (LCA) in terms of CO_2 emissions (Table 9) and a measure of the lifetime of storage of CO_2 in the product (Table 10). The use of these factors is motivated by one of the key motivations for CCU: the need to reduce the impact of CO_2 emissions on the global climate. The life storage measure for carbon dioxide in a product is included to mitigate the lack of comparable LCA data in some cases. For example, methanol may only have temporary storage due to its use as a combustible fuel, whereas polyalkylene presents a long term store for the CO_2. The scores from the two factors are combined with equal weighting after scaling to yield a score in the range [1,10].

3.5 Visualising the 4As

The 4As framework described above generates a quantitative assessment of the individual factors. Displaying this multi-dimensional information can be done in

Table 8 Affordability score system

Catalyst to revenue ratio	Score
0–0.001	10 points
0.001–0.0025	9 points
0.0025–0.005	8 points
0.005–0.01	7 points
0.01–0.025	6 points
0.025–0.05	5 points
0.05–0.1	4 points
0.1–1	3 points
1–10	2 points
10–20	1 points
20+	0 points

Table 9 LCA score system based on the ratio of amount of CO_2 emissions to the amount of CO_2 utilised in the process

Ratio	Score
0–0.5	10 points
0.5–1	9 points
1–1.5	8 points
1.5–2	7 points
2–2.5	6 points
2.5–3	5 points
3–3.5	4 points
3.5–4	3 points
4.5–5	2 points
5+	1 point

a variety of ways. We have chosen to use star charts, otherwise known as radar charts and spider charts, to present this multi-dimensional information. This is a simple graphical representation that facilitates the comparison of different alternatives and is illustrated in Fig. 1.

Each of the 4 measures is itself an indicator of the security of supply for a given process, but it is the overall combination of these factors that needs to be compared when looking at alternative catalysts and products. A perfect rhombus is an illustration of a technology that would be considered to have a secure supply.

3.6 Assumptions

There have been 5 key assumptions made in this framework:

(1) The catalyst is the limiting factor for the success of the process. Given that overcoming the thermodynamic constraints of carbon dioxide based reactions is the key, this assumption is reasonable.

(2) A second assumption is that the availability and the affordability are dependent solely on the catalyst. We assume that the general plant cost and material sourcing is comparatively less volatile given available knowledge of developing chemical process. For example, in processing polyalkylene carbonates, the majority of the processing units will be similar if not the same for a process without CCU and one with CCU. However, the catalytic reactor and the capture cost are the key differences in the processes.

(3) The turnover frequency (TOF) is assumed to be the same for different catalysts. This is not true in general but we believe is sufficient for an initial comparison. A more accurate approach would involve a full economic analysis, including product revenue and catalyst costs. These data are often proprietary and therefore difficult to obtain generally.

Table 10 CO_2 life storage in the final product

Storage life	Score
No storage	1 point
Temporary storage	2 points
Permanent storage	3 points

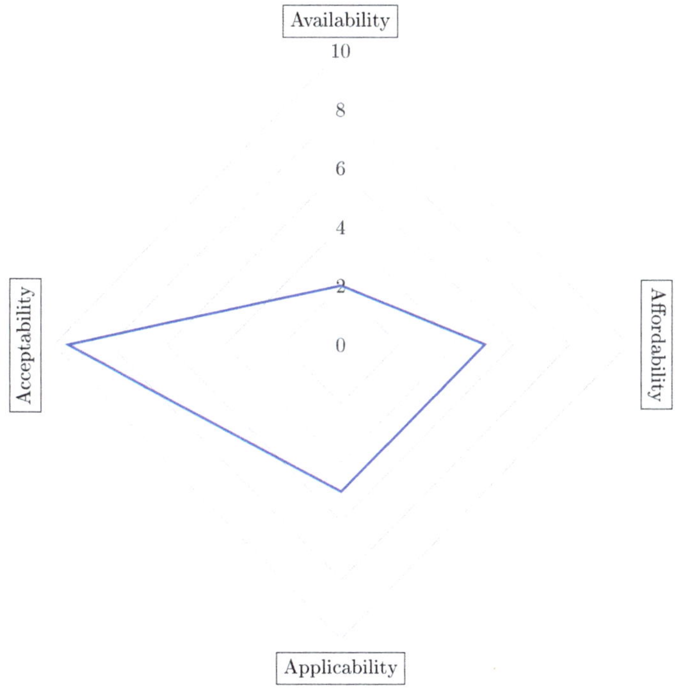

Fig. 1 Example radar chart for visualising the 4A measure showing a case with high acceptability (score of 9.5), low availability (2), average applicability (5) and affordability (5).

(4) Equal weighting is given to sub-items within each category, *e.g.* corruption *versus* stability within the availability category. For specific cases, it may be useful to have non-equal weightings. This would be straightforward to implement should it be desirable.

(5) The results presented below are based on current estimates for each of the categories with no attempt at projecting into the future. For instance, the political situation in relevant countries may change, new sources of raw materials may be discovered, or improved mining operations could change the affordability of a specific catalyst. Furthermore, historical context could be useful in estimating the values of some of the sub-items, especially in terms of the impact of stability or corruption on availability. However, updating individual inputs to the framework is straightforward.

It is also worth noting that the framework need not be limited to 4 categories. In fact, it is highly likely that further economic considerations for the particular product, *e.g.* process cost and market demands, would be included to define a 5As framework. We have not included such elements as the data required are often sensitive and company specific. The overall methodology, however, does not preclude such an extension.

4 Carbon utilisation targets and processes

The information required for the assessment of each target, on the basis of the methodology described above, is given in this section. The various processes, including both chemistry and actual processing, are described, and any catalysts required are specified.

4.1 Urea

Urea accounts for approximately 50% of global nitrogen fertiliser production. Urea has the highest concentration of nitrogen of all solid nitrogenous fertilisers which are widely used in agriculture.[18] Ammonia, a key reactant for the production of urea, and urea plants are often combined as one plant.[19]

The basic synthesis of urea has been established since 1922, using a process known as the Bosch–Meiser process.[4] This process consists of two main equilibrium reactions. The first is carbamate formation in a fast exothermic reaction which is then followed by urea conversion, a slow endothermic decomposition of the ammonium carbamate into urea and water.

$$2NH_3 + CO_2 \rightleftharpoons NH_2COONH_4$$

$$2NH_2COONH_4 \rightleftharpoons H_2O + NH_2CONH_2$$

The synthesis of urea is a non-catalytic process and therefore does not require any catalyst material.

Interestingly, historically the production of ammonia has often exceeded the amount required stoichiometrically when compared with the amount of CO_2 readily available for the production of urea. Combined ammonia and urea plants would sell the surplus ammonia because the cost of CO_2 to meet the deficiency was not economically justified. However, with the potential increase in CCS, there would be an opportunity to boost urea production economically in these plants. This is known as the KM-CDR process, which is used to enhance the yield in urea production.[19] This aligns well with the desire to increase CCU.

4.2 Polyalkylene carbonate

Polyalkylene carbonates are polymers that have a range of uses from biodegradable polymers in medical use to high temperature tolerance polymers. Polypropylene carbonate, a polyalkylene derivative, is amongst the most promising polymer products. The synthesis of polyalkylene carbonates is through the reaction between epoxides and CO_2. The process conditions determine the generated product. Conditions can vary from 30–40 °C for polymers obtained from polypropylene oxide and up to temperatures of 110–120 °C for polymers from cyclohexene epoxide. Pressure also has an impact on the reactions.[20] Current methods utilise biomass feedstocks. If CO_2 were used to generate these polymers, competition with food production would be reduced (which highlights that security of supply is also an issue for food and water, but out of scope for this paper).

Various catalysts have been researched for the co-polymerisation process. Inoue *et al.*[21] discovered that the combination of $ZnEt_2$ and water allowed viable catalyst performance for the co-polymerisation of carbon dioxide and propylene oxide. Further validation on the performance of a zinc catalyst was made when Kawachi *et al.*[22] demonstrated high catalytic activity for polymerization of epoxides and CO_2. Previous varieties include aryl, alkyl, diimines, Schiff bases and zinc compounds. More recently, researchers have focused on 3 main catalysts: chromium, cobalt and zinc. For example, Noh *et al.*[23] discuss the reaction between carbon dioxide and propylene oxide in the presence of Co(salen). The cobalt catalyst showed a superior turnover number at 22 000 g per gcat (ref. 23) whilst zinc only yields a turnover value of 1441 g per gcat. The catalyst proved to be highly active in the reaction for CO_2 and propylene oxide co-polymerization although commercial viability of the process has yet to be demonstrated.

4.3 Formic acid

Formic acid, HCOOH, has a wide range of uses including silage, additives to pharmaceutical intermediates, and as a fundamental feedstock in the chemicals industry to produce aldehydes, ketones and carboxylic acids.[24] More recently there has been strong interest in utilising formic acid within fuel cells due to its strong electrochemical oxidation ability for Pt–Ru electrodes. Formic acid is traditionally produced through the reaction of methanol and carbon monoxide which produces an intermediate, methyl formate:

$$CH_3OH + CO \rightarrow HCOOCH_3$$

Hydrolysis then produces formic acid and methanol:

$$HCOOCH_3 + H_2O \rightarrow HCOOH + CH_3OH$$

However, the hydrogenation of CO_2 is also possible using both transition and non-transition metal compounds as catalysts, with a resulting benefit in the reduction of the cost of raw materials:

$$CO_2 + H_2 \rightarrow HCOOH \tag{2}$$

Farlow[25] investigated the reaction in the presence of a nickel catalyst under high pressure, 20–40 MPa, and high temperature, 353–423 K. This eventually progressed to favourable conditions (298 K) through catalyst complexes of Ru or Pd combined with halides and hydrides.[26] Simultaneously, there has been research in homogeneous catalysts, such as supercritical CO_2, water and ionic liquids. This can achieve comparable reaction rates to transition complexes in the supercritical phase. However, the heterogeneous catalysts proved more attractive as the separation of the formic acid from the catalyst is easier. An activated carbon supporting ruthenium through impregnation is preferred as it does not use the hazardous and expensive reagents used in other approaches.[8] Using ruthenium as the active component results in high activity and selectivity.

4.4 Methanol

Methanol is widely used as a fuel for transport and as a chemical feedstock.[27] There are over 90 process plants worldwide producing 75 million tons of methanol annually.[9] Methanol has traditionally been produced through fossil fuels *via* syngas chemistry. Therefore, elements of energy security of supply come into play in global methanol production from C_1 feedstocks. However, recently new pathways and catalysts have been developed.

Olah[27] devised an approach to produce methanol through what is known as "metgas". Metgas is composed of CO and H_2 at a 1 : 2 ratio through a single step by reacting CO_2, $3CH_4$ and steam. This is known as *bireforming*. The temperature is high, 1073–1273 K, whilst pressure is held between 0.5–4 MPa in the presence of a nickel based catalyst:

$$3CH_4 + CO_2 + 2H_2O \rightarrow 4CO + 8H_2 \rightarrow 4CH_3OH \quad (3)$$

The hydrogenation of CO_2 has become an increasingly promising solution given an appropriate catalyst. The conversion has favourable thermodynamics although the high activation energy barrier remains a challenge.[28] Heterogeneous catalysts have therefore been widely investigated. Copper has proven to be the most favourable catalyst. With the use of copper catalysts, such as Cu/ZnO, CuO/ZnO and CuO–ZnO/ZrO_2, the hydrogenation proceeds at lower temperatures and under higher pressures.

4.5 Cyclic carbonates

Cyclic carbonates are commonly used as degreasing agents, polar aprotic solvents and electrolytes for lithium ion batteries.[7,29] Cyclic carbonates can also be converted into dimethyl carbonate which is used as a quality oxygenating additive for both petrol and aviation fuel. The main chemical pathway for cyclic carbonates is the reaction between epoxides and carbon dioxide.

The development of the catalysts and the mechanisms for this reaction has been well documented over several permutations of catalysts, including organic bases,[30] zeolites, metal oxides,[31] alkali metal halides[32] and metal complexes.[7] Whilst the synthesis has been applied to industry with a variety of such catalysts, the recovery and stability of the catalyst itself has yet to be improved. More recently, research has focused on using ionic liquids as the catalyst due to advantageous negligible vapour pressures.[33] The use of ionic liquids as a clean catalytic form has proven to improve reaction rate and reaction selectivity. However, the use of ionic liquids exhibits low catalyst stability and activity and requires a co-catalyst. Whilst Dai *et al.*[34] suggested that the performance was substantiated through the use of Lewis acidic compounds as co-catalysts, it still held industrial limitations due to challenges in separation. Recent research has begun to tackle this by immobilising the ionic liquid onto solid supports.

A further form of catalyst proposed for this reaction has been through using metal complexes such as Co, Cr, Ni, Al, Zn and Re mounted onto supports. North & Young[7] have proposed a catalyst based on bimetallic aluminium(acen) complexes. This allows the reaction to take place at room temperature and at atmospheric pressure.

5 Assessing carbon utilisation targets

The 4As methodology has been applied to the processes described above. Table 11 summarises the results which are also presented using radar charts in Table 12 for easy comparison. Data from a number of sources,[6,14–16,18,27] along with those cited in the discussion above and below, were used to determine these scores. For all the categories, except for acceptability, the scores cover almost the full range of values possible, indicating that the scoring system for these categories is able to discriminate between alternatives. The acceptability scores cover only half of the range. This is mostly due to the energy requirements of each process. Currently, energy sources are typically not carbon neutral, and so there is a significant impact on emissions in each case. What these scores show, besides the possibility of comparing the alternatives, is that there is scope for improvement, in this category in particular.

The results can be used to identify which processes show the most promise, in comparison with the others, in the context of security of supply. For each individual case, the different factors can help focus the attention on any aspects that could be addressed by stakeholders to reduce supply risks. The benefit of the 4As model for a process is that the nature of future and current challenges can be efficiently identified.

Some discussion about each target process follows.

5.1 Urea

Overall, urea is the most promising target with maximum values for three of the criteria. First of all, the applicability is high because the process has been used for a long time and is well established. With respect to availability and affordability, the process requires no catalyst. It is useful to compare this process for the production of urea with an alternative that is based on the use of ruthenium as a catalyst.[35]

In either case, the acceptability score ranked second overall. This is due to unfavourable CO_2 emissions by these processes.

5.2 Polyalkylene carbonate

Two different catalysts, cobalt and zinc, have been considered for the production of polyalkylene carbonates. The radar charts for the two options (second row of Table 12) show significant differences, particularly in availability and affordability. Zinc is more easily available and also more affordable when compared with cobalt (see Table 13). In both cases, the relatively low score for acceptability was mostly due to CO_2 emissions.

For availability, a zinc catalyst is preferred for both economic and geo-political factors. Cobalt is 3 times less abundant than zinc and the production levels of

Table 11 Summary of scores for each process/catalyst combination. The maximum score in each category is 10

Product	Catalyst	Avail.	Afford.	Applic.	Accept.	Average
Cyclic carbonates	Al	5	9	2	4	5
Formic acid	Ir	5	1	3	2	3
Formic acid	Ru	5	2	3	2	3
Methanol	Cu	7	8	6	4	6
Polyalkylene carbonates	Co	3	5	9	4	5
Polyalkylene carbonates	Zn	7	9	9	4	7
Urea	Ru	5	2	10	5	5
Urea	None	10	10	10	5	9

Table 12 Star charts for the 8 CCU process alternatives

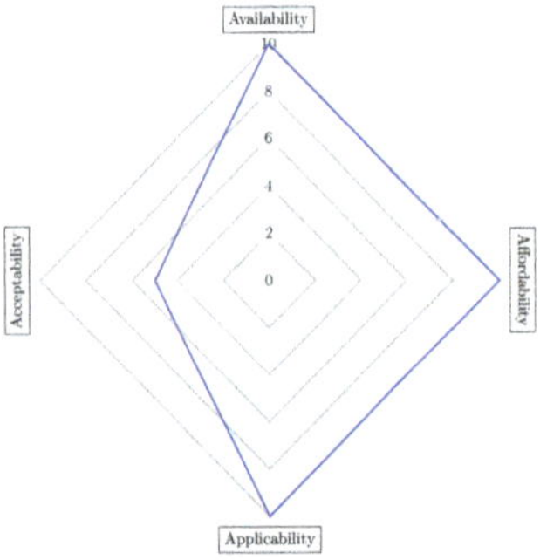

Urea with no catalyst

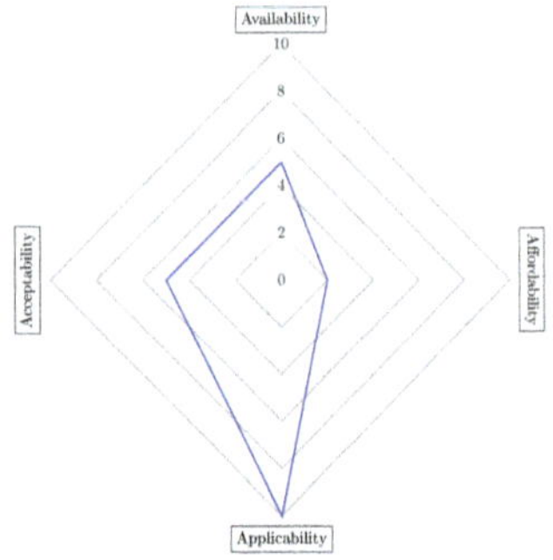

Urea with ruthenium

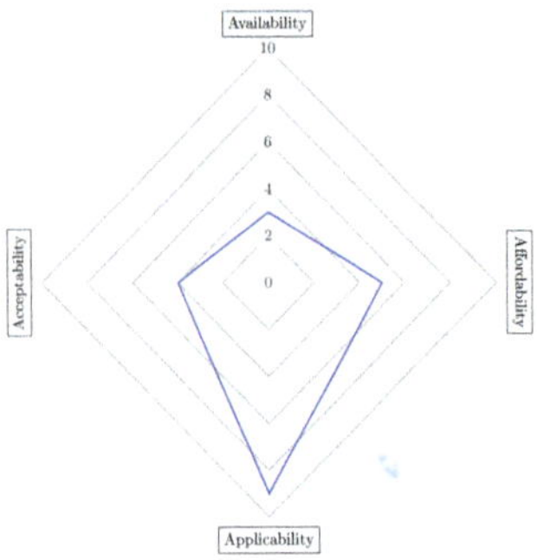

Carbonate with cobalt

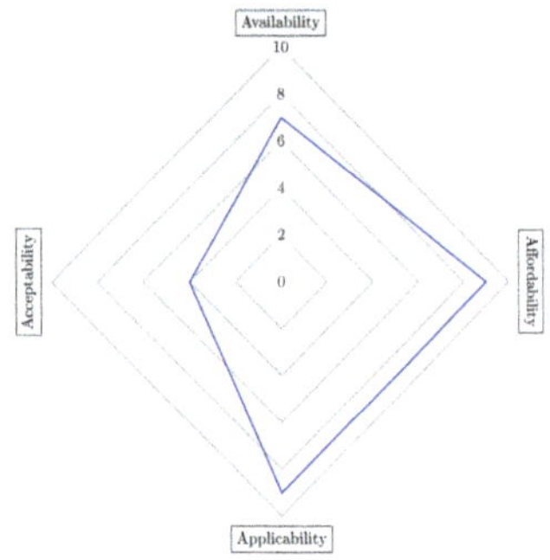

Carbonate with zinc

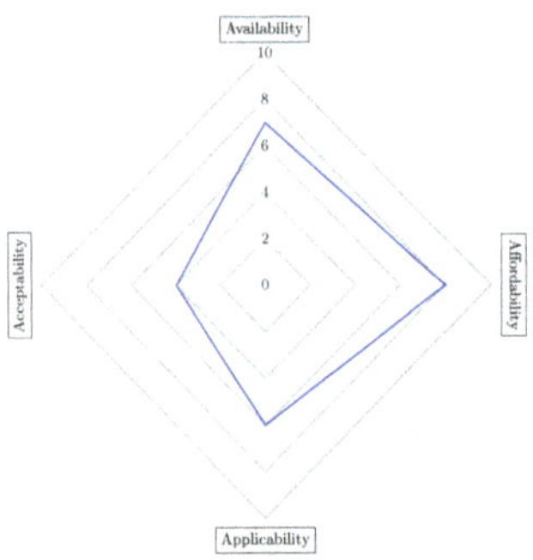

Methanol with copper

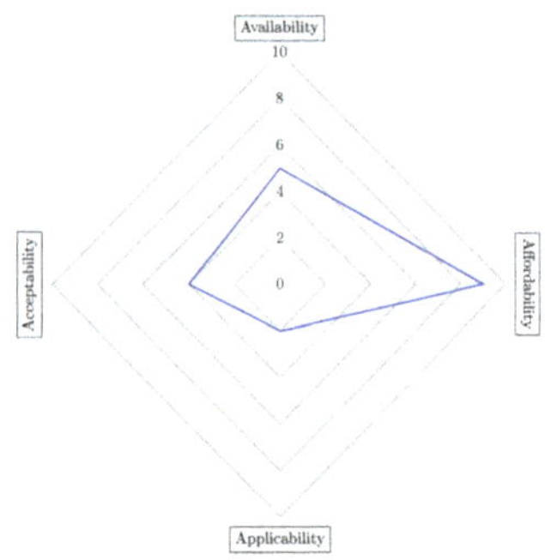

Cyclic carbonates

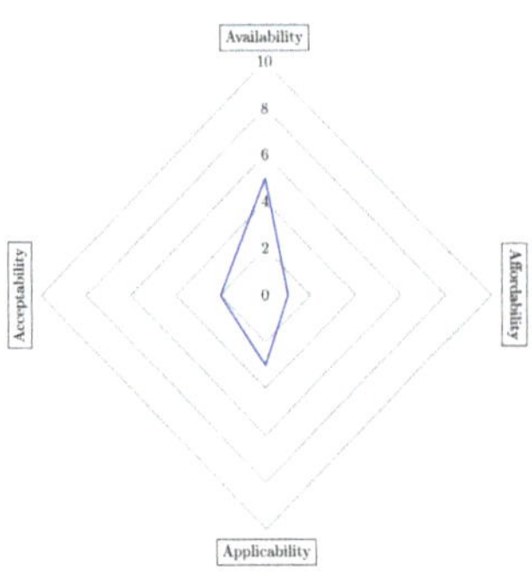

Formic acid with iridium

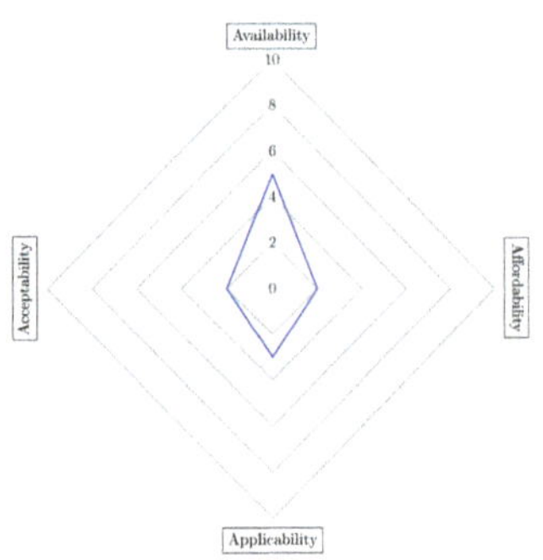

Formic acid with ruthenium

zinc are 84 times that of cobalt. The ability to source the catalyst is also a consideration: 81% of cobalt production is concentrated in 3 producing countries, whilst only 49% of zinc production is in its top 3 producing countries. The risk to supply is lower for zinc, *i.e.* were there to be instability in one of the top producing regions, a large amount will be available from other sources. The political considerations of production and of reserve distribution led to the low score for cobalt: the top producing country for cobalt (Democratic Republic of the Congo) is ranked high in global corruption levels whereas, for zinc, the top producing country (P. R. China) has a better ranking for both corruption and political stability.

The high score for applicability is based on the fact that existing technologies can be used for the production of these polymers. Also, current production is based on food based feedstocks such as corn. Replacing these with CO_2 will help address food security and therefore increase the applicability of this CCU alternative.

5.3 Methanol

Methanol rates reasonably well in availability and affordability, with applicability average. Acceptability is low, again mostly due to emissions of CO_2 due to energy requirements for the process. There are no particular concerns beyond this and so methanol can be considered a good choice as a target, especially if CO_2 neutral energy sources are available.

5.4 Cyclic carbonates

The key factor in the production of cyclic carbonates is the low applicability score. This is primarily due to the low TRL value of 2. Although there has been significant research into catalysts for the production of cyclic carbonates, no commercial implementations exist.[7] The existing processes for cyclic carbonates, not based on CO_2 utilisation, perform well enough that the incentive to develop CO_2 based processes is not there currently. As pressure, financial and regulatory, on utilising CO_2 increases, there will be increasing incentives to take existing lab-scale processes into pilot and eventually plant scale. This will increase the applicability score and, depending on the energy requirements, may also lead to an increase in the acceptability score.

The availability score of 5 is the same as for urea with ruthenium. Although aluminium has a much higher abundance, the overall score is the same due to the political and corruption assessments of the countries with greatest production

Table 13 Price of catalyst relative to the price of aluminium

Material	Relative cost
Al	1
Zn	1.14
Cu	4
Co	18
Ru	1529
Ir	11 418

and reserves. The overall security of supply is deemed to be similar, although this is a function of the weighting of the different criteria that form the basis for the availability score.

5.5 Formic acid

The progress of formic acid production through CO_2 utilisation is the least mature of all the processes presented, leading to low applicability scores. Furthermore, both of the catalysts considered are costly, having a significant impact on the affordability scores. However, because of the potential applications, *e.g.* the use of formic acid in fuel cells, there is an incentive to develop and to improve these processes; see, for instance, the press release from Market Wired.[36]

6 Conclusion

This paper presents a framework for the analysis of the security of supply for carbon dioxide utilising processes. Development in CCU has concentrated on catalysis to ameliorate the energy requirements for reactions involving CO_2. At these early stages of CCU development, gaining insight into any potential limitations that arise from security of supply issues is of value to various stakeholders, including industry, governments and potential end-users of the products.

The 4As approach, proposed by the Asian Pacific Energy Research Centre for energy security, has been adapted to CCU, combining micro- and macro-economic criteria with process considerations. The framework enables us to compare and contrast alternative products and processes through the analysis of the impact of catalyst choice. It also highlights those aspects which could benefit from further development. The cases considered show that urea production is currently the most secure, while formic acid is at the other end of the scale. The reasons for the differences amongst the various target products range from geopolitical, through to the stage of development of the individual processes.

A number of assumptions have been made. These are all subject to change as the framework is fundamentally extensible. Of primary concern to industrial users of the framework would be the addition of process and market economics beyond the impact of the catalysts. However, economics will necessarily trade-off with security of supply and eventual decisions will be based on the stakeholders' own perceptions of relative importance.

References

1 C. Creutz and E. Fujita, *Carbon Management: Implications for R&D in the Chemical Sciences and Technology: A workshop Report to the Chemical Sciences Roundtable*, 2001.
2 T. E. Graedel, J. Allwood, J.-P. Birat, M. Buchert, C. Hagelueken, B. K. Reck, S. F. Sibley and G. Sonnemann, *J. Ind. Ecol.*, 2011, **15**, 355–366.
3 T. Paul, Process for producing sodium carbonate, *US Pat.*, US25256763A, 1965.
4 N. MacDowell, N. Florin, A. Buchard, J. Hallett, A. Galindo, G. Jackson, C. S. Adjiman, C. K. Williams, N. Shah and P. Fennell, *Energy Environ. Sci.*, 2010, **3**, 1645–1669.

5 A. Lindsey and H. Jeskey, *Chem. Rev.*, 1957, **57**, 583–620.
6 P. Zakkour, Implications of the reuse of captured CO_2 for European climate action policies, 2013, http://setis.ec.europa.eu/setis-deliverables/setis-workshops-hearings/workshop-co2-re-use-technologies.
7 M. North and C. Young, *Catal. Sci. Technol.*, 2011, **1**, 93–99.
8 G. Peng, S. J. Sibener, G. C. Schatz and M. Mavrikakis, *Surf. Sci.*, 2012, **606**, 1050–1055.
9 Methanol Institute, http://www.methanol.org/Methanol-Basics/The-Methanol-Industry.aspx, retrieved 22 October 2014.
10 S. Moolji, Global Polymer Business – Trends & Market Dynamics, Presented at IOCL Petrochem Conclave, http://www.petrochemconclave.com/presentation/2013/Mr.SMoolji.pdf, 2013, retrieved 15 April 2015.
11 Asahi Kasei Chemicals Corporation, Phosgene-free polycarbonate process, http://www.asahi-kasei.co.jp/chemicals/en/license/page01.html, retrieved 15 April 2015.
12 Ernst & Young, http://iffcocan.com/iffco/IFFCO-CF_FinalΩReport-w-App-20130904.pdf, retrieved 22 October 2014.
13 Asia Pacific Energy Research Centre,http://aperc.ieej.or.jp/file/2010/9/26/APERC_2007_A_Quest_for_Energy_Security.pdf, retrieved 24 October 2014.
14 R. L. Rudnick and S. Gao, in *The Crust*, ed. K. Turekian and H. Holland, Elsevier, 2013, vol. 3, ch. 1.
15 T. J. Brown, A. S. Walters, N. E. Idoine, R. A. Shaw, C. E. Wrighton and T. Bide, *World Mineral Production 2006–2010*, British Geological Survey, Keyworth, Nottingham, 2012.
16 World Bank,http://info.worldbank.org/governance/wgi/index.aspx#home, retrieved 24 October 2014.
17 J. C. Mankins, Technology Readiness Levels, A White Paper, Advanced Concepts Office, Office of Space Access and Technology, NASA technical report, 1995.
18 Ceresana, http://www.ceresana.com/en/market-studies/agriculture/urea/, retrieved 22 October 2014.
19 M. Iijima, Mitsubishi Heavy Industries, http://www.gpcafertilizers.com/pdf/2011/12.pdf, retrieved 23 October 2014.
20 V. Macho, M. Králik, M. Olšovský, J. Jorniaková and D. Mravec, *Pet. Coal*, 2008, **50**, 30–38.
21 S. Inoue, H. Koinuma and T. Tsuruta, *J. Polym. Sci., Part B: Polym. Lett.*, 1969, **7**, 287–292.
22 H. Kawachi, S. Minami, J. N. Armor, A. Rokicki, B. K. Stein and Mitsui Petrochemical Industries Ltd., Pulverized reaction product of zinc oxide and dicarboxylic acid, *US Pat.*, 4,981,948, 1989.
23 E. K. Noh, S. J. Na, S. Sujith, S.-W. Kim and B. Y. Lee, *J. Am. Chem. Soc.*, 2007, **129**, 8082–8083.
24 C. Hao, S. Wang, M. Li, L. Kang and X. Ma, *Catal. Today*, 2011, **160**, 184–190.
25 M. W. Farlow and H. Adkins, *J. Am. Chem. Soc.*, 1935, **57**, 2222–2223.
26 A. Behr, P. Ebbinghaus and F. Naendrup, *Chem. Eng. Technol.*, 2004, **27**, 495–501.
27 G. A. Olah, *Angew. Chem., Int. Ed.*, 2005, **44**, 2636–2639.
28 R. Raudaskoski, E. Turpeinen, R. Lenkkeri, E. Pongracz and R. L. Keiski, *Catal. Today*, 2009, **144**, 318–323.

29 J. A. Castro-Osma, M. North and X. Wu, *Chem.–Eur. J.*, 2014, **20**, 15005–15008.
30 C. R. Gomes, D. M. Ferreira, C. J. Leopoldo Constantino and E. R. Perez Gonzalez, *Tetrahedron Lett.*, 2008, **49**, 6879–6881.
31 T. Yano, H. Matsui, T. Koike, H. Ishiguro, H. Fujihara, M. Yoshihara and T. Maeshima, *Chem. Commun.*, 1997, 1129–1130.
32 H. Zhu, L. B. Chen and Y. Y. Jiang, *Polym. Adv. Technol.*, 1996, 7, 701–703.
33 L. A. Blanchard, D. Hancu, E. J. Beckman and J. F. Brennecke, *Nature*, 1999, **399**, 28–29.
34 W.-L. Dai, B. Jin, S.-L. Luo, X.-B. Luo, X.-M. Tu and C.-T. Au, *Catal. Sci. Technol.*, 2014, **4**, 556–562.
35 R. Voorhoeve, L. Trimble and D. Freed, *Science*, 1978, **200**, 759–761.
36 Market Wired, Mantra Signs Technology Development Cooperation Agreement for the Production of Green Chemicals in Korea Using Proprietary Carbon Capture and Recycling (CCR) Technology, 2010, http://www.marketwired.com/printer_friendly?id=1305650, retrieved 15 April 2015.

Faraday Discussions

PAPER

Investigating public perceptions of carbon dioxide utilisation (CDU) technology: a mixed methods study†

C. R. Jones,*[ac] D. Kaklamanou,[b] W. M. Stuttard,[a] R. L. Radford[a] and J. Burley[a]

Received 5th May 2015, Accepted 10th June 2015
DOI: 10.1039/c5fd00063g

Carbon dioxide utilisation (CDU) technologies hold promise for helping to limit atmospheric releases of CO_2 while generating saleable products. However, while there is growing investment in the research and development required to bring CDU to the market, to date there has been very little systematic research into public perceptions of the technology. The current research reports upon the findings of a series of six qualitative focus groups (and an associated questionnaire) held with members of the UK public in order to discuss the perceived benefits and risks of CDU technology. The findings reveal that public awareness of CDU is currently very low and that there is a desire to learn more about the technology. While our participants did, on average, appear to develop an overall positive attitude towards CDU, this attitude was tentative and was associated with a number of caveats. The implications for the findings in terms of the development of communication and broader strategies of public engagements are outlined.

Introduction

Anthropogenic emissions of carbon dioxide (CO_2) are a primary cause of current global warming and climate change.[1] Carbon dioxide utilisation (CDU) technologies have the potential to help mitigate the release of CO_2 to the atmosphere by making use of some of the emissions from carbon intensive processes like fossil-fuel power generation. By utilising the CO_2 as a carbon source for the manufacture of saleable chemical products (*e.g.* polymers) and fuels, or through direct use in other industries (*e.g.* enhanced oil recovery); CDU also holds promise for

[a]*Department of Psychology, University of Sheffield, Western Bank, Sheffield, UK. E-mail: c.r.jones@sheffield.ac.uk; Tel: +44 (0)114 222 6592*
[b]*Department of Psychology, Sociology and Politics, Sheffield Hallam University, Collegiate Crescent, Sheffield, UK*
[c]*UK Centre for Carbon Dioxide Utilisation (CDUUK), University of Sheffield, Western Bank, Sheffield, UK*

† The informational video used within the current research is available at http://www.co2chem.co.uk/research-clusters/public-perception.

generating economic revenue. This revenue could help to offset some of the costs associated with CDU/CCS (carbon capture and storage) processes and present a viable alternative to fossil-fuel based feedstocks in the manufacture of these products.[2,3] As such, there is growing interest into the research, development and deployment (RD&D) of CDU technology – exemplified by this Faraday discussion.

Social acceptability of CDU

A key consideration in the RD&D of CDU should be the systematic assessment of the *social acceptability* of the technology. Social acceptability (*i.e.* the extent to which a phenomenon, like CDU, is endorsed or rejected by key social actors, *e.g.* politicians, financiers, the general public, *etc.*) is now recognised as being necessary for the successful implementation of new technologies.[4]

As key groups of actors are known to affect the social acceptability of emerging technologies at a number of levels (*e.g.* household, community, national), understanding and responding to the opinions of the general public (*i.e.* examining *public acceptability*) should be a priority consideration for CDU proponents.[5] However, with the exception of a preliminary pilot study conducted by the current authors, to date there has been no systematic research in this field.[6]

Assessing public perceptions of CDU

Public engagement is a diverse term covering any attempt to contact members of the public in order to inform decision making.[5] Research shows that more deliberative, participatory forms of engagement – which involve affected publics earlier (*i.e.* upstream) and in a sustained and transparent way – will tend to yield better outcomes for those behind the engagement activity (*e.g.* increased public trust and decreased objection to decisions, *etc.*).[7,8]

While there is an emerging precedent for upstream engagement, there are evident challenges and risks to realising this in any meaningful sense with emerging technologies, like CDU. Not only will a lack of awareness of the technology likely prove to be a barrier to people's willingness to engage, but once engaged there are risks that the opinions registered towards the technology could be misleading if appropriate forms of attitude assessment are not employed. Reference to literature on the formative assessment of public opinion to CCS, for example, indicated the potential for registering pseudo-opinions (or pseudo-attitudes) if traditional questionnaire-based survey methods were used.[9,10] Pseudo-opinions are, in essence, uniformed judgements that people provide on issues which they have given little or no thought and are problematic as they tend to be weak, unstable and not very predictive of later thought and behaviour.[9,11]

The prospect of registering pseudo-opinions is increased when using traditional questionnaire-based surveys because they provide limited contextual information on the issues being discussed and are often self-completed, thereby offering little opportunity to clarify misunderstanding. In the context of understanding public perceptions of other emerging technologies (*e.g.* CCS, hydrogen), the spectre of recording pseudo-opinions has been addressed through the use of non-traditional survey methods (*i.e.* information choice questionnaires [ICQs]) and qualitative research techniques (*e.g.* focus groups, interviews).[9,10,12,13]

Focus groups (FGs), for instance, provide a good forum for exploring controversial, unfamiliar and/or complex issues, by offering a setting within which information can be presented to and discussed by participants, and where responses and understanding can be probed.[14] If facilitated carefully, FGs provide a useful context for establishing: (a) 'why' people feel the way they do about issues and; (b) how such issues become socially represented and shared.[15]

Comparative case study: public perceptions of CCS

The importance of seeking to understand and appropriately assess the opinion of *the public* towards emerging carbon mitigation technologies is exemplified in work into the public perception of CCS. As a sister technology of CDU, such research provides an appropriate analogue for communicating the value of conducting similar work into CDU. For instance, public opinion research conducted over the last decade or so in a number of countries (*e.g.* USA,[10] UK,[12] Europe,[16] and Japan[17]) has proven invaluable in elucidating the roots of subjective concerns about CCS at a national, regional and local level; leading to guidance on how best to tailor education, communication and development practices to more appropriately address public concerns.[18–20]

Together, these studies have illustrated the multifaceted nature of lay (and expert) opinion of CCS, revealing that public attitudes are not simply a sum of anticipated technical risks but are also influenced by myriad social and economic considerations (*e.g.* mistrust in the proponents of the technology).[18,21]

The current research

We argue that forging a better understanding of emerging public opinion towards CDU is timely and should be seen as an integral accompaniment to the ongoing RD&D of the technology. In view of the current dearth of research into the public opinion of CDU technology, our team is conducting a series of studies with the dual objectives of (1) learning more about public perceptions of the perceived benefits, risks, utility and relevance of CDU; and (2) identifying appropriate means of communicating with the lay public about the science and technology behind CDU (*i.e.* the 'What a Waste!' programme).

We feel that appropriate engagement and communication efforts *should* be predicated on developing a systematic understanding of public attitudes towards the technology. As such, the current research builds upon that reported in a recently published communication article[6] by detailing the results and implications of six qualitative FGs and an associated survey-based activity designed with these objectives in mind.‡

In addition to providing insight into people's opinions of CDU, these FGs also provided a forum to 'market test' a pilot informational video about CDU being developed by the CO2Chem Network (www.co2chem.org).

‡ The two FGs mentioned as part of the communication article do also feature within the present article. However, the current article presents new systematic analysis of these FGs alongside 4 new FGs, details of which have not previously been published.

To our knowledge this study is the first to formally investigate and assess public perception of CDU. While a relatively small qualitative study, this research should be considered as part of a preliminary but growing body of research in this novel and important field.

Methods

Participants

Six focus groups (FGs), each comprising 6–8 participants (44 participants total: 14 female, 30 male; 15–54 years) were convened. All participants were offered a monetary incentive for participating.§ Further details of the participants comprising each FG can be found in Table 1. FGs 1–4 took place at the University of Sheffield in June or December 2013. Participants were recruited *via* a university volunteers list. FG4 also included members of the general public recruited from the part-time workplace of one of the authors. FGs 5 and 6 were convened in December 2013 and comprised year 11 pupils from a local high school. Staff at the school selected students based upon their interest and ability in science and/or their presence on outreach schemes previously run by the University of Sheffield. All participants were aged 15–16 years; both groups comprised a mix of genders.

Materials

Focus group information sheet. Provided details of the research team and sponsor; an outline of what to expect from the research activity; and a very brief introduction to CDU. Participants were told that CDU can make use of the CO_2 emitted from carbon intensive processes like fossil fuel power generation. They were informed that the CO_2 could be used in things like plastic manufacture, meaning that CDU could help to limit atmospheric CO_2 emissions and provide a use for an otherwise 'waste' greenhouse gas.

Pre-discussion questionnaire. Recorded participants' age, gender and occupation; their awareness of CDU and CCS ("*Have you heard of Carbon Capture & Storage/Carbon Dioxide Utilisation?*" Yes/No/Don't Know); their self-reported level of knowledge about CDU and CCS ("*How much do you think you know about …?*" Not a lot/A little/A fair amount/A lot); their attitudes to CDU and CCS ("*Overall, what is your attitude to…?*" 5-point Likert scale: very positive *to* very negative, plus a 'Don't Know' option) and their attitude certainty for both technologies ("*How certain or uncertain are you of your attitude to…?*" 5-point Likert scale: very certain *to* very uncertain, plus a 'Don't know' option).

Pre-discussion presentation. Contextualised the FG discussion by presenting participants with some background information on CDU *via* PowerPoint. This presentation expanded on the information sheet by verbally introducing the research team and outlining the central aims for the focus group (*i.e.* to gather public opinions on CDU and to aid the creation of a video for the CO2Chem Network).

§ Monetary incentives varied by group. All participants age 18+ received a personal monetary incentive. Members of FGs 1 and 2 each received £20 on account of the fact they also took part in a secondary research task following the FG. Members of FGs 3 and 4 each received £5. The high school students did not receive individual payment but the school received a lump-sum of £80 as reimbursement for the students' time.

Table 1 Focus group descriptive characteristics

Grp	Date	Participant profile	No.	Age (years)	Gender	Aware of CDU	Mean CDU knowledge[a]	Aware of CCS	Mean CCS knowledge[a]
1	June 2013	University students & non-academic university staff	8	Mean = 25.6 SD = 7.6 Range = 20–43	3 Female 5 Male	1 Yes 5 No 2 DK	1.00 (0)	3 Yes 3 No 2 DK	1.38 (0.52)
2	June 2013	University students & non-academic university staff	8	Mean = 26.6 SD = 11.6 Range = 19–54	3 Female 5 Male	0 Yes 8 No 0 DK	1.00 (0)	5 Yes 3 No 0 DK	1.88 (0.99)
3	Dec 2013	University students & non-university support workers	7	Mean = 32.4 SD = 13.4 Range = 20–53	4 Female 3 Male	1 Yes 6 No 0 DK	1.14 (0.38)	2 Yes 5 No 0 DK	1.29 (0.49)
4	Dec 2013	University students & academic/non-academic university staff	6	Mean = 26.5 SD = 13.4 Range = 19–53	1 Female 5 Male	1 Yes 4 No 0 DK	1.00 (0)	3 Yes 3 No 0 DK	1.33 (0.52)
5	Dec 2013	High school students (year 11, England)	7	Mean = 15.4 SD = 0.5 Range = 15–16	1 Female 6 Male	2 Yes 5 No 0 DK	1.14 (0.38)	5 Yes 2 No 0 DK	1.57 (0.53)
6	Dec 2013	High school students (year 11, England)	8	Mean = 15.4 SD = 0.5 Range = 15–16	2 Female 6 Male	0 Yes 6 No 2 DK	1.13 (0.35)	5 Yes 2 No 1 DK	1.88 (0.99)
		Totals:	44	Mean = 23.5 SD = 10.8 Range = 15–54	14 Female 30 Male	5 Yes 34 No 4 DK	1.07 (0.26)	23 Yes 18 No 3 DK	1.57 (0.73)

[a] "How much do you think you know about CDU/CCS?" (1 = not a lot; 2 = a little; 3 = a fair amount; 4 = a lot).

Participants were briefly talked through a diagram of the CCS process associated with a coal-fired power station. The CCS concept was used as a counterpoint for introducing two often cited benefits of CDU: (a) the value of CDU in offsetting some of the costs associated with CCS by creating saleable chemical products; and (b) the value of CDU in reducing the current reliance on fossil-fuel derived carbon as a feedstock for these products.

Participants were then shown Fig. 1 and informed of some of the products that CO_2 could be converted to *via* CDU. It was noted that many of the depicted conversion processes would require energy and that this would necessarily have to come from renewable sources to mitigate the release of additional CO_2 during the manufacture of the products. The presentation ended with a slide outlining a protocol for the remainder of the session. This told participants they would first watch and then comment on a video about CDU before being asked to talk more generally about their opinions of CDU.

Informational video about CDU. A short (75 seconds) informational video combining a mix of cartoon animation and cutaways to real life industrial CDU operations. This video was being developed for the CO2Chem Network in order to communicate fundamental details of CDU technology to an interested, lay audience.¶ People watching the video were first introduced to the CO2Chem Network and its purpose in furthering the research and development of CDU. The video then spoke of the relationship between CO_2 emissions and climate change. CCS was mentioned as a way of achieving reductions in CO_2 emissions and the process of separating and storing the CO_2 in geological reservoirs was illustrated. Making use of captured CO_2 to create chemical products *via* CDU was then introduced and framed as a means of offsetting some of the costs associated with

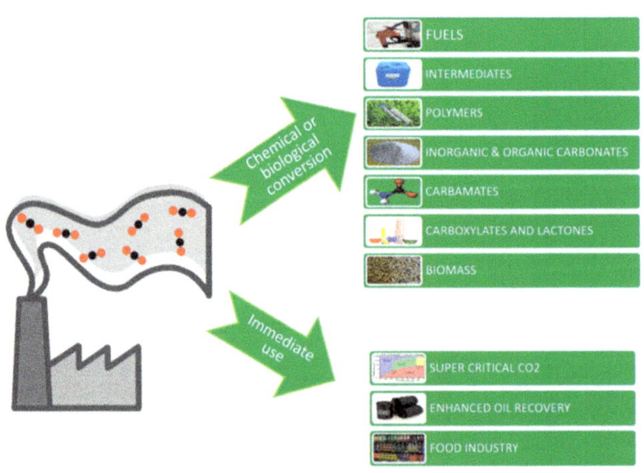

Fig. 1 Some products that CO_2 can be converted to *via* carbon dioxide utilisation (CDU) processes. Source: CO2Chem Network, available at http://www.co2chem.co.uk. Reproduced with the permission of CO2Chem Media and Publishing Limited.

¶ There was a problem with the video in FG6, which meant that it did not run smoothly. This issue was taken into consideration when analysing responses towards the video in this group.

CCS. CDU was also registered as a way to reduce reliance on fossil fuels as a feedstock for producing these chemical products. The video ended by noting that CDU would need energy to produce the chemical products and confirmed that this would necessarily need to come from renewable sources to avoid the release of more CO_2 emissions (note: the video is available to view at: www.co2chem.co.uk/research-clusters/public-perception).

Post-discussion questionnaire. Asked for participants' opinion about 26 risks and benefits of CDU technology ("*To what extent would you agree or disagree with each of the following statements relating to CDU?*" 5-point scale: strongly disagree *to* strongly agree) (see Appendix 1 for a full list of statements); their self-claimed knowledge, attitude and attitude certainty towards CDU (assessed as outlined in pre-discussion questionnaire); their environmental worldview (revised New Ecological Paradigm [NEP] scale);[22] and their 'green' identity (4-item scale).[23]

FGs 1 and 2 completed the questionnaire online 1–2 weeks after the FGs. This was necessary as the questionnaire was partially developed on the basis of their responses within the FGs. The remaining FGs (3–6) completed a paper-pencil version of the questionnaire immediately following the focus group discussion.‖

Procedure

All groups were audio-recorded for later transcription and analysis. Upon arrival participants were provided with refreshments and asked to: (a) read the *information sheet*; (b) provide their written consent for their participation; and (c) complete the *pre-discussion questionnaire*.

The FG then began with participants being invited to first provide their names and occupation in order to acquaint themselves with one another. The *pre-discussion presentation* and *informational video* were then provided and participants were invited to provide feedback on the video – focusing upon both issues of style and content (*e.g.*, how engaging, informative and understandable it was). Discussion about the video lasted approximately 20 minutes, at which point participants re-viewed the video and were invited to provide any final comments. Participants were then asked to discuss their general opinions about CDU and to comment on: (a) any perceived risks and benefits of the technology; (b) the utility of CDU in tackling climate change and; (c) comparative preferences for CDU *vs.* other carbon mitigation options. This discussion lasted approximately 20 minutes and took a semi-structured format.

Having completed the FG discussion, participants spent the last part of the session completing the *post-discussion questionnaire*. They were finally invited to ask any final questions or make any final comments before being debriefed, thanked, paid and dismissed.

Data transcription and analysis

The FG audio-recordings were fully transcribed and analysed using an exploratory thematic analysis approach.[25] All transcripts were first-coded by one of the authors (WS) who was not present during the FGs. Two additional members of the research

‖ Additional questions were included in the post-discussion questionnaire; however, due to small differences in how these questions were asked in FGs 1–2 *versus* FGs 3–6, these data are not reported on further.

team (CJ and DK) then independently second-coded one FG transcript using the coding manual created by WS. All coders then convened to discuss and confirm the emergent themes from the FG and to check the reliability of the initial coding scheme created by WS. Any missed coding or disagreement was discussed, before relevant adaptations were made to the coding manual. CJ and DK then independently analysed a further three FGs before convening a second meeting. Within this meeting any disagreements or missed coding were again discussed, before any final, relevant changes were made to the coding manual. WS then used the revised coding manual to recode (where relevant) all the FG transcripts.

Results

Focus group findings

The thematic analysis[24] of the FG data is presented and discussed in accordance with participants' evaluation of: (1) the style and content of the informational video; and (2) the perceived risks and benefits of CDU. In order to aid interpretation of the comments relating to the video, the analysis is structured according to the issues of source, message and audience.[25]

Informational video

Source factors. Participants noted that it was unclear who the source of the video was. This led to questions about who was behind the video (and CDU more generally) and what their motivation was. The lack of clear authorship, in combination with the perceived "simplistic" nature of the video, negatively affected perceptions of its scientific credibility:

...it definitely wasn't a scientific backed-up video. It could've been an advert for anything. (FG4)

Participants suggested that this issue could be resolved if the video were to include interviews with visible, neutral, expert sources. It was suggested that this would put a face to the technology, which should help to engender more trust in the message content and more generally CDU.

Message factors. Opinions were shaped by the perceived intent of the video (*i.e.* whether it was designed to entertain or inform) and the groups discussed what level of entertainment might be needed in order to keep peoples interest. Participants agreed that more visually and emotionally engaging video content was needed and they criticised the video for being quite rushed, lacking a consistent visual style and for being quite dull.

Participants questioned whether the information in the video contained sufficient detail and clarity of expression to effectively describe the technology, its purpose and how it differs from CCS.

It [the video] doesn't necessarily very well convey the difference between CCS and CDU. I think you need to make clear that CCS proposes to store it [CO_2]; you are proposing to do something else. On reflection I don't think that comes over particularly well or easily. (FG1)

Some participants suggested that the central message behind the video was not apparent and that the explanation provided in the video needed to follow a more logical, narrative structure in order to appropriately engage with the audience.

...actually seeing what the problem is and explaining the problem, and what is the solution that you are looking for, that is the main focus of the research, and that was not very deeply touched upon. (FG4)

Comments were also made about some of the technical language (or jargon) used within the video. The following exchange highlights how a number of scientific terms used within the video promoted confusion and misunderstanding among some of our participants, also leading them to question the viability of the video for a general, lay audience.

P1: *...no-one cares about carbonates, I'm probably one of the only people in the university who cares about them, no one knows what they are.*

P2: *I don't know what synth oil is?*

P3: *It's synthetic oil.*

P2: *If [the video] is for a general audience then ...*

P4: *What does feed-stock mean? When I hear that I think of animals. (Laughter) I don't have a background in chemistry.* (FG2)

The video was also perceived to be lacking a balanced critique of CDU. Participants suggested that the potential risks of CDU were not fully addressed and therefore the video came across as one-sided and as an effort to persuade people to like the technology. This imbalance negatively affected the perceived credibility of the message and led to suspicion as to why CDU was being presented in such a positive light.

P1: *Like you said, there is no debate [about the risks] so you think well 'what are you not saying'.*

P2: *It is just like one sided, they are trying to sell you something.* (FG3)

Audience factors. Participants commented that it was unclear as to who the intended audience was for the video and agreed that establishing this was a high priority for understanding the purpose of the video and determining the appropriateness of the style and message content.

I don't understand the point of the video, or whether it was trying to tell me to take action or to improve something or to go on the website, I don't know what the point was. (FG1)

Participants tended to agree that the video provided a reasonable basic introduction to CDU but that it was lacking in depth and detail if it were to be used for any other purpose than a basic introduction to the concept. This led to a tension among our participants, who desired more detail (to fully engage in the focus group) but recognised that such detail would increase the length and complexity of the video and thus negatively affect audience interest outside of the experimental context.

Having more facts or figures might make your video altogether a bit boring because it really wouldn't make sense to the wider audience who are not involved in the research. A little bit of it [more detail] would definitely help, giving more examples, actually seeing what the problem is and explaining the problem, and what is the solution that you are looking for, that is the main focus of the research, and that was not very deeply touched upon. (FG4)

Participants' age appeared to shape evaluations of the adequacy of the video. While our adult participants tended to feel that the video was too simplistic and lacked seriousness (bearing in mind the seriousness of the issue it was trying to resolve), our high school groups tended to be less critical on these grounds. It was suggested that developing multiple, tailored videos intended for different age groups would be very useful in the future.

I think it [the video style/content] depends on the audience, because you were trying to appeal to everyone by having facts and stuff in as well as the cartoons and the music and stuff, so they should split it up ideally, one for a younger audience and one for older audience. (FG5)

Perceived risks and benefits of CDU

Three principal areas were discussed by participants, relating to the *conceptual issues, technical issues* and *societal issues* associated with CDU. Conceptual issues related to the general underlying principles of the CDU concept and its position relative to other carbon mitigating options (*i.e. should we do this*); technical issues focused on the technological and market feasibility of CDU (*i.e. can we do this*); and societal issues related to the implications that might result from an investment in the technology (*i.e. what are the consequences*).

Conceptual issues. Participants saw CDU to be a technology that would not provide a long term solution to CO_2 emissions but would simply stall an inevitable release of CO_2 into the atmosphere.

...I like it [CDU] because it is doing something, but it shouldn't be seen as a long term fix, because you are not really going anywhere you are just hiding it [CO_2] right? (FG2)

Some examples of CDU were particularly susceptible to this criticism (*e.g.* synthetic fuels) and tended to be negatively evaluated by participants. In contrast, CDU options that implied a longer-term storage of CO_2 option (*e.g.* plastics, concrete) tended to be more positively evaluated.

I think also a lot of what you think about this technology will also depend on its application, [...] if you are getting carbon dioxide from a coal fired power plant and turning that carbon dioxide into polymers that go into plastic, you have created kind of a legitimate carbon sink where it is fixed and it is not going into the atmosphere [...]. But if you are turning it into, somehow managing to turn it into a fossil fuel, that you can use to run on a car, train, whatever, then all the effort that you are going to put into turning that CO_2 into some sort of fuel it is still going to end up as carbon dioxide in the atmosphere. (FG1)

While 'delaying the inevitable release of CO_2' was considered problematic, participants did note the pragmatic value of CDU as a 'stop-gap' technology option (*i.e.* something which could 'buy us time' as we transition to a low-carbon economy) and as something symbolic of efforts being made to combat climate change.

I just feel that it [CDU] is a step in the right direction, providing that [...] if you can do this [capture and use CO_2] and it works then brilliant. (FG3)

There was also a sense that investing in current CDU technologies could also expedite the development of other CDU options that would not suffer as much from the prospect of re-releasing captured carbon (*e.g.* using CO_2 from the air).

I think if there was potential in the future of just not using CO_2 from power plants and just using CO_2 from the atmosphere then I might feel like the power plant one might be a step on the way and maybe that would swing it [the participant's opinion]. (FG1)

CDU was conceptually criticized for presenting an 'end of pipe' solution to the problem of CO_2 emissions; a solution that did not address the root cause of the problem (*i.e.* the activities that were producing CO_2 in the first place). In short, CDU was seen as treating the symptoms of the problem as opposed to the cause.

...they [CDU technologies] are trying to fix something but they are not going to the root of the problem, that there is more cars, more population, more pollution, more everything. So they are trying to fix that but not the actual problem that humans are creating more and more pollution. (FG2)

Participants outlined an array of alternative supply and demand side options that they felt would more appropriately address the CO_2 problem at source (*e.g.* promotion of more sustainable living practices, direct investment in renewables). These points are noteworthy bearing in mind some participants believed CDU to be a barrier to necessary lifestyle changes and questioned why renewable energy was being used in the conversion of CO_2, rather than being used to more directly power the economy (see below).

Technical issues. High investment costs and cheaper alternatives (*e.g.* unmitigated emission) were thought to be an economic obstacle to CDU (particularly in a climate of austerity). Participants questioned as to whether CDU would ever become cost-effective without some kind of market intervention.

...there is also a question of cost-effectiveness. Kind of sticking a chimney up and spewing out CO_2 I imagine is going to be a whole lot cheaper than the capital investment needed to build either a carbon capture and storage facility or kind of a CDU facility. So there would have to be some sort of pricing mechanism in place. (FG1)

The value of CDU was calculated in more than just economic terms. Many participants suggested that they would endorse the economic cost of investment in CDU if there were significant environmental benefits in doing so. However, there was uncertainty about how readily demonstration CDU operations could be scaled-up and what magnitude of environmental benefit would be realised by CDU.

It [CDU] might be significant but we don't know how significant it might be. General logic says that it should be, because CO_2 emissions would increase, we will have more cars, more people, carbon dioxide and utilizing them would help. But I don't know what impact or how much of an impact it could make for the future generations. (FG4)

This uncertainty was related to the fact that participants felt ill-informed about the relative technical and economic feasibility of CDU *vs.* alternatives. Indeed, while participants appeared to have a generally favourable attitude to CDU, this opinion was evidently conditional upon CDU performing well against these other options.

The question is what alternatives are there, because I'm all for 'we'll spend a little bit more if it has benefits' [CDU]. But if we spend a little bit more on this and there is actually something out there that will work better I'd probably rather spend my money on that. (FG3)

Debate of the likely impact of CDU was also tied to perceptions about the timeframes for bringing the technology to market. There was tension between the seemingly long period of time needed to develop CDU into an economically competitive technology option and the urgency of addressing climate change. However, it was recognised that financial investment in CDU would be necessary for it to become economically competitive. Parallels were drawn with the photovoltaic industry, where investment in solar had eventually made it competitive with more traditional energy sources.

P1: *Well that [economic cost] is an argument that they had against early solar but as oil production starts to come lower and lower, prices do go up and eventually the argument could be that if they develop the technology to do this [CDU] then it will become cost effective…*

P2: *By the time that happens it will be probably too late.*

P1: *I don't know; solar got there, solar is cost-efficient now, competitive with oil.* (FG2)

Participants were sceptical about whether CDU would result in a net reduction in CO_2 emissions across the whole lifecycle. The sense was that emissions associated with the energy needed to convert CO_2 into commodity chemicals would undermine any savings resulting from utilisation. Participants drew upon other purportedly 'green' initiatives (*e.g.* early solar) which turned out to emit more CO_2 than they would save to back up this concern.

…we have had too many cons, I think especially some of the early solar panels and things like that when they were so inefficient that […] once you had it in its box it was saving carbon dioxide, but to produce the sucker and especially if you went back to the mines to mine the silicon […] you were causing so much more damage than anything that you were saving. (FG1)

This issue was deemed particularly important when considering CDU for fuel synthesis. For some participants it seemed counter intuitive (and thermodynamically infeasible) to burn a fossil fuel only to then capture the CO_2 produced and expend significant amounts of energy to convert it into another 'fossil fuel'.

Participants' recognition that CDU processes were energy intensive also highlighted the importance to them of using renewables to power the processes. The prospect of using large amounts of renewable energy in CDU, however, led participants to consider whether or not there would be more benefits from just using the renewable energy more directly.

…I like the fact that you show that you use renewable energy to do it. So it is not as if we are going to produce 20 tons of CO_2 to get the energy to use up 1 ton of CO_2. That to me was a crucial message. (FG1)

…if you are using renewable energy to convert carbon dioxide into something else, couldn't you use the renewable energy sources to make energy [electricity]. (FG2)

Societal issues. There was concern that as an 'end of pipe' solution CDU might be used as an excuse for people to continue their environmentally-damaging lifestyles. Participants therefore tended to believe that CDU should only be considered alongside demand-side CO_2 reduction strategies.

…people might sort of think like 'great we can, you know, keep going and use loads of cars and doing this that and the other because we've got all this green stuff now'. It's not quite as it might seem. (FG3)

It was also feared that CDU would propagate a 'business as usual' approach to the use of fossil fuels in powering the economy and it was felt that the technology might create societal complacency towards tackling climate change.

…sometimes these things [CDU] can get used to justify more and more coal power stations, 'ah we can capture, you know, a bit of the CO_2 from them and make a plastic cup' […] if it was like that then it wouldn't be worth it. (FG3)

The belief that CDU might produce ostensibly 'unsustainable products' was also of concern to some participants. Plastics and chemicals, even produced from captured CO_2, were deemed to run counter to a drive to reduce anthropogenic environmental impact. This led some to devalue the products of CDU.

...most of the things that are mentioned [in the video] do look like they have a bit of, they don't look exactly environmentally friendly, things like chemicals, you know people don't look at chemicals and think that is good for the environment. Plastic, cars, fuels are not things that people associate with environmentally friendliness. (FG2)

Finally, there was a sense that there might be unknown chemical risks and localised environmental impacts from CDU processes (*e.g.* acidification of soil or chemical explosions).

There must be dangers involved in like the manipulation of carbon dioxide I would think [...] I'm not sure of the process so... (FG2)

However, in the absence of a full outline of the CDU process, participants felt that they could not comment on these 'capture' risks with certainty. Instead, when considering the risks of CDU, the discussions principally focused on the issue of CO_2 sequestration (*e.g.* CO_2 leakage) as opposed to specific concerns with utilisation *per se*.

Overall evaluation of CDU. Overall, participants appeared to be generally favourable towards CDU. They knew that there were drawbacks but could see value in the idea of trying to recycle CO_2. There was also recognition that with new industry would come new jobs, and it was acknowledged that CDU could produce useful products. However, this positivity was caveated by participants' realisation that they still knew very little about CDU, leaving some requiring more convincing of its value.

The idea of recycling CO_2 sounds like a good idea in theory but I don't know enough about this process at all, to say whether the process is a good idea. (FG4)

I'm more favourable to capture than to utilisation [...] I believe that the CDU, it is a bit bizarre, it is trying to, well you know it is making plastic that... I'm not convinced by CDU basically. (FG4)

Also, participants only appeared willing to entertain the prospect of investing in CDU alongside investment in other mitigation options.

P1: *I think that it [CDU] is good because they are looking at another [option to mitigate climate change], it is just one of the things that they are looking at...*

P2: *Yes, it is good to consider them all.* (FG3)

Quantitative survey findings

Statistical analysis of some of the key questions in the pre- and post-discussion questionnaires was conducted. This analysis focused on identifying participants' attitudes to CDU and the factors underlying these attitudes. The analysis also indicated the presence of any initial pseudo-opinions.

Pre-discussion questionnaire

Pseudo-opinions. Of 44 participants, 5 stated that they had heard of CDU before beginning the FG. The remaining 39 participants stated that they had 'not heard' of CDU ($n = 34$) or that they 'didn't know' ($n = 5$). Congruently, self-reported knowledge of CDU was low, with just 2 participants holding 'a little' knowledge of the technology. Factoring out those who had heard of the technology and/or stated holding 'a little' knowledge of CDU ($n = 6$), we investigated the stated pre-discussion attitudes of the participants. While the majority of these participants stated that they held a neutral attitude ($n = 9$) or that they 'didn't know' what their attitude was towards CDU ($n = 18$); 11 participants registered holding either a fairly ($n = 8$) or very positive ($n = 3$) attitude. We feel that this can be taken as reasonable evidence of these participants (25% of our sample) having registered pseudo-opinions before beginning the study and, as such, as a justification for using FGs within the current research activities.

Post-discussion questionnaire

CDU belief statements. Responses to the 26 belief items were assessed by comparing the mean score for each statement with the scale midpoint (*i.e.* 'neutral') using one-sample *t*-tests. Items where there was a significant deviation from the midpoint were indicative of emerging agreement on the positive or negative attributes of CDU among our participants. Six items showed a significant positive deviation (t values ≥ 3.85, p values < 0.001) from the midpoint, with six showing a significant negative deviation (t values ≥ 3.60, p values ≤ 0.001). Details of these items can be found in Table 2. The remaining items were statistically comparable to the midpoint using a Bonferonni-corrected alpha level of 0.002 (t values ≤ 3.15, p values ≥ 0.003).

The six *positive* items related to three key issues: (1) the value of CDU as an example of efforts being made to combat climate change; (2) the positive delaying potential for CDU in helping to address climate change; and (3) the potential for

Table 2 CDU belief statements showing significant positive or negative deviation from the scale midpoint[a]

	N	Mean	SD
Positive deviation from scale midpoint			
CDU is a step in the right direction for combating climate change	41	3.78	0.85
CDU will help to delay the negative effects of having too much CO_2 in the atmosphere	41	3.59	0.97
CDU will create new employment opportunities	41	4.05	0.77
CDU will produce useful products	43	3.93	0.77
CDU indicates a commitment to tackling climate change	42	3.69	0.90
CDU will 'buy us time' as we aim to tackle climate change	42	3.52	0.86
Negative deviation from scale midpoint			
CDU will promote a 'business as usual' approach to current wasteful lifestyle practices	39	2.56	0.85
CDU will have a limited impact on CO_2 emissions	37	2.35	0.95
CDU should only be considered alongside other technologies for tackling climate change	41	1.81	0.90
CDU will draw funding from other technologies better suited to tackling climate change	33	2.21	0.82
CDU will undermine efforts to promote behaviour change among the general public	40	2.43	1.01
CDU will only delay the inevitable release of CO_2 at high economic cost	38	2.42	0.91

[a] Notes: negatively worded items were reverse coded such that higher scores for all statements reflected a more pro-CDU opinion. All means discount missing data and respondents who answered 'Don't Know' when responding to the item. Significance *vs.* scale midpoint (3.00) using one-sample *t*-tests, calculated using Bonferroni-corrected alpha value of $p = 0.002$. Statement 1 ("CDU will help to slow the negative effects of climate change") was removed from the analysis due to the misspelling of the word *slow* in the surveys distributed to FGs 3–6. A full list of the 26 belief statements can be found in Appendix 1.

CDU to create useful products and employment opportunities. The retained *negative* items also related to three key issues: (1) the potential for CDU to undermine necessary behaviour and/or lifestyle change; (2) the limited impact of CDU on CO_2 emissions; and (3) a concern that investment in CDU might affect other, more preferred, options for addressing climate change.

Post-discussion knowledge, attitudes and attitude certainty. Forty-three participants completed the post-discussion questions relating to their CDU knowledge, attitude and attitude certainty. Self-claimed knowledge of CDU improved markedly from pre-discussion levels, with 41 participants stating that they now knew either 'a little' (n = 24) or 'a fair amount' (n = 17) about the technology after the FG. On the basis of these findings, we can be fairly certain that our participants had developed a basic understanding of CDU.

Overall, post-discussion attitudes towards CDU were fairly positive, with the mean attitude (mean = 3.35, SD = 0.84) differing significantly from the scale midpoint, t (42) = 2.72, p = 0.010. Overall, post-discussion attitude certainty (mean = 3.47, SD = 0.80) was also found to differ from the scale midpoint in an affirmative direction, t (42) = 3.83, $p < 0.001$. This is indicative that participants were on average 'fairly certain' of their opinions about CDU post-discussion.

Post-discussion attitudes, green identity and ecological worldview. With the emerging ambivalence in the perceived 'green credentials' of CDU within our sample (*e.g.* CDU was seen as a delaying solution for climate change but a threat to lifestyle change), we investigated how participants' green identity and ecological worldview related to their post-discussion attitudes towards CDU. Two of the 44 participants were omitted from these analyses as they did not provide useable response data.

Spearman's *rho* correlations (two-tailed, pairwise deletion) confirmed the expected significant positive relationship between participants' green identity (mean = 3.92, SD = 0.61) and NEP (mean = 3.61, SD = 0.48) scores, r (42) = 0.31, $p < 0.045$; and indicated that there was a significant negative relationship between ecological worldview and attitudes (mean = 3.36, SD = 0.85), r (42) = −0.31, p = 0.048. Participants with a stronger pro-ecological worldview tended to hold less favourable attitudes towards CDU. The correlation between green identity and attitude was not statistically significant, r (42) = −0.07, p = 0.665.

Discussion

This study combined focus group (FG) and survey methods to (a) establish more about public perceptions of CDU; and (b) help identify appropriate means of communicating with the lay public about CDU. While there are limitations to the current research design; we believe we have fulfilled both aims and that our findings offer pioneering insight into the emerging nature of public opinion towards CDU. The remainder of this article will seek to summarize the main findings from the study in relation to public engagement and communication efforts before outlining some of the limitations and key future directions for research in this field.

Main research findings

The findings indicate that by the end of the research process our participants had, on average, formed a tentative positive attitude towards CDU. This attitude appeared to stem principally from the 'delaying potential' offered by CDU in combating climate change, its symbolic status as an attempt to address climate change and its potential to generate useful products and employment opportunities. This positivity was, however, firmly caveated by participants' recognition that they knew little about CDU (related to perceived inadequacies in the informational video) and by several conceptual, technical and societal tensions.

Lack of awareness. Before participating, only 5 of 44 participants stated that they had heard of CDU and even then self-claimed knowledge among those 5 participants was low. Despite claiming to have no awareness or little knowledge of CDU, 11 participants (25%) claimed to hold (very) positive attitudes towards the technology. While it should not be assumed that these participants were being deceitful, these data confirm the potential for registering 'pseudo-opinions'[9,11] in the current context and thus arguably justify our choice of a focus group method for our research.

The lack of awareness and knowledge of CDU negatively affected participants' ability and willingness to comment on the perceived risks, benefits and applications of the technology. While evidently posing problems for maintaining fluid FG discussion, we feel that this confirms the opportunity facing CDU proponents at the present time. Specifically, not only is there growing evidence of the benefits of upstream public engagement (if done correctly) in helping to foster the success of emerging technology[5,8] but it is recognised that the optimum time to shape opinion towards new phenomena is when awareness is low and attitudes have yet to form.[18] CDU evidently fulfils these criteria and confirms that now is the time to begin a dialogue with the public about CDU.

Importantly, our results also point to the importance of considering the purpose and adequacy of any planned communication in order to lessen the potential for misunderstanding or misrepresentation of the technology. Indeed, one of the key findings from the FGs related to how our participants evaluated the adequacy of the informational video used as an aide to discussion. While many felt that this video could reasonably act to spark public interest in CDU, they questioned the sufficiency of the information in providing the depth of coverage required to debate the technology in full. In short, the perceived quality of the video was tied to beliefs about its intended purpose (and the intended audience). Some participants were also seen to question *why* they were being asked to discuss CDU at all, which is to say they were unclear as to the purpose of the engagement activity (*e.g.* what implications there would be for their comments). While we did attempt to clarify the purpose of the research activity, we feel that both these comments underline the same issue: the importance of communicating the purpose of engagement activities and careful selection of communication tools. This conclusion is not novel – the importance of identifying and communicating the goals of planned engagement is well-established[26,27] – but we feel that the point is illustrated well in the present context, in that a brief informational video was deemed incongruent with the apparent substantive goals of the FG and hence was more negatively evaluated by participants.[28]

There were a number of other stylistic and content concerns that affected participants' evaluations of the adequacy of the video. Issues of message clarity (*e.g.* words used, structure of narrative) were important and it appeared that trust in the video was undermined by its 'facelessness' and the lack of discussion of risk. These factors led participants to speculate over who would stand to benefit from the technology, what risks had gone unmentioned and whether the video had positive persuasive intent. To the extent that trust is used as a heuristic in guiding decision-making has been found to be important in shaping perceptions of similar technologies (*e.g.* CCS[21]). If the intent of future communications is to provide impartial information so as to allow people to make an informed judgement about CDU technology, then including a fuller description of the anticipated risks and more clearly identifying the source (and beneficiaries) would appear prudent.

Conceptual, technical and social tensions. There was a desire for more information among our participants and it is possible that many of the registered *technical concerns* (*e.g.* issues of technical and economic feasibility; lifecycle CO_2 emissions and energy critique, *etc.*), might have been addressed by the presence of more detail on these matters. Arguably, future correspondence should build upon our findings in order to formally address these concerns and counter the emerging gaps and misperceptions in lay understanding of the technology. Importantly though, while more information is perhaps needed, one should not assume that the simple provision of this information alone will guarantee acceptance of CDU. Not only are there known limitations to interventions centred solely on presumed knowledge deficit[29] but there is evidence within our study that attitudes were shaped by more than a simple lack of technical understanding. Rather, attitudes were also governed by more subjective considerations of the *conceptual* (*e.g.* end of pipe critique) and *societal* (*e.g.* encouraging wasteful life-styles) implications of investing in CDU.

Further research into how these *conceptual* and *societal* concerns might shape perceptions of CDU is a key avenue for future research. Not only will they likely shape public opinion of CDU in their own right but they may also impact upon how any provided technical information is interpreted and used.[30] A particular focus of future research might be placed upon the apparent conflict forming over the pro-environmental credentials of CDU. For instance, while we found that participants with a stronger environmental worldview tended to be less favourable to CDU; it cannot be inferred that more pro-ecological individuals will automatically reject CDU outright. Rather, whilst they might see CDU as making a direct (*e.g.* locking away CO_2) or indirect (*e.g.* raising the profile of CO_2 reduction attempts) contribution to tackling climate change, it is possible that such individuals might show a reluctant acceptance of the technology – akin to that shown in the responses to the recent reframing of nuclear power as a low-carbon energy option.[31,32]

Agnosticism on CDU attributes. While a large number of interesting issues were raised and discussed within the FGs, relatively few were clearly evaluated as positive or negative. Rather, participants remained largely agnostic about many perceived attributes of the technology. These findings are remarkably similar to those from a study by Flynn and colleagues[13] into public perceptions of hydrogen energy technologies (HET) and help to confirm the challenges faced by engaging in upstream discussions about a new technology. We feel that as more information on the relative costs and benefits of CDU becomes available, systematic investigation of how this information affects public agnosticism on some of the identified issues will be important. Thus it should help to clarify whether the tentative positivity seen towards CDU in our study will likely become strengthened and less caveated, or undermined and more negative over time.

Limitations & future directions

While the current research has succeeded in providing some initial insight into public perceptions of CDU technology; when seeking to transfer our research findings to other groups or contexts, one should carefully consider the limitations relating to this study.

Transferability of findings. The present research was conducted on a convenience sample of participants recruited principally *via* a university mailing list. While few of the university participants were engineers and/or pure scientists; the self-selected, well-educated nature of our participants presents limitations to the direct transferability of our research findings. This argues in favour of repeating the research – perhaps with more purposive sampling – on participants from more diverse backgrounds. This should help to identify the extent to which the opinions of our participants are socially shared.

Future research could continue to have a qualitative focus although confirming our findings *via* quantitative methods would also be useful. One option would be to conduct a nationally representative survey of public opinion; however, such activity would need to recognise the issues presented by the low levels of public awareness (*e.g.* the prospect of registering pseudo-opinions). Distributing an Information-Choice Questionnaire (ICQ)[9] could present one solution to this problem and formative efforts to pilot a CDU ICQ have already been made by the current authors.[6]

Framing of materials. The perceived imbalance in the description of CDU present within the informational video (*i.e.* the absence of considerations of risk) indicates that the technology was *positively* framed in this research. While this did not prevent participants engaging in considering potential drawbacks of CDU, it does have implications for the strength of the conclusions that can be drawn. Specifically, studies show that the manner in which information is presented or framed, can exert an impact upon people's decisions and preferences.[33] Thus, one could hypothesise that the positive tone of the video may have yielded more favourable opinions of the technology than would have arisen in a context where the potential drawbacks of CDU were more explicitly considered (or were the focus of the video). While the deliberative nature of the FG context (*i.e.* where both the benefits and risks of CDU were debated) should have lessened the impact of this positive framing in the current context; we contend that a systematic investigation of the impact of purposive framing on comparative preferences for CDU (or different CDU options) presents an important, empirical question for future research.

Conclusions

With a growing recognition of the impact that public opinion can have in shaping the social acceptance (and likely success) of emerging technologies,[4] investing time and appropriate resources for developing public engagement and communication strategies is essential.[5] In the context of CCS, an awareness of the value of public engagement has not only promoted invaluable social scientific research into the factors underlying public perceptions of the technology but has given rise to best practice guidelines designed to inform more effective engagement and education programmes.[18,19]

Consistent with this precedent, the current study has provided formative insight into the beliefs that are likely to underlie emerging public opinion of CDU; helping to shed light on the current low level of awareness of the technology and how this might feed technical misunderstanding and shape perceptions about conceptual fit and societal implications. While we found that participants generally valued the idea of recycling CO_2, this general-level support masked differences in the favourability of different CDU options and was strongly qualified. We feel that now is the time to work with the findings and limitations of the current study to engage in a fuller programme of research in order to investigate how this qualified support of CDU holds up to further scrutiny and which CDU options are most preferred.

Appendices

Appendix 1

The full list of the 26 CDU risk and benefit statements presented to participants in the post-discussion questionnaire (*"To what extent would you agree or disagree with each of the following statements relating to CDU?"* 5-point scale: strongly disagree *to* strongly agree). CDU is/will/should/has: (1) a step in the right direction for combating climate change; (2) help to delay the negative effects of having too much CO_2 in the atmosphere; (3) create new employment opportunities; (4) be good for the environment; (5) be good for the UK economy; (6) a cost-effective way of tackling climate change; (7) promote a 'business as usual' approach to current wasteful lifestyle practices; (8) have a limited impact on CO_2 emissions; (9) only be considered alongside other technologies for tackling climate change; (10) the wrong solution for tackling climate change; (11) produce useful products; (12) be accepted by the general public; (13) indicates a commitment to tackling climate change; (14) be negatively evaluated by the general public; (15) draw funding from other technologies better suited to tackling climate change; (16) undermine efforts to promote behaviour change among the general public; (17) promote an unwelcome continuing use of fossil fuels; (18) only delay the inevitable release of CO_2 at high economic cost; (19) alleviate the storage risks associated with Carbon Capture and Storage (CCS); (20) only delay the inevitable release of CO_2 at high energy cost; (21) a 'green' technology; (22) many unknown risks; (23) more risks than benefits; (24) 'buy us time' as we aim to tackle climate change; (25) not become a commercial reality in my lifetime; (26) help to slow the negative effects of climate change.

Acknowledgements

The authors would like to thank Louise Essam for her help in producing the CDU informational video and the CO2Chem network (http://www.co2chem.co.uk) for funding this research (EPSRC grants EP/H035702/1 and EP/K007947/1). The authors would also like to thank Prof. Peter Styring and Katy Armstrong for their assistance in the formative stages of this study.

Notes and references

1 IPCC, Climate Change 2014: Synthesis Report, Contribution of Working Groups I, II and III to the Fifth Assessment Report of the Intergovernmental Panel on Climate Change, 2014, IPCC, Geneva, Switzerland.

2 P. Styring, D. Jansen, H. de Coninck, H. Reith and K. Armstrong, *Carbon Capture and Utilisation in the Green Economy*, UK, 2011, Centre for Low Carbon Futures.
3 *Carbon Dioxide Utilisation: Closing the Carbon Cycle*, ed. P. Styring, E. A. Quadrelli and K. Armstrong, Elsevier, 2014.
4 R. Wüstenhagen, M. Wolsink, and M. J. Bürer. *Energy Policy*, **35**, 2683.
5 Energy Research Partnership (ERP), Engaging the Public in the Transformation of the Energy System, 2014, available at: http://erpuk.org/wp-content/uploads/2014/12/ERP-Public-Engagement-Report-May-2014.pdf.
6 C. R. Jones, R. L. Radford, K. Armstrong and P. Styring, *J. CO_2 Util.*, 2014, **7**, 51.
7 *Renewable Energy and the Public: from NIMBY to Participation*, ed. P. Devine-Wright, Routledge, 2014.
8 J. Wilsdon and R. Willis, *See-through science: Why public engagement needs to move upstream*, Demos, 2004.
9 M. de Best-Waldhober, D. Daamen and A. Faaij, *Int. J. Greenhouse Gas Control*, 2009, **3**, 322.
10 E. L. Malone, J. J. Dooley and J. A. Bradbury, *Int. J. Greenhouse Gas Control*, 2010, **4**, 419.
11 G. F. Bishop, A. J. Tuchfarber and R. W. Oldendick, *Publ. Opin. Q.*, 1986, **50**, 240.
12 S. Shackley, C. McLachlan and C. Gough, *Clim. Pol.*, 2004, **4**, 377.
13 R. Flynn, P. Bellaby and M. Ricci, The limits of 'upstream' public engagement: citizens' panels and deliberation over hydrogen energy technologies, in *Renewable Energy and the Public: from NIMBY to Participation*, ed. P. Devine-Wright, Routledge, 2014.
14 V. Price and P. Neijens, *Int. J. Publ. Opin. Res.*, 1997, **9**, 336.
15 L. Litosseliti, *Using Focus Groups in Research*, A&C Black, 2003.
16 P. Upham and T. Roberts, *Int. J. Greenhouse Gas Control*, 2011, **5**, 1359.
17 K. Tokushige, K. Akimoto and T. Tomoda, *Int. J. Greenhouse Gas Control*, 2007, **1**, 101.
18 P. Ashworth, N. Boughen, M. Mayhew and F. Millar, *Int. J. Greenhouse Gas Control*, 2010, **4**, 426.
19 P. Ashworth, J. Bradbury, S. Wade, C. Y. Feenstra, S. Greenberg, G. Hund and T. Mikunda, *Int. J. Greenhouse Gas Control*, 2012, **9**, 402.
20 W. Bruine de Bruin, L. A. Mayer and M. G. Morgan, *J. Risk Res.*, 2015, 1, ahead-of-print.
21 B. W. Terwel, E. Ter Mors and D. D. Daamen, *Int. J. Greenhouse Gas Control*, 2012, **9**, 41.
22 R. E. Dunlap, K. D. Van Liere, A. G. Mertig and R. E. Jones, *J. Soc. Issues*, 2000, **56**, 425.
23 L. Whitmarsh and S. O'Neill, *J. Environ. Psychol.*, 2010, **30**, 305.
24 V. Braun and V. Clarke, *Qual. Res. Psychol.*, 2006, **3**, 77.
25 C. I. Hovland, I. L. Janis and H. H. Kelley, *Communication and Persuasion: Psychological Studies of Opinion Change*, Yale University Press, 1953.
26 M. C. Powell and M. Colin, *Sci. Comm.*, 2008, **30**, 126.
27 R. Wooden, *Social Research: An International Quarterly*, 2006, vol. 73, p. 1057.
28 A. Stirling, *Sci. Tech. Hum. Val.*, 2008, **33**, 262.
29 P. Sturgis and N. Allum, *Publ. Understand. Sci.*, 2004, **13**, 55.
30 J. C. Stephens, J. Bielicki and G. M. Rand, *Energy Procedia*, 2006, **1**, 4655.

31 K. Bickerstaff, I. Lorenzoni, N. F. Pidgeon, W. Poortinga and P. Simmons, *Publ. Understand. Sci.*, 2008, **17**, 145.
32 A. Corner, D. Venables, A. Spence, W. Poortinga, C. Demski and N. Pidgeon, *Energy Policy*, 2011, **39**, 4823.
33 D. Chong and J. N. Druckman, *Annu. Rev. Polit. Sci.*, 2007, **10**, 103.

Faraday Discussions

DISCUSSIONS

Wider Impacts: general discussion

Christopher Jones, Katy Armstrong, Ian Ingram, Elizabeth Bay, Jennifer Dodson, Pedro Abrantes, Michael North, Peter Styring, Niall MacDowell, Dina Churina, Katie Lamb, Michele Aresta, George Dowson, Richard Heyn, Elsje Alessandra Quadrelli, Renata Silva, Aryane Marciniak, Niklas Meine, Kusman Dossumov, Alexander Navarrete, André Bardow, Anthony Coogan and Eric Fraga

DOI: 10.1039/C5FD90081F

Michael North opened the discussion of the paper by André Bardow: It is an interesting paper, but you do not discuss the lifetimes of the polymers? One application of polycarbonates is in mattresses and we all sweat at night. Carbonate bonds are more prone to hydrolysis than ether bonds, so would polycarbonate derived polyurethanes be more prone to hydrolysis, which would affect their life time so they wouldn't last as long, and would this affect the life cycle analysis?

André Bardow replied: In our analysis we used a cradle-to-gate approach, *i.e.*, we assumed that the CO_2-based polyethercarbonate polyols have the exact same life cycle from gate-to-grave as conventional polyether polyols. This implies that we assumed that the CO_2-based polymers would last as long as conventional polymers. This assumption was made due to a lack of data on the performance. One point I am aware of that changes from gate-to-grave is the end-of-life treatment; usually, mattresses are burnt to generate heat. Since the heating values are lower for CO_2-based polyethercarbonate polyols than for conventional polyether polyols, less heat is generated.

Michael North said: It will be interesting to see real data on polymer life times when mattresses become available

André Bardow responded: I fully agree. The actual demonstration plant for the CO_2-based polymers is being built at this moment. I hope that we will get the opportunity to include real data on the polymer life time and performance in the future.

Michael North asked: When applying polyurethane foam to building insulation, where its life time would be much longer, what effect would this have on the life cycle analysis?

André Bardow replied: Thank you for giving me the opportunity to discuss the importance of storage duration for the climate impact of carbon utilization. In my view, the effect is very often double-counted. Often, researchers assume a CO_2-footprint of −1 kg $CO_{2,eq}$ per kg CO_2 used. First of all, this value is too optimistic since it would imply that CO_2 from the atmosphere is used directly in the process without any further climate impacts for capture, transport, *etc.*[1] In addition, it should be noted that the climate impact of the feedstock CO_2 already implies the storage effect. This can be easily seen from the analysis of a simple storage process (Fig. 1a). This 'miracle process' is able to store atmospheric CO_2 for infinite time without any other climate-related impacts. The carbon balance on this 'miracle process' is simple: The global warming impact is −1 kg $CO_{2,eq}$ per kg $CO_{2,stored}$. It is noteworthy that there is no additional credit for storage. If there is no additional credit for infinite storage, there can clearly be no additional credit

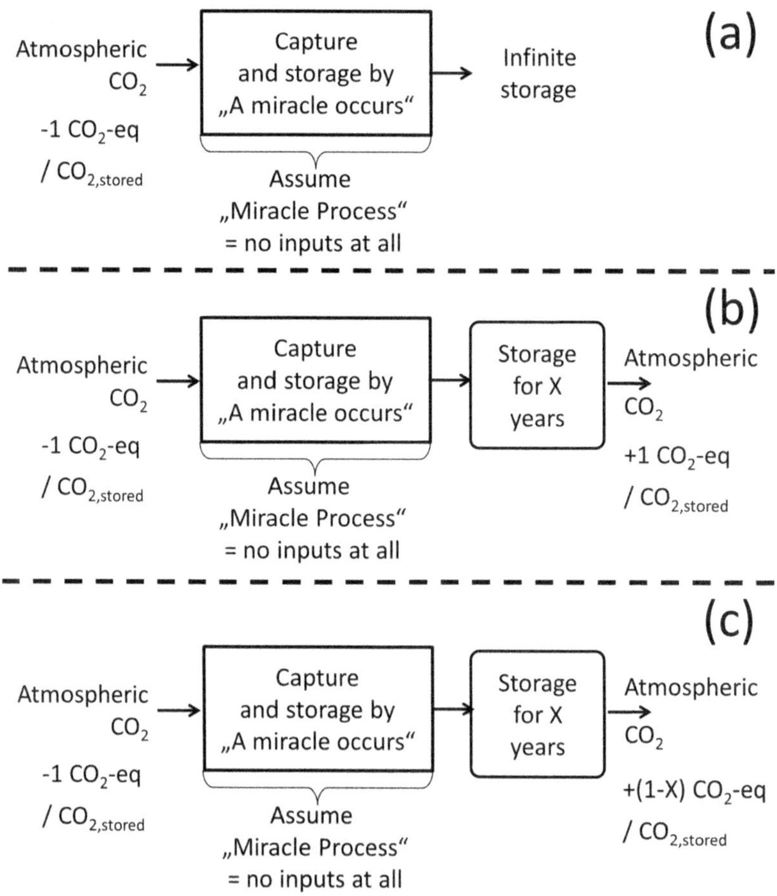

Fig. 1 Outline of the impact of storage duration on the carbon footprint (CF) for a simple CO_2 storage process working by a miracle, *i.e.* requiring no inputs other than atmospheric CO_2. (a) Infinite storage process leading to CF = −1 kg $CO_{2,eq}$ per $CO_{2,stored}$. (b) Process with Storage for *X* years and use of classical global warming potential (GWP) leading to CF = 0, *i.e.* no benefit from storage. (c) Process with Storage for *X* years and use of time-corrected global warming potentials (GWP) leading to CF = −*X*% kg $CO_{2,eq}$ per $CO_{2,stored}$.

for finite storage. This is demonstrated in Fig. 1b. Here, the stored CO_2 is released back to the atmosphere after X years. In this case, the total carbon balance is zero according to classical life-cycle metrics for climate impact. Thus, there is no benefit at all from temporary storage. This view is dominant in the life cycle assessment (LCA) community. In recent years, however, several researchers have promoted time-corrected metrics for climate impacts which account for emissions timing.[1] These metrics quantify the climate impacts within a specified time horizon, *e.g.*, within the next 100 years. A greenhouse gas emission released today is then considered to absorb radiation for the entire duration of 100 years, whereas a delayed greenhouse gas emission released 40 years later should be considered to absorb radiation only for the following 60 years. All emissions over the next 100 years are then related to equivalent emission at time $t = 0$. With such a time-corrected LCA metric, a 'miracle process' with storage for more than 100 years would reduce the climate impacts (within the next 100 years) by -1 $CO_{2,eq}$ per $CO_{2,stored}$. If CO_2 is stored for less than 100 years as shown in Fig. 1c, we have derived a – slightly optimistic – rule-of-thumb that shifting a CO_2 emission by one year decreases the corresponding global warming impact by 1% for a time horizon of 100 years.[1] If the analysis of this 'miracle process' is transferred to carbon utilization processes, it is important to keep in mind that the emissions shifted are only the end-of-life emissions; emissions from raw material extraction and production occur at time $t = 0$ and have thus to be accounted for by 100%. Emissions from raw material extraction and production often dominate the end-of-life emissions leading to small benefits of storage duration even when using a time-corrected global warming impact metric.

1 Niklas von der Assen, Johannes Jung and André Bardow, *Energy Environ. Sci.*, 2013, **6**, 2721–2734.

Ian Ingram enquired: For many years, Ford have been using polyols derived from soybean oil in their foam seat cushions. Have you done an environmental assessment of these or other plant-based polyols? If so, how do they compare to the polyol systems you have presented here?

André Bardow responded: Thank you for this question and the interesting suggestion. I am aware of plant-based polyols but we have not yet compared them to CO_2-based polyols. I believe that such a comparison would be highly interesting. In my view, it would be of particular importance to also look beyond CO_2 emissions to the broader environmental impacts.

Peter Styring commented: Soybean polyols have been used to produce polymers for use as insulating foams in Ford cars. However, when you look at the soybean route, you have to consider land use in any environmental impact studies.

Niall MacDowell asked: I would have expected there to be a delay between the sequestration of CO_2 and accruing a climate benefit. Can you please elaborate on the methodology you employed in this aspect of your work?

André Bardow replied: Thank you for allowing me to clarify the time perspective of global warming impact metrics. The global warming impact in LCA measures the cumulative absorbed radiation of a greenhouse gas emission over a given time horizon, typically 100 years. Thus, the reported values for a climate benefit are integrated over the considered time horizon. It would certainly have been more accurate to state clearly that we consider the climate benefit over the next 100 years.

Niklas Meine said: Can you exclude CCU approaches that don't make sense from a thermodynamic point of view? What are the most promising uses of CO_2 from today's view for the next 10 years? If you distinguish between different CO_2 uses, which of them should be preferably investigated for CCU? Can you also distinguish which are the most feasible CO_2 sources?

André Bardow replied: Thermodynamics allows us to study the lower bounds for the environmental impacts of potential carbon dioxide utilization pathways. Studying these lower bounds would allow us to identify routes which will never make sense. We are currently conducting such an analysis.

The most promising uses of CO_2 for the next 10 years are for me the ones that do not require hydrogen produced from carbon-free electricity. I do not expect that this 'green hydrogen' will be abundantly available soon. Thus, we will have a competition. Hereby, we have to acknowledge the fact that pathways for using renewable electricity *via* CO_2 utilization usually have a severe efficiency penalty when compared to other pathways.[1] Regarding the CO_2 sources, we have analyzed which CO_2 sources are environmentally most preferential.[2] The basic conclusion is that CO_2 purity is the most influential parameter. Thus, we should go for CO_2 sources with the highest purity. This purity effect strongly dominates over impacts from transportation. The other conclusion is that there is and will be sufficient CO_2 available. Thus, CO_2 is usually not the limiting resource. From this viewpoint, asking which CO_2 utilization option is the best one is the wrong question: Any CO_2 utilization option which reduces environmental impacts over the life cycle should be employed. However, we have to be careful if other limited resources are required, such as hydrogen from renewable electricity.

1 André Sternberg and André Bardow, *Energy Environ. Sci.*, 2015, **8**, 389–400.
2 Niklas von der Assen, Leonard Müller, Annette Steingrube, Philip Voll and André Bardow, Selecting CO_2 sources for CO_2 utilization by environmental-merit-order curves, 2015, submitted.

Alexander Navarrete communicated: I am working on what I think is a new family of reactors to locally activate the catalytic materials with electromagnetic fields. There is lots to do yet in terms of optimization and therefore scale up studies will need a couple of years more. Thus, I am curious (and admittedly really ignorant) about any estimate procedure to approximate the loss in efficiency upon scale up of a given process. That would be very useful for me during this early development stage to discard/improve certain process concepts as suggested by you during the conference. Would you suggest any given procedure (or rule-of-thumb) considering the CO_2 conversion framework?

André Bardow communicated in reply: Thank you for your interest in our work. Uncertainties are certainly very high in the early stages of development. However, you can usually state already whether certain aspects will get worse or better during scale up. Thus, you could already use your data as either the upper or lower bound on the process performance to be expected, which you can benchmark against existing processes. In addition, I would recommend process simulations to study the full flowsheet to identify potential bottlenecks in downstream processing. First process simulations can usually be based on very little information (ideal thermodynamics, equilibrium reactions, *etc.*) but can already provide valuable insight.

Michele Aresta asked: When one starts an LCA of a new process, one has serious problems with the availability of data. If one has process data, also at the pilot scale, one can work with much more reliable figures. Have you had the possibility of comparing database data with pilot process data for any of the processes you have investigated? Can you comment on how they differ?

André Bardow replied: Data availability for new processes is indeed an enormous problem. In my experience, values on raw material use can often be estimated quite well; the actual energy use in reactions and separations are more difficult to predict and are often even site-specific, *e.g.*, due to heat integration. In our experience, early values are often too optimistic. In particular, product separation is often even neglected but can have a significant impact.

Peter Styring opened a general discussion of the paper by Eric Fraga: One of your metrics state that TOF is a constant. TOF is not a constant and depends on many variables including whether the catalyst is used in a batch or continuous process. Turnover number (TON) again is not a constant, and its value is also difficult to predict. So how do you account for inclusion of these non-constants in the metrics?

Eric Fraga answered: Yes, these values are not constant and this is recognised in the paper: "This assumes an equal turnover frequency for all catalysts which is not typically the case...However, the measure should be appropriate as an indication of affordability." The TOF was used as a proxy for the affordability of the catalyst and this factor could and should be improved using often confidential data.

Peter Styring asked: When considering availability, you rank catalysts based on crustal abundance of the metals. Where did you get the data from? There are some inconsistencies from a number of sources. A good starting point would be the RSC Periodic Table App that gives supply risk on a decimal scale.

Eric Fraga replied: The data were obtained from a number of sources, primarily the RSC periodic table on the web (http://www.rsc.org/periodic-table/element/13/aluminium) and the British Geological Survey minerals statistics site (http://www.bgs.ac.uk/products/minerals/statistics.html). The RSC site does have its own measure of supply risk which may not be consistent with our measure.

Elsje Alessandra Quadrelli remarked: You present a framework helping asses the parameters influencing the security of supply (of catalyst) on the overall deployment of CO_2 technologies as a chemical feedstock. Just to focus on catalyst supply (and keeping aside as you rightly mention, that it is a multi-criteria problem, where catalyst supply is only one of the criteria coming into play): how does your system take into consideration the fact that substantial efforts are being devoted to the development of an alternate catalyst, shunting most of the critical points you bring into focus in your present study? For example, development of an iron based catalyst, that will most likely be cheaper, easier to access and overall safer catalysts than the current precious metal ones.

Eric Fraga responded: The framework has two main objectives: to enable the comparison of alternatives and to motivate research into the generation of better alternatives. If a more available catalyst, together with an appropriate route, would make the overall assessment better, then this would be a good choice. In the example you suggest, an iron based catalyst would improve the scores in all the A factors we have included in the framework.

Jennifer Dodson asked: Regarding the discussion on catalyst availability, there has been a large number of studies, including some by the UN and the EU on the issue of critical elements, which have examined a wide range of factors affecting elemental availability. It would be useful to draw on this for your work.

Eric Fraga answered: Thank you for bringing this to our attention. The availability factor in the 4As framework currently takes into account a number of factors, including abundance, production concentration, reserve concentration and geo-political stability. It would straightforward to either add the factors identified by the UN and the EU to these, or to enhance existing factors with any data available from these organisations.

Dina Churina communicated: 1) What is the stability of the catalyst for utilising CO_2 as a chemical feedstock? 2) What is the cost of this catalyst?

Eric Fraga communicated in reply: We did not include the cost of the catalyst as this would then have required bringing in the process cost as well to provide a value for the affordability factor. To avoid this, we used the turnover frequency. In practice, however, it would be better to use full costing and this is something a commercial organisation would be better placed to do, having access to their own internal confidential cost models. The stability of the catalyst will typically be reflected in the cost of the process, requiring either fresh supplies of catalyst or process steps for regenerating the catalyst, as appropriate.

Pedro Abrantes asked: In technological start-up matters, I think there are two things very important to keep in mind. The chance to have a great project is very small, so when you work on a project, remember that investments are made if your product, service or process is simple and cheap. If your solution is too complex to remove the existing one, or if it's very expensive, it makes your work inefficient at becoming a great solution.

Eric Fraga replied: This is true. Simplicity in design is a key criterion in the selection when alternatives are available. In terms of the framework we have proposed, this feature of a design of a process will come into the applicability measure. A simpler process design will move up the TRL scale more easily and quickly.

Michael North commented: I found your paper confusing to read – you give data for products, but it is not always clear what production route this refers too. Table 3 shows some data still in a research stage, that are actually in practice.

Eric Fraga responded: The paper did not attempt to be exhaustive in the processes that are available for the products considered. The examples were chosen to illustrate the framework. The process routes actually used in the results summarised are presented in each product description.

Regarding Table 3, the boundaries between the different categories of maturity are by necessity fuzzy. Nevertheless, the aim of this table is to summarise the majority of the effort for each product domain so it is indeed possible that there are examples of processes beyond the research stage for some products, while most of the activity is still in the research stage.

For the framework, what is important is the TRL assessment for the individual processes considered, not the categorisation done in Table 3.

George Dowson commented: While your affordability analysis considers catalyst performance in its inclusion of a catalyst to revenue ratio factor, the availability analysis does not. I think this omits a key consideration of the sustainability of rarer earth catalysts and risks encouraging premature blanket bans on the use of certain elements.

Crustal abundance is indeed very important for calculating whether a rare earth catalyst and its associated element are suitable, but the performance, activity, recoverability and recyclability of a catalyst must also be considered. This is often process-specific, for example a solid-state precious metal catalyst that is used to generate a gas-phase product is going to be entirely recoverable and desirable, unlike a precious polymerisation catalyst that will likely be unrecoverable.

Using a tiny amount of a recoverable rare earth catalyst which is selective towards the desired product will surely be more desirable than using much larger quantities of a much less selective earth-abundant catalyst.

Eric Fraga responded: The point is not to look at only one attribute to the exclusion of the others. The visualisation we propose enables one to look at the whole combination of factors. If only availability were to be considered, in isolation, it is true that decisions would be taken that would lead to the scenario you propose. This would be a mistake. It is important, in considering complex systems, that all factors be considered simultaneously. As mentioned previously, there is always a desire to combine factors into a single economic value for assessment. If this can be done to the satisfaction of the decision maker, this is fine; if it cannot, then a multi-criteria decision making support system is helpful. The framework we propose, along with the visualisation, aims to provide this decision support.

Katie Lamb communicated to André Bardow and Eric Fraga: I want to query something about metal availability in these studies. I understand that we must take care when choosing metals for our processes in terms of their abundance, so that we do not deplete our entire resources of precious metals entirely. I was however wondering if it would be possible to consider factors such as recyclability or mass required for some processes? Many electrochemical methods for example use platinum based electrodes, and although platinum is a precious metal, it can be used over and over again because is acts as an inert electrode. If some highly efficient catalytic process also only required a small catalytic loading of a precious metal, which could be used over and over again, we should surely utilise this process? I therefore was wondering what your opinions are on this matter?

André Bardow communicated in reply: An LCA study would account for recyclability; simply stated, less fresh catalyst would be needed per product. Thus, all catalyst-related environmental impacts would be reduced accordingly. One would expect a trade-off between the efficiency gain from using the catalyst *versus* the additional environmental impact from catalyst production. This trade-off is case dependent and can be different for different environmental impact categories (*e.g.*, climate impact *versus* metal depletion). Thus, there is no simple answer.

Eric Fraga communicated in response: Decision support is about providing the information decision makers need. Ideally, when comparing alternatives, the less factors that need to be considered, the easier the decision making process becomes. For this reason, we often try to bring everything down to economics and the most economically efficient alternative will win out, although even here the modelling of economics will vary from organisation to organisation (*e.g.* assessments based on return on investment, net present value, revenue, profit, *etc.*). The catalyst and its properties with respect to the process in which it is used is a key element. The stability and recyclability of the catalyst will affect the economics.

However, in the absence of costing information, the framework we propose still enables a comparison of alternatives. The different A factors address aspects which can be looked at simultaneously, noting that there is overlap between them.

Katie Lamb communicated to Christopher Jones, André Bardow and Eric Fraga: I noticed in this discussion that each of the speakers used different data sources, when conducting research into LCA factors and values such as metal availability, affordability and acceptability, and members of the audience have highlighted that there are even more data sources available. Not only has the diverse range of data sources lead to some disagreements over conclusions reached in this discussion, but it has also caused the speakers to reach different conclusions in their assessments. Do you therefore think there is an urgent need for some type of overall agreed global database of some sort, when people are considering data for their LCA models or assessing different chemical processes, or do you think this would not be possible?

Christopher Jones communicated in reply: My opinion would be that having an overall database with agreed 'impact' figures for different processes, *etc.*, would be a very valuable resource. From the perspective of my own research, I think that having access to such data would be invaluable in helping to communicate with lay peoples about the relative pros and cons of different CDU options. This could be helpful in understanding which features of CDU options (*e.g.* energy requirements, carbon use potential) have the most impact on people's perceptions.

André Bardow communicated in response: I agree that the acceptance of LCA is suffering from the fact that different databases report different values and that there are no globally accepted values. I therefore welcome initiatives like the EU directives which specify how to apply LCA in a certain area. I would be glad to contribute to such an effort for carbon dioxide utilization.

Currently, I am very careful when using absolute numbers on environmental impacts. In my experience, LCA study results are much more robust for relative impacts when comparing technologies. In our practice, we commonly employ several databases to test our conclusions for robustness.

Eric Fraga communicated in reply: Yes, an agreement on the data sources that should be used would be welcome. However, in a commercial organisation, the sources that are most appropriate will already be in use. We are proposing a framework which can be applied with any data source the user wishes and so is agnostic with respect to the data sources. The results obtained will, of course, depend directly on these choices.

Pedro Abrantes opened a general discussion of the paper by Christopher Jones: For further comprehension and appreciation of your first results, can you present us your work level, and how many people did you contact? Do these people see CO_2 as a waste or a raw material?

Christopher Jones answered: Our study presents a formative, qualitative investigation of emerging public opinions employing focus group methodology. Focus groups are commonly used in exploratory work as they are valuable for providing insights into how people think about things and help to determine the range of responses you are likely to receive in response to discussions about a particular topic or issue. Focus group studies typically comprise between 4 and 6 groups, with each group containing 6–10 participants. In the present study we interviewed a total of 44 participants spread across 6 focus groups. In terms of whether people see CO_2 as a 'waste' or a 'raw material', well, I think that people principally think of CO_2 as a negative waste product of industrial processes. Indeed, the prospect of utilising CO_2 as a raw material was seen as a novel concept to most of the people we spoke to. Importantly, most participants did value the prospect of making use of CO_2 rather than just simply disposing of it through CCS. That said, to some extent participants evaluations of CCS *vs.* CDU depended upon the overarching frame used to evaluate these technologies. Specifically, when thinking about CDU/CCS in terms of mitigating climate change, most of our participants appeared to value CCS over CDU, except those CDU options that ostensibly could capture and store CO_2 in the long term (*e.g.* cement). This stemmed from a belief that CDU generally only delays the inevitable release of CO_2 into the atmosphere at high cost and so does not really offer a solution to the climate change problem. Conversely, if framing CO_2 as a raw material, then CDU options tended to be seen as preferable to CCS. This is logical, inasmuch as we'd be foolish to simply dispose of something we have spent time, energy and money capturing if it could be used as a feedstock for the things we need. My sense is that climate change, at least among those that we spoke to, remains a primary concern and so the perceived limitations of CDU in this regard is something that needs addressing if meaningful public support for the technology is to be garnered. That said, systematic research into the impacts of different framing on people's preferences for CDU is something that still requires research.

Aryane Marciniak communicated: It is very important to know about the public perceptions of CDU and CCS. In your study, you interview people from university (students and non-academic university staff), and also students from a local high school, with different ages. I would like to know if you thought to do the same interview but in others schools or with people from different social groups? Because in my country (Brazil) we have some similar studies but in different schools and lay public in general and the results are very interesting.

Christopher Jones communicated in reply: Thank you for your interest in our research. I think the short answer to your question is 'yes'; we would be keen to look at emerging public opinion in different social groups. I agree that we might get different and interesting viewpoints if we were to speak with different groups and we indicate that the use of a more purposive (rather than convenience) sampling method in future research might be a route to achieving this. Also, I think that moving beyond the focus group research method we employed towards the development of an online Information Choice Questionnaire (ICQ) could be a key means of reaching more people. This could provide a route to getting more representative and general opinions of CDU within a given population, and testing the relative importance of the risks and benefits highlighted in our groups in shaping people's attitudes.

Ian Ingram remarked: I was recently involved in running an RSC funded public dialogues project (Entitled "Green Reactions") which focused on sustainable technologies, and I was quite surprised to encounter some members of the public with an attitude of "environmental absolutism". At events where we were talking about cleaner and biomass-derived plastics, fuels and lubricants, members of the audience sometimes expressed the strong opinion that we ought to just stop using cars, plastics *etc.* altogether instead of developing cleaner technologies. Does this resonate with the sorts of opinions you encountered in your research?

Christopher Jones responded: Yes, although perhaps not to the same extent that you are insinuating from your own research. We did find evidence of people questioning whether or not CDU might be committing us to a 'business and usual' future and whether it might be seen to undermine efforts to stimulate appropriate behaviour change. For these reasons people were cautious about supporting the concept. Also, some people viewed some of the potential products from CDU processes as being perceptively 'unsustainable' (*e.g.* plastic bottles, transport fuel) and argued that this conflicted with the current, necessary societal zeitgeist of edging closer to sustainability. We also found some negative correlations between 'green identity' and 'environmental world views' which suggested that people with stronger environmental values tended to be more skeptical of the technology. To some extent this makes sense as CDU is perceptively viewed as a technological solution to a behavioural problem (of over consumption) by some. These individuals would likely prefer efforts to promote behaviour change, efficiency savings and direct use of renewable energy over investment in CDU and we have found some evidence for this. I think that this makes an important point regarding the promotion of the technology (if that is what one wishes to do). It is evident that many people question the pro-environmental credentials of the technology and so perhaps marketing on such grounds might fall short of convincing people that CDU is something that is required. Maybe the use value of CDU in creating a perpetual feedstock for the things we need and use everyday might be a better tack.

Peter Styring enquired: It's a small sample you've used. One way to increase the sample size would be to do an online survey, using for example Survey Monkey. How would you handle the trolls in this environment? I am thinking of the people who say the obvious things like you should just ride a bike, irrelevant of the circumstances, or use the forum to perpetuate uninformed views.

Christopher Jones replied: The reason we have utilised a focus group method in this study is to counteract (to some extent) the potential for registering so-called 'pseudo-opinions' in the research. There are, of course, limitations to this method (*e.g.* it tends not to produce findings that can be easily generalised) and thus we are considering the next steps for the research. One such 'next step' could be to take the research online in the form of an Information Choice Questionnaire (ICQ). Your point about the potential for 'trolling' is something that we should consider. To some extent you could argue that the so-called 'trolls' are a group of the public that we should be hearing from in such research. Such vehement opponents of a technology can be quite an important counterweight to runaway optimism. Also, it is likely that these opponents will be the most vocal (*e.g.* attending public consultations, signing petitions) when it comes to deploying instances of the technology, and so it is important to identify what their concerns might be, such that they can be adequately addressed (if possible). It will often be possible to categorize strong opponents (like strong supporters) by their responses within a survey. As such, it should be possible to control for their influence in any analysis that it performed on the data. Indeed, there are other important trends in responding that should be looked for when seeking to analyse survey data conducted at a distance, *e.g.* people simply rushing through the survey (*e.g.* by ticking only the middle option on everything) in order to make themselves a viable candidate for any proposed incentive (*e.g.* voucher, cash payment); or those who do not answer a lot of questions. Thankfully, modern survey software and data analysis tools allow us to easily spot and adjust for such individuals in our analysis.

One group that pose a different issue than the so-called 'trolls' that you mention, are those who seek to respond to questions in a 'socially desirable' way (*i.e.* in accordance with perceived social norms). These individuals might mask their 'true' feelings for things behind more socially palatable responses, so that they don't get negatively evaluated by others for the opinions they hold. This can be an issue with survey data, although this issue is somewhat lessened by the anonymity offered by self-completion surveys. Interestingly, we have scales that we can introduce into studies which are there to identify the extent to which people engage in socially desirable responding, such that this tendency can be controlled for in any analysis that is performed. This could be important in our research as CDU has tended to be framed in terms of 'climate change mitigation'. To the extent that it is generally agreed that climate change is a bad thing, this might lead people to say that they favour investment in CDU (as it is socially right to say this), without necessarily believing this to be true.

Renata Silva communicated: Do you think that 44 participants would be enough to represent public opinion about such a new subject and so many differing opinions, as is the CDU?

Christopher Jones communicated in reply: Your question would appear to relate to the issue of generalizability (*i.e.* can the findings from our study be generalized to the wider population)? The short answer to your question is 'No'; a convenience sample of 44 people is not likely to be representative of the UK population as a whole and therefore it would be ill-advised to automatically assume that the opinions of our participants will necessarily be shared by the average Briton. While I do think that many of the perceived risks and benefits discussed within our groups will be indicative of how other lay people's might think about CDU, it is important to think about our rationale for using focus group methods in this research (stated within our article) and what you can and cannot conclude based upon the findings of such research (outlined below).

Focus groups tend to be used in exploratory work where you are generating hypotheses and identifying concepts of interest through discussion, as opposed to quantifying the extent to which a particular viewpoint or concern is shared within a population. Focus groups tend to look at the 'why' questions, as opposed to looking at how many people believe X or Y about something. As such, the results of focus groups are rarely generalizable or representative but are more indicative and illustrative of the kinds of opinions that can be raised when discussing a social phenomenon.

As such, when we are talking about how to use the findings of data from focus groups we tend to talk about the 'transferability of findings' as opposed to the 'generalizability of findings'. What does this mean? Well, generalization can be considered to be an automatic process. To the extent that as you have accrued a representative sample of your chosen population you can essentially automatically assume that the opinions shared within your sample will be reflective of the population at large. These kinds of generalizable data will be the data you get from things like national surveys. When using the term 'transferability' we are talking about more of an active process. That is, to the extent that a new group of people (Group Y) share similar characteristics to your sample (Group X), you can reasonably hypothesize that their should be an overlap in their opinions. However, it is incumbent on you as the researcher to actively evaluate whether and to what extent you can/should transfer the findings from one group to another. That said, if you had a representative sample of the population within your focus groups, then generalizability becomes a possibility; however, getting a representative sample is often very difficult to achieve.

There is a good quote from Krueger[1] that sums up focus group research: *"...the intent of focus groups is not to infer but to understand, not to generalize but to determine the range, not to make statements about the population but to provide insights about how people perceive a situation"* At the moment we are in the process of using focus groups to develop the kind of insight Krueger talks about. The next step for us is to think about piecing together a survey (based on an Information Choice Questionnaire [ICQ] design) to identify the extent to which opinions are shared within more representative samples.

R. A. Krueger, *Focus groups: a practical guide for applied research*, 1994, Sage Publications.

Michael North asked: Given that the video was not well received, would it help to have props as well *i.e.* a mattress or a rubber duck *etc.* to show people some 'normal' examples?

Christopher Jones responded: Yes, I think that props of this kind would be helpful in illustrating what is an abstract and 'hard to believe' concept for some. There a examples of educational programmes relating to carbon capture and storage that seek to illustrate on a desktop scale the process of storing captured CO_2. These educational tools help to counter misperceptions and uncertainties, as well as providing a useful talking point for discussion. Of course, one should not assume that making the process more real through the use of props will guarantee acceptance of CDU (or individual technology/product options). As we have shown in our paper, even if people are convinced that the process works, it does not necessarily mean that they will think that we should be investing in this technology. However, I think that having tools to illustrate the process is a good start in helping people to understand the potential for the technology, perhaps allowing for a more considered discussion of the relative pros and cons.

Katy Armstrong stated: The video was created as an infomercial as part of a CO_2Chem summer student project. It was always intended to be a very short 90 s introduction to what CO_2 utilisation is; introducing viewers to the idea that CO_2 could be used to produce useful products. As such, the comments from the focus groups are unsurprising, the video is promotional in essence, it is short, very positive and does not raise issues surrounding the implementation of these technologies. It was intended to raise awareness that CO_2 could be useful in the chemical industry, not provide a more detailed explanation of how CDU technologies would fit within the industrial landscape and renewable energy sector. What we have learnt is that it is complex to explain CDU and that people would prefer a longer more rounded discussion in a video. As such we will be looking to make new longer films which explain CDU more fully.

Christopher Jones commented: I think that in retrospect it might have been preferable within our study to have formally developed a more balanced and slightly longer video to act as a stimulus for discussion; however, we were to some extent constrained by what was available to us as the time. As I mentioned previously, I think that the video we used did do its job in communicating the basics of CDU to our participants and did stimulate productive discussion. The key thing now is to look forward and work on producing an more in-depth video that will hopefully serve to test whether the tentative acceptance for CDU we have found in our work to date holds up to further scrutiny.

It was evident from the findings presented within our article and some of the points made in the discussion that the infomercial used had a number of flaws (*e.g.* the source, message, audience and purpose of the video was deemed to be unclear by our participants). That said, the video did still serve its purpose in getting participants to debate openly the potential risks and benefits of the concept (presented within the article). The purported issues with the video perhaps reflect a discrepancy in the 'depth' and 'content' of the video relative to the perceived purpose of our research. Specifically, participants saw the brevity of the video as being okay if the video was intended as an 'advert', but somewhat lacking if the purpose was for them to provide a more substantive commentary on the pros and cons of CDU. Going forward I would be keen to piece together a longer video (or series of videos) that sum up the CDU process in more depth and detail – akin to those which have been developed in order to inform people about CCS – which would be more consistent with the research angle that I am pursuing. If people are interested in assisting me with this, then I would encourage you to contact me.

Michael North asked: What is the ideal length for a video to provide sufficient information to inform people, but not so long as to bore them?

Christopher Jones responded: The ideal length for a informational video about CDU will vary depending on what the purpose of the video is and the intended audience. Within our research we were utilising a short, 75 s infomerical about CDU as a stimulus for discussion. While our participants tended to think that the video was okay (there were of course many reservations) in terms of 'advertising' CDU to a lay population, they felt that they needed more information if the purpose of the video was to allow them to consider the pros and cons of CDU in real depth. As such, I think that producing a more detailed and more balanced (set of) video(s) would be useful as we take our research forward. Exactly what length these should be is an interesting question. In some work that I did a few years ago around public perception of CCS in North America, we utilised a couple of videos produced by the National Energy Technology Laboratory (NETL) that were available on YouTube. These were approximately 5–8 min long and did a good job of stimulating in depth discussion of the benefits and risks of CCS in affected communities (*i.e.* those facing the prospect of CCS in their communities). Producing something similar for CDU might be the next step and I would be keen on speaking to people who might be interested in helping to create these.

Michele Aresta said to Christopher Jones: You and your coworkers are doing quite unique research. As you have pointed out, it is essential that both the public is correctly informed on the fundamentals (I am not saying that they have to be specialists), and the interviewers have the correct sensibility. Therefore, much work must be done in both producing some information (printed and visual) and in bringing it to the public. It is too risky to reach the wrong conclusions, in whatever direction, pro or con.

You are doing an excellent job considering all the factors and difficulties and I am sure that all of us, starting from myself, will be willing to collaborate for the success of your initiative.

Anthony Coogan commented: I am impressed by the way your group takes on so many challenges. It is clear that politicians, financiers, the general public *etc.* must be informed of the major issues for the successful implementation of new technologies. The importance of educating children is essential too, not merely because of their potential, but because any misconceptions that they may have may be typical of those of many adults.

Your video introduced CCS and led on to CDU. Some may think that this new area may replace CCS. A video,[1] which is available for download, featuring Prof. Mike Stephenson of the British Geological Survey, states that 'The rocks under the North Sea could absorb about 22 billion tonnes of CO_2 which is 180 years of the UK's twenty largest point sources. This is a really hefty reduction in emission'. With respect, it seems that even if CDU is as successful as we hope, it can only make a small dent on the problem. Another CSS issue: If a CDU project fails to make a significant profit, it may be economically viable for CSS projects to subsidise it.

In your future videos, global warming issues may be raised. Methane is likely to cause more severe ecological problems in the future than carbon dioxide. In a video[2,3] by Dr. Peter Cox of Exeter University, he states that it may be better to tackle methane emissions rather than carbon dioxide, and still operate within the restrictions set by the Kyoto Protocol.

Other issues: Jason Cox made videos[4,5] about the melting ice in Greenland and the Arctic. This problem is of worldwide importance and cannot be understated but the general public may not understand how CDU can help.[6] Can your future video(s) convince the public at large that global warming is truly global, and that CDU is a small but important part? Hopefully yes, and although I am retired, I am still hopeful for society and wish I could help.

Is there any consensus of a time frame for tackling these problems? The Kyoto Protocol specifies 2030 as a deadline, but Arctic problems may become critical long before then. People, especially children, are likely to ask about the likely times that problems will arise. It will be difficult for you to balance the issues: people must be informed of the other issues,[7,8] without underrating their significance, yet you still need to promote the vital nature of CDR for future generations.

1 "What is carbon capture and storage CCS 2013", https://youtu.be/oMkTHFqUUo8.
2 "Arctic Methane Catastrophe from The Eco-Logic Channel", https://youtu.be/MVwmi7HCmSI.
3 "Climate change: thinking outside the low carbon box: Peter Cox at TEDxExeter", https://youtu.be/V6gAJXcEx5o.
4 "Greenland Ice Sheet: Starting to Slip", https://youtu.be/qkpFNteryX8.
5 "Jason Box at MAD3: Dark Snow", https://youtu.be/khs-2Yr0vtk.
6 "Climate change will be catastrophic for mankind after study reveals methane leaking from the Arctic Ocean, scientist warns", Tuesday June 16th 2015, Mail Online.
7 "How does Carbon Capture & Storage work?", https://youtu.be/lyEt3lGQVWw.
8 "Climate Change is a real problem. NewGenCoal is all about solutions", https://youtu.be/lyEt3lGQVWw.

Elizabeth Bay asked: When trying to look at people's attitudes and the information they have at hand you want to see how that impacts the decisions they make. Therefore do you find that attitudes have a larger impact on decision making than economics?

Christopher Jones answered: There is no denying that financial cost is a prominent driver of consumer decision-making, however, there are a number of complexities to your question that need addressing. First, this is a difficult question to answer inasmuch as a person's attitudes are likely to shape, to some extent, their perception as to whether or not an advertised price is deemed 'too costly'. That is, value is partly subjectively determined. Also, in some situations any attitudinal influence will be trumped by circumstance. By this I mean that if someone actually cannot physically afford to buy something (*e.g.* an electric car) then whether or not they have a positive attitude towards doing so is a moot point. I believe that the thrust of your question, though, relates to circumstances where a sustainable option is in the price bracket of a consumer but is perhaps at a higher price point than a less sustainable option. In such circumstances will people take the financial hit in order to act in accordance with their pro-environmental attitudes or will they just opt for the cheaper of the two options? I guess the answer to this question is, unfortunately, 'it depends'. Certain individuals (*e.g.* those with particularly strong pro-environmental values) will likely opt for the more costly sustainable option on the grounds that this is consistent with their world view and sense of identity. For those who are less motivated to be pro-environmental, they will probably just opt for the cheaper option. But let's say for the sake of argument that we are speaking about someone who values the idea of being pro-environmental but also has a number of competing (and equally desirable) demands on their finances; how would they respond? Well, there is certainly evidence to suggest that while cost is likely to remain at the forefront of peoples thinking in retail situations (people like a good deal), there is also evidence that people will factor in normative considerations (what is appropriate in this situation) and hedonic considerations (what will make me feel good) when making purchase decisions. To the extent that our individual is likely to recognise that being pro-environmental (*i.e.* buying sustainable) is socially desirable, this may make them likely to consider forgoing the absolutely cheapest option and spending a little bit more on their purchase. However, this will likely be reliant on the price differential being fairly narrow and on the sustainable product being similarly desirable to the less sustainable one. In sum, economics are a very important consideration when looking at consumer behaviour and will exert a large influence on decision making, however, they are not the only predictor of behaviour, and a person's attitudes (amongst other psychological constructs) will also exert an influence on these decisions.

Kusman Dossumov communicated: Will chemical or biological methods fundamentally solve the problem of CO_2 utilization in terms of social acceptability, or you can offer other more effective methods?

Christopher Jones communicated in reply: At present we just don't know. Social acceptance is an area that does need more attention at the present time and drilling into these more specific questions would be worthwhile. My sense is that some individuals will have a preference for biological or chemical options, while others will see both as harbouring certain risks. It might come down to the phrasing used and what existing knowledge people go to when trying to gain information about these unfamiliar technologies. For example, some lay people might – depending on phrasing – associate terms like biological or chemical with dangerous things like warfare, while others might think of more benign analogues like washing powder. Evidently CDU is different to these subjective analogues but these analogues might affect preferences for different CDU options or the concept more generally.

Michele Aresta asked: You have been working for a number of years on the topic of public acceptance. Have you noticed any change of attitude in this time? Or is it too short a period for analysis of this kind?

Christopher Jones replied: I think the short answer is 'No'. I have not seen any real attitude change in the time that I have been conducting my research. I think that this is for a couple of reasons. First, there is not a lot of social scientific work being produced that is looking at the nature of public perceptions at the current time. Indeed, I am one of very few people specifically working on this topic at the current time. Thus, tracking emerging public perception is difficult at present. As more people do research in the area it will become easier to map and track opinions. Second, public awareness of CDU remains very low at the present time. CDU, unlike CCS which is more widely known nowadays, has yet to really capture the public's attention. If and when it does, then there is clearly the potential for more rapid, evident and marked shifts in opinion to occur. For the time being though, we tend to find that the introduction we provide to CDU in our research is the first that most people have ever heard of it. Moreover, the comments and opinions that are subsequently expressed tend to centre on common issues (*i.e.* those reported in our Faraday Discussions article). I think that now is genuinely the time to be conducting research into the likely risks, benefits and uses that the general public (or more accurately general publics) see for CDU. We have noticed, for example with hydraulic fracturing (AKA fracking), that opinions of new technologies – particularly when they are perceived to have (unknown) risks and/or conflict with moves towards a more sustainable economy – can quickly become controversial, divisive and politically-charged. In such circumstances, social scientists are often bought in to assess what has gone wrong and why; however, by this time the 'battle lines' are drawn and opinions are more crystallized. Being more pro-active and assessing emerging perceptions of new technology before things potentially turn sour should, for me, become normative. While it is possible, of course, that the results of such work could be utilised to help proponents of a technology to promote their business interests; for me (a necessary CDU agnostic in a research capacity), the most important thing is that such research could be used to help develop the means by which affected populations might be able to make a more balanced, informed judgement about the relative benefits and risks of CDU and associated technology options.

Elsje Alessandra Quadrelli asked: While your paper is about 'public perception', some of the questions you have answered revolve around 'acceptability' and increase thereof. How does your research make the distinction between these two different aspects?

Christopher Jones responded: I think that the key distinction to draw here is between the generation of knowledge and the subsequent application of that knowledge for some intended purpose. We do not deny that the findings of our research are useful to those who wish to promote (or battle against) the development and deployment of CDU; however, we see our role as social scientists working in this field to be aligned with the 'knowledge generation' side of this divide. This is to say that we do not have a professional vested-interested in the fate of CDU – if it succeeds it succeeds, if it fails it fails – and I think that this impartiality is important if we are to get a true appreciation of emerging perceptions of the technology.

Perhaps the distinction between 'generation' and 'application' was blurred a little in the current context due to the fact that we were working with technologists who do have a professional interest in seeing CDU succeed. This was potentially reflected in some of the feedback we received about the video we used as a discussion prompt within our study. Again, though, we recognise this to be a limitation to the current study and we wish to create discussion materials for future research that present a more balanced critique of CDU (both in general and individual technology options). I guess a follow-up question to the one you ask would be "Why bother doing the research if you do not have a vested interested in the future of the technology?" Well, aside from the fact that hearing people discuss new, emerging technologies is interesting; we feel that developing a better appreciation of public perception is an important step in making investment and deployment decisions more transparent, democratic and fair. You could argue that this kind of advocacy counters our previous assertions about the importance of being impartial. However, the important distinction here is that we are advocating more democratic decision-making, as opposed to backing any individual technology/policy option.

Richard Heyn remarked: A couple of summers ago, the online version of a technical magazine in Norway had an article about Bayer's (Covesto) Dream Process and incorporation of CO_2 into the polyurethane in mattresses. The online comments to the article suggested that several readers thought it was gaseous CO_2 in the mattress. Can you provide any suggestions about how we can educate the public regarding CDU, *i.e.*, that the CO_2 is covalently bound into the polymer as a substitute for carbon from fossil feedstocks, and not simply gaseous CO_2?

Christopher Jones responded: Firstly, this confusion is logical. For the lay person, the term "CO_2" is likely to be associated with thoughts about greenhouse gases, memories of school disco dry-ice machines or health and safety videos about fire extinguishers. The point here is that, for most people, CO_2 is something that has only ever really been considered in a gaseous or liquid form. If you add to this the fact that people are probably all too familiar with the lilos, blow-up camp beds and water beds, then the prospect of having a gaseous-CO_2 filled mattress becomes a fairly logical, if erroneous, conclusion to draw. This is similar, perhaps, to some of the misperceptions about CCS and the prospect that we might get sparkling water out of our kitchen taps if stored CO_2 doesn't stay where it is supposed to.

We know from other social scientific research into new technologies that when something is unfamiliar, people will tend to try and find analogues of that technology that they are familiar with in order to help process and understand the incoming information. In some cases the analogue will be suitable and will lead people to draw reasonable conclusions about the strengths and weaknesses of the new technology – indeed, the use of natural analogues is a technique that has been used to help people understand CCS. For example, using descriptions about natural underground stores of CO_2 has been used as a technique for helping people to appreciate that storage of CO_2 underground for long periods is feasible. In other cases, the conclusions drawn can be objectively erroneous (as illustrated by your anecdote) but can still strongly shape people's opinions. An example from my own research relates to public perceptions of CCS. It was evident in some of my discussions with people living in affected communities that certain people equated the disposal of waste CO_2 underground as being conceptually similar to the prospect of storing spent nuclear fuel underground. This led to less favourable opinions of the technology as a whole. Also, it was clear that some people confused CO_2 with CO, again having a strong negative impact on attitudes. To the extent that CDU is an extremely novel concept for most people, I think that comments along the lines of that which you describe are to be expected. In terms of what we should do about this, well, firstly investing in more social scientific study of emerging perceptions is important. By engaging with people at an early stage it might be possible to identify which issues, concerns and misperceptions are most (or most likely to become) prevalent and then appropriate steps can be taken to ensure that communication materials appropriately address them. Secondly, I think it pays to be tolerant and respectful of such misperceptions. People are not stupid. Indeed, within our focus groups the people we spoke to asked very logical and insightful questions about how CDU processes might operate and what the implications for these technologies might be. There is no such thing as a stupid question and I think that it is incumbent on those who would wish to promote a technology to ensure that all (serious) questions are dealt with in a sincere manner. Finally, I think that being willing to meet with and engage people (particularly affected populations) in two-way discussions about CDU is important for proponents of the technology. It is evident that meaningful public participation in decision-making processes can be a good means of exchanging knowledge (from an expert to a lay person and *vice versa*) but can be a route to fostering trust.

Michael North commented: The misunderstanding is not that people are stupid – CO_2 is used to blow foams in polyurethanes – so maybe this is just an example of how a little knowledge can be a dangerous thing.

Jennifer Dodson addressed André Bardow, Eric Fraga and Christopher Jones: How do you want researchers developing technologies, and policy-makers, to engage with your studies?

Christopher Jones answered: At the current stage I think that technologists and policy-makers can help by paying attention to the rationale that is driving our research and keeping a watching brief on our findings. Specifically, our research is an illustration of the need for – and the benefit that can be derived from – upstream engagement with members of the general public (or publics) regarding new and emerging technologies. Our research remains at an early stage and I think that it is incumbent upon us to complete a few more studies before engaging in more formal consultations with technologists and policy-makers about the 'dos and don'ts' of CDU communication. That said, I think that our current findings (as discussed here) do provide an initial indication of the possible issues and misconceptions that might be at the forefront of people's minds as discussion about CDU at a national, regional and community level picks up pace.

Something I have mentioned here though, and something that technologists can help with now, is helping to develop the next stimulus materials for use in our work (*e.g.* developing a new, balanced informational video). As such, I would invite people who are interested in finding out what the public think about their technology or CDU in general, to get in touch with me.

André Bardow replied: Regarding the engagement with researchers developing technologies, we have been very fortunate in our work. Often, life cycle assessment (LCA) is performed as a posteriori analysis for the final product. In contrast, we have been involved with LCA from the start of the development. Thereby, we could analyze the first data available and provide guidance to the development. In our experience, this input from an LCA perspective has been very helpful in the development since today's scientists and engineers still lack an intuition for environmental impacts. I would therefore strongly recommend to involve LCA studies very early on in a development project.

With respect to policy-makers, I am glad to see that the importance of environmental impact studies is being more and more realized. This can be seen, *e.g.*, from the emphasis on this topic in the Horizon 2020 program.

Eric Fraga answered: The framework we have proposed aims to provide a mechanism by which alternative catalysts, combined with specific process routes, can be compared. The comparisons can be used to motivate research into alternative catalysts and routes, to inform policy makers about which alternatives may be more sustainable and to enhance dissemination for decisions taken.

In the other direction, the various stake-holders can help improve the framework through the provision of more data, although it is recognised that some fraction of the desired data may be confidential, and through the suggestion for further measures that would enhance or complement the 4 different attributes used in the framework. The visualisation suggested is extensible to more than 4 attributes easily and each attribute is already a combination of factors.

Christopher Jones commented: Research into the public perception of CDU is at a very early stage. I am perhaps one of the only people in the world who is systematically investigating emerging perceptions of these technologies and even my work is at an early stage. Public perception work into affiliated technologies, notably CCS, is in a much more advanced stage than CDU. The integration of the social sciences into the research agenda for this technology has provided many dividends; *e.g.* we now know much more about the perceived conceptual, technical and societal risks and benefits that general 'publics' see in this technology. This understanding has proven to be invaluable in shaping communication and public engagement efforts relating to this technology. To the extent that there is an increasingly salient role for the social sciences in the RD&D of new technologies (indeed, this is exemplified by the fact that the EPSRC often now require social scientific input into large grant funding-initiatives), perhaps now is the time for greater investment in research into the social acceptability of CDU (both conceptually and for individual technology options). Note: By 'social acceptability' I refer not only to public perception but also market/commercial acceptance and socio-political acceptance.

Faraday Discussions

PAPER

Extraction of $Mg(OH)_2$ from Mg silicate minerals with NaOH assisted with H_2O: implications for CO_2 capture from exhaust flue gas

Silvia Madeddu,*[a] Michael Priestnall,†[b] Erik Godoy,[c] R. Vasant Kumar,[d] Sugat Raymahasay,[e] Michael Evans,[f] Ruofan Wang,[a] Seabelo Manenye[a] and Hajime Kinoshita*[a]

Received 1st May 2015, Accepted 5th June 2015

DOI: 10.1039/c5fd00047e

The utilisation of $Mg(OH)_2$ to capture exhaust CO_2 has been hindered by the limited availability of brucite, the $Mg(OH)_2$ mineral in natural deposits. Our previous study demonstrated that $Mg(OH)_2$ can be obtained from dunite, an ultramafic rock composed of Mg silicate minerals, in highly concentrated NaOH aqueous systems. However, the large quantity of NaOH consumed was considered an obstacle for the implementation of the technology. In the present study, $Mg(OH)_2$ was extracted from dunite reacted in solid systems with NaOH assisted with H_2O. The consumption of NaOH was reduced by 97% with respect to the NaOH aqueous systems, maintaining a comparable yield of $Mg(OH)_2$ extraction, *i.e.* 64.8–66%. The capture of CO_2 from a CO_2–N_2 gas mixture was tested at ambient conditions using a $Mg(OH)_2$ aqueous slurry. $Mg(OH)_2$ almost fully dissolved and reacted with dissolved CO_2 by forming $Mg(HCO_3)_2$ which remained in equilibrium storing the CO_2 in the aqueous solution. The CO_2 balance of the process was assessed from the emissions derived from the power consumption for NaOH production and $Mg(OH)_2$ extraction together with the CO_2 captured by $Mg(OH)_2$ derived from dunite. The process resulted as carbon neutral when dunite is reacted at 250 °C for durations of 1 and 3 hours and CO_2 is captured as $Mg(HCO_3)_2$.

[a]*Department of Materials Science and Engineering, The University of Sheffield, Sheffield, UK. E-mail: mtq11sm@sheffield.ac.uk; h.kinoshita@sheffield.ac.uk*

[b]*Innovate UK, Swindon, UK*

[c]*Polarcus DMCC, Dubai, United Arab Emirates*

[d]*Department of Materials Science and Metallurgy, University of Cambridge, Cambridge, UK*

[e]*WRK Design & Engineering Ltd, Birmingham, UK*

[f]*Cambridge Carbon Capture Ltd, Cambridge, UK*

† Previously, Cambridge Carbon Capture Ltd.

Introduction

Carbon Capture and Storage (CCS) is a portfolio of technologies developed for the abatement of anthropogenic CO_2 emissions *via* permanent isolation of CO_2 from the atmosphere.[1] CCS includes the separation of CO_2 from exhaust flue gases and its storage either underground, in seawater or through the fixation into stable mineral carbonates *via* mineral carbonation.[1]

$Mg(OH)_2$ can capture CO_2 in a wide range of conditions: it can permanently and safely store CO_2 into solid Mg carbonate minerals upon exposure to gaseous or supercritical CO_2 or *via* dissolution in H_2O with purged CO_2.[2–4] $Mg(OH)_2$ aqueous slurries have been successfully tested to separate CO_2 from gas mixtures *via* liquid–gas scrubbing, and the technology has been proposed to separate CO_2 from exhaust flue gases.[5] $Mg(OH)_2$ and CO_2 dissolved in H_2O can react to form $Mg(HCO_3)_2$ which remains in solution as a soluble phase under a controlled pH.[5,6] This chemistry is suited to industrial applications such as in the coastal industry and power plants or ships, where CO_2 can be separated from exhaust flue gases using $Mg(OH)_2$ slurries and permanently stored in seawater as dissolved $Mg(HCO_3)_2$.[7] This technology combines the separation and storage of CO_2 into a single stage, which is favourable because it avoids the capture and conversion of CO_2 into a pure CO_2 stream, as well as the compression and transport operations, all of which are expensive and energy demanding.[8,9] Despite the wide range of possible applications and potential advantages, $Mg(OH)_2$ utilisation for CO_2 capture has been hindered due to its rare occurrence in natural outcrops.[1,10]

Mg silicate minerals, on the other hand, in particular serpentine and forsterite have been the preferred feedstock materials for CO_2 sequestration due to their vast availability in natural deposits which provide large storage capacity.[1,11,12] The main challenge of Mg silicate mineral carbonation is the slow kinetics of the carbonation reaction which requires acceleration through mechanical and chemical pre-treatment of the materials and the application of high pressure and temperatures.[1,13]

The extraction of $Mg(OH)_2$ from Mg silicate minerals *via* chemical processing benefits from the large availability of Mg silicate minerals. The overall efficiency of the process is also improved as $Mg(OH)_2$ has faster kinetics in carbonation than these Mg silicate minerals and allows greater flexibility in the design of the CCS technologies.[4] There are technologies currently available to obtain $Mg(OH)_2$ from Mg silicate minerals, *e.g.* a solid state reaction with ammonium salts at 400–500 °C or dissolution with HCl at 150 °C.[4,14] These technologies involve a second step where the pH of the system is increased by introducing another reactant to favour the precipitation of $Mg(OH)_2$ under alkaline conditions.[4,14]

In a previous study we investigated the alkaline digestion of Mg silicate minerals to obtain $Mg(OH)_2$ using highly concentrated NaOH aqueous systems.[15] The alkaline digestion of dunite, an ultramafic rock composed of Mg silicate minerals, at 180 °C for 6 hours, resulted in the near-complete digestion and formation of $Mg(OH)_2$. The technology is advantageous because it is a one-step reaction and involves a single reactant, although the high consumption of NaOH was identified as a possible obstacle for the implementation.[15]

The present study proposes a new technique for the extraction of $Mg(OH)_2$ from Mg silicate minerals, aiming to reduce the NaOH consumption by introducing NaOH into the system in the solid state. After studying the basic reaction of the system at 180 °C with the aid of thermodynamic consideration, the investigation focused on the effect of H_2O on the efficiency of $Mg(OH)_2$ extraction from the dunite–NaOH solid mixtures. The effects of temperature and duration of reaction were also studied for a dunite–NaOH–H_2O system with a fixed composition reacted at 130, 180 or 250 °C for 1, 3 and 6 hours. The obtained results are compared with those from the alkaline digestion of dunite with NaOH aqueous systems previously investigated. The feasibility of CO_2 sequestration using $Mg(OH)_2$ was also demonstrated using a Downflow Gas Contactor (DGC) reactor. CO_2 was separated from a flow of CO_2–N_2 mixture at ambient conditions using an aqueous slurry of reagent grade $Mg(OH)_2$, and the implications of using $Mg(OH)_2$ derived from dunite for CO_2 sequestration are discussed.

Materials and methods

Dunite

The dunite mined in Åheim, Norway, was provided by Sibelco Ltd. The material was analysed *via* X-ray fluorescence (XRF) and was composed of 48.3 wt% MgO, 45.35 wt% SiO_2, and 6.16 wt% Fe_2O_3, while other oxides were present at <1 wt%. The X-ray diffraction (XRD) analysis revealed that dunite was mainly composed of forsterite, Mg_2SiO_4, the Mg-rich member of the olivine group. The minor components were also Mg-bearing minerals, *i.e.* clinochlore, serpentine, enstatite, talc, hornblende and spinel. The Rietveld Refinement Quantitative Phase Analysis (QPA) conducted on the XRD pattern estimated the amount of forsterite to be 73 ± 2 wt%. Thermogravimetric analysis (TGA) detected the presence of $Mg(OH)_2$ at 0.42 wt% in the dunite.

The powdered dunite was sieved to <63 μm and analysed by laser diffraction with dry dispersion, which showed an average distribution of the particle size to be around 25 μm, with 90% of the particles <58 μm. These results were confirmed *via* SEM analysis.

Extraction of $Mg(OH)_2$

NaOH pearl, reagent grade provided by Fisher Chemical, was mixed with powdered dunite, with or without addition of distilled H_2O, in an agate mortar with a pestle. The mole ratio of the reactants in the samples used in the present study is summarised in Table 1. Three series were prepared at different dunite : NaOH mole ratios, *i.e.* 1 : 1, 1 : 1.5, and 1 : 2, which are referred to as Series 1,

Table 1 Normalised mole of reactants in samples

	Dunite[a] (mole)	NaOH (mole)	H_2O[b] (mole)				
Series 1	1	1	—	0.25	0.5	1	2
Series 1.5	1	1.5	—	0.375	0.75	1.5	3
Series 2	1	2	0	0.5	1	2	4

[a] Approximate mole of dunite was estimated based on the chemical formula of forsterite, Mg_2SiO_4. [b] Different amounts of H_2O were tested at NaOH : H_2O mole ratios of 4 : 1, 2 : 1, 1 : 1 or 1 : 2 for each series. Series 2 also includes a *dry* system without H_2O.

Series 1.5, and Series 2, respectively. The moles of dunite were estimated based on the chemical formula of forsterite, Mg_2SiO_4. For all series, the samples were prepared at different NaOH : H_2O mole ratios of 4 : 1, 2 : 1, 1 : 1, and 1 : 2. Series 2 also includes a sample reacted without H_2O addition.

When the reactants were well-blended, the mixture was transferred in a steel vessel lined with Teflon which was sealed and heated at 180 °C for 6 hours in a Carbolite electric oven, series PF30. The system with dunite : NaOH : H_2O mole ratio of 1 : 2 : 0.5 was replicated at different durations of reaction, *i.e.* 1 and 3 hours, and different temperatures, *i.e.* 130 and 250 °C.

After the reaction, the products were ground and washed with distilled water and the solid component was separated from the liquid phase *via* vacuum filtration. The solid filtrate was dried for 1 hour at 90 °C and then analysed *via* XRD and TGA.

CO_2 capture with $Mg(OH)_2$ suspension

A Downflow Gas Contactor (DGC) reactor, designed by WRK Design & Engineering Ltd, with a liquid volume capacity of 10 litres was used as the liquid–gas reactor to capture CO_2 from a flowing gas-mixture. In this reactor, the gas-mixture at atmospheric pressure enters at the top of the bubble column, is entrained (at 0.8 atm) in the downward circulating liquid flow, and exits at 1 atm from the bottom of the column after reaction and separation from the recirculating liquid phase. The DGC reactor provides a large interfacial area between the gas bubbles and liquid which enhances the reactivity of the phases involved. The rate of liquid recirculation was typically 10 L minute^{-1} to maintain a stable bubble–liquid interface.

A gas-mixture of N_2 and 4–5% CO_2 was injected at the top of the column at a fixed flow-rate of 2.25 ± 0.07 L minute^{-1} and bubbled through the column at approximately 10 °C and under ambient pressure. When the solution was saturated with CO_2, *i.e.* the concentration of CO_2 in the outlet gas became the same in the inlet gas, 50.8 g of powdered $Mg(OH)_2$ was added to the circulating H_2O.

The $Mg(OH)_2$ with 98% purity used in the study was provided by Lehmann & Voss & Co. The material had an average particle size of 7 μm, and also contained 0.5% CaO and smaller amounts of SiO_2 and Fe_2O_3 as impurities.

The CO_2 concentration in the inlet and outlet gas was periodically monitored using a calibrated inline infrared data logger, and the monitoring continued until the concentration of CO_2 in the outlet gas had increased and become equal to that in the inlet gas, indicating that no further reaction was occurring. The experiment took approximately 12 hours to complete and was run over the course of two days with an overnight shut-down. Samples of the circulating liquid were also taken periodically to monitor the change in pH.

Products and data analysis

XRD was conducted for phase analysis of the reaction products using a Siemens D5000 with Cu Kα X-ray source, $\lambda = 1.54$ Å. The powdered samples were placed in a plastic sample holder, and the scans were run from 10° to 70° 2θ with a step size of 0.05° 2θ.

TGA of the reaction products was also conducted using a Perkin Elmer Pyris 1. The samples were placed in an alumina crucible and heated at a rate of 10 °C minute^{-1} from room temperature to 1000 °C under a N_2 flow. The first derivative of the TG curve, DTG, was also determined to assist the analysis. The amount of $Mg(OH)_2$ was calculated based on the dehydroxylation of $Mg(OH)_2$ (reaction (1)) taking place at 350–450 °C.[16]

$$Mg(OH)_{2(s)} \rightarrow MgO_{(s)} + H_2O_{(g)} \quad (1)$$

The analysis was also assisted with the thermodynamic consideration. The change in the Gibbs free energy and enthalpy for the possible reactions were calculated using the SGTE Substances Database (SSUB5) in Thermo-Calc.[17] The calculation was also performed for the minor mineral phases in the dunite. The calculation was not performed for clinochlore and hornblende due to the lack of available data in the SSUB5.

The carbon content in the solid residue recovered from the CO_2 capture test was determined using a Perkin Elmer 2400 CHNS/O Series II Elemental Analyser. The sample was combusted at 975 °C under an oxygen environment. The gases released were reduced by copper and separated through a chromatographic column. The amount of carbon was obtained from the gases eluting off the column based on their thermal conductivity, which were converted into CO_2 wt% to assess the CO_2 captured in the solid phase.

Results and discussion

Extraction of $Mg(OH)_2$

The reaction of dunite with NaOH usually resulted in the formation of brucite, Na_2SiO_3 and natrite, *i.e.* Na_2CO_3, together with the mineral components from the unreacted dunite and remaining NaOH. Fig. 1 shows the XRD patterns of selected reaction products from the solid systems with a dunite : NaOH mole ratio of 1 : 2 and a dunite : NaOH : H_2O mole ratio of 1 : 2 : 0.5, before and after being washed with distilled H_2O, respectively. The by-products, Na_2SiO_3 and natrite are usually detected before washing the reaction products as demonstrated in Fig. 1. These

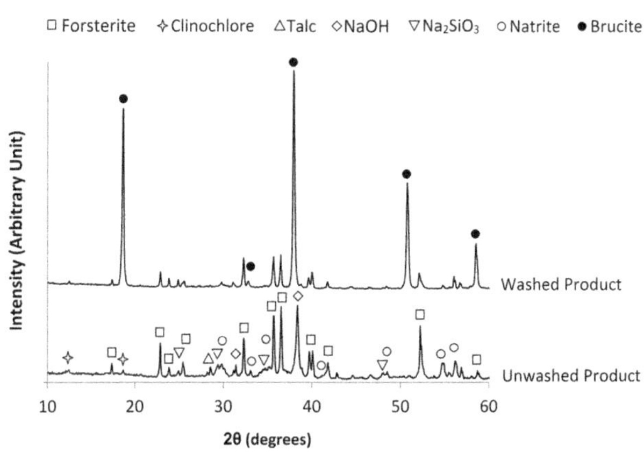

Fig. 1 XRD patterns of selected reaction products from solid systems with a dunite : NaOH mole ratio of 1 : 2 and a dunite : NaOH : H_2O mole ratio of 1 : 2 : 0.5 before and after washing, respectively.

by-products are soluble in water and can be removed together with the remaining NaOH by washing the reaction products. After washing and successive filtration, the samples are typically left with a solid fraction composed of $Mg(OH)_2$ and remaining dunite components (Fig. 1).

It is expected to be difficult to obtain Na_2SiO_3 from the reaction of dunite and NaOH. Reactions (2)–(4) are examples of reactions for Mg_2SiO_4 and NaOH which assume Na_4SiO_4, Na_2SiO_3 and $Na_2Si_2O_5$ as reaction products, respectively. Mg_2SiO_4 was chosen as a representative of dunite because it is the main mineral component. The calculation of the Gibbs energy change (ΔG^o) for reactions (2)–(4) showed that only the formation of Na_4SiO_4 is thermodynamically favoured, as indicated by the negative ΔG^o of the reaction in Fig. 2, whereas, the formation of Na_2SiO_3 or $Na_2Si_2O_5$ is not possible because $\Delta G^o > 0$.

$$Mg_2SiO_4 + 4NaOH \rightarrow 2Mg(OH)_2 + Na_4SiO_4 \quad (2)$$

$$Mg_2SiO_4 + 4NaOH \rightarrow 2Mg(OH)_2 + Na_2SiO_3 + Na_2O \quad (3)$$

$$Mg_2SiO_4 + 4NaOH \rightarrow 2Mg(OH)_2 + \frac{1}{2}Na_2Si_2O_5 + \frac{3}{2}Na_2O \quad (4)$$

However, when H_2O is involved in the reactions (reactions (5)–(7)), the formation of these Na silicate species becomes possible because their ΔG^o are all negative, as shown in Fig. 2.

$$Mg_2SiO_4 + 2NaOH + H_2O \rightarrow 2Mg(OH)_2 + \frac{1}{2}Na_4SiO_4 + \frac{1}{2}SiO_2 \quad (5)$$

$$Mg_2SiO_4 + 2NaOH + H_2O \rightarrow 2Mg(OH)_2 + Na_2SiO_3 \quad (6)$$

$$Mg_2SiO_4 + NaOH + \frac{3}{2}H_2O \rightarrow 2Mg(OH)_2 + \frac{1}{2}Na_2Si_2O_5 \quad (7)$$

It should be noted that the formation of Na_2SiO_3 in the presence of H_2O becomes thermodynamically more favourable than that of Na_4SiO_4 in a *dry*

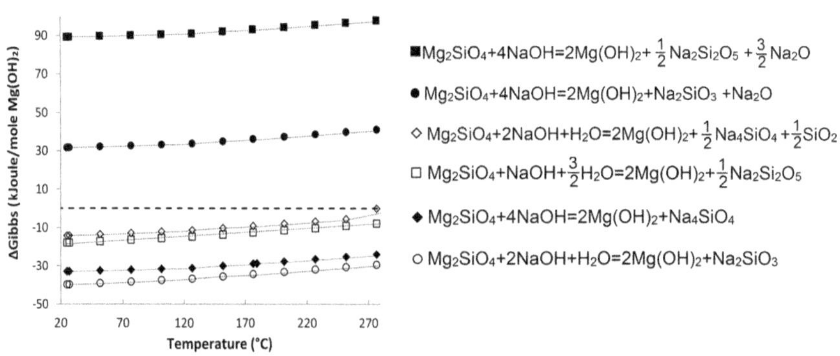

Fig. 2 ΔG^o per 1 mole of $Mg(OH)_2$ produced calculated for possible reactions for Mg_2SiO_4 and NaOH. Closed data points are for the *dry* reactions, and open data points for those involving H_2O.

reaction. These data suggest the involvement of H_2O in the studied reaction and its thermodynamical advantage over the *dry* reaction. The carbonation of NaOH could also aid the involvement of H_2O in the reaction. As seen in the XRD data in Fig. 1, natrite has been formed in the system. Because the samples are in contact with air during the preparation and the reaction (in a closed vessel with trapped air), some of the NaOH in the system can be carbonated, and this reaction releases H_2O through the following reaction:

$$2NaOH + CO_2 \rightarrow Na_2CO_3 + H_2O. \quad (8)$$

The ΔG^o for the reactions of NaOH and minor phases in the dunite *i.e.*, serpentine ($Mg_3Si_2O_5(OH)_4$), talc ($Mg_3Si_4O_{10}(OH)_2$), and enstatite ($MgSiO_3$) are shown in Fig. 3 and suggests that the formation of $Mg(OH)_2$ and Na_2SiO_3 from these Mg silicate minerals is also thermodynamically possible ($\Delta G^o < 0$). The thermodynamical feasibility of $Mg(OH)_2$ extraction from serpentine with NaOH is particularly relevant because serpentine is widely available in natural deposits and contains a high wt% of Mg which makes it a suitable alternative feedstock material to dunite.[4]

The enthalpy change (ΔH^o) for the possible reactions occurring in the dunite–NaOH system was also calculated at the temperatures of reaction investigated in this study, *i.e.* 130, 180 and 250 °C. All the reactions showed negative ΔH^o in the temperature range of interest and the lowest was found at 250 °C. The values of ΔH^o at this temperature are shown for each reaction in Table 2. These reactions are all exothermic and although their balance is not known, overall the extraction of $Mg(OH)_2$ from dunite should be energetically favoured.

Effect of added H_2O

The introduction of H_2O into the system had a significant effect on the extraction of $Mg(OH)_2$. Fig. 4 compares the XRD patterns of the reaction products from the *dry* reaction and the reaction with added H_2O in Series 2. These samples were washed with distilled water and Na_2SiO_3, Na_2CO_3 had been removed. The extraction of brucite ($Mg(OH)_2$), in the *dry* dunite–NaOH system appeared to be

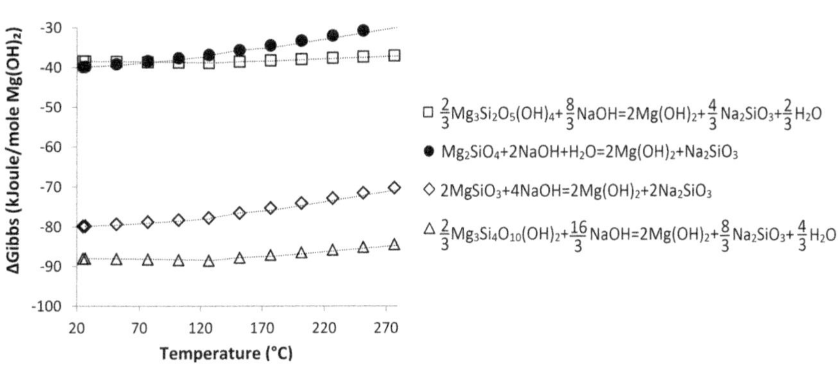

Fig. 3 ΔG^o per 1 mole $Mg(OH)_2$ produced calculated for possible reactions for serpentine ($Mg_3Si_2O_5(OH)_4$), talc ($Mg_3Si_4O_{10}(OH)_2$), and enstatite ($MgSiO_3$) with NaOH in comparison with that for forsterite (Mg_2SiO_4).

Table 2 Standard enthalpy change per 1 mole of $Mg(OH)_2$ produced at 250 °C

Reactions	ΔH^o (kJ)
$Mg_2SiO_4 + 2NaOH + H_2O \rightarrow 2Mg(OH)_2 + Na_2SiO_3$	−58
$\frac{2}{3}Mg_3Si_2O_5(OH)_4 + \frac{8}{3}NaOH \rightarrow 2Mg(OH)_2 + \frac{4}{3}Na_2SiO_3 + \frac{2}{3}H_2O$	−44
$2MgSiO_3 + 4NaOH \rightarrow 2Mg(OH)_2 + 2Na_2SiO_3$	−98
$\frac{2}{3}Mg_3Si_4O_{10}(OH)_2 + \frac{16}{3}NaOH \rightarrow 2Mg(OH)_2 + \frac{8}{3}Na_2SiO_3 + \frac{4}{3}H_2O$	−99
$2NaOH + CO_2 \rightarrow Na_2CO_3 + H_2O$	−176

limited. The only reflection peaks identified are for the mineral components present in the dunite, while the reflection peaks for brucite were not observed. On the other hand, the systems with added H_2O all indicated the presence of brucite. The two main reflection peaks for the brucite, were detected at 18.6° and 38° 2θ. These peaks partially overlap with those for clinochlore and forsterite, respectively, but are distinguishable based on the proportion of the other peaks intensity. The introduction of H_2O is beneficial and the intensity of the reflection peaks for brucite considerably increases in systems with 0.5, 1 and 2 moles of H_2O, although, the intensity significantly decreases in the system with 4 moles of H_2O. A similar trend was observed also in Series 1.5 and 1.

The TGA and DTG curves of the reaction products from Series 2 are shown in Fig. 5(A) and (B), respectively. The sample reacted in *dry* conditions had a small weight loss between 300 and 450 °C for the dehydroxylation of $Mg(OH)_2$, whereas those reacted in the presence of water showed a larger weight loss in this temperature region, indicating that more $Mg(OH)_2$ was produced when H_2O was added to the system.

The calculated concentration of $Mg(OH)_2$ in the reaction product was 19 wt% for the *dry* dunite–NaOH mixture, and 57.6, 55.4, 51.2 and 28 wt% with 0.5, 1, 2 and 4 moles of H_2O added per 1 mole of dunite, respectively. Thus, the amount of Mg(OH) in the products decreases with the increase in H_2O content.

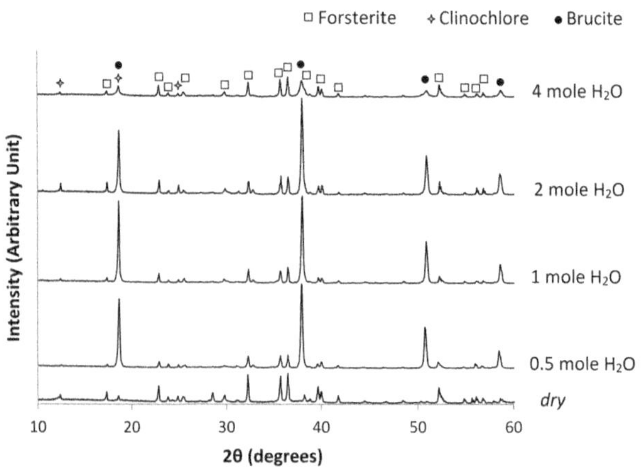

Fig. 4 XRD patterns of reaction products from Series 2 with different amounts of H_2O.

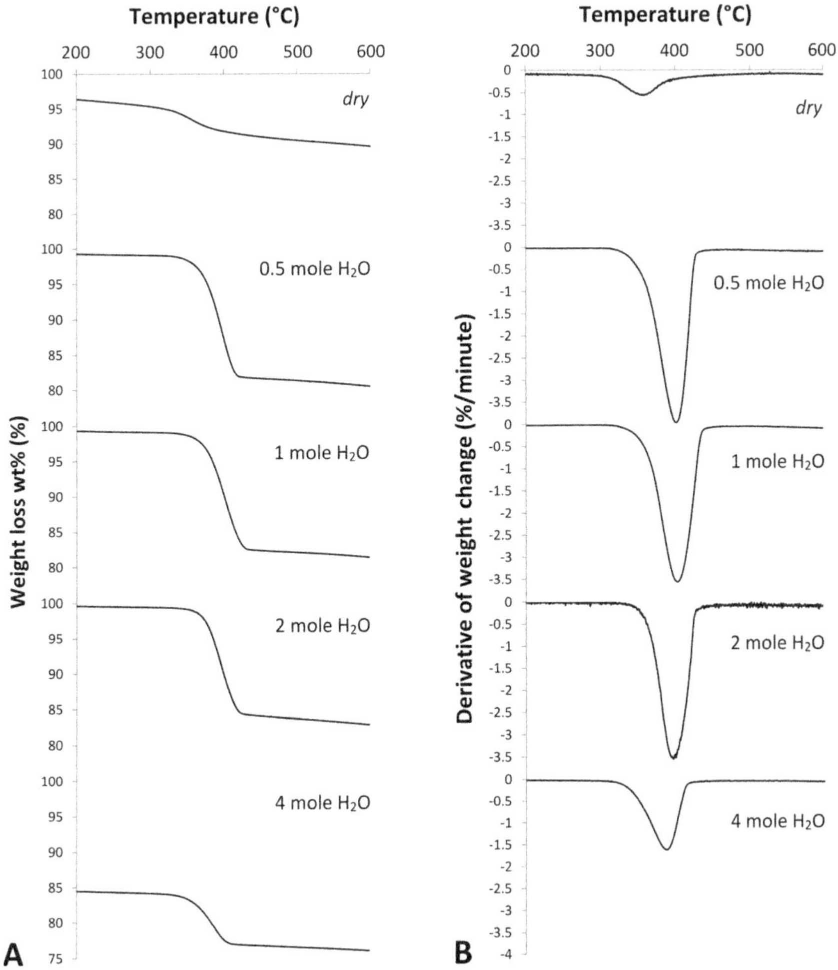

Fig. 5 (A) TGA and (B) DTG curves of the reaction products from Series 2.

The TGA results are in agreement with the XRD data and indicate that the addition of H_2O was beneficial compared to *dry* conditions, which is likely because the dissolution of NaOH in H_2O favoured the ion exchange and the diffusivity of materials involved in the reaction.[18,19] The addition of smaller amounts of H_2O, *i.e.* a higher NaOH : H_2O mole ratio, was a more preferable condition for $Mg(OH)_2$ formation, which suggests that the concentration of NaOH in H_2O played a decisive role for the extraction of $Mg(OH)_2$ from dunite.

Effect of NaOH concentration

The $Mg(OH)_2$ content in the reaction products were estimated *via* TGA for Series 2, 1.5 and 1, and are compared in Fig. 6. The data are plotted against the NaOH : H_2O mole ratio to study the effects of NaOH concentration in H_2O.

The reaction products from Series 2 showed the highest concentration of $Mg(OH)_2$, followed by those from Series 1.5 and 1, indicating the advantage of

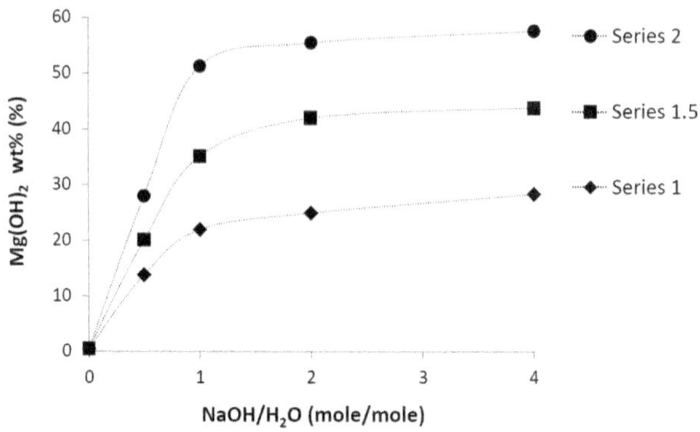

Fig. 6 $Mg(OH)_2$ concentrations estimated *via* TGA in reaction products from Series 1, 1.5, and 2.

having a higher NaOH content in the system. The three series showed the same trend in the presence of H_2O, and the wt% of $Mg(OH)_2$ in the reaction products increased with the increase of the NaOH : H_2O mole ratio. However, this effect becomes less significant at a higher NaOH : H_2O mole ratio. In fact, at higher NaOH : H_2O mole ratios the concentration of NaOH in H_2O is higher but the amount of H_2O in the system is lower, and thus the positive effect of H_2O, such as the improvement of the reactants diffusion is reduced.

The data shown in Fig. 6 are also plotted against the amount of H_2O added to the system in Fig. 7(A), grouped by the same NaOH : H_2O mole ratio.

The gradients of the linear fitting for these data sets are proportional to the NaOH : H_2O mole ratio as shown in Fig. 7(B). Based on these results, it is possible to estimate the approximate wt% of $Mg(OH)_2$, $Mg(OH)_2$ (%), expected in the reaction product under the conditions studied, using the following empirical equation with the amount of NaOH (X_{NaOH}) and H_2O (X_{H_2O}):

$$Mg(OH)_2\ (\%) = 30.79(X_{NaOH}) - 7.34(X_{H_2O}). \tag{9}$$

These data together with those previously discussed for Series 2, indicate that the extraction of $Mg(OH)_2$ from dunite in NaOH–H_2O solid systems is the result of a combined effect of NaOH concentration in H_2O and the amount of liquid phase present in the system. At 180 °C, the NaOH in the system should be fully dissolved at a NaOH : H_2O mole ratio of 1 : 2 and 1 : 1, or constitute a solid system with partially dissolved NaOH at a NaOH : H_2O mole ratio of 2 : 1 and 4 : 1.[20] The optimal conditions for the extraction of $Mg(OH)_2$ appears to be found when H_2O is present in a sufficient quantity to help the diffusion of the reactants but not in excess to reduce the relative concentration of NaOH in the liquid.

Effect of temperature and time

Amongst the compositions investigated, the system with a mole ratio of dunite : NaOH : H_2O = 1 : 2 : 0.5 gave the highest $Mg(OH)_2$ extraction. The effect of the reaction conditions was further investigated for this system, using three

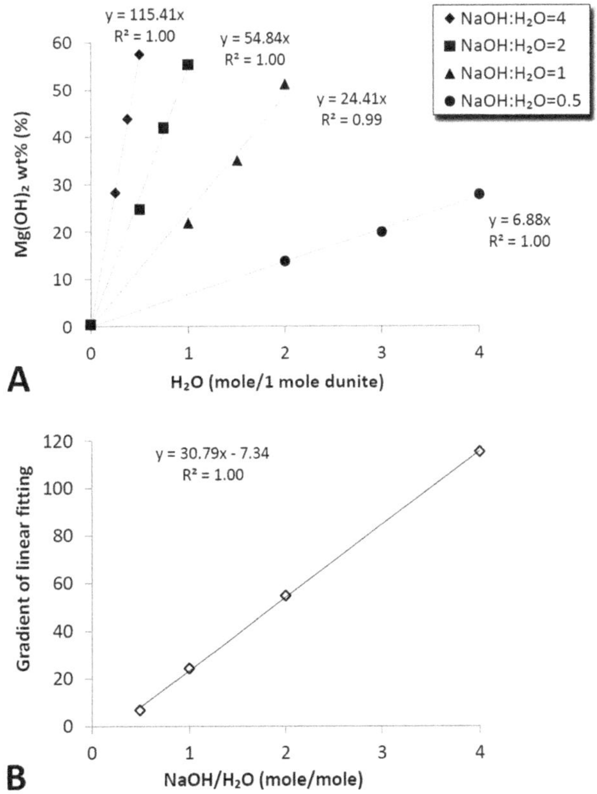

Fig. 7 $Mg(OH)_2$ concentrations estimated *via* TGA of the reaction products: (A) effect of H_2O content for different NaOH : H_2O mole ratios, and (B) the gradient of the linear fitting for the datasets shown in (A).

durations of reaction, *i.e.* 1, 3 and 6 hours, at different temperatures, *i.e.* 130, 180 and 250 °C. The reaction products were analysed *via* TGA, and the wt% of $Mg(OH)_2$ in the reaction products are compared in Fig. 8.

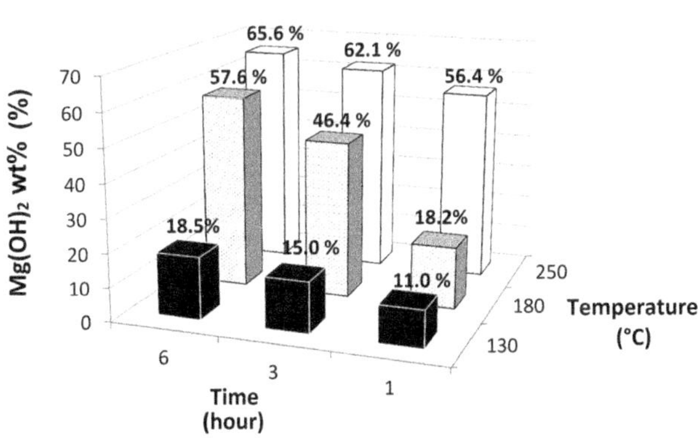

Fig. 8 $Mg(OH)_2$ concentration estimated *via* TGA in the reaction products of dunite : NaOH : H_2O systems with a 1 : 2 : 0.5 mole ratio at different temperatures and times of reaction.

As expected, increasing the duration and temperature of the reaction was beneficial for the extraction of $Mg(OH)_2$. At 130 °C the formation of $Mg(OH)_2$ was limited, achieving only 18.5 wt% after 6 hours which is comparable with 18.2 wt% obtained at 180 °C in 1 hour. The reaction conducted at 180 °C resulted in a significant $Mg(OH)_2$ extraction of 46.4 wt% after 3 hours of reaction. Increasing the temperature to 250 °C significantly improved the extraction of $Mg(OH)_2$, and a concentration of 56.4 wt% $Mg(OH)_2$ was achieved after 1 hour, which increased to 65.6 wt% in 6 hours. The $Mg(OH)_2$ extraction obtained within 1 hour at 250 °C is comparable with the 57.6 wt% attained at 180 °C over 6 hours of reaction. The reaction at 250 °C is faster than at the other temperatures, and the majority of the reaction appears to have taken place in the first hour. This is highly advantageous for the industrial application of the process, as it would allow a reduction of the reaction time from 6 hours to 1 hour to achieve the same degree of $Mg(OH)_2$ extraction obtained at 180 °C.

Fig. 9 shows the NaOH–H_2O phase diagram.[20] The diagram has been adapted to show the temperatures and NaOH concentration used in this study. The NaOH : H_2O mole ratio of 4 : 1 corresponds to a NaOH concentration in H_2O of approximately 90% by weight. The diagram shows that there could be a solid portion remaining at 130 and 180 °C in the NaOH–H_2O system at 90 wt% NaOH whereas, only liquid or gas can be present at 250 °C, similar to molten salt systems.[20] Consequently, at 250 °C dunite reacted with a highly alkaline melt instead of a solid system with a limited amount of NaOH dissolved in H_2O, which should have favoured the diffusion of reactants,[21] contributing to the much faster kinetics of $Mg(OH)_2$ extraction observed within the first hour of reaction at this temperature. Although the increase of pressure in the closed vessel at higher temperatures may shift the melting temperature, based on the obtained results it appears to be still below 250 °C.

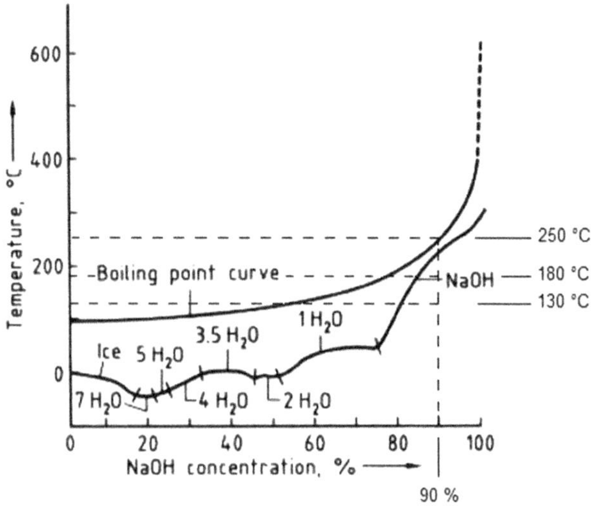

Fig. 9 NaOH–H_2O phase diagram, modified from Kurt and Bittner, 2003.[20]

Efficiency of the reaction

The efficiency of $Mg(OH)_2$ extraction is estimated through the yield of the reaction which calculates the percentage of $Mg(OH)_2$ produced with respect to the maximum amount of $Mg(OH)_2$ theoretically producible from dunite.[15,22] The procedure adopted for the yield calculation is described by Madeddu *et al.*[15] In the present study, the yield was calculated using eqn (10) and (11) from which the amount of $Mg(OH)_2$ extracted from dunite was obtained, based on the wt% estimated *via* TGA.

The wt% of $Mg(OH)_2$ in the reaction products determined *via* TGA, $Mg(OH)_2$ (%), can be expressed according to eqn (10), where $Mg(OH)_{2(rp)}$ and $dunite_{(rp)}$ are the amount of $Mg(OH)_2$ and dunite in the reaction products, respectively. Eqn (11), on the other hand calculates the amount of MgO involved in the reaction. $Dunite_{(i)}$ is the initial feed of dunite, and 48.3 is the wt% of MgO in $dunite_{(i)}$. 48.21 is the wt% of MgO in dunite excluding the MgO initially presented as 0.42 wt% of $Mg(OH)_2$, and 40.3 and 58.3 are the molecular weights of MgO and $Mg(OH)_2$, respectively.

$$\mathrm{Mg(OH)_2(\%)} = \frac{100\mathrm{Mg(OH)_{2(rp)}}}{\mathrm{Mg(OH)_{2(rp)}} + \mathrm{dunite_{(rp)}}} \tag{10}$$

$$\frac{48.3 \times \mathrm{dunite_{(i)}}}{100} = \frac{48.21 \times \mathrm{dunite_{(rp)}}}{100} + \frac{40.3 \times \mathrm{Mg(OH)_{2(rp)}}}{58.3} \tag{11}$$

With the values of $Mg(OH)_2$ (%) and $dunite_{(i)}$, it is possible to obtain $Mg(OH)_{2(rp)}$ and $dunite_{(rp)}$ by solving the system of eqn (10) and (11).

For the three series tested at 180 °C in the present investigation, a highest yield of 66% for the reaction was achieved in the reaction product of Series 2, with a NaOH : H_2O mole ratio of 4 : 1 reacted for 6 hours. The same yield was obtained from the alkaline digestion of dunite with NaOH 50 mol kg^{-1} aqueous system at the same temperature and same duration of reaction.[15] The NaOH consumed in the solid system reaction was 0.57 g per 1 g of dunite, whereas 20.6 g of NaOH was demanded in the aqueous system to process 1 g of dunite, which corresponds to the reduction of NaOH usage by 97%.[15] The H_2O consumed was also reduced by 99%.[15] When the system was reacted at 130 °C, the yield of the reaction was 24% after 6 hours, whereas, at 250 °C the yield was 64.8, 70 and 73% after 1, 3, and 6 hours, respectively. As expected the yield achieved after a 1 hour reaction at 250 °C is comparable with that achieved at 180 °C after 6 hours.

The NaOH was reduced by 97.9% in Series 1.5 and 98.6% in Series 1, although the yield of the reaction was reduced to 53% and 36%, respectively, after a 6 hour reaction at 180 °C.

These results show that the extraction of $Mg(OH)_2$ *via* the reaction of dunite with NaOH in a solid system is preferable to the alkaline digestion with NaOH aqueous systems as the amount of NaOH required can be significantly reduced while maintaining the same efficiency of $Mg(OH)_2$ extraction.

Feasibility of CO_2 capture with $Mg(OH)_2$

Fig. 10 shows the CO_2 concentration in the flow gas measured at the inlet and outlet of the DGC reactor during the experiment. The CO_2 concentration at the

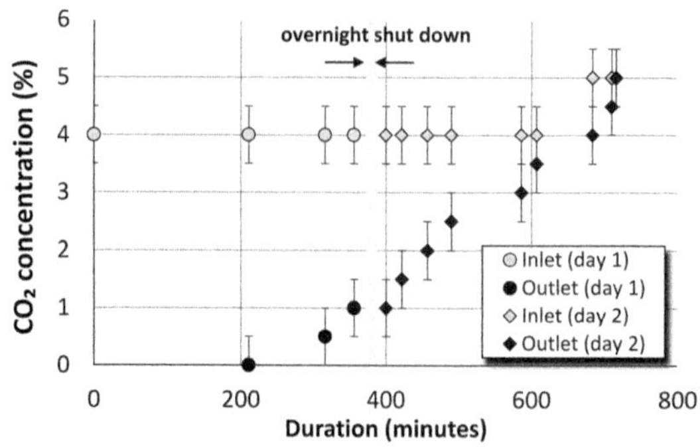

Fig. 10 CO_2 concentration in the flow gas at the inlet and outlet of the DGC reactor.

outlet was initially negligible, and all CO_2 in the flow gas appeared to be retained in the circulating solution by reacting with $Mg(OH)_2$ in the system. The CO_2 concentration started increasing after approximately 200 minutes due to the consumption of $Mg(OH)_2$ in the system, and became equal to that in the inlet gas at 720 minutes where the system no longer captures CO_2. The overnight shut down period did not cause any obvious change in the system. The obtained results clearly show the CO_2 capture capability of this system.

The amount of CO_2 captured from the flow gas was estimated using eqn (12) where CO_2i % and CO_2o % are the concentration of CO_2 in the inlet and outlet gas, respectively. The volume of gas under atmospheric pressure at 10 °C was estimated as 23.2 L mole^{-1}.

$$CO_2(\text{mole}) = \frac{2.25\ (\text{L minute}^{-1})}{23.2\ (\text{L mole}^{-1})} \times \frac{CO_2i(\%) - CO_2o(\%)}{100} \times \text{time (minute)} \quad (12)$$

The amount of captured CO_2 is plotted in Fig. 11 together with the pH of the circulating solution. CO_2 was steadily captured up to around 400 minutes, and then the rate of capture slightly decreased. The significant reduction in pH at

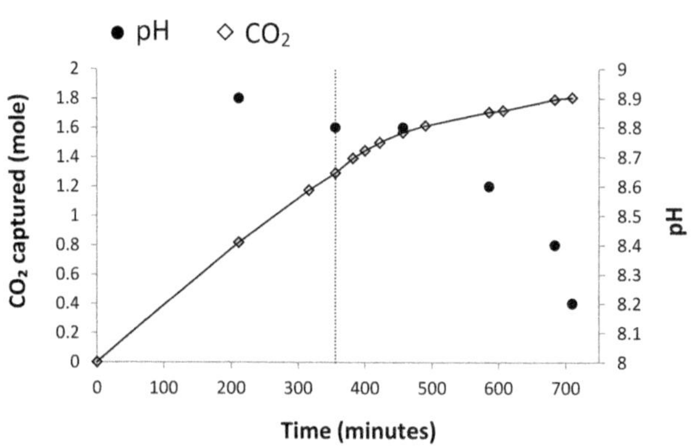

Fig. 11 Amount of CO_2 captured and pH of the circulating solution in the liquid–gas scrubbing with $Mg(OH)_2$–H_2O slurry.

this period suggests the exhaustion of the dissolved $Mg(OH)_2$ available for reaction.

In total, 1.80 ± 0.19 moles of CO_2 were sequestered over the 2 days of the experimental run.

Since the circulating solution used in the present study contained approximately 0.85 mole of $Mg(OH)_2$, the maximum CO_2 to be captured is 1.71 mole (1.72 mole if CaO presented as an impurity is also counted) assuming the reaction product is $Mg(HCO_3)_2$, and 0.86 mole of $MgCO_3$ was produced. Although our estimation, 1.80 ± 0.19 mole, contains a significant deviation, it is clear that the reaction product in the tested system was mainly $Mg(HCO_3)_2$. It is known that the formation of $Mg(HCO_3)_2$ is favoured when the pH of the solution is around 8.4 where HCO_3^- ions are the predominant species of CO_2 dissolution, whereas $MgCO_3$ preferentially forms when the pH of the solution is above 10 as the CO_3^{2-} ions are the predominant species.[5,23,24] The obtained results confirm that $Mg(HCO_3)_2$ can form at a pH ranging between 8.2 and 8.9.

The amount of solid material collected upon completion of the experiment was 2 g. The XRD analysis showed that the solid residue constituted of brucite and poorly crystalline hydromagnesite ($Mg_5(CO_3)_4(OH)_2 \cdot 4H_2O$) and another poorly crystalline phase which was not identified. The amount of CO_2 in the collected solid was determined *via* CHN analysis and estimated as 0.01 mole. These results indicate that $Mg(OH)_2$ almost fully dissolved and captured CO_2 by forming $Mg(HCO_3)_2$ which remained in equilibrium in solution while only a small fraction reacted forming Mg carbonate phases.

The exact quantification of CO_2 captured by the reaction with $Mg(OH)_2$ is challenging and requires further analysis. Nevertheless, these preliminary results suggest that at least 93.6% of CO_2 potentially capturable by the $Mg(OH)_2$ suspension was effectively captured.

Implications of $Mg(OH)_2$ extraction for CCS

The consumption of NaOH was identified as one of the major drawbacks for the extraction of $Mg(OH)_2$ in the NaOH aqueous system[15] because the production of NaOH involves significant energy consumption and thus CO_2 emissions. Therefore, the reduction in NaOH achieved in the present study is significant and it would reduce the environmental impact of the process.

In the present study, the CO_2 balance was evaluated based on the emissions associated with the NaOH, $Mg(OH)_2$ extraction and the sequestration potential of $Mg(OH)_2$ derived from dunite. The assessment is based on the mass balance of the materials used assuming that 1 kg of dunite is processed with the corresponding amount of NaOH used for Series 1, 1.5 and 2. The amount of $Mg(OH)_2$ extracted was estimated from the yield of reaction achieved. The CO_2 emission from NaOH production was calculated assuming that 2.9 k W h of electricity is required to produce 1 kg of NaOH, which corresponds to the emission of at least 250 g of CO_2 per 1 k W h, when natural gas is used.[25,26] The CO_2 emission from the extraction of $Mg(OH)_2$ is based on the use of an electric oven with a capacity of 28 litres to process 1 kg of dunite. The oven consumes 0.17, 0.26 and 0.385 k W h of energy to operate at 130, 180 and 250 °C, respectively, according to the manufacturer's specification.[27] The CO_2 sequestration was calculated based on the full conversion of $Mg(OH)_2$ into either $MgCO_3$ or $Mg(HCO_3)_2$.

Fig. 12 compares the CO_2 potentially captured by the $Mg(OH)_2$ extracted at 180 °C in 6 hours for Series 1, 1.5, and 2, together with that achievable at a 100% yield of the reaction. The dotted lines represent the sum of the CO_2 emissions from the NaOH production and $Mg(OH)_2$ extraction. The graph shows that the potential CO_2 capture naturally increases with the improvement of the yield of $Mg(OH)_2$ extraction, but the estimated CO_2 emission from NaOH production and $Mg(OH)_2$ extraction is still larger, indicating the importance of process optimisation. The achievement of a 100% yield at 180 °C in 6 hours would effectively reduce CO_2 emission for all series when $Mg(OH)_2$ is fully converted into $Mg(HCO_3)_2$. On the other hand, the CO_2 captured into $MgCO_3$ would not be sufficient to offset the CO_2 emission even at the maximum yield of $Mg(OH)_2$ extraction. According to reaction (6), 1 mole $Mg(OH)_2$ requires the consumption of 1 mole NaOH, and from the amount of $Mg(OH)_2$ in the reaction products it was estimated that only 55, 59 and 60% NaOH reacted with dunite in Series 2, 1.5 and 1, respectively, suggesting that there is still a wide margin for improvement of $Mg(OH)_2$ extraction.

Fig. 13(A)–(C) show the CO_2 balance for Series 2 extracted at 130, 180 and 250 °C, respectively. As shown in Fig. 13(A), the yield of $Mg(OH)_2$ extraction at 130 °C is too low and the CO_2 emission from the power consumption is considerably larger. Fig. 13(B) shows that the CO_2 balance is significantly improved at 180 °C. In fact, at this temperature the potential CO_2 capture comes close to the CO_2 produced in the case of 3 hours of extraction, when CO_2 is captured as $Mg(HCO_3)_2$. Thus, a small improvement of the extraction conditions could result in a carbon neutral process. Extending the duration of the reaction to 6 hours is not energetically beneficial as more CO_2 is emitted. The CO_2 balance for the extraction at 250 °C is much better. As shown in Fig. 13(C), at 1 and 3 hours extraction, the CO_2 capture exceeds the CO_2 produced from the NaOH production and $Mg(OH)_2$ extraction, and CO_2 is successfully reduced. At this temperature, a shorter duration is clearly more beneficial. It should be noted that the CO_2 capture exceeds the CO_2 emission only when $Mg(HCO_3)_2$ is produced as a reaction product. The achievement of a 100% yield of extraction at 130, and 180 °C would

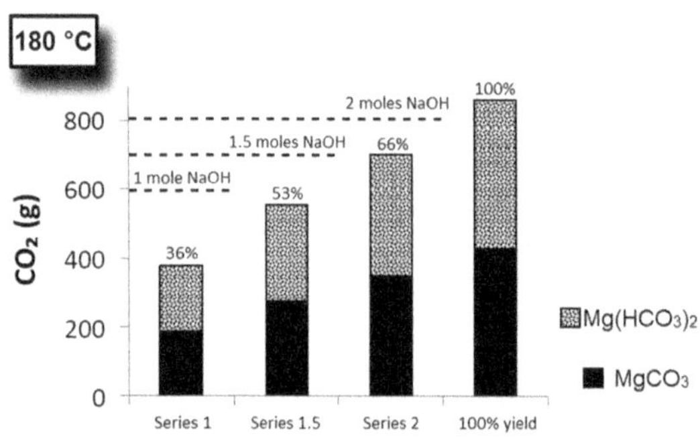

Fig. 12 Potential CO_2 capture by $Mg(OH)_2$ conversion into $MgCO_3$ or $Mg(HCO_3)_2$ (columns). The estimated CO_2 emission associated with the NaOH production and $Mg(OH)_2$ extraction at 180 °C for 6 hours are also shown (dotted lines) for Series 1, 1.5 and 2 for comparison. The yield of $Mg(OH)_2$ extraction is reported on top of the columns for each series.

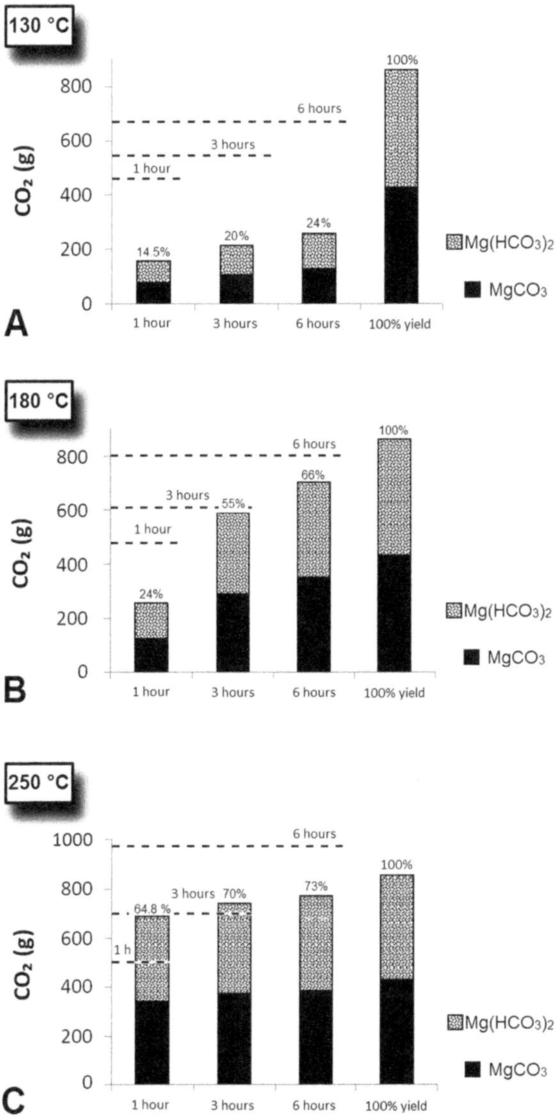

Fig. 13 Potential CO_2 capture by $Mg(OH)_2$ conversion into $MgCO_3$ or $Mg(HCO_3)_2$ (columns) for Series 2 at (A) 130 °C, (B) 180 °C and (C) 250 °C. The estimated CO_2 emission associated with the NaOH production and $Mg(OH)_2$ extraction for 1, 3 and 6 hours are also shown (dotted lines) for comparison. The yield of $Mg(OH)_2$ extraction is reported on top of the columns for each series.

reduce CO_2 under the studied conditions when $Mg(HCO_3)_2$ is formed, while at 250 °C, this occurs only for the extraction at 1 and 3 hours, and with 6 hours CO_2 is emitted. When $MgCO_3$ is produced as the reaction product, the amount of CO_2 to be captured is less than the CO_2 produced during the process, even at 100% yield, and thus the production of $Mg(HCO_3)_2$ is a key factor to offset the CO_2 produced during the extraction.

Conclusions

The present study demonstrated that the NaOH required to extract $Mg(OH)_2$ from dunite can be reduced to a stoichiometric ratio with the reaction in the solid system assisted with H_2O.

H_2O considerably improved the extraction of $Mg(OH)_2$ compared with the reaction in *dry* conditions, favouring the diffusion of the reactants involved. Both NaOH concentration in H_2O and the amount of liquid phase in the system played a significant role in the extraction of $Mg(OH)_2$ from dunite and the best results were achieved in systems with a dunite : NaOH : H_2O 1 : 2 : 0.5 mole ratio. Decreasing the NaOH below the stoichiometric ratio resulted in less $Mg(OH)_2$ extraction.

At 180 °C, the maximum yield of $Mg(OH)_2$ extraction was 66% over 6 hours reaction for solid systems with a dunite : NaOH : H_2O 1 : 2 : 0.5 mole ratio. The NaOH consumption was reduced by 97% without affecting the efficiency of extraction with respect to the NaOH aqueous systems used in the previous study reacted at the same temperature and time. The H_2O consumption was also reduced by 99%. At 250 °C, the extraction of $Mg(OH)_2$ was significantly accelerated and a 64.8% yield was achieved in 1 hour for solid systems with a dunite : NaOH : H_2O mole ratio of 1 : 2 : 0.5.

CO_2 was successfully separated from a gas mixture composed of 4–5% CO_2 and N_2 *via* liquid–gas scrubbing using a $Mg(OH)_2$ aqueous slurry and stored in the solution as $Mg(HCO_3)_2$. At least 93.6% capture efficiency was achieved over 12 hours at ambient conditions. This technology integrates the separation and capture of CO_2 in one single step and may improve the efficiency of the overall CCS process.

The CO_2 balance of the process was estimated from the emission associated with the power consumed for NaOH production and $Mg(OH)_2$ extraction together with the CO_2 captured by $Mg(OH)_2$ derived from dunite. At 130 and 180 °C, the process is carbon neutral only at yields of reaction higher than those achieved in the present study when $Mg(OH)_2$ is converted into $Mg(HCO_3)_2$. The CO_2 balance becomes negative when dunite is processed at 250 °C for 1 or 3 hours and CO_2 is captured as $Mg(HCO_3)_2$. This is promising for the possible application of $Mg(OH)_2$ derived from dunite for CO_2 separation from flue gases and storage in H_2O.

Acknowledgements

The authors thank Regione Autonoma della Sardegna, Agenzia Regionale per il Lavoro, Italy, and the EU-European Social Fund for their contribution to the work presented which is part of the PhD project funded by the Master and Back Programme 2011, Percorsi di Alta Formazione, P.O.R. FSE 2007–2013.

References

1 *IPCC: IPCC Special Report on Carbon Dioxide Capture and Storage*, ed. B. Metz, O. Davidson, H. C. De Coninck, M. Loos and L. A. Meyer, Prepared by Working Group III of the Intergovernmental Panel on Climate Change, Cambridge University Press, Cambridge, United Kingdom and New York, NY, USA, 2005.

2 L. Zhao, L. Sang, J. Chen, J. Ji and H. H. Teng, *Environ. Sci. Technol.*, 2010, **44**, 406–411.

3 H. T. Schaef, C. F. Windisch Jr, B. P. McGrail, P. F. Martin and K. M. Rosso, *Geochim. Cosmochim. Acta*, 2011, **75**, 7458–7471.

4 J. Fagerlund, E. Nduagu, I. Romão and R. Zevenhoven, *Energy*, 2012, **41**, 184–191.

5 K. S. Jung, T. C. Keener, S.-J. Khang and S.-K. Lee, *Clean Technol. Environ. Policy*, 2004, **6**, 201–212.

6 H. K. Bharadwaj, J.-Y. Lee, X. Li, Z. Liu and T. C. Keener, *J. Hazard. Mater.*, 2013, **250–251**, 292–297.

7 P. Renforth and T. Kruger, *Energy Fuels*, 2013, **27**, 4199–4207.

8 A. Sanna, M. R. Hall and M. Maroto-Valer, *Energy Environ. Sci.*, 2012, **5**, 7781–7796.

9 X. Wang, M. Maroto-Valer, G. Shiwang and X. Shisen, *Energy Procedia*, 2013, **37**, 2529–2535.

10 G. J. Simandl, S. Paradis and M. Irvine, *Geosci. Can.*, 2007, **34**, 57–64.

11 K. S. Lackner, C. H. Wendt, D. P. Butt, E. L. Joyce and D. H. Sharp, *Energy*, 1995, **20**, 1153–1170.

12 M. G. Best, *Igneous and metamorphic petrology*, Blackwell Science Ltd, Oxford, UK, 2nd edn, 2003.

13 W. J. J. Huijgen and R. N. J. Comans, *Carbon dioxide sequestration by mineral carbonation, literature review update 2003-2004*, 2005, ECN-C–05–022, ECN.

14 P.-C. Lin, C.-W. Huang, C.-T. Hsiao and H. Teng, *Environ. Sci. Technol.*, 2008, **42**, 2748–2752.

15 S. Madeddu, M. Priestnall, H. Kinoshita and E. Godoy, *Miner. Eng.*, 2014, **59**, 31–38.

16 M. Földvári, *Handbook of thermogravimetric system of minerals and its use in geological practice*, Occasional Papers of the Geological Institute of Hungary, Geological Institute of Hungary, Budapest, 2011, p. 213.

17 *Thermo-Calc Software User's Guide, Version S*, ed. P. Shi and B. Sundman, Royal Institute of Technology, Sweden, 1995.

18 H. B. Boubaker, M. Mhamdi, E. Marceau, S. Khaddar-Zine, A. Ghorbel, M. Che, Y. B. Taarit and F. Villain, *Microporous Mesoporous Mater.*, 2006, **93**, 62–70.

19 J. Zhang, M. Huang, K. Yanagisawa and S. Yao, *Ceram. Int.*, 2015, **41**, 5439–5444.

20 C. Kurt and J. Bittner, Sodium Hydroxide, in *Ullmann's Encyclopedia of Industrial Chemistry*, Wiley-VCH, Weinheim, Germany, 2003.

21 A. R. West, *Solid state chemistry and its applications*, Wiley, 2nd edn, 2014.

22 K. W. Whitten, R. E. Davis, M. L. Peck and G. G. Stanley, *Chemistry*, Brooks/Cole, Cengage Learning, Belmont, USA, 9th edn, 2010.

23 W. Stumm and J. J. Morgan, *Aquatic Chemistry, Chemical Equilibria and Rates in Natural Waters*, John Wiley & Sons, 3rd edn, 1970.

24 M. A. Shand, *The chemistry and technology of magnesia*, John Wiley & Sons Inc., Hoboken, New Jersey, USA, 2006.

25 Y. Kiros and M. Bursell, *Int. J. Electrochem. Sci.*, 2008, **3**, 444–451.

26 Parliamentary Office of Science and Technology, Carbon footprint of electricity generation, postnote, October 2006, Number 268, United Kingdom.

27 Carbolite, Installation, operation & maintenance instructions, Ovens Peak Series, MF15 – 3.14, available at: http://www.carbolite.com/downloads/operating-manuals/.

Faraday Discussions

PAPER

CO_2 capture and electrochemical conversion using superbasic $[P_{66614}]$-[124Triz]

Nathan Hollingsworth,*[a] S. F. Rebecca Taylor,*[b] Miguel T. Galante,[c] Johan Jacquemin,[b] Claudia Longo,[c] Katherine B. Holt,[a] Nora H. de Leeuw[ad] and Christopher Hardacre*[b]

Received 22nd May 2015, Accepted 3rd July 2015

DOI: 10.1039/c5fd00091b

The ionic liquid trihexyltetradecylphosphonium 1,2,4-triazolide, $[P_{66614}]$[124Triz], has been shown to chemisorb CO_2 through equimolar binding of the carbon dioxide with the 1,2,4-triazolide anion. This leads to a possible new, low energy pathway for the electrochemical reduction of carbon dioxide to formate and syngas at low overpotentials, utilizing this reactive ionic liquid media. Herein, an electrochemical investigation of water and carbon dioxide addition to the $[P_{66614}]$[124Triz] on gold and platinum working electrodes is reported. Electrolysis measurements have been performed using CO_2 saturated $[P_{66614}]$[124Triz] based solutions at −0.9 V and −1.9 V on gold and platinum electrodes. The effects of the electrode material on the formation of formate and syngas using these solutions are presented and discussed.

1 Introduction

Since the industrial revolution, the harnessing of fossil fuels by combustion for energy and transportation has seen levels of CO_2 in the atmosphere steadily rise.[1,2] One concern is the increasing concentration of CO_2 in the atmosphere is connected with rising global temperatures and sea levels.[3] Given the potential implications of global warming, strategies to capture and/or convert CO_2 provide the subject of much scientific endeavour.[4]

Aside from a greenhouse gas, CO_2 can be considered as an abundant, renewable and low cost C1 building block for the production of fuels, or as a feedstock for industrial chemical synthesis. The synthesis of fuels is particularly attractive given the carbon neutrality of the product.

[a]Department of Chemistry, University College London, 20 Gordon Street, London, WC1H 0AJ, UK. E-mail: n.hollingsworth@ucl.ac.uk

[b]School of Chemistry and Chemical Engineering, Queen's University Belfast, Belfast, N. Ireland, BT9 5AG, UK. E-mail: r.taylor@qub.ac.uk; c.hardacre@qub.ac.uk

[c]Institute of Chemistry, University of Campinas – UNICAMP, Campinas, SP, Brazil

[d]School of Chemistry, Cardiff University, Cardiff, CF10 3XQ, UK

Electrochemical reduction offers a viable route to this goal due to the potential to reduce CO_2 at high efficiencies and selectivity, as well as the ability to harness electrical energy from renewable sources.[5,6] The main barriers are the thermodynamic stability and the kinetic inertness of CO_2, due to the fact that the first electron reduction of linear CO_2 to form the bent $[^{\bullet}CO_2]^-$ radical anion requires a high reduction potential, −1.9 V *vs.* NHE.[7] Attempts have been made to reduce this potential by employing various catalysts, such as the recently reported reductions on Fe_3S_4 surfaces.[8]

Ionic liquids (ILs) have been employed as CO_2 capture agents and electrochemical conversion media due to their intrinsic ionic conductivity, wide potential windows and CO_2 solubility levels.[9–12] ILs generally physically absorb CO_2 into the solution free volume. Reports have shown that ILs can act to stabilise the first electron reduction of the physically absorbed CO_2 by complexation of the reduction product with the IL cation.[12] This significantly reduces the energy required to form the $[^{\bullet}CO_2]^-$ radical anion. Initial CO_2 electrochemical reduction experiments in ILs employed 1-alkyl-3-methylimidazolium ($[C_n\text{mim}]^+$) based ILs with a range of non-coordinating anions such as $[BF_4]^-$, Br^-, Cl^-. The electrochemically generated anion radical, $[^{\bullet}CO_2]^-$, was then reacted with alcohols forming dialkyl carbonates.[13–15]

Further studies used hydrated ILs, to provide a source of H^+. CO_2 physically absorbed within the IL was reduced with Ag, as the working electrode, further enabling the reduction of the IL stabilised $[^{\bullet}CO_2]^-$ to CO, as shown in the Scheme 1.[12]

CO_2 reduced to CO in a hydrated IL $[C_2\text{mim}][BF_4]$ on a Ag working electrode was reported to have overpotentials as low as 0.2 V.[12] H_2 production was also suppressed due to the cation forming a monolayer on the electrode surface.[16] Further improvements have been reported by employing a MoS_2 working electrode, resulting in overpotentials as low as 0.054 V.[17]

Product selectivity has also been observed upon altering the IL anion. For example, using $[C_2\text{mim}]^+$ based ILs on Pb electrodes, oxalate is favoured in the presence of $[NTf_2]^-$ and formate is favoured when $[TFA]^-$ is used.[9,11]

Recently, we have reported an alternative route for CO_2 reduction in ILs, by using $[P_{66614}][124\text{Triz}]$.[18] $[P_{66614}][124\text{Triz}]$ is able to capture significantly more CO_2 than ILs such as $[C_4\text{mim}][NTf_2]$ or $[C_2\text{mim}][OAc]$, due to the chemical reactivity of the 1,2,4-triazolide anion.[19] The anion can chemically bind CO_2 allowing up to equimolar amounts of CO_2 to be captured, as illustrated in the Scheme 2.[19]

Crucially, upon binding, the CO_2 geometry is altered from linear to bent. A lower energy pathway to CO_2 reduction was demonstrated in this IL, which may be related to the chemical binding of the CO_2 by the anion.[18] $[P_{66614}][124\text{Triz}]$ is also able to absorb CO_2 through physical exothermic absorption in the solution free

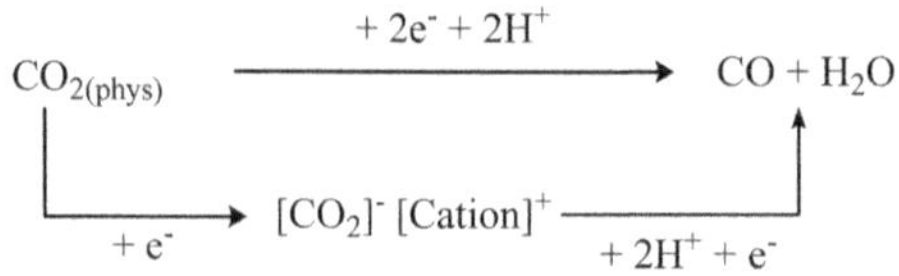

Scheme 1 Reduction of CO_2 to CO in an IL, which absorbs CO_2 physically.[12]

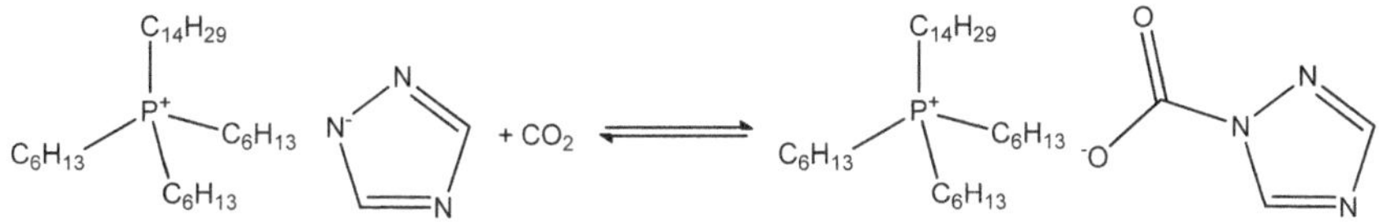

Scheme 2 CO_2 binding to the $[124Triz]^-$ anion in the IL $[P_{66614}][124Triz]$.[19]

volume that is driven by an entropic process of solvation. The physically absorbed CO_2 may also undergo reduction stabilised by the $[P_{66614}]^+$ cation. Formate was produced by the lower energy route at −0.7 V *vs.* Ag/$AgNO_3$ with 95% Faradaic efficiency on the Ag electrodes. In addition, reduction of the physically absorbed CO_2 at −1.9 V *vs.* Ag/$AgNO_3$ produced CO with a 41% Faradaic efficiency.[18] In comparison, on using $[P_{66614}][NTf_2]$ or $[P_{66614}][BF_4]$, which do not support chemical CO_2 binding, little or no product was formed at −0.7 V suggesting that the reaction occurring at less negative potentials (*i.e.* −0.7 V) in $[P_{66614}][124Triz]$ is related to the reduction of chemically bound CO_2 (Scheme 2). Products were formed at −1.9 V, *i.e.* the same potential as found in $[P_{66614}][124Triz]$, indicating that they are formed from the reduction of physically absorbed CO_2 stabilised by the $[P_{66614}]^+$ cation (Scheme 1).[18] Herein, the study is expanded to investigate the influence of the electrode material on the CO_2 reduction potential, product distribution and Faradaic efficiency.

2 Experimental

2.1 Materials and IL synthesis

Trihexyltetradecylphosphonium chloride ($[P_{66614}]Cl$) was received from Cytec (97.7%) and 1,2,4-triazole (98%) was purchased from Sigma-Aldrich. The Au foil (0.1 mm thick 99.9975+%) and Pt wire (0.3 mm dia., 99.99%) were purchased from Alfa Aesar. Gaseous nitrogen (99.998%) and carbon dioxide (99.99%) were obtained from BOC and passed through drying columns before contact with the IL samples. The water was purified using a Milli-Q 18.3 MΩ water system.

$[P_{66614}][124Triz]$ was prepared using a previously reported method[20] and a two step synthesis procedure; $[P_{66614}][OH]$ was synthesized using an anion exchange resin from $[P_{66614}]Cl$ followed by addition of the superbase. The water content of the ILs was measured using a Metrohm 787 KF Titrino Karl Fischer and was found to be <0.1 wt% for all ILs. Halide content was below the detectable limit by testing with $AgNO_3$.

2.2 Electrochemical experiments

Cyclic voltammograms (CVs) were recorded using an Ecochemie Autolab Potentiostat/Galvanostat (PGSTAT302) and carried out using a three-electrode arrangement with a Au or Pt working electrode (1.6 mm), a platinum coil as the counter electrode. All potentials were measured with respect to a 0.01 mol L^{-1} Ag/Ag^+ reference, with $AgNO_3$ dissolved in $[C_4mim][NO_3]$ and separated from the bulk solution *via* a glass frit. Electrolysis experiments were carried out using a three electrode set up in a sealed 50 cm^3 cell with a gas tight syringe attached. The working electrodes were a coiled Pt wire or a Au foil, a Pt coil was used as the counter electrode (contained within a counter compartment) and the potentials were measured with respect to a 0.01 mol L^{-1} Ag/Ag^+ reference, with $AgNO_3$ dissolved in $[C_4mim][NO_3]$ and separated from the bulk solution *via* a glass frit. A 0.1 mol L^{-1} $[P_{66614}]$[124Triz], 0.7 mol L^{-1} H_2O in acetonitrile solution (8 cm^3) was added to the cell and bubbled with CO_2 (25 cm^3 min^{-1}) for 60 min, during this time the gas syringe was gradually opened. The cell was then sealed, connected up to a potentiostat (Biologic VMP3 controlled by EC-Lab) and a potential applied to the cell for the time required to pass 10 C of charge. It should be noted that no ionic liquid decomposition or reaction was observed under any of the conditions tested in this study.

2.3 Solution phase detection

Solution samples were analysed by obtaining ^1H NMR spectra using a Bruker Avance III 600 with a DCH cryoprobe. Quantitative analysis was performed using the ERETIC2 quantification tool within Topspin 3.2.

2.4 Gas phase detection

The gas phase of the reaction was analysed using a Perkin-Elmer Clarus 500 Gas Chromatograph equipped with a TCD and FID (with methanizer) fitted with a packed column (stainless steel, 30 ft, 1/8 inch OD, 2.0 mm ID packed with Haysep DB 100/120 mesh). The gaseous products were transferred to the GC by connecting up the cell and injecting the contents of the gas tight syringe attached to the cell into the GC. H_2 was quantified using the TCD and CO using the FID.

3 Results and discussion

3.1 Voltammetry study of CO_2 reduction in $[P_{66614}]$[124Triz] using various working electrodes

CVs have been recorded at both Au and Pt electrodes in 0.1 mol L^{-1} $[P_{66614}]$-[124Triz] in acetonitrile (MeCN). CVs were recorded at a scan rate of 100 mV s^{-1} at 22 °C. The reference electrode was a previously reported 0.01 mol L^{-1} Ag/Ag^+ reference, with $AgNO_3$ dissolved in $[C_4mim][NO_3]$ and separated from the bulk solution *via* a glass frit.[21,22] 0.1 mol L^{-1} $[P_{66614}]$[124Triz] in MeCN was first saturated with argon and a baseline CV taken. Thereafter, water was added (0.7 mol L^{-1}), as a source of H^+ and the influence on the CV investigated. The H^+ is required to make protonated reduction products and CO through the reaction of reduced CO_2 and H^+ (Scheme 1). Finally, the hydrated IL mixture was purged with CO_2 for 30 min and a CV taken to assess CO_2 reduction processes.

3.1.1 Au working electrode. CVs relating to the use of a polished Au working electrode (0.16 cm^2) are displayed in Fig. 1. The CV, after purging with Ar, is shown in blue. A small increase in reduction current is observed at *ca.* −1.1 V with a rapid increase in reduction current at *ca.* −2.0 V. Upon the addition of H_2O to the IL/MeCN mixture, a small increase in current is observed at −1.0 V and a larger increase at *ca.* −1.75 V (Fig. 1, red). Although the small increase in current observed at −1.0 V has a slight anodic shift compared with that observed in the Ar saturated sample (−1.1 V), this feature may be attributed to the reduction of 1,2,4-triazole formed from the reaction of the $[124Triz]^-$ anion and H_2O (Scheme 3).

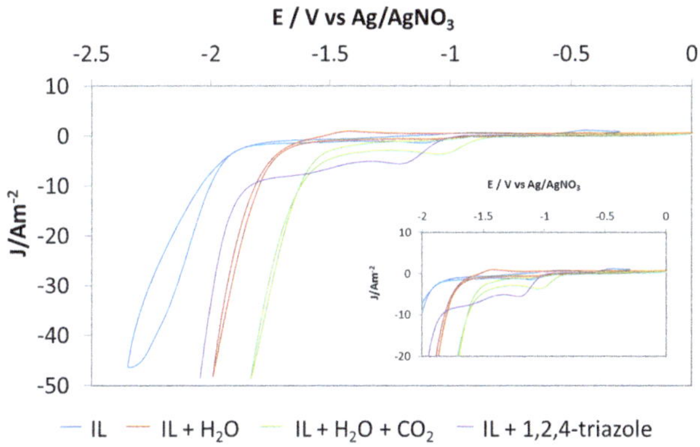

Fig. 1 Cyclic voltammograms of 0.1 mol L^{-1} $[P_{66614}][124Triz]$ (IL) in acetonitrile (MeCN) at a gold electrode; (blue) IL in MeCN; (red) IL in MeCN + 0.7 mol L^{-1} H_2O; and (green) IL in MeCN + 0.7 mol L^{-1} H_2O + CO_2, IL in MeCN + 0.5 mol L^{-1} 1,2,4-trizole; only the cathodic linear sweep is shown for clarity. Magnification is shown inset.

To test this theory, 1,2,4-triazole was deliberately added to a solution of hydrated 0.1 mol L^{-1} $[P_{66614}][124Triz]$ in MeCN. The CV showed that, upon addition of 1,2,4-triazole, the current from −1.0 V increased, supporting this proposal. The presence of the small current increase at −1.1 V in the Ar saturated sample could, therefore, either be due to adventitious water or trace 1,2,4-triazole impurities from the synthesis process of $[P_{66614}][124Triz]$, which is synthesised from the addition of 1,2,4-triazole to $[P_{66614}][OH]$.

The large reduction current at −1.8 V, observed upon the addition of H_2O, shows a large anodic shift compared to the onset of the large reduction current in the Ar saturated sample. This is proposed to be due to the reduction of H_2O to H_2. This region also increases in current when 1,2,4-triazole is deliberately added to the electrolyte, suggesting that generation of H_2 from the reduction of 1,2,4-triazole may also contribute to this current. In addition, a small oxidative stripping peak is observed at −1.5 V, following the rapid increase of current at −1.8 V, suggesting that the product of the reduction may interact with the electrode surface. Recent reports employing $[C_4mim][BF_4]$ on Au surfaces have shown the hydrogen evolution reaction to occur at *ca.* −2.0 V *vs.* Ag/Ag^+,[23] suggesting that the use of chemically co-ordinating IL provides a pathway to less negative reduction potentials.

Saturation of the hydrated IL mixture with CO_2 (Fig. 1 green) resulted in an increase in current at −0.9 V, followed by a rapid increase in current at −1.6 V. The peak at −0.9 V shows an increased current and greater anodic shift when compared to the hydrated sample. The large increase in current observed at −1.6 V shows an anodic shift in comparison to Ar saturated and hydrated samples. Previous reports on Ag electrodes have suggested that the reduction at less negative potentials (−0.7 V) could be assigned to the reduction of CO_2

$$[124Triz]^- + H_2O \longrightarrow [124Triz\text{-}H] + OH^- \quad ...(1)$$

Scheme 3 H_2O reaction with the $[124Triz]^-$ anion in the $[P_{66614}][124Triz]$.

chemically bound to the $[124Triz]^-$ anion, while reductions at more negative potentials (−1.9 V) could be assigned to the reduction of physically absorbed CO_2 (Scheme 4 eqn (2)–(5), respectively).[18] The reduction of physically absorbed CO_2 at −1.6 V is in good agreement with CO_2 reduction reported on Au electrodes in $[C_4mim][BF_4]$. Reduction of CO_2 in $[C_4mim][BF_4]$, an IL that only exhibits physical CO_2 absorption, on the Au surface was reported at −1.8 V *vs.* NHE (*i.e.* −1.6 V *vs.* Ag/Ag^+).[23] Interestingly, the potential for the reduction of CO_2 over Au reported, herein, at −0.9 V is the lowest potential, to the best of our knowledge, reported to date using ionic liquid systems showing the potential of reactive IL to enhance the electrochemical activity.

Assuming similar mechanisms take place at the Au electrode, this suggests that chemically bound CO_2 can be reduced at less negative potentials on Ag than Au (−0.7 V compared to −0.9 V), but that the reduction of physically bound CO_2 occurs at less negative potentials on Au than Ag (−1.6 V *vs.* −1.9 V). One plausible explanation would be that Ag is more catalytically active than Au and, therefore, the IL is more strongly bound to the Ag electrode than the Au. This may suggest that the chemically bound CO_2 on Ag can be reduced at less negative potentials due to the increased contact with the electrode but the physically absorbed CO_2 is blocked by the IL requiring more negative potentials for reduction. Reduced binding in the Au system suggests the physically absorbed CO_2 has easier access to the electrode and, therefore, physically absorbed reduction occurs at a less negative potential than on Ag, but that the reduction of the chemically absorbed CO_2 requires a more negative potential than in the Ag system.

3.1.2 Pt working electrode. CVs relating to the use of a polished Pt working electrode (0.16 cm^2) are displayed in Fig. 2, the CV after purging with Ar is shown in blue. An increase in current is observed at −1.6 V followed by the start of a rapid current increase at *ca.* −2.1 V. Upon the addition of H_2O to the IL/MeCN mixture (Fig. 2 red), a rapid increase in current is observed at −1.5 V with no smaller current increase observed at lower potentials. The rapid increase in current at −1.5 V shows an anodic shift and enhanced current compared to the Ar saturated sample. This peak is plausibly associated with the reduction of H_2O to H_2. On comparison with the Au electrode the reduction of H_2O to H_2 takes place at a less negative potential on Pt than Au (−1.5 V *vs.* −1.8 V). This is consistent with Pt being well known as an excellent catalyst for water splitting.

$$[124TrizCO_2]^- + H^+ + 2e^- \longrightarrow [HCOO]^- + [124Triz]^- \quad ...(2)$$

$$[124TrizCO_2]^- + 2H^+ + 2e^- \longrightarrow CO + H_2O + [124Triz]^- \quad ...(3)$$

$$CO_2 + e^- \longrightarrow [CO_2]^- [P_{66614}]^+ \xrightarrow{+ 2H^+ + e^-} CO + H_2O \quad ...(4)$$

$$[CO_2]^- [P_{66614}]^+ \xrightarrow{+ H^+ + e^-} [HCOO]^- \quad ...(5)$$

Scheme 4 Suggested CO_2 reduction processes occurring on the working electrode.

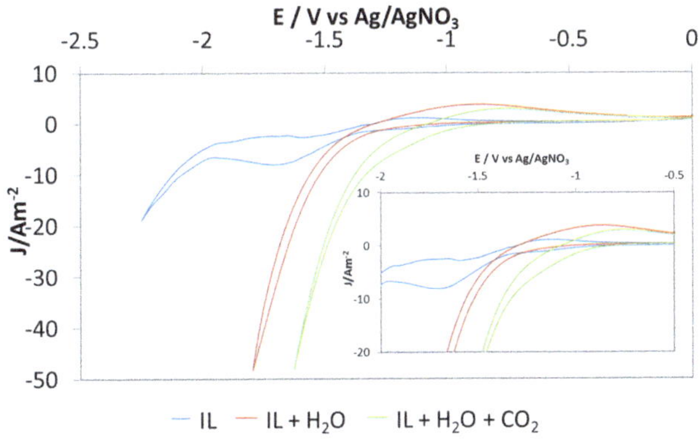

Fig. 2 Cyclic voltammograms of 0.1 mol L^{-1} [P_{66614}][124Triz] (IL) in acetonitrile (MeCN) at a platinum electrode; (blue) IL in MeCN, (red) IL in MeCN + 0.7 mol L^{-1} H_2O, and (green) IL in MeCN + 0.7 mol L^{-1} H_2O + CO_2. Magnified CVs are inset.

Saturating the hydrated IL mixture with CO_2 (Fig. 2 green) results in a slow increase in current at −0.9 V. The increase becomes rapid at −1.4 V. The slow increase in current is plausibly due to the reduction of CO_2 bound to the $[124Triz]^-$ anion, while the rapid increase in current is associated with the reduction of physically absorbed CO_2.

Both Au and Pt display similar reduction potentials for CO_2 bound to the $[124Triz]^-$ anion (−0.9 V), which are more negative than reduction on Ag (−0.7 V). However, the reduction of physically bound CO_2 is less negative for Pt than Au (−1.4 V and −1.6 V respectively), both of which are less negative than Ag (−1.9 V). It is important to note that the Pt is more catalytically active than both the Au and Ag and, therefore, should show a higher activity. However, as a result of this increased activity Pt will also bind the IL more strongly than the Ag, therefore, the results suggest that a balance between activity and adsorption strength is required for CO_2 reduction at lower potentials. As CO is also formed during these experiments and the binding of CO to Pt will be much stronger than on Ag or Au, site blocking from the CO may also be contributing to the observed differences between the electrode performance.

3.2 Electrolysis study

Electrolysis was performed in CO_2 saturated and hydrated (0.7 mol L^{-1} H_2O) 0.1 mol L^{-1} [P_{66614}][124Triz] in MeCN (8 cm^3). Electrolysis was continued until a fixed charge of 10 C had passed, after which the solution phase was tested by quantitative ^{1}H NMR and in the gas phase by gas chromatography (GC). The reported values for the analogous electrolysis using Ag electrodes is displayed in Fig. 3. While Ag electrodes are reported to display Faradic efficiencies of 95% at −0.7 V, with an applied potential of 0.17 V, electrolysis on Au and Pt at this potential resulted in no detected reduction product. Therefore, electrolysis was also performed at −1.9 V and −0.9 V to test for reduction of physical and chemical absorbed CO_2, respectively, using these electrodes. The experiments were also performed at these potentials with Ag for comparison (Fig. 3 and 4), although it should be noted that electrolysis at −0.9 V is not the optimum potential for formate production on Ag.[18]

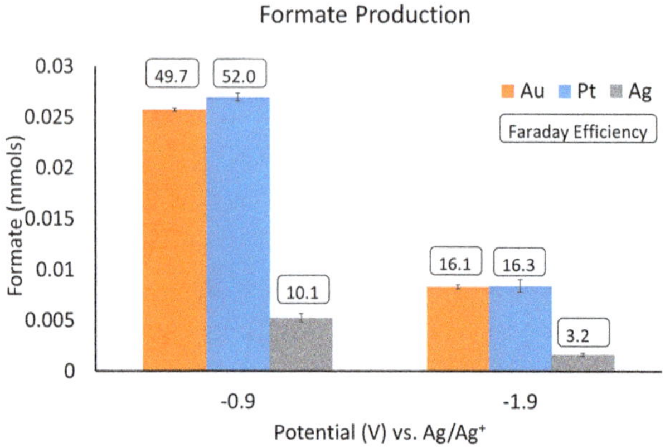

Fig. 3 Quantitative ^{1}H NMR analysis of formate production at the Au, Pt, and Ag electrodes after 10 C of charge is passed. Error bars represent ±1.5 standard deviations. Faraday efficiencies for each potential and electrode are shown in the boxes.

3.2.1 Au electrode. 0.5 cm^2 of Au foil was employed as a working electrode. Quantitative ^{1}H NMR of the electrolyte after electrolysis showed the presence of formate for both investigated potentials (−0.9 V and −1.9 V). The formate produced at −0.9 V is proposed to be formed from the reduction of CO_2 bound to the $[124Triz]^-$ anion (Scheme 4 eqn (2)), while the formate produced at −1.9 V is proposed to be generated from the reduction of physically absorbed CO_2 stabilised by $[P_{66614}]^+$ cation (Scheme 4 eqn (5)). A greater quantity of formate was detected at −0.9 V (0.026 mmol) compared with −1.9 V (0.0083 mmol) with Faradaic efficiencies of 49.7 and 16.1%, respectively. This is in contrast to previous reports using $[C_2mim][TFA]$, whereby the Faradaic efficiency for formate production increases with potential.[9] The open circuit potential (OCP) prior to electrolysis, but after CO_2 saturation, was measured at −0.45 V, suggesting an applied potential of −0.45 and −1.45 V for potential holds of −0.9 and −1.9 V, respectively. A comparison with the reported values for the analogous electrolysis

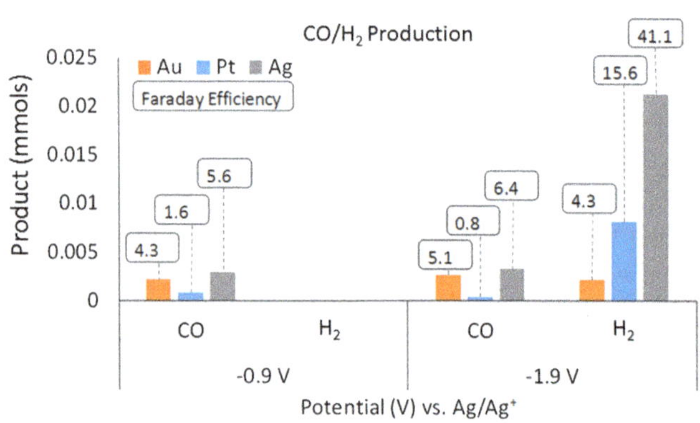

Fig. 4 GC analysis of CO and H_2 production at the Au, Pt, and Ag electrodes after 10 C of charge is passed. Faraday efficiencies (%) for each potential and electrode are shown in the boxes.

using Ag electrodes is displayed in Fig. 3. Electrolysis on Ag resulted in considerably less formate at both −0.9 and −1.9 V, and it should be noted that the maxima for formate production is −0.7 V, at which potential considerably more formate is produced than on the Au electrode.

GC analysis of the headspace gas mixture after electrolysis showed that both CO and H_2 were formed with 0.0022 mmol and 0.0027 mmol of CO produced at −0.9 V and −1.9 V, respectively. This is assigned to the reduction of CO_2 bound to the $[124Triz]^-$ anion (Scheme 4 eqn (3)) and the reduction of CO_2 stabilised by the $[P_{66614}]^+$ cation (Scheme 4 eqn (4)). In addition, hydrogen was only detected at −1.9 V, which is consistent with H_2O being reduced to H_2.

3.2.2 Pt electrode. Quantitative ^1H NMR of the electrolyte after electrolysis showed a similar trend for formate production on Pt compared to Au. A greater quantity of formate was detected at −0.9 V (0.027 mmol) compared with −1.9 V (0.008 mmol) with Faradaic efficiencies of 52.0 and 16.3%, respectively. However, formate production at both potentials was not significantly different from the amount produced using the Au electrode. The OCP prior to electrolysis, but after CO_2 saturation, was more negative than that previously found on Au. It was measured at −0.60 V, giving smaller applied potentials of 0.3 and 1.3 V for electrolysis at −0.9 and −1.9 V, respectively. CO was detected at both −0.9 and −1.9 V, however, the amount of CO detected was significantly reduced at both potentials compared with the analogous reductions over Au and Ag. No H_2 was detected at −0.9 V while significantly more H_2 was detected at −1.9 V on Pt than Au, as expected. This corresponds well with previous reports showing that the order in ability to catalyse the hydrogen evolution reaction is Pt > Au > Ag.[24] This in turn relates to the observed reduction potentials after the addition of H_2O on the Pt, Au and Ag electrodes (−1.5, −1.8 and −1.9 V, respectively). However, more H_2 is detected for Ag electrodes at −1.9 V than over Pt, this is surprising given that Ag has the most negative onset potential for reduction of H_2O to H_2. Previous reports have stated that the IL cation may suppress hydrogen production by blocking H_2O binding. As such, the disparity in hydrogen production may indicate a differing affinity for the cation to the three electrode surfaces, of the order Ag < Pt < Au, as well as the ability of the electrode material to activate water. As such, the production of H_2 and CO may be a balance between catalytic activity (through onset potentials) and binding energy of the reduced species.

Overall, for all three electrodes (Ag, Au and Pt) the trend in the species formed during the reduction is the same, *i.e.* less negative potentials favour formate whereas more negative potentials favour CO and H_2. The main difference between the electrodes is the potential at which the maxima/optimum for each species occur. For example, for formate production, on the Ag electrode this occurs at −0.7 V, which is less negative than found for the Au and Pt electrodes. In addition, it is clear that as the potential becomes more negative the Faradaic efficiency for the species detected decreases for all three electrodes.

4 Conclusions

We have shown that the CO_2 saturated superbasic IL $[P_{66614}][124Triz]$ can be reduced on Au and Pt electrodes to produce formate and syngas. Two reaction pathways appear to exist at −0.9 V and −1.9 V. Formate is formed in increased quantities at −0.9 V over −1.9 V on both electrode surfaces with Faradaic efficiencies of *ca.* 50% for both electrode surfaces. The production of CO is higher on Au electrodes than Pt electrodes for both electrolysis potentials of −0.9 and −1.9 V. The production of hydrogen, only observed at −1.9 V, is *ca.* 3 times higher on Pt electrodes, enabling the ratio of CO : H_2 in the syngas mixture to be influenced by electrode choice. The reduction of physically bound CO_2 is shown to occur at similar potentials on Au electrodes to those reported for ILs that only physically absorb CO_2, such as [C_4mim][BF_4], albeit at lower Faradaic efficiencies.[23] Formate production in ionic liquids has been previously reported to occur on indium surfaces using hydrated CO_2 saturated [C_2mim][TFA].[9] The $[TFA]^-$ anion only very weakly binds to CO_2, which results in little significant activation of the CO_2 and formate production at low Faradaic efficiencies (*ca.* 10%) at 1.55 V *vs.* Ag/Ag^+, rising to 90% efficiency at −1.95 V *vs.* Ag/Ag^+. This behaviour is in good agreement with our proposed onset of reduction of physically bound CO_2. In the present study, the binding of CO_2 is much stronger and, thus, formate production on Au and Pt electrodes, from the chemically bound CO_2, is observed at applied potentials lower than any previously reported ILs. These results show the same trends as found over the previously reported Ag electrode, albeit with some differences. For example, the maximum formate production is found at −0.7 V over Ag at which potential no reduction was observed on Pt and Au.

5 Abbreviations

$[P_{66614}]^+$	Trihexyltetradecylphosphonium
$[C_2mim]^+$	1-Ethyl-3-methylimidazolium
$[C_4mim]^+$	1-Butyl-3-methylimidazolium
$[124Triz]^-$	1,2,4-Triazolide
$[BF_4]^-$	Tetrafluoroborate
$[NTf_2]^-$	Bis(trifluoromethylsulfonyl)imide
$[TFA]^-$	Trifluoroacetate
$[OAc]^-$	Acetate
$[NO_3]^-$	Nitrate

Acknowledgements

This work was carried out as part of the "4CU" programme grant, aimed at sustainable conversion of carbon dioxide into fuels, led by The University of Sheffield and carried out in collaboration with The University of Manchester, Queen's University Belfast and University College London. The authors acknowledge gratefully the Engineering and Physical Sciences Research Council (EPSRC) for supporting this work financially (Grant No. EP/K001329/1). S. F. R. Taylor and N. Hollingsworth wish to thank the ELSOL workshop, Researcher Links, FAPESP and Newton fund for funding. M. T. Galante thanks the RSC and CNPq for funding.

Notes and references

1 S. Solomon, D. Qin, M. Manning, R. B. Alley, T. Berntsen, N. L. Bindoff, Z. Chen, A. Chidthaisong, J. M. Gregory, G. C. Hegerl, M. Heimann, B. Hewitson, B. J. Hoskins, F. Joos, J. Jouzel, V. Kattsov, U. Lohmann,

T. Matsuno, M. Molina, N. Nicholls, J. Overpeck, G. Raga, V. Ramaswamy, J. Ren, M. Rusticucci, R. Somerville, T. F. Stocker, P. Whetton, R. A. Wood and D. Wratt, *Technical Summary, in: Climate Change 2007: The Physical Science Basis. Contribution of Working Group I to the Fourth Assessment Report of the Intergovernmental Panel on Climate Change*, ed. S. Solomon, D. Qin, M. Manning, Z. Chen, M. Marquis, K. B. Averyt, M. Tignor and H. L. Miller, Cambridge University Press, Cambridge, United Kingdom, New York, NY, USA, 2007.

2 T. R. Karl and K. E. Trenberth, *Science*, 2003, **302**, 1719–1723.

3 *Climate Change and Water, Technical Paper of the Intergovernmental Panel on Climate Change*, ed. B. C. Bates, Z. W. Kundzewicz, S. Wu and J. P. Palutikof, IPCC Secretariat, Geneva, 2008, p. 210.

4 P. Styring, E. A. Quadrelli and K. Armstrong, *Carbon Dioxide Utilisation: Closing the Carbon Cycle*, Elsevier Science, 2014.

5 D. T. Whipple and P. J. A. Kenis, *J. Phys. Chem. Lett.*, 2010, **1**, 3451–3458.

6 N. S. Spinner, J. A. Vega and W. E. Mustain, *Catal. Sci. Technol.*, 2012, **2**, 19–28.

7 J. Schneider, H. Jia, J. T. Muckerman and E. Fujita, *Chem. Soc. Rev.*, 2012, **41**, 2036–2051.

8 A. Roldan, N. Hollingsworth, A. Roffey, H. U. Islam, J. B. M. Goodall, C. R. A. Catlow, J. A. Darr, W. Bras, G. Sankar, K. B. Holt, G. Hogarth and N. H. de Leeuw, *Chem. Commun.*, 2015, **51**, 7501–7504.

9 J. D. Watkins and A. B. Bocarsly, *ChemSusChem*, 2014, **7**, 284–290.

10 M. C. Buzzeo, O. V. Klymenko, J. D. Wadhawan, C. Hardacre, K. R. Seddon and R. G. Compton, *J. Phys. Chem. B*, 2004, **108**, 3947–3954.

11 (*a*) L. Sun, G. K. Ramesha, P. V. Kamat and J. F. Brennecke, *Langmuir*, 2014, **30**, 6302–6308; (*b*) J. L. Anthony, J. L. Anderson, E. J. Maginn and J. F. Brennecke, *J. Phys. Chem. B*, 2005, **109**, 6366–6374; (*c*) L. A. Blanchard, D. Hancu, E. J. Beckman and J. F. Brennecke, *Nature (London, U. K.)*, 1999, **399**, 28–29.

12 B. A. Rosen, A. Salehi-Khojin, M. R. Thorson, W. Zhu, D. T. Whipple, P. J. A. Kenis and R. I. Masel, *Science*, 2011, **334**, 643–644.

13 Q. Feng, S. Liu, X. Wang and G. Jin, *Appl. Surf. Sci.*, 2012, **258**, 5005–5009.

14 X. Yuan, B. Lu, J. Liu, X. You, J. Zhao and Q. Cai, *J. Electrochem. Soc.*, 2012, **159**, E183–E186.

15 F. Liu, S. Liu, Q. Feng, S. Zhuang, J. Zhang and P. Bu, *Int. J. Electrochem. Sci.*, 2012, **7**, 4381–4387.

16 B. A. Rosen, J. L. Haan, P. Mukherjee, B. Braunschweig, W. Zhu, A. Salehi-Khojin, D. D. Dlott and R. I. Masel, *J. Phys. Chem. C*, 2012, **116**, 15307–15312.

17 M. Asadi, B. Kumar, A. Behranginia, B. A. Rosen, A. Baskin, N. Repnin, D. Pisasale, P. Phillips, W. Zhu, R. Haasch, R. F. Klie, P. Král, J. Abiade and A. Salehi-Khojin, *Nat. Commun.*, 2014, **5**, 4470.

18 N. Hollingsworth, S. F. R. Taylor, M. T. Galante, J. Jacquemin, C. Longo, K. B. Holt, N. H. de Leeuw and C. Hardacre, *Angew. Chem., Int. Ed.*, 2015, DOI: 10.1002/anie.201507629.

19 (*a*) C. Wang, X. Luo, H. Luo, D.-e. Jiang, H. Li and S. Dai, *Angew. Chem., Int. Ed.*, 2011, **50**, 4918–4922; (*b*) S. Seo, M. A. DeSilva and J. F. Brennecke, *J. Phys. Chem. B*, 2014, **118**, 14870–14879.

20 S. F. R. Taylor, C. McCrellis, C. McStay, J. Jacquemin, C. Hardacre, M. Mercy, R. Bell and N. H. de Leeuw, *J. Solution Chem.*, 2015, **44**, 511–527.

21 L. Aldous, D. S. Silvester, C. Villagran, W. R. Pitner, R. G. Compton, M. Cristina Lagunas and C. Hardacre, *New J. Chem.*, 2006, **30**, 1576–1583.
22 N. S. A. Manan, L. Aldous, Y. Alias, P. Murray, L. J. Yellowlees, M. C. Lagunas and C. Hardacre, *J. Phys. Chem. B*, 2011, **115**, 13873–13879.
23 J. H. Koh, H. S. Jeon, M. S. Jee, E. B. Nursanto, H. Lee, Y. J. Hwang and B. K. Min, *J. Phys. Chem. C*, 2015, **119**, 883–889.
24 F. Zhou, S. Liu, B. Yang, P. Wang, A. S. Alshammari and Y. Deng, *Electrochem. Commun.*, 2014, **46**, 103–106.

Faraday Discussions

PAPER

Highly optimized CO_2 capture by inexpensive nanoporous covalent organic polymers and their amine composites

Hasmukh A. Patel and Cafer T. Yavuz*

Received 30th May 2015, Accepted 16th June 2015
DOI: 10.1039/c5fd00099h

Carbon dioxide (CO_2) storage and utilization requires effective capture strategies that limit energy penalties. Polyethylenimine (PEI)-impregnated covalent organic polymers (COPs) with a high CO_2 adsorption capacity are successfully prepared in this study. A low cost COP with a high specific surface area is suitable for PEI loading to achieve high CO_2 adsorption, and the optimal PEI loading is 36 wt%. Though the adsorbed amount of CO_2 on amine impregnated COPs slightly decreased with increasing adsorption temperature, CO_2/N_2 selectivity is significantly improved at higher temperatures. The adsorption of CO_2 on the sorbent is very fast, and a sorption equilibrium (10% wt) was achieved within 5 min at 313 K under the flow of simulated flue gas streams. The CO_2 capture efficiency of this sorbent is not affected under repetitive adsorption–desorption cycles. The highest CO_2 capture capacity of 75 mg g^{-1} at 0.15 bar is achieved under dry CO_2 capture however it is enhanced to 100 mg g^{-1} in the mixed gas flow containing humid 15% CO_2. Sorbents were found to be thermally stable up to at least 200 °C. TGA and FTIR studies confirmed the loading of PEIs on COPs. This sorbent with high and fast CO_2 sorption exhibits a very promising application in direct CO_2 capture from flue gas.

Introduction

The issue of global warming resulting from increased atmospheric concentrations of carbon dioxide (CO_2) is arguably the most important environmental challenge facing the world today.[1,2] Effective strategies of CO_2 remediation have to include capture and utilization in close connection with each other. To begin such an endeavour, capturing of CO_2 generated by large point sources, such as fossil-fuel-fired power gasification plants, is essential.[3]

Liquid amine scrubbing technology (20% to 30% aqueous solutions of alkanolamines) has been commercialized industrially for decades and it is the most suitable for high volume flue gas streams among the CO_2 separation technologies in the present scenario.[4] The major shortcoming of the aqueous amine systems is

Graduate School of EEWS, Korea Advanced Institute of Science and Technology (KAIST), Daejeon 305-701, Republic of Korea. E-mail: yavuz@kaist.ac.kr

the high amount of energy required for the regeneration step due to the chemisorptive mechanism, degradation of amines, corrosion of equipment, and sluggish adsorption–desorption kinetics. Nanoporous solid adsorbents have been proposed to counter these shortcomings associated with aqueous amine systems. Organic, inorganic and organic–inorganic hybrid porous materials with high surface areas are used for the sorption, purification, and storage of gases due to their plentiful quantities, composition, pore topology and morphology.[5] Solid adsorbents—activated carbons, zeolites, mesoporous silica, metal–organic frameworks and nanoporous polymers—have been utilized for CO_2 separation.[3,6–8] Although these sorbents show high CO_2 adsorption capacities (under pure, dry CO_2 feeds), most of them don't demonstrate selective separation of CO_2 from the mixed gas streams at warmer temperatures and under flue gas conditions.

Amines supported on porous materials are highly selective towards CO_2 by an effective chemisorptive mechanism and their capture capacity is preserved in the presence of moisture unlike the sorbents based on physisorption.[9–11] Additionally, the disadvantages associated with the traditional amine scrubbing technique, *i.e.*, high-energy sorbent regeneration and corrosion, could easily be overcome if amines are immobilized on porous solid supports. Amine immobilization is usually carried out through either chemical or physical binding on various porous supporting materials.[12,13] Since the chemically tethered aliphatic amines may not be able to withstand harsh reaction conditions and is thus an expensive route, a facile synthetic route involving physical impregnation of polyamines on porous supports could be an alternative, cost-effective technique for CO_2 capture technology.[14]

Most of the studies oriented towards physically immobilized polyamines used a mesoporous silica support because of the presence of hydroxyl groups on the surface (facilitates hydrogen bonding with amines) and large pore channels (hold a large amount of amines). Goeppert *et al.* showed polyethylenimine (PEI) loaded fumed silica and found a CO_2 uptake capacity of 206 mg g^{-1} at 298 K and 1 bar.[15] In fact, polyamine impregnated silicas are the materials which demonstrated a CO_2 capture capacity of up to 349 mg g^{-1} at warmer sorption conditions under a humid atmosphere however very few reports studied the CO_2/N_2 selectivity for silica–polyamines hybrids.[14] Furthermore, the synthesis of mesoporous silica requires costly surfactants and raw materials, leading to costly CO_2 capture technologies.

Here, we have developed a low cost nanoporous covalent organic polymer (COP) for loading polymeric amines and presented an optimized concentration of amines required for CO_2 uptake for this type of nanoporous COP support. CO_2 and N_2 sorption characteristics of an amine impregnated sorbent are also studied with their CO_2/N_2 selectivity trends. Apart from static sorption experiments, we have also demonstrated regeneration behaviour and CO_2 sorption under a continuous flow of humid CO_2 mixed gas stream at warmer temperatures to establish the sorbent's functioning in simulated flue gas conditions.

Experimental

Materials

Branched polyethylenimine (M_w = 800, PEI-800), branched polyethylenimine (M_w = 25 000, PEI-25 000), melamine, terephthalaldehyde, dimethyl sulfoxide

(DMSO), acetone, tetrahydrofuran and methanol were obtained from Aldrich, USA and used as received.

Synthesis of covalent organic polymer-19 (COP-19)

A slightly modified procedure by Schwab *et al.* was used.[16] Melamine (4.0 g, 0.032 mol) was dissolved in 150 mL dimethyl sulfoxide and terephthalaldehyde (6.44 g, 0.048 mol) was added after the complete dissolution of melamine. The solution was heated to 180 °C under vigorous stirring in a nitrogen environment for 3 days. The reaction was allowed to cool to room temperature, and the solid obtained was filtered, washed with dimethyl sulfoxide (20 mL, 3×), water (20 mL, 4×), acetone (20 mL, 3×), tetrahydrofuran (15 mL, 3×) and CH_2Cl_2 (15 mL, 3×). The resulting off-white solid was dried under vacuum at 150 °C for 10 h. Product: 7.2 g.

Synthesis of PEI impregnated COP-19

The calculated amount of PEI-800 or PEI-25 000 was dissolved in 10 mL of methanol. To this solution, COP-19 was dispersed and stirred for 24 h at room temperature. The dispersion was subjected for centrifugation to separate out precipitates and dried at 80 °C for 24 h under vacuum. The amount of PEI and COP-19 used for each impregnation is listed in Table 1 with sample designation.

Characterisation

FTIR spectra were obtained on KBr disks using a Perkin-Elmer FT-IR spectrometer. Thermogravimetric analysis (TGA) was performed using a NETZSCH-TG 209 F3 instrument by heating the samples up to 800 °C at 10 °C min^{-1} under a N_2 atmosphere. Nitrogen adsorption isotherms were obtained with a Micromeritics Triflex accelerated surface area and porosimetry analyser at 77 K after the samples were degassed at 120 °C for 5 h under vacuum. The surface area of the samples was calculated using the Brunauer–Emmett–Teller (BET) method. Low-pressure CO_2 and N_2 adsorption–desorption isotherms were measured at 273 K and 298 K using a Micromeritics Triflex system. IAST calculations were carried out by fitting adsorption isotherms with a single-site Langmuir model or a dual-site Langmuir model. These models were fitted only on the basis of deriving the best fit with adjusted r^2 values exceeding 0.999. The Origin Pro v8.5 program has been used to calculate IAST selectivity.[17] The CO_2 sorption used 15% CO_2, 3.8% H_2O, 81.2% He gas mixtures under flow conditions at 313 K up to 20 min. The gas mixture flow rate was 80 mL min^{-1}. The sorbent was loaded in a column and the concentration of inward and outward gas mixture was measured using GC-MS. The sorbent was degassed in He flow at 100 °C before analysis.

Table 1 Experimental composition used for synthesis of PEI impregnated samples

Samples	COP-97A	COP-97B	COP-97C	COP-98A	COP-98B	COP-98C
COP-19	0.25 g	0.25 g	0.25 g	0.25 g	0.25 g	0.25 g
PEI-800	0.25 g	0.5 g	1.0 g	—	—	—
PEI-25 000	—	—	—	0.25 g	0.5 g	1.0 g

Results and discussion

It is a prime requisite for CO_2 sorbents to be of minimal cost and this can only be possible if the sorbents can be produced using low cost raw materials and facile synthetic methodologies. Nanoporous COP-19 is synthesized through a one pot poly-condensation reaction of melamine and terephthalaldehyde (Scheme 1). COP-19 is utilized as a support for the immobilization of branched aliphatic amines of two different molecular weights (800 and 25 000 g mol^{-1}) and three different concentrations (Table 1), to demonstrate the effect of chain length and loading on the CO_2 uptake capacity. Thermogravimetric (TGA) and FTIR studies of COP-19 and amine impregnated COP-19 were carried out to obtain the amount of amine loaded on COP-19 and to show the structural stability of the support polymers.

The mass loss patterns of COP-19 and amine impregnated COP-19 were obtained from a TGA study under an inert atmosphere (Fig. 1a and b). The amount of amine loaded on COP-19 is calculated from the mass loss of amine impregnated COP-19 (COPs-97A,B,C and COPs-98A,B,C) normalized with the mass loss of COP-19. The calculated PEI-800 in COP-97A, COP-97B, and COP-97C is 24.1%, 36.4%, and 46.2% respectively while PEI-25 000 in COP-98A, COP-98B, and COP-98C is 22.4%, 32.3%, and 44.8% respectively. PEI-800 shows a slightly higher loading efficiency over PEI-25 000 because of the smaller steric size of the low molecular weight aliphatic amines. The TGA study also showed a thermal stability up to 200 °C for PEI impregnated COP-19.

FTIR spectra show evidence of amine impregnation and also demonstrate the stability of COP-19 after amine immobilization (Fig. 1c). The C–N stretching vibration is observed at 1150 cm^{-1} and N–H stretching is clearly visible at 3500–3300 cm^{-1}, which is from merged peaks of primary and secondary amines. The broad peak and small intensity within this region is due to the presence of aliphatic amines. Broad peaks at 1640–1560 cm^{-1} and 800 cm^{-1} are due to a N–H

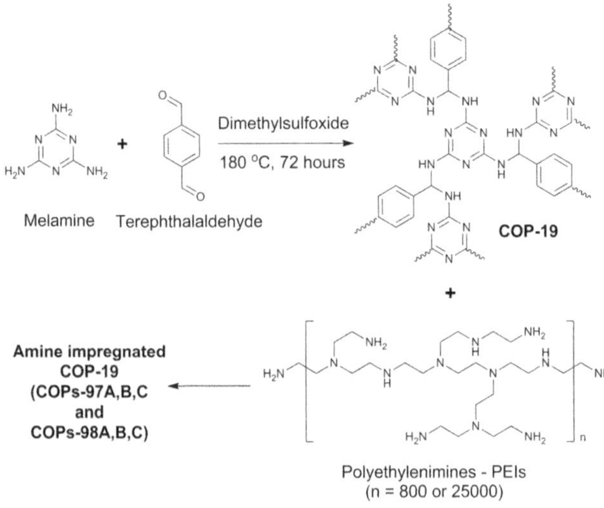

Scheme 1 Synthetic route for nanoporous COP-19 and PEI impregnation of COP-19.

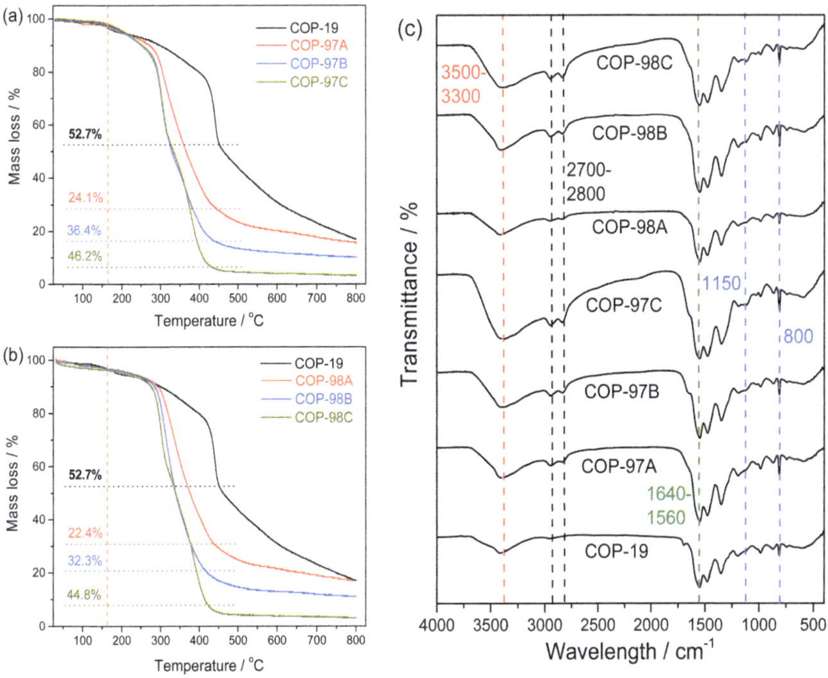

Fig. 1 (a and b) Thermogravimetric analysis of COP-19 and amine impregnated COP-19 up to 800 °C under a N_2 atmosphere. (c) FTIR spectra of COP-19 and amine impregnated COP-19 obtained as a KBr pellet.

in-plane-bend for primary amines. The vibration band at 2700–2800 cm^{-1} is observed for C–H stretching of aliphatic chain of PEI and the intensity of this mode is higher than COP-19, clearly suggesting the presence of aliphatic amines.

The effect of amine impregnation on the porosity of the amine impregnated COP-19 is studied using N_2 adsorption–desorption isotherms at 77 K (Fig. 2). N_2 uptake at low relative pressure for COP-19 reveals the presence of micropores however it also shows a large uptake at higher relative pressure which is characteristic of meso–macro porous interconnected porosity in networked polymeric materials. COP-19 shows a Brunauer–Emmett–Teller (BET) surface area of 640 m^2 g^{-1} which is substantially reduced after amine impregnation (Table 2). COP-97C and COP-98C (PEI loading ≈ 45% wt) became non-porous, mainly caused by blocking of accessible porosity of COP-19. The BET surface area of PEI supported COP-19 reduced upon an increase in amount of loading as well as the molecular chain length of the PEI. The BET surface area of COP-97A, COP-97B, COP-98A, and COP-98B is 118, 59, 88, and 47 m^2 g^{-1} respectively. These results provide direction toward focusing on the linear low molecular weight aliphatic amines for a support material, which contain largely micro and meso porosity. It should also be noted that low molecular weight CO_2-philic amines may show lower thermal stability and thus degradation and loss of active species could be realised over a prolonged time.[18]

The moderate surface area and presence of CO_2-philic basic amine moieties on COP-19 are favourable to high CO_2 adsorption, particularly at very low pressure, a

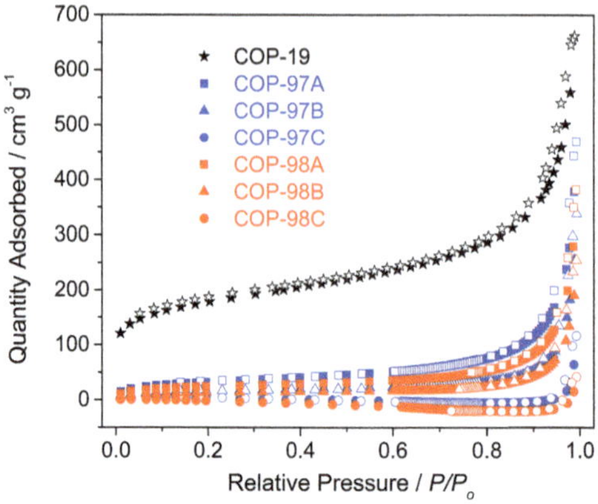

Fig. 2 N_2 adsorption–desorption isotherms of COP-19 and amine impregnated COP-19 at 77 K up to 1 bar. Filled and empty symbols correspond to adsorption and desorption respectively. The sorbents were activated at 120 °C under vacuum for 5 hours before each isotherm.

necessity for post-combustion CO_2 capture where the partial pressure of CO_2 ranges from 0.08–0.15 bar in flue gas streams. We studied CO_2 and N_2 sorption of COP-19 and amine impregnated COP-19 under static conditions at 273 and 298 K and up to 1 bar (Fig. 3 and Table 2).

Amine impregnated COP-19 (COP-97A,B,C and COP-98A,B,C) show very fast CO_2 uptake at low pressure compare to COP-19. PEI-800 loaded COP-97A, COP-97B, and COP-97C show a CO_2 uptake capacity of 95.5, 102.5 and 72 mg g^{-1} at 273 K and 80.6, 96.7 and 62 mg g^{-1} at 298 K, respectively at 1 bar. This suggests that PEI-800 loading of 24.1% wt is the optimum concentration for COP-19. Lower than this loading results in low capacity due to insufficient CO_2-philic groups while higher loading causes pore blocking and thus restricts accessibility of the amines for CO_2 molecules. Furthermore, the formation of carbamate species upon reaction of CO_2 and amines also can block accessible CO_2-philic sites.[19] CO_2 isotherms at studied temperatures show fast CO_2 uptake for amine impregnated

Table 2 Surface area, CO_2 uptake and CO_2/N_2 selectivity of COP-19 and amine impregnated COP-19

Sorbent	Surface area, m^2 g^{-1}	CO_2 uptake at 273 K, mg g^{-1} 0.15 bar	1 bar	CO_2 uptake at 298 K, mg g^{-1} 0.15 bar	1 bar	CO_2/N_2 selectivity 273 K	298 K
COP-19	640	45.4	107.4	17.7	58.3	141.5	131.2
COP-97A	118	63.6	95.5	50.6	80.6	574	965.6
COP-97B	59	75	102.5	69	96.7	445.3	779.2
COP-97C	7	47.5	72	37.6	62	1505	2511.4
COP-98A	88	70.4	115	53.3	99.2	—	—
COP-98B	47	63.2	95	58.2	90	452.6	852
COP-98C	2	43	68	38	63.6	813.7	1428

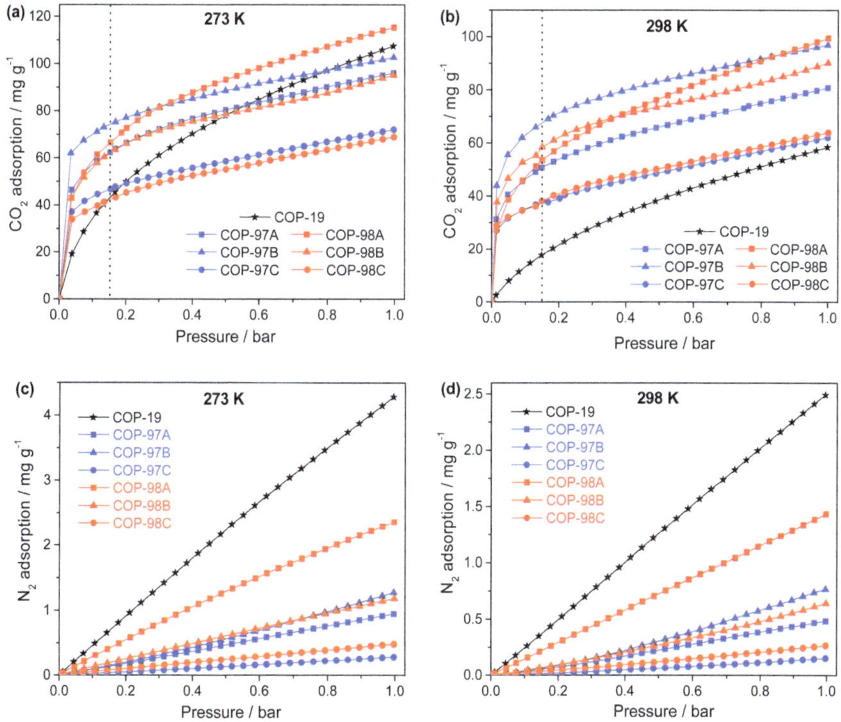

Fig. 3 CO_2 and N_2 adsorption isotherms of COP-19 and amine impregnated COP-19 at 273 K and 298 K up to 1 bar. The sorbents were activated at 120 °C under vacuum for 5 hours before each isotherm. Desorption profiles were omitted for clarity.

COP-19. The CO_2 sorption capacity at 0.15 bar for COP-97A, COP-97B, and COP-97C is 63.6, 75 and 47.5 mg g^{-1} at 273 K and 50.6, 69 and 37.6 mg g^{-1} at 298 K, respectively, however the CO_2 capture capacity of COP-19 is 45.4 and 17.7 mg g^{-1} at 273 and 298 K respectively. COP-97B demonstrated 1.65 and 3.9 times higher CO_2 uptake compare to COP-19 at 273 and 298 K. PEI-25 000 impregnated COP-98A, COP-98B and COP-98C also revealed similar trends of higher CO_2 uptake compare to COP-19 at 0.15 bar but lower than COP-97A,B,C (Fig. 3 and Table 2). N_2 sorption isotherms at 273 and 298 K show a decrease in the N_2 uptake with respect to an increase in the PEI loading. Owing to the non-reactivity of the N_2 molecule with amine functionality the N_2 sorption only depends on the availability of microporous porosity, which is filled and/or blocked by amines in impregnated samples. Indeed, lower N_2 sorption could be useful for attaining high CO_2 selective sorption from the mixture of gases.

CO_2/N_2 selectivity is one of the significant parameters for real applications of any sorbent material in CO_2 capture from flue gas streams, especially post-combustion CO_2 capture where N_2 concentration is as high as 85% in gas mixtures. Ideal adsorbed solution theory (IAST) is used in this study to illustrate CO_2/N_2 selectivity by a single-site Langmuir model or a dual-site Langmuir model fitting calculation of CO_2 and N_2 sorption isotherms.[20] CO_2/N_2 selectivity for CO_2 and N_2 (15 : 85) gas mixture against pressure is plotted in Fig. 4.

COP-19 shows CO_2/N_2 selectivity of 141.5 and 131.2 at 273 and 298 K respectively at 1 bar for a $CO_2 : N_2$ (15 : 85) mixture. A significant increase in CO_2/N_2 selectivity is observed for amine-impregnated samples, attributed to higher CO_2 uptake in CO_2-philic PEI loaded sorbents. Although the CO_2 capture capacity of COP-97C is relatively low compared to COP-97B, it showed the highest CO_2/N_2 selectivity of 1505 and 2511.4 at 273 and 298 K. COP-97B demonstrated a CO_2/N_2 selectivity of 445.3 and 779.2 at 273 and 298 K. Other PEI impregnated samples also showed excellent CO_2/N_2 selectivity, ranging from 452.6–1428 at 273 and 298 K (Table 2). The increase in the CO_2/N_2 selectivity with increase in temperature is a significant characteristic since post-combustion CO_2 capture requires CO_2 sorption at high temperatures. A similar phenomenon can be observed for azo linked covalent organic polymers (azo-COPs) where the azo functionality was found to be N_2-phobic.[21] Here, however, the increment in selectivity at high temperatures resulted predominantly from CO_2-philic functionalities. It is also noticeable that higher CO_2 capture and CO_2/N_2 selectivity is obtained in amine impregnated sorbents even though their surface area ranges from moderate to

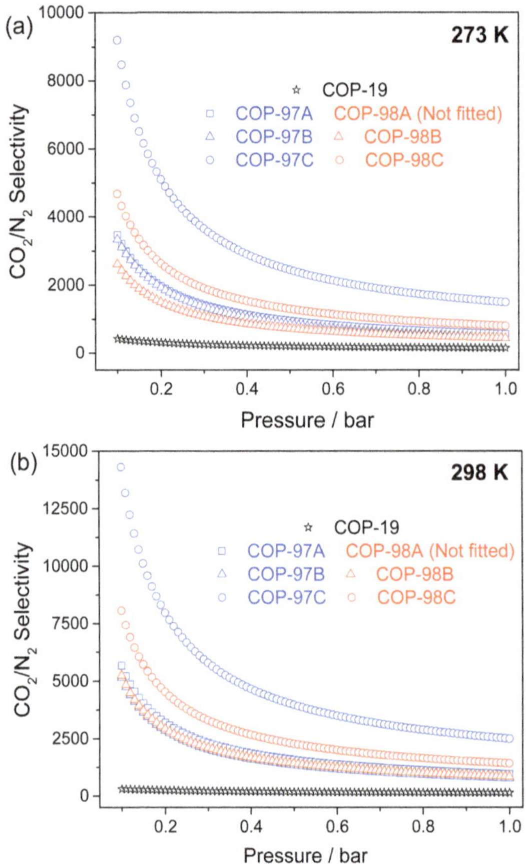

Fig. 4 IAST CO_2/N_2 selectivity of COP-19 and amine impregnated COP-19 at (a) 273 K and (b) 298 K up to 1 bar. The CO_2 and N_2 sorption data weren't satisfied for the fitted parameters for COP-98B at 273 and 298 K.

very low. These results evidently recommend the importance of CO_2-philic functionality within the sorbent materials rather than exceptionally high surface areas—higher surface area is clearly not the imperative parameter for higher CO_2 capture capacity.

Since the CO_2 capture capacity and CO_2/N_2 selectivity should be governing factors for post-combustion CO_2 capture capacity, we studied COP-97B for regeneration under continuous flow of mixed gas experiments. The stability of CO_2 sorbent materials under repetitive cycles of CO_2 adsorption–desorption was examined in static conditions for COP-97B up to 10 cycles at 273 K (Fig. 5a). COP-97B was subjected to regeneration after each cycle at 80 °C under vacuum for 2 hours. CO_2 adsorption–desorption cycles show no loss of CO_2 capture efficiency and COP-97B demonstrated an average CO_2 capture capacity of 100 mg g^{-1} (10% wt). The heat of adsorption for CO_2 adsorption in COP-97B is 46 kJ mol^{-1}, which facilitates desorption at quite a low temperature (80 °C) compared to the high temperatures (>120 °C) used in traditional aqueous amine scrubbing technology.

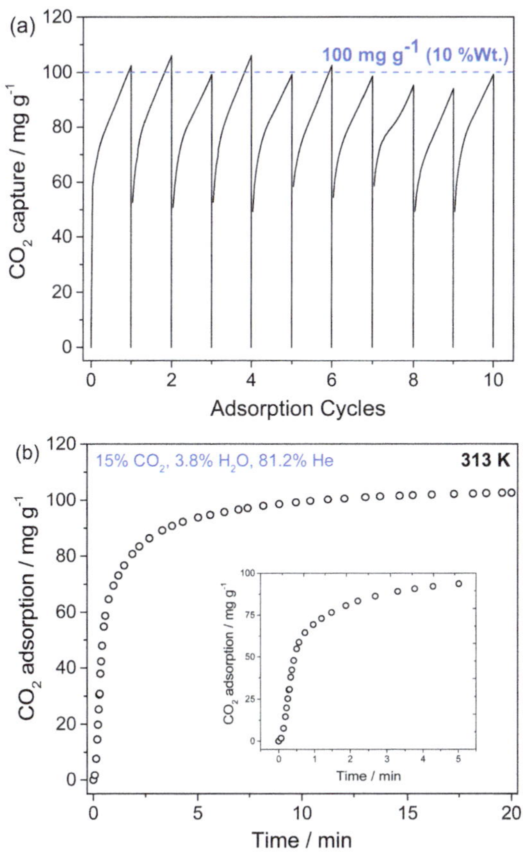

Fig. 5 (a) CO_2 adsorption cycles of COP-97B in static conditions at 273 K up to 1 bar. Desorption was carried out at 80 °C under vacuum for 2 hours. (b) CO_2 sorption of COP-97B for a 15% CO_2, 3.8% H_2O, 81.2% He gas mixture under flow conditions at 313 K up to 20 min. The gas mixture flow rate was 80 mL min^{-1}.

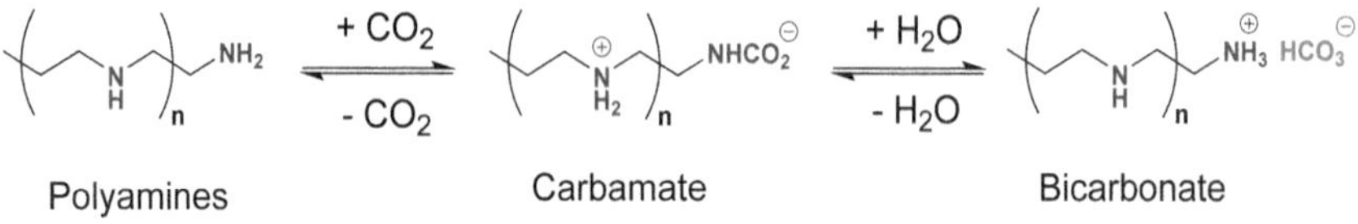

Scheme 2 Interaction of CO_2 with amines, forming carbonate in dry conditions and bicarbonate in the presence of water.

To emulate with post-combustion CO_2 capture, we studied CO_2 sorption under the flow of a moist gas mixture (15% CO_2, 3.8% H_2O, 81.2% He) at 313 K (Fig. 5b). We found that CO_2 sorption isn't affected in the presence of water vapour and the CO_2 capture kinetics are also extremely fast: 97 mg g^{-1} of CO_2 adsorbed within 5 min (Fig. 5 inset) and the plateau was reached (100 mg g^{-1}; 10% wt) at 10 min. Obviously, the fast kinetics and higher capacity of CO_2 sorption are achieved mainly because of aliphatic amines loaded on the porous support. In static and dry CO_2 sorption experiments, COP-97B uptakes 102.5 and 96.7 mg g^{-1} at 273 and 298 K which shows a decrease in CO_2 uptake at higher temperature. CO_2 capture for COP-97B under a humidified gas mixture under continuous flow demonstrated 100 mg g^{-1} at 313 K.

The enhanced CO_2 uptake is related to the formation of bicarbonate in the presence of water. Indeed, the improvement in the CO_2 capture capacity in the presence of water vapour has been documented in several reports where polyamines immobilized on a porous substrate are subjected to dry and humid CO_2 adsorption.[22–24] The primary and secondary amino groups in polyamines form carbamates upon dry CO_2 interaction which subsequently transform into bicarbonate if water is present during adsorption (Scheme 2). Two amino functionalities are needed for each CO_2 molecule to be adsorbed in the dry CO_2 capture, however theoretically one amino functionality is required for a CO_2 molecule in the humid environment. Therefore, the CO_2 capture capacity of COP-98B is greater under humid conditions at a warmer sorption temperature.

Conclusions

Sorbents with high CO_2 uptake with optimized kinetics are developed by PEI impregnation into nanoporous COP-19. The amount of loaded amine and the coverage throughout the support play a significant role and greatly affect the CO_2 sorption performance. Moderate to low surface areas are found to be enough for high CO_2 uptake at low pressure and the CO_2 sorption kinetics are remarkably fast with a heat of adsorption right below 50 kJ mol^{-1}. Experimental investigations revealed that the highest IAST CO_2/N_2 selectivity of 2511.2 was obtained upon amine impregnation for COP-97C at 298 K and it is enhanced at higher temperatures. After repeated adsorption–desorption cycles, the sorbent retained its CO_2 capture capacity (10% wt) over 10 cycles. COP-97B also showed exceptional CO_2 sorption under flow of humid mixed gas stream at 313 K. The higher CO_2 uptake capacity in moist CO_2 is suspected to be from the formation of bicarbonate where only one amine functionality is used for each CO_2 molecule. Since the higher CO_2 sorption in humid conditions was observed for COP-97B, we concluded that the sorbent is not deteriorated in the presence of water vapour unlike well-known high CO_2 adsorptive materials. In summary, we developed and optimized, a commercially viable CO_2 sorbent by combining a wealth of knowledge set forth by the leaders in the field and our own design principles. Future studies will include the continuous development of low cost, robust porous COPs with a higher pore size (20–50 nm) to increase amine loading and thus even higher CO_2 capture capacities.

Acknowledgements

We acknowledge the financial support by grants from Korea CCS R&D Centre, Basic Science Research Program through the National Research Foundation of Korea (NRF) funded by the Ministry of Science, ICT & Future Planning (2013R1A1A1012998), IWT (NRF-2012-C1AAA001-M1A2A2026588), and KAIST EEWS Initiative.

Notes and references

1 C. Mora, A. G. Frazier, R. J. Longman, R. S. Dacks, M. M. Walton, E. J. Tong, J. J. Sanchez, L. R. Kaiser, Y. O. Stender, J. M. Anderson, C. M. Ambrosino, I. Fernandez-Silva, L. M. Giuseffi and T. W. Giambelluca, *Nature*, 2013, **502**, 183–187.

2 J. G. Canadell, C. Le Quere, M. R. Raupach, C. B. Field, E. T. Buitenhuis, P. Ciais, T. J. Conway, N. P. Gillett, R. A. Houghton and G. Marland, *Proc. Natl. Acad. Sci. U. S. A.*, 2007, **104**, 18866–18870.

3 J. Y. Wang, L. Huang, R. Y. Yang, Z. Zhang, J. W. Wu, Y. S. Gao, Q. Wang, D. O'Hare and Z. Y. Zhong, *Energy Environ. Sci.*, 2014, **7**, 3478–3518.

4 M. E. Boot-Handford, J. C. Abanades, E. J. Anthony, M. J. Blunt, S. Brandani, N. Mac Dowell, J. R. Fernandez, M. C. Ferrari, R. Gross, J. P. Hallett, R. S. Haszeldine, P. Heptonstall, A. Lyngfelt, Z. Makuch, E. Mangano, R. T. J. Porter, M. Pourkashanian, G. T. Rochelle, N. Shah, J. G. Yao and P. S. Fennell, *Energy Environ. Sci.*, 2014, **7**, 130–189.

5 S. Tedds, A. Walton, D. P. Broom and D. Book, *Faraday Discuss.*, 2011, **151**, 75–94.

6 Z. J. Zhang, Z. Z. Yao, S. C. Xiang and B. L. Chen, *Energy Environ. Sci.*, 2014, **7**, 2868–2899.

7 H. A. Patel, F. Karadas, A. Canlier, J. Park, E. Deniz, Y. Jung, M. Atilhan and C. T. Yavuz, *J. Mater. Chem.*, 2012, **22**, 8431–8437.

8 M. Sevilla, C. Falco, M. M. Titirici and A. B. Fuertes, *RSC Adv.*, 2012, **2**, 12792–12797.

9 H. Zhang, A. Goeppert, M. Czaun, G. K. S. Prakash and G. A. Olah, *RSC Adv.*, 2014, **4**, 19403–19417.

10 J. J. Vericella, S. E. Baker, J. K. Stolaroff, E. B. Duoss, J. O. Hardin, J. Lewicki, E. Glogowski, W. C. Floyd, C. A. Valdez, W. L. Smith, J. H. Satcher, W. L. Bourcier, C. M. Spadaccini, J. A. Lewis and R. D. Aines, *Nat. Commun.*, 2015, **6**, 6124.

11 M. B. Yue, Y. Chun, Y. Cao, X. Dong and J. H. Zhu, *Adv. Funct. Mater.*, 2006, **16**, 1717–1722.

12 W. G. Lu, M. Bosch, D. Q. Yuan and H. C. Zhou, *ChemSusChem*, 2015, **8**, 433–438.

13 C. Chen, S. T. Yang, W. S. Ahn and R. Ryoo, *Chem. Commun.*, 2009, 3627–3629, DOI: 10.1039/b905589d.
14 C. Chen, J. Kim and W. S. Ahn, *Korean J. Chem. Eng.*, 2014, **31**, 1919–1934.
15 A. Goeppert, H. Zhang, M. Czaun, R. B. May, G. K. S. Prakash, G. A. Olah and S. R. Narayanan, *ChemSusChem*, 2014, **7**, 1386–1397.
16 M. G. Schwab, B. Fassbender, H. W. Spiess, A. Thomas, X. L. Feng and K. Mullen, *J. Am. Chem. Soc.*, 2009, **131**, 7216–7217.
17 J. Byun, S. H. Je, H. A. Patel, A. Coskun and C. T. Yavuz, *J. Mater. Chem. A*, 2014, **2**, 12507–12512.
18 A. Sayari, A. Heydari-Gorji and Y. Yang, *J. Am. Chem. Soc.*, 2012, **134**, 13834–13842.
19 A. Goeppert, M. Czaun, R. B. May, G. K. S. Prakash, G. A. Olah and S. R. Narayanan, *J. Am. Chem. Soc.*, 2011, **133**, 20164–20167.
20 H. A. Patel, S. H. Je, J. Park, Y. Jung, A. Coskun and C. T. Yavuz, *Chem.–Eur. J.*, 2014, **20**, 772–780.
21 H. A. Patel, S. H. Je, J. Park, D. P. Chen, Y. Jung, C. T. Yavuz and A. Coskun, *Nat. Commun.*, 2013, **4**, 1357.
22 G. G. Qi, L. L. Fu, B. H. Choi and E. P. Giannelis, *Energy Environ. Sci.*, 2012, **5**, 7368–7375.
23 Z. H. Chen, S. B. Deng, H. R. Wei, B. Wang, J. Huang and G. Yu, *ACS Appl. Mater. Interfaces*, 2013, **5**, 6937–6945.
24 A. Goeppert, S. Meth, G. K. S. Prakash and G. A. Olah, *Energy Environ. Sci.*, 2010, **3**, 1949–1960.

Faraday Discussions

PAPER

Solar energy utilization in the direct photocarboxylation of 2,3-dihydrofuran using CO_2

Michele Aresta,*[ab] Angela Dibenedetto,[bc] Tomasz Baran,[bd] Szymon Wojtyła[d] and Wojciech Macyk[d]

Received 23rd April 2015, Accepted 1st July 2015
DOI: 10.1039/c5fd00040h

The conversion of CO_2 into high energy products (fuels) and the direct carboxylation of C–H bonds require a high energy input. Energy cannot be derived from fossil carbon, in this case. Solar energy can be used instead, with a low environmental impact and good profit. We have studied the use of white light or solar energy in the photoreduction of CO_2 and in photocarboxylation reactions, using different semiconductors modified at their surface. Two examples of reduction of CO_2 to methanol and CO will be shortly discussed, and two cases of carboxylation of organic substrates. The case of carboxylation of 2,3-dihydrofuran will be discussed in detail.

Introduction

Today, the introduction of a carboxylic group into an organic molecule occurs *via* synthetic methodologies characterized by low Carbon Fraction Utilization (CFU) and high Carbon Footprint (CF). Some years ago, we undertook a critical analysis of such issues and suggested that the direct carboxylation of C–H bonds (eqn (1)) represents one of the most effective ways to reduce CO_2 emissions and save fossil carbon.[1,2] Such new syntheses may require a quite different approach than thermal reactions. In fact, a critical barrier that can occur for the reaction, represented by C–H activation, must be overcome. The use of the photochemical technology may be of great help, if photocatalysts able to activate the C–H bond using visible light are developed (eqn (1)).

$$R - H + CO_2 \xrightarrow[\text{cat}]{h\nu} R - CO_2H \tag{1}$$

[a]ChBE, NUS, 4 Engineering Drive 4, Singapore 117585, Singapore. E-mail: cheam@nus.edu.sg
[b]IC²R, Tecnopolis, via Casamassima km3, Valenzano 70018, Bari, Italy
[c]CIRCC, Via Celso Ulpiani 27, Bari 70126, Italy
[d]Faculty of Chemistry, Jagiellonian University in Kraków, Ingardena 3, 30-060 Kraków, Poland

On the other hand, CO_2 can be used as a source of carbon for the production of C_1 and C_n energy rich molecules, a process that would also be of great interest if fossil carbon was not used to provide the energy required for the reduction to occur, and if the hydrogen was sourced from water or waste organics. Furthermore, the use of solar energy as a primary energy source would be of fundamental importance.[3]

We have started a programme on the photochemical conversion of CO_2 using newly synthesized semiconductor-based photocatalysts that are able to operate under visible and/or solar light and have applied such systems to a number of reactions implying CO_2 conversion: either fixation into organic substrates of the entire molecule, or reduction to C_1 and C_n molecules using water or abundant and cheap polyols as a source of hydrogen.

Fig. 1 shows the matching of some semiconductor band gaps and CO_2 reduction potentials in $2ne^-$ transfer processes in water compared to the higher energy $1e^-$ transfer.[4]

In this paper we recall the concept on which the catalysts are based on and present some very recent applications in both carboxylation reactions and CO_2 reductions. We discuss new results relevant to the carboxylation of a quite intriguing substrate, 2,3-dihydrofuran (2,3-DHF).

The new-concept photocatalysts and their application

The photocatalysts have been engineered by carrying out a surface modification of semiconductors such as TiO_2, ZnS and several others. As Fig. 1 shows, TiO_2 has a band gap energy of 3.2 eV that makes it not particularly suited for visible light utilization in CO_2 conversion. In fact, TiO_2 alone is better used under UV irradiation in oxidation processes that convert recalcitrant pollutant organics either in water or in air, thus producing CO_2. Moreover, due to a high oxidation potential of holes photoinduced in the valence band of titanium dioxide, reoxidation of the reduction product is plausible. The use of UV-light in CO_2 conversion would not be an efficient use of solar light, making the process both economically not viable and environmentally not acceptable. Nevertheless, it is

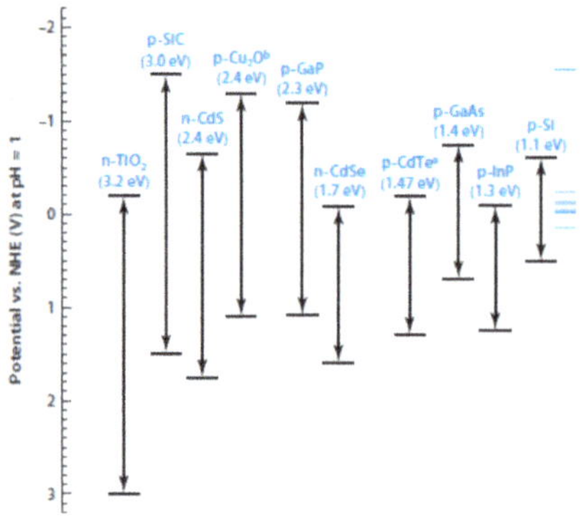

$CO_2 + e^- = CO_2^{\cdot-}$ -2.21 in aprotic solvents;	-1.90 in water
$CO_2 + 2H^+ + 2e^- \rightarrow CO + H_2O$	-0.53
$CO_2 + 2H^+ + 2e^- \rightarrow HCO_2H$	-0.61
$CO_2 + 4H^+ + 4e^- \rightarrow CH_2O + H_2O$	-0.48
$CO_2 + 6H^+ + 6e^- \rightarrow CH_3OH + H_2O$	-0.38
$CO_2 + 8H^+ + 8e^- \rightarrow CH_4 + 2H_2O$	-0.24

Fig. 1 Comparison of the band gap energies of some semiconductors with the reduction potentials of CO_2 in aqueous solutions (pH = 1) for selected one- and multi-electron transfer processes.[4]

possible to modify the spectral properties of titanium dioxide in several modes, as shown in Fig. 2.

Modifications of semiconductors with species that are either "*hole*" (to the VB) or "*electron*" (to the CB) injectors may sensibly extend the spectral response of the photocatalysts and enhance their use with exploitation of solar light. This concept may be applied to a number of semiconductors which do not absorb in the visible region, but more in the UV. So, TiO_2 modified with fluorochromates has been advantageously used in the reduction of NAD^+ into NADH used as an "e^- and H^+" vector to enzymes which reduce CO_2 to methanol.[5–7]

Another approach to enhance solar light utilization is to increase the quantum efficiency of processes driven by near-UV light (*e.g.* within the range of 320–400 nm) instead of questing for broadening the spectral response of the wide band gap semiconductors. This goal can be achieved for instance by surface functionalization of semiconductors with co-catalysts improving efficiency of charge separation and interfacial electron transfer processes. For example, nano-particles of elemental platinum deposited on the surface of TiO_2 act as electron sinks and enhance photoreduction of the adsorbed electron acceptor (*e.g.* O_2).[8–11] A similar methodology was successfully applied by us to zinc sulfide photocatalysts. As mentioned above, ZnS is an interesting photomaterial having a band gap energy larger than TiO_2 (3.6 *vs.* 3.2 eV). Due to its redox properties, it may have a good potential for C–H activation. Unfortunately, ZnS alone does not show good photoactivity. However, we have synthesized ZnS decorated with Ru nanoparticles[12] and used it for the carboxylation of C–H bonds with a different activation energy: either activated CH_2 moieties as in $CH_3COCH_2COCH_3$ (acac) or terminal CH_3, as shown in eqn (2).

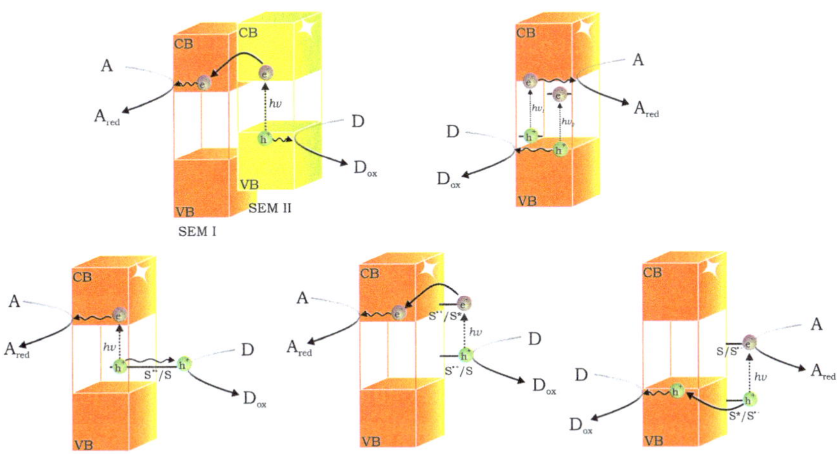

Fig. 2 Mechanisms of wide band gap semiconductor photosensitization. (A) Formation of composite semiconductors, (B) bulk doping resulting in the formation of acceptor or donor levels, (C) direct photosensitization (optical charge transfer), (D) dye-photosensitization involving an electron injection to the conduction band from the excited photosensitizer, (E) dye-photosensitization involving a hole injection to the valence band from the excited photosensitizer.

$$CH_3COCH_2COCH_3 \xrightarrow[cat]{h\nu} CH_3COCH(COOH)COCH_3 + CH_3COCH_2COCH_2COOH \quad (2)$$

In this photochemical process the formation of the radical anion $CO_2^{\bullet-}$ has been demonstrated to be the key issue. The photomaterial used (ZnS) has the correct band gap energy for its generation, while Ru enhances the light harvesting capacity and drives the carboxylation reaction under sunlight irradiation. It is worth noting that acetylacetone can be carboxylated using chemical catalysts (typically a base such as the phenate anion) and affords only $CH_3COCH(COOH)COCH_3$. The use of the photocatalyst enables the activation of the C–H bond of the terminal methyl group, affording the terminal carboxylic acid $CH_3COCH_2COCH_2COOH$. It is noteworthy that the C–H bonding energy is quite different in the CH_2 and CH_3 moieties. The former is activated due to tautomerism that brings the hydrogen onto the oxygen atom (eqn (3)). In the presence of bases, the OH moiety is deprotonized and prompted to interact with CO_2. The terminal-CH_3 moiety is much less activated and almost does not react in thermal reactions, but it does in photochemical processes.

$$CH_3COCH_2COCH_3 \rightleftarrows CH_3C(OH){=}CHCOCH_3 \quad (3)$$

ZnS decorated with ruthenium and p-type photocatalysts built on the same concept have been found to be active in the photoreduction of carbon dioxide to formic acid and carbon monoxide[13] in protic solvents. p-Type semiconductors, in contrast to n-type materials, have rarely been studied for carbon dioxide conversion. We have found that wide-band gap semiconductors (like CuI and NiO) characterized by a band gap energy of 3.6–4.0 eV and a very low potential of the conduction band edge (in the range of −3.0 to −3.4 V *vs.* SHE) or also materials having a band gap energy equal to 3.1 eV and the potential of conduction band edge of −2.1 V *vs.* SHE can be usefully applied in the photoreduction of CO_2.[14] For example, under full solar light irradiation using isopropanol as the hole scavenger, CO_2 was reduced to CO, CH_4 and/or HCOOH. The preparation method (annealing temperature) plays a key role in determining the activity of the catalyst and its selectivity towards one of the products.

Results and discussion

Characterization of the catalysts

The properties of Ru@ZnS-A and Ru@ZnS-B used in this work have already been published.[12,13] Here, we report some new additional properties that have recently been determined, namely the diffuse reflectance spectra and the photocurrent measurement, which are useful for determining the spectral region within which the catalysts are active, and the EDX analysis. The band gap energies were estimated from diffuse reflectance spectroscopy. Obtained spectra (reflectance, R, *vs.* wavelength, λ) have been transformed to Kubelka–Munk function (KM) according to the equation: $KM = (1 - R)^2/2R$ (Fig. 3). Spectra of bare ZnS and materials decorated with ruthenium are similar, *i.e.* ruthenium does not influence the absorption onset. The band gap energy was calculated according to Tauc theory as published elsewhere.[12,13] ZnS-A shows a smaller bandgap energy (3.47 eV) than

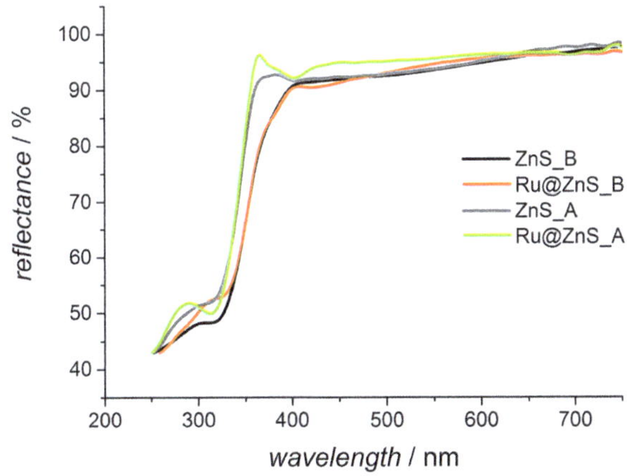

Fig. 3 Diffuse reflectance spectra of the synthesized photocatalysts transferred to Kubelka–Munk function.

ZnS-B (3.64 eV). Materials decorated with ruthenium (0.5% of Ru) have band gap energies of 3.57 and 3.65 eV for Ru@ZnS-A and Ru@ZnS-B, respectively. The prepared materials offer strong reduction properties (for example, the conduction band edge of ZnS-B is *ca.* −2.5 V *vs.* SHE) as learnt from spectroelectrochemical measurements published elsewhere.[13]

UV-Vis analysis nicely corresponds to the results of photocurrent measurements. Irradiation of the ZnS-A-covered electrode with near-UV-Vis light (solar light) results in anodic photocurrent generation (375 nm track), as shown in Fig. 4. The use of only visible light decreases the intensity of the photocurrent (see wavelength 405 and 465 nm in Fig. 4). Application of light of longer wavelengths, 520, 595 and 630 nm, did not result in photocurrent generation.

The characterization of materials has been done by EDX analysis. Energy-dispersive X-ray spectra of ZnS-A and 1%Ru@ZnS-A are shown in Fig. 5. Both materials show typical signals of zinc and sulphur. Moreover, the spectrum of 1%

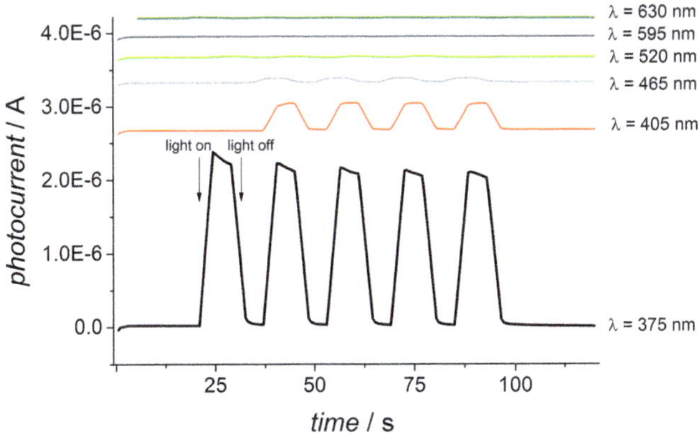

Fig. 4 Photocurrent measurements: ZnS-A electrode irradiated with chopped light of 375, 405, 465, 520, 595, 630 nm wavelength.

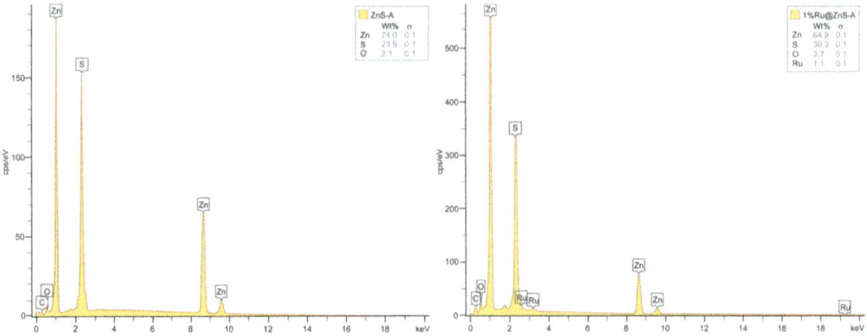

Fig. 5 Energy-dispersive X-ray spectra of ZnS-A and 1%Ru@ZnS-A.

Ru@ZnS-A also shows signals assigned to ruthenium (2.69 and 19.27 keV). Raman spectroscopy excluded the formation of ruthenium(IV) oxide, since bands at 515 and 626 cm^{-1}, characteristic for this compound (RuO stretching modes) were not found. This is in accordance with the XRD features of the materials,[12] previously published, which confirm the presence of Ru(0).

The SEM image of Ru@ZnS-A (Fig. 6) shows that the material is composed of irregular aggregates of various sizes ranging from several nanometers up to 4 micrometers. DLS measurements prove that the average particle sizes are in the range of hundreds of nanometers. SEM-EDX analysis proves the chemical composition of the material. All materials have relatively high specific surface areas (BET measurements) in the range of 55–105 $m^2\ g^{-1}$, as published elsewhere.[12] The specific surface area of the ZnS samples decorated with ruthenium is reduced when compared to bare materials.

Fig. 6 SEM of Ru@ZnS-A.

The target molecule and the observed reactions

In this paper, we focus on the discussion of the photoconversion of intriguing molecules, such as 2,3-dihydrofuran (2,3-DHF) (Fig. 7), at room temperature under inert gas or in the presence of CO_2 dissolved in aprotic or protic solvents.

DHF is an interesting molecule as it contains a multiplicity of bonds having quite different energies. Table 1 shows the nature of the bonds and their average energies.

Thus, this molecule has several reactive sites. Despite such complexity, it is foreseeable that the most reactive position will be the allylic $-CH_2-$ moiety, corresponding to the lower energy of the C–H bond, while C–C and C–O bond splitting may occur as secondary events (Table 1). However, in principle, H-extraction should occur, and occurs, at C-4, affording an allylic radical (Fig. 8). In order to check whether the photocatalyst would be able to promote C–H bond cleavage, we carried out a cyclic-voltammetry (CV) analysis. The CV measurements on a solution of 2,3-DHF are shown in Fig. 9. The voltammogram shows an irreversible peak at 1.35 V *versus* the non-aqueous silver electrode (Ag/Ag^+), corresponding to the oxidation of 2,3-DHF. The oxidation potential of 2,3-DHF (1.55 V *vs.* SHE) is lower than the potential of the ZnS valence band edge (1.8 V *vs.* SHE[20]), thus the photoinduced oxidation of 2,3-DHF by VB holes is thermodynamically possible.

Such species can evolve in two different ways (Fig. 10), affording either 2,5-dihydrofuran (2,5-DHF) or, *via* a C–O splitting and structural rearrangement, the formyl-cyclopropane molecule (F-Cyp), or even the linear crotonaldehyde (C-ald). Such photoinduced-isomerization products have been observed when 2,3-DHF was directly irradiated using UV-light.[18]

We now show that white (solar) light in the presence of a photocatalyst can induce the isomerization. This adds a complexity to the system as one can ask if the interaction of the substrate with CO_2 will occur prior to or after isomerization of the substrate and, thus, whether carboxylated forms of the rearrangement products can also be expected. Therefore, we have investigated the reactivity of the pure substrate and the reactivity of it in the presence of CO_2 under the same experimental conditions (solvent, radiations, irradiation time, temperature, concentration). The study was carried out at a double scale: at a micro-scale in a pressurized NMR tube and in a larger photoreactor having a volume of 10 mL, allowing sample withdrawal for GC analysis. Moreover, in order to establish if proton donors may influence the reaction, two different solvents were used, chloroform and methanol. We shall discuss the behaviour of 2,3-DHF in the two solvents separately, comparing the role of Ru@ZnS-A and Ru@ZnS-B.

Photocarboxylation in chloroform

Fig. 11a shows the modification of the ^{13}C NMR spectrum of 2,3-DHF during irradiation under an Ar atmosphere, in chloroform, in presence of Ru@ZnS-A.

Fig. 7 The 2,3-dihydrofuran molecule.

Table 1 Bonds in 2,3-dihydrofuran and their average energy in kJ mol^{-1}

Bond	Number of bonds	Average energy kJ mol^{-1}	Ref.
$sp^3C–sp^3C$	1	346	15
$sp^3C–sp^2C$	1	*386–390*	16
$sp^2C–sp^2C$	1	610	15
$sp^3C–H$	4	411	17
$sp^2C–H$	2	422	18
$sp^3C–O$	1	358	15
$sp^2C–O$	1	418	19

Fig. 11b shows the same system when Ru@ZnS-B is used. It is noteworthy that within 2 h of irradiation, in the presence of Ru@ZnS-A, all of the starting substrate is converted (disappearance of resonances at 146, 99.5, 69.5 and 29.3 ppm typical of 2,3-DHF). Two major products are formed, namely, 2,5-DHF (signals at 104 and 63 ppm) and formylcyclopropane, which are identified through the signals at 205, 31 and 14 ppm. Crotonaldehyde, the alternative isomerization product derived from ring-opening, is absent as its resonances at 194.0, 154.3, 134.6 and 18.6 ppm are not found in the spectrum.[21] Other very minor products are also formed (*ca.* 1% yield), while signals of ethanol are present from the beginning, most probably an impurity derived from the work-up of the photocatalysts. Interestingly, the use of Ru@ZnS-B increases the amount of formylcyclopropane, while producing the same isomerization products. It is worth noting, that the NMR spectra do not show any change with respect to starting 2,3-DHF in the absence of irradiation and in the presence of the photocatalyst.

Fig. 11 presents the products formed in $CHCl_3$ when the photoconversion is carried out under 0.7 MPa of CO_2 in a high pressure NMR-sapphire tube using Ru@ZnS-B as the photocatalyst. Ru@ZnS-A gives the same products, but at lower concentrations. As well as the signals due to the isomerization products, it is quite evident that the insurgence of some new signals, some of which can be ascribed to species formed upon incorporation of CO_2. In particular, the signals at 178, 144, 100, 62 and 50 ppm are due to the new species. A careful comparison of the ^{13}C-spectra of the reaction solution with those of the photoisomerization products found under Ar, and then of the expected carboxylation products, brings us to the conclusion that the major photocarboxylation product is 2,5-dihydrofuran-2-

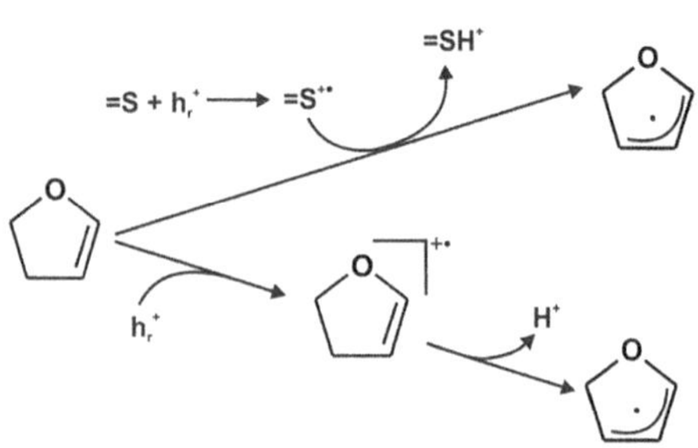

Fig. 8 Radical formation in 2,3-DHF upon irradiation.

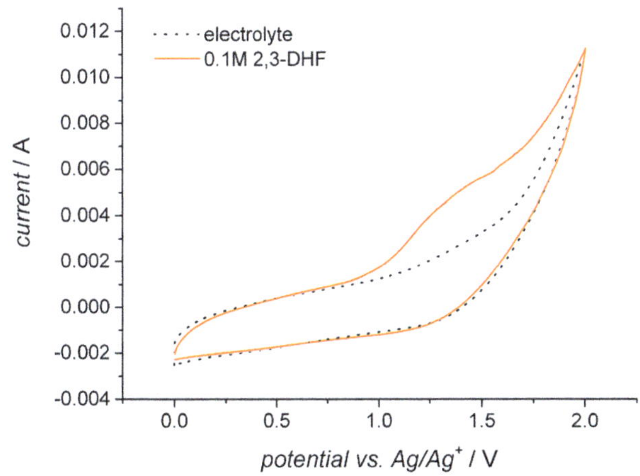

Fig. 9 Cyclic voltammogram of 0.1 mol L^{-1} 2,3-dihydrofurane in acetonitrile with lithium perchlorate (0.1 mol L^{-1}) as the electrolyte. Scan rate: 50 mV s^{-1}.

carboxylic acid (2-COOH-2,5-DHF). Such species have ^{13}C-NMR signals at 184 (*C*OOH), 104 (*C*=*C*), 63 (*C*H2) and 60 (*C*H–COOH) ppm. The absence of a signal at 140 ppm (characteristic of the O-bonded *C* in the O–*C*H=CH– moiety) confirms the isomerization of the starting material. A simulation of the ^{13}C-NMR spectrum made using CS Chem Draw Ultra 6 (Cambridge Software Corporation) shows that 2-COOH-2,5-DHF (Fig. 12A) has signals at 176, 129, 89 and 68 ppm, while 4-COOH-2,3-DHF (Fig. 12B) has computed signals at 177, 144, 100, 80 and 32 ppm. Finally, 2-COOH-1-formyl-cyclpopropane has computed peaks at 204, 184, 36 and 6–15 ppm. The signals found in the ^{13}C-spectrum of the mixture irradiated with visible light in the presence of Ru@ZnS-B shows, in addition to the signals of 2,5-DHF and 1-formyl-cyclopropane, two series of new signals at: 184, 126, 81 and 68 ppm due to 2-COOH-2,5-DHF and lower signals attributable to 2-COOH-2,3-DHF at 184, 145, 102, 80 and 29 ppm. The absence of signals in the high part of the spectrum excludes the formation of the carboxylated form of 1-formyl-cyclopropane. The formation of the two carboxylated isomers can be explained by the coupling of the radical shown in Fig. 6 with the $CO_2^{\bullet-}$ radical anion. The new C–C bond is formed at C-2, and the double bond migrates to the 3–4 carbons or to the 4–5 carbons, the latter with less favourable energetics as it implies H-migration. As reported above, 2-COOH-2,3-DHF is present at a lower concentration. Besides the signals of the products reported above, weak resonances due to THF and furan are also found, which also occur in solutions irradiated in the absence of CO_2. A new signal due to methanol also appears (50 ppm), which is not found in the

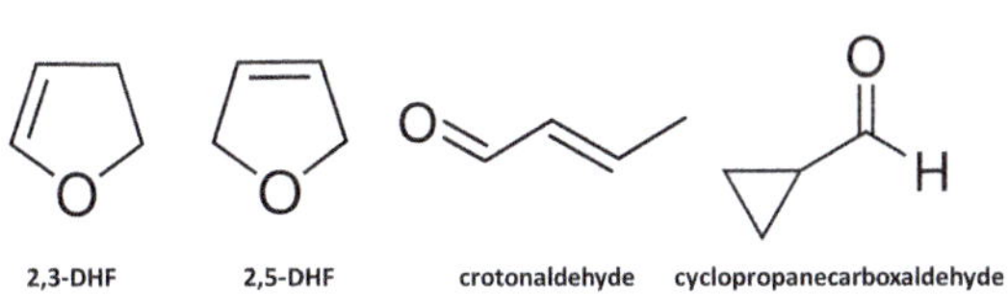

Fig. 10 2,3-DHF and the products of its photocatalytic isomerization (inert atmosphere).

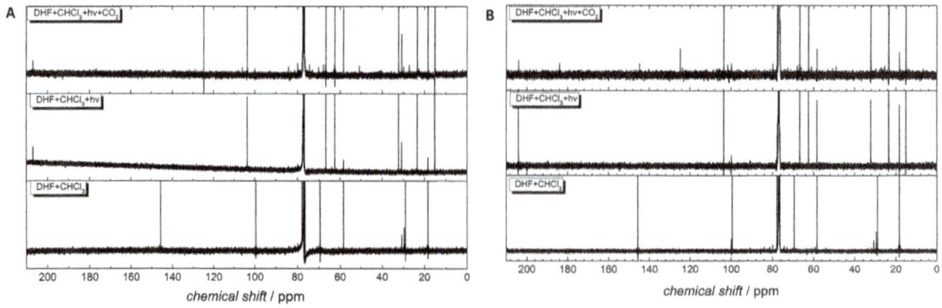

Fig. 11 ^{13}C-NMR spectra of 2,3-DHF and after 2 h of using Ru@ZnS-A (A) and Ru@ZnS-B (B) as photocatalysts: the starting product is quantitatively converted into isomerization products. In absence of CO_2, carboxylated products are not formed.

absence of CO_2, and this is quite interesting but not surprising as we have shown that CO_2 can be reduced to methane, formic acid or methanol under irradiation in the presence of the ZnS photocatalysts or others (such as CuI).[13]

In order to obtain further information on the compounds formed, we have produced a ^{1}H-NMR spectrum of the mixture. The resulting spectrum is quite complex due to the many signals of the various isomers. The only significant signals are those of the aldehydic (9.75 ppm in $CDCl_3$) and carboxylic groups (11.3 ppm in $CDCl_3$, not seen in CD_3OD most probably due to the hydrogen bonding and H–D exchange that causes broadening).

In order to confirm the structural features of the carboxylated forms obtained through the multinuclear NMR, a GC-MS analysis of the reaction solution in methanol produced on a large scale was carried out, allowing the identification of various products. Several peaks were found in the GC of the solution irradiated in the absence and presence of CO_2, corresponding to the isomers described above. Under CO_2, the 2-COOH-2,3-DHF peak appeared, attributed on the basis of its fragmentation showing *m*/*z* values at: 114 (molecular peak), 85, 69, 45 (carboxylic group) and 41.

Carboxylation in methanol

The carboxylation in methanol has features different from those of the same reaction in $CHCl_3$. A single carboxylated isomer is formed in this solvent, namely 2-COOH-2,3-DHF. This product is also formed in a higher concentration than in chloroform, reaching over 25% of the total (Fig. 14) of the starting product, after 12 h with 1%Ru@ZnS-A as the photocatalyst (Fig. 13C compares the various photocatalysts: ZnS-A, 1%Ru@ZnS-A, ZnS-B and 1%Ru@ZnS-B). The accumulation of the carboxylated products over 10 h can be explained assuming that:

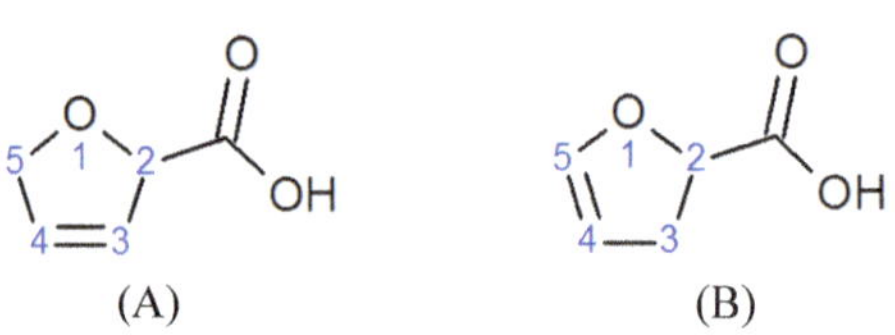

Fig. 12 (A) 2-COOH-2,5-DHF and (B) 2-COOH-2,3-DHF.

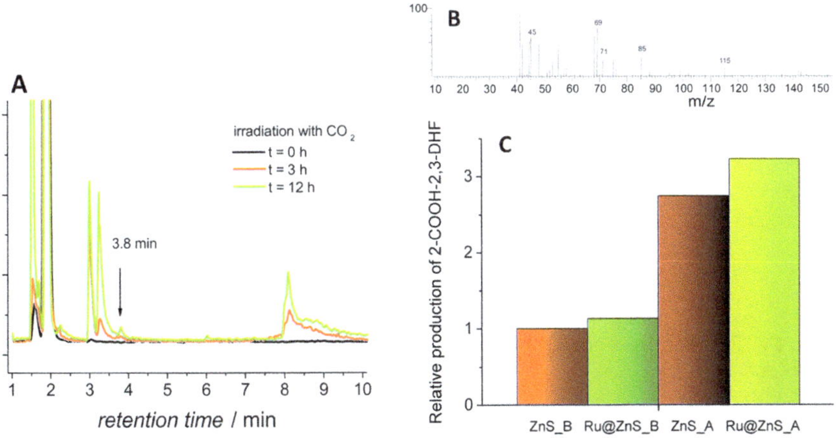

Fig. 13 GC-MS analysis of the reaction solution during photocatalytic carboxylation of 2,3-DHF in methanol. (A) Gas-chromatograms of samples withdrawn at different times during the photocatalytic tests in methanol using 1%Ru@ZnS-A as a photocatalyst with irradiation times of 0, 3 and 12 h, (B) the MS spectrum on the peak at 3.8 min, and (C) the relative formation of 2-COOH-2,3-DHF in the presence of ZnS-A (set equal to 1), 1% Ru@ZnS-A, ZnS-B and 1%Ru@ZnS-B after 10 hours of irradiation of a 2,3-DHF solution in methanol.

1. The isomerisation is fast and the isomerized product can then be carboxylated. This is not the case as discussed in the next sub-chapter.

2. The isomerisation rate in methanol is slower than in chloroform or the lifetime of the radical shown in Fig. 8 is longer in methanol, so that a more extended carboxylation is observed with time, with higher selectivity.

In order to collect data on the formation rate of the carboxylated product and on the activity of the catalysts, we have carried out extended irradiations of the solution up to 12 h and compared the catalysts. Fig. 13 reports an example of the evolution of the chromatograms with time, the MS spectrum of the peak at 3.8 min due to 2-COOH-2,3-DHF, and the relative activity of the different photocatalysts. Fig. 13C clearly shows that the Ru-decorated ZnS materials are more active than bare sulphides. Using Ru@ZnS-A as the photocatalyst and collecting data at fixed time intervals (30 min) over 12 h, it was possible to obtain preliminary information about the kinetics of the conversion of the starting material. Fig. 14 shows the percentage of conversion of 2,3-DHF into 2-COOH-2,3-DHF with time. It is interesting to note that at the time of 12 h all the starting materials was converted either into the isomerisation product or its carboxylated form, as demonstrated by ^{13}C NMR. Dimers of 2,5-DHF were also identified.

Does the carboxylation occur concurrently with the photoisomerisation or after it?

As reported above, the only carboxylation products observed in $CHCl_3$ are 2-COOH-2,3(or 2,5)-DHF, while no carboxylated forms of F-Cyp were found. The question came to our mind of whether the carboxylation is concurrent with the isomerisation or whether it follows this step. We have carried out a simple experiment to answer such a question. We first irradiated the 2,3-DHF solution

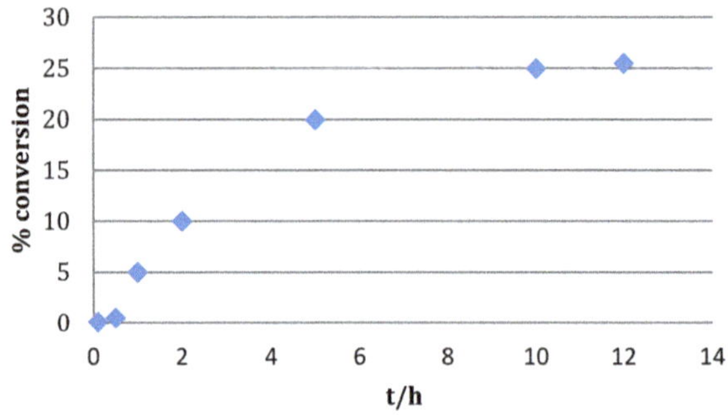

Fig. 14 Photoconversion of 2,3-DHF into 2-COOH-2,3-DHF using Ru@ZnS-A as the photocatalyst.

under argon in $CHCl_3$ and in the presence of the photocatalyst Ru@ZnS-B or Ru@ZnS-A and, when the photoconversion of 2,3-DHF into 2,5-DHF and 1-formyl-cyclopropane was complete, CO_2 was admitted and the irradiation continued. The analysis of the reaction mixture after irradiation (2 h) showed that there was no further conversion of 2,5-DHF nor 1-formyl-cyclopropane, thus inferring that the carboxylation occurs on the radical species (shown in Fig. 8) more than on the products of photoisomerization. The carboxylation of the substrates formed upon isomerisation of 2,3-DHF takes place under quite different conditions than those described in this paper.

Conclusions

The irradiation of a solution of 2,3-DHF in $CHCl_3$ in the presence of Ru@ZnS-B under inert atmosphere causes isomerization of the starting product into 2,5-DHF and 1-formyl-cyclopropane. This reaction proceeds *via* the formation of allylic radicals (hydrogen abstraction from 2,3-DHF), identical to those formed in the reaction of 2,5-DHF oxidation by valence band holes generated upon the irradiation of ZnS and CdS.[22–25] Since the radical is stabilized by its two resonance structures, its lifetime is sufficient to enable bimolecular reactions. When methanol is used, the photoisomerization reaction requires a longer time.

When the irradiation is carried out under CO_2 atmosphere in $CHCl_3$, the carboxylation of the substrate occurs affording two isomeric products, namely 2-COOH-2,3-DHF and 2-COOH-2,5-DHF. The carboxylation occurs by coupling of a $CO_2^{\bullet-}$ radical anion with the allylic radical. The formation of the $CO_2^{\bullet-}$ radical under the operative conditions has been confirmed by our recent studies involving spin-trapping and EPR measurements.[26] The generation of this very reactive radical is thermodynamically feasible due to the high reduction abilities of electrons excited to the conduction band of zinc sulphide. Moreover, the presence of ruthenium nanoparticles enhances adsorption of carbon dioxide and improves the charge separation, which is reflected in the diminished recombination efficiency and the pronounced $CO_2^{\bullet-}$ formation.[26] When the photocarboxylation is carried out in methanol, 2-COOH-2,3-DHF is formed as the only product. This allowed us to carry out preliminary kinetic studies and calculate the photocarboxylation yield to be 25% after 12 h.

The photocarboxylation process discussed here is another example of a possible application of photocatalytic systems in the synthesis of chemicals. Although the overall efficiency of the process can be improved, the actual photoconversion is quite interesting (25% of 2,3-DHF converted after 12 h, or 0.025 M) and mimics a reductive path of photosynthesis (reduction of carbon dioxide) utilizing the UV part of the solar radiation.

Experimental part

Synthesis of photocatalyst

Zinc sulphide (ZnS-A) was prepared according to methods published elsewhere.[12] Under an argon atmosphere, using Schlenk's system, a solution of Na_2S (0.1 mol) in 25 mL of distilled water was added dropwise to the aqueous solution of $ZnSO_4$ (0.1 mol, 25 mL). The mixture was stirred for twenty four hours, filtrated, washed with water to neutrality and dried at room temperature under vacuum. Another sample of zinc sulphide (ZnS-B) was also prepared under a nitrogen atmosphere. A solution of sodium hydroxide (10 g) in distilled water (20 mL) was added dropwise to a solution of $ZnSO_4$ (0.01 mol, 20 mL), as already described.[12] The primarily formed solid $Zn(OH)_2$ was dissolved as $[Zn(OH)_4]^{2-}$. Afterwards, 40 mL of a thiourea solution (1.52 g) was added. The mixture was heated to 353 K for 48 h. After separation *via* centrifugation, the powder was washed with H_2O and dried at 333 K. Ruthenium nanoparticles were deposited as follows: 1 g of ZnS was suspended in an aqueous solution of $RuCl_3 \cdot xH_2O$ (10 mL, 10 mM) under an argon atmosphere. A concentrated ethanol solution of $NaBH_4$ was added dropwise (in a big excess) under ultrasonic agitation. Then the powder was filtered off and washed with water under argon, dried and stored under a nitrogen atmosphere.

Characterization of materials

Diffuse reflectance spectra of photocatalysts were measured using a UV-Vis spectrophotometer (UV-3600 type, Shimadzu), equipped with an integrating sphere (15 cm diameter). 25 mg of analysed samples were ground with 0.5 g of $BaSO_4$. Barium sulphate was also used as the reference material. Hydrodynamic particle diameters were measured using Zetasizer NanoZS (Malvern). The scanning electron microscope (Vega 3 LMU, Tescan) was equipped with a LaB_6 cathode. EDX spectra were recorded using the EDX detector (Oxford Instruments, X-act, SDD 10 mm^2). The photoelectron spectra were measured using a monochromatized aluminium AlKα source (E = 1486.6 eV) and a low energy electron flood gun (FS40A-PS) to compensate the charge on the surface of the non-conductive samples. A typical three-electrode set-up was employed for photocurrent analysis. The electrolyte solution was 0.1 M KNO_3. Platinum and Ag/AgCl were used as auxiliary and reference electrodes, respectively. The working electrodes were prepared by casting the samples onto ITO conductive foil. Light emitting diodes of various wavelengths (*Instytut Fotonowy*) were used as light sources. The measurements were collected using the potentiostat (PGSTAT 302N, Autolab).

Cyclic voltammetry

CV experiments were carried out using a BioLogic SP-150 potentiostat. The measurements were collected at a scan rate of 100 mV s^{-1}. The electrochemical cell consisted of three electrodes: the glassy-carbon electrode (working electrode), the nonaqueous Ag/Ag^+ (reference electrode) and platinum wire (counter electrode). The reference electrode was prepared by placing a clean silver wire into an electrolyte (0.01 M tertrabutylammonium perchlorate in acetonitrile) containing silver ions (0.01 M $AgNO_3$). The electrodes were placed in a Teflon cuvette filled with 0.1 mol dm^{-3} $LiClO_4$ solution in anhydrous acetonitrile. Oxygen was thoroughly removed from the electrolyte by purging with argon before (15 min) and during the experiment.

Photocatalytic tests

2,3-Dihydrofuran, a Sigma-Aldrich product, was carefully distilled under reduced pressure prior to photocatalytic tests. It was stored at 4 °C and used for up to 30 days after distillation. Chloroform was dried with alumina, distilled and stored under a nitrogen atmosphere. Methanol was distilled under a nitrogen atmosphere, dried using molecular sieves and distilled again. The photocatalyst (1 g dm^{-3}) was suspended (by sonication) in the solvent (5 or 10 mL) and the organic substrate (0.1 mol dm^{-3} unless otherwise stated) was added. CO_2 was bubbled through the suspension for 3 minutes to remove oxygen and afterwards the mixture was closed under an elevated pressure of carbon dioxide (7–8 bar) in a sapphire tube reactor, unless otherwise stated. Photocatalytic tests were carried out in a closed reactor using a XBO-150 lamp as the light source equipped with a cut-off filter (360 nm). Samples of the reaction mixture were collected periodically, filtered through syringe filters and analysed. The progress of the reaction was monitored by GC-MS (Shimadzu QP5050) equipped with a DB 5 MS 30 m capillary column. Nuclear Magnetic Resonance (NMR) experiments were carried out using a 400 MHz Varian INOVA apparatus and 600 MHz Bruker Advance.

Acknowledgements

IC^2R srl is acknowledged for having made available the photocatalytic equipment. The authors wish to thank Anna Regiel-Futyra for SEM measurements. The support from the Foundation for Polish Science within the VENTURES Project (2011-8/1) and TEAM Project (2012-9/4), co-financed by the European Union, Regional Development Fund, is highly acknowledged by TB and SW, respectively. Some equipment was purchased with the financial support of the European Regional Development Fund in the framework of the Polish Innovation Economy Operational Program (contract no. POIG.02.01.00-12-023/08).

References

1 *Carbon Dioxide as a Source of Carbon*, ed. M. Aresta and G. Forti, Springer Netherlands, Dordrecht, 1987.

2 *Developments and Innovation in Carbon Dioxide (CO2) Capture and Storage Technology*, ed. M. M. Maroto-Valer, Woodhead Publishing, 2010, vol. 1, pp. i–iii.

3 E. Barton Cole and A. B. Bocarsly, in *Carbon Dioxide as Chemical Feedstock*, ed. M. Aresta, Wiley-VCH Verlag GmbH & Co. KGaA, 2010, pp. 291–316.

4 M. Aresta, A. Dibenedetto and E. Quaranta, *Reaction Mechanisms in Carbon Dioxide Conversion*, Springer-Verlag, Berlin Heidelberg, 2015.
5 M. Aresta, A. Dibenedetto, W. Macyk, T. Baran, *Pat. IT* MI20131135, 2015
6 A. Dibenedetto, P. Stufano, W. Macyk, T. Baran, C. Fragale, M. Costa and M. Aresta, *ChemSusChem*, 2012, **5**, 373–378.
7 M. Aresta, A. Dibenedetto, T. Baran, A. Angelini, P. Łabuz and W. Macyk, *Beilstein J. Org. Chem.*, 2014, **10**, 2556–2565.
8 D. W. Bahnemann, J. Moenig and R. Chapman, *J. Phys. Chem.*, 1987, **91**, 3782–3788.
9 J. Lee and W. Choi, *J. Phys. Chem. B*, 2005, **109**, 7399–7406.
10 A. A. Ismail and D. W. Bahnemann, *Green Chem.*, 2011, **13**, 428–435.
11 A. A. Ismail and D. W. Bahnemann, *J. Phys. Chem. C*, 2011, **115**, 5784–5791.
12 T. Baran, A. Dibenedetto, M. Aresta, K. Kruczała and W. Macyk, *ChemPlusChem*, 2014, **79**, 708–715.
13 T. Baran, S. Wojtyła, A. Dibenedetto, M. Aresta and W. Macyk, *Appl. Catal., B*, 2015, **178**, 170–176.
14 T. Baran, S. Wojtyła, A. Dibenedetto, M. Aresta and W. Macyk, 2015, data not published.
15 L. G. Warde, *Organic Chemistry, Hardcover*, 2009.
16 R. Morrison and R. Boyd, *Organic chemistry, Hardcover*, 1992.
17 S. J. Blanksby and G. B. Ellison, *Acc. Chem. Res.*, 2003, **36**, 255–263.
18 K. M. Ervin, S. Gronert, S. E. Barlow, M. K. Gilles, A. G. Harrison, V. M. Bierbaum, C. H. DePuy, W. C. Lineberger and G. B. Ellison, *J. Am. Chem. Soc.*, 1990, **112**, 5750–5759.
19 G. da Silva, C.-H. Kim and J. W. Bozzelli, *J. Phys. Chem. A*, 2006, **110**, 7925–7934.
20 H. Kisch, *Semiconductor Photocatalysis: Principles and Applications*, Wiley-VCH, 2015.
21 F. Osada, H. Furukawa, M. Matsubara, T. Imai and K. Tsukiyama, Fukuoka, 2012, http://sdbs.db.aist.go.jp, access date: 11.09.2015.
22 R. Künneth, C. Feldmer, F. Knoch and H. Kisch, *Chem.–Eur. J.*, 1995, **1**, 441–448.
23 H. Keck, W. Schindler, F. Knoch and H. Kisch, *Chem.–Eur. J.*, 1997, **3**, 1638–1645.
24 A. Reinheimer, A. Fernández and H. Kisch, *Z. Phys. Chem.*, 1999, **213**, 129–133.
25 G. Hörner, P. Johne, R. Künneth, G. Twardzik, H. Roth, T. Clark and H. Kisch, *Chem.–Eur. J.*, 1999, **5**, 208–217.
26 T. Baran, A. Dibenedetto, M. Aresta, K. Kruczała and W. Macyk, *ChemPlusChem*, 2014, **79**, 708–715.

Faraday Discussions

PAPER

CO_2 capture systems based on saccharides and organic superbases†

G. V. S. M. Carrera,* N. Jordão, L. C. Branco* and M. Nunes da Ponte

Received 27th April 2015, Accepted 29th June 2015
DOI: 10.1039/c5fd00044k

In this report, novel systems, based on highly abundant saccharides, D-mannose, D-glucose, β-cyclodextrin, alginic acid and mannitol, in combination with an organic superbase, tetramethylguanidine (TMG) or 1,8-diazabicyclo[5.4.0]undec-7-ene (DBU), are studied for carbon dioxide capture. With D-mannose and D-glucose, several ratios of equivalents of alcohol groups of saccharide : superbase were tested: 1, 0.625, 0.5 and 0.25. High wt% values of CO_2 uptake were obtained with TMG-based systems. However, TMG itself can react directly with CO_2, and, in the presence of D-mannose, competition between carbonate and carbamate based products was established. In order to circumvent this competition and obtain exclusively the carbonate-based product, DBU was used instead as an organic superbase. In the D-mannose series the highest result was obtained with a D-mannose : DBU ratio eq. = 0.625 (13.9% CO_2 uptake, 3.3/5 alcohol groups converted into carbonates). A more effective stirring system, designed to overcome the high viscosity of the products, allowed the use of a D-glucose : DBU = 1 : 1 ratio with 11.5 wt% of CO_2 uptake and 2.47/5 alcohol groups converted into carbonates. Additionally a DSC thermal study was performed in order to study the stability/reversibility of the CO_2 loaded systems.

Introduction

Carbon dioxide capture and utilization is a fundamental concept to develop in the current environmental and energetic context.[1] The benchmark scrubbing systems for CO_2 capture that have been available in the market for more than 60 years are aqueous solutions of alkanolamines, which have a drawback of the requirement of dilution of the capture agent in water (in order to avoid corrosion and mitigate excessive release of heat during reaction), leading to poor performances in CO_2 capture (7 wt% of CO_2 uptake in 30% aqueous solution of ethanolamine) and high energy demand for CO_2 stripping, due to the high heat capacity of water. Additionally, the solvent is lost during operations.[2,3] A competitive system should fulfil the criteria of high capacity to store this gas and sustainability from an energetic,

LAQV, REQUIMTE, Departamento de Química, Faculdade de Ciências e Tecnologia, Universidade Nova de Lisboa, 2829-516 Caparica, Portugal. E-mail: goncalo.carrera@fct.unl.pt; l.branco@fct.unl.pt

† Electronic supplementary information (ESI) available. See DOI: 10.1039/c5fd00044k

economic and environmental point of view. The possibility to avoid water as a reaction solvent made task specific ionic liquids (TSILs) interesting alternatives for CO_2 capture.[4–6] Ten years ago, Jessop and collaborators[7] developed the concept of a reversible ionic liquid, using a mono-alcohol, an organic superbase and CO_2. These reagents led to an ionic liquid, formed by an alkylcarbonate and the protonated superbase cation. The reaction could be reverted by bubbling an inert gas, like nitrogen or argon. The initial goal of this pioneering work was to develop solvents where polarity could be tuned by the introduction or displacement of an acid gas such as CO_2. Following it, other studies with modification of several parameters such as the nucleophile (amines,[8,9] amino-esters,[10] amino-alcohols,[11] amino-acids[12] and mono-saccharides[13]), type of organic superbase,[9] element of reversibility[14] and number of functionalities in the same molecules able to react with CO_2 [9,13] have been reported. In parallel, several applications based on the same principle of reversibility, such as media to promote reactions and extractions,[15] dissolution of biomolecules,[16] quenching of fluorescence,[17] reversible manipulation of colour[14] and CO_2 capture have also been described. Regarding this last aspect, it is important to highlight that Heldebrant *et al.*[2] could achieve a maximum of 19 wt% of CO_2 uptake using a system of hexanol : diazabicyclo[5.4.0]-undec-7-ene (DBU), an excellent performance compared with the conventional scrubbing system. The authors also state that alkyl carbonates are less stable than bicarbonates or carbamates, due to hydrogen bond interaction decrease. Naturally-occurring molecules with multiple alcohol groups could therefore become suitable platforms to reversibly capture CO_2. In agreement with these conjectures, Zhang *et al.* have recently reported a dramatic enhancement of the solubility of cellulose in DMSO after admission of CO_2 to cellulose : superbase systems.[16] The authors suggested that the reason for this behavior is related to the conversion of the alcohol groups of cellulose into carbonates. On the other hand, the order of alcohol reactivity is primary > secondary > tertiary, which may decrease the effectiveness of carbon capture by saccharides in comparison with primary alcohols. Herein, we present a study of saccharides and derived sugar structures containing multiple alcohol groups: D-glucose, D-mannose, alginic acid, mannitol and β-cyclodextrin, in combination with one of two organic superbases, diazabicyclo[5.4.0]undec-7-ene (DBU) or tetramethylguanidine (TMG). Considering reactivity constraints due to diffusion of CO_2 from the gas to liquid phase associated to stirring efficiency[2,13] different ratios of saccharide to superbase, from equivalent amounts to large excesses of superbase–alcohol equivalents in saccharide/mol of superbase from 1 to 0.25 were tested in the present work. Additionally, magnetic coupled stirring was tested as alternative method to a conventional magnetic bar coupled system. Contrarily to our previous report,[13] no solvents were used.

Experimental

Materials

All chemicals were used as purchased, with the exception of 1,8-diazabicyclo [5.4.0]undec-7-ene (DBU), with a purity of at least 99%, provided by Fluka and 1,1,3,3-tetramethylguanidine (TMG), supplied by Sigma-Aldrich (99%), where water was removed using molecular sieves. The remaining chemicals used were

D-(+)-mannose (99%), D-(+)-glucose (99%), D-mannitol (99%), alginic acid, β-cyclodextrin from Alfa Aesar and carbon dioxide (Air Liquid, 99.998 mol%).

Methods

The prepared compounds were characterized by ^1H and ^{13}C-NMR, HMBC (heteronuclear multiple bond correlation) and HSQC (heteronuclear single quantum coherence) and in specific cases by TOCSY (total correlated spectroscopy) on a Bruker AMX400 spectrometer. Chemical shifts are reported downfield in parts per million from a tetramethylsilane reference, using d_6-DMSO as the deuterated solvent. IR spectra were recorded on a Perkin-Elmer FTIR Spectrometer, Spectrum 1000. The samples were prepared in a KBr matrix. DSC analysis was carried out by using a TA Instruments Q-series TM Q2000 DSC with a refrigerated cooling system. The sample was continuously purged with 50 mL min^{-1} nitrogen gas and 2–20 mg of salt was crimped into an aluminum standard sample pan with a lid. The samples were submitted to an isothermal step (40 °C, 1 minute), cooled to −90 °C (20 °C min^{-1}, isothermal step 1 minute at −90 °C) and then heated to 150 °C (20 °C min^{-1}, isothermal step 1 minute at 150 °C). Followed by another cycle of cooling (−90 °C, 20 °C min^{-1}) and heat (150 °C, 20 °C min^{-1}). The glass transition temperature, the melting point and the decomposition temperature were determined on the heating process of the first cycle.

Synthetic procedures

The syntheses were performed in a cylindrical high pressure steel reactor (11 mL) with sapphire windows at both ends (Fig. 2), allowing a full view of the contents of the cell. In each experiment, 2.5 g of superbase (DBU or TMG) was used, alone or in combination with the proper number of equivalents (alcohol or carboxylic acid groups) of each saccharide. In the case of D-mannose and D-glucose, the ratios (in equivalents) of saccharide : DBU were 1, 0.625, 0.5 and 0.25, whereas for β-cyclodextrin, alginic acid and mannitol, they were 0.5, 1 and 0.5 respectively. Stirring of the saccharide + superbase mixtures was done with a small magnetic bar, coupled to a common laboratory plate, for at least 20 min until complete dissolution of the saccharide in the (liquid) base. CO_2 was then introduced into the cell, at room temperature, normally until a pressure of 5 MPa was reached (exceptions were 2 MPa for TMG alone, and 4 MPa for TMG in combination with D-mannose). The pressure dropped continuously until reaching a stable minimum, usually in less than 3 hours. In some situations it was necessary to refill the cell with CO_2 in order to complete the reaction. Decompression of the reactor was then carried out slowly, in order to avoid the release of any fine powder. The reactor was weighed prior to CO_2 supply and after decompression, in order to measure the carbon dioxide uptake. The products were stored in a freezer at temperatures below 0 °C for their analysis.

Visual observations of the reactions as they proceeded indicated that a solid progressively formed and separated from the liquid mixture. In the cases of higher concentrations of saccharide in the superbase (1 : 1 ratio), the solid could even block the magnetic bar and stop stirring. In order to check whether in these cases the kinetics of the reaction were significantly decreased; one experiment was carried out using different apparatus, with a more effective stirring system. A 250 mL reactor with a magnetically coupled stirring system, built by Thar, was used. This stirring system is often used in high-pressure reactions, and consists of a powerful external electromagnet that rotates, and couples its movement with permanent magnets attached to the top of a rod, located inside the reactor, with a stirring helix at the bottom. The nominal torque exerted by this configuration is at least 5 times higher than what we could obtain with the magnetic bar in the reactor of Fig. 2.

Although several disadvantages would not allow the systematic use of that experimental apparatus for our purposes, continuous stirring throughout the whole time of reaction was obtained for a (1 : 1) DBU + glucose mixture, consisting of 20.36 g of DBU combined with an equivalent amount of D-glucose. The mixture was stirred 35 min prior to the reaction with CO_2.

Results and discussion

Saccharides or derived sugar structures with different numbers of OH groups were studied (Fig. 1). In the cases of D-glucose and D-mannose (five OH groups per molecule), two highly abundant natural compounds, several numbers of equivalents of OH groups in mono-saccharides per number of superbase molecules ratios (nr. –OH eq./nr. SB: 0.25, 0.5, 0.625 and 1) were tested. Considering this approach, mixtures of monosaccharide : superbase with different viscosities were achieved for CO_2 capture studies. The other saccharide-based structures were alginic acid (two OH groups and one carboxylic acid group per monomer), β-cyclodextrin (three OH groups per unit and twenty-one per molecule) and D-mannitol (six OH groups per molecule). For these last saccharide scaffolds, the saccharide : superbase ratios (nr. –OH and COOH eq./nr. SB) were 1, 0.5 and 0.5 respectively. In the case of alginic acid, the carboxylic group was also considered to be deprotonated by the superbase, along with the alcohol groups, in order to determine the ratio under discussion.

In this context it is fundamental to determine the percentage of CO_2 uptake respective to the capture system (superbase alone and superbase + saccharide) as well as the number of alcohol groups of the saccharide converted to carbonates.

The superbase TMG can by itself react with carbon dioxide, forming carbamates according to the work of Pereira *et al.*[18] Considering this, in an initial stage of our studies, we measured the selectivity of the carbonate *vs.* carbamate in a mixture of D-mannose : TMG (ratio 1), using pressures of CO_2 up to 4 MPa, leading to 18 wt% of CO_2 uptake. The reaction was repeated, leading to 22.8 wt% of CO_2 uptake (Table 1) corresponding to 2.32 –OH groups of D-mannose (from 5 possible to be functionalized) being converted into carbonates (determined by ^1H-NMR with aid of HMBC). 1.58 equivalents of TMG (from 5 possible) were converted into carbamates, as determined by weighing the mass of CO_2 uptake and considering the fraction converted into carbonates. This result is supported by the work of Ozturk *et al.*[19] who performed a kinetic study of the reaction between CO_2, TMG and 1-hexanol, with the formation of a carbamate intermediate product (reaction with TMG), stabilized by the alcohol.

The authors proposed two alternative products: $TMGCO_2^-$ + $TMGH^+$ + hexanol and the other configuration $HexCO_3^-$ + $TMGH^+$ + hexanol. According to the authors, the second path is favoured and the carbonate product is preferred when compared with carbamate. In this context, we checked the reactivity of TMG towards CO_2, leading to 29.5% of CO_2 uptake (77.5% yield). It is important to note

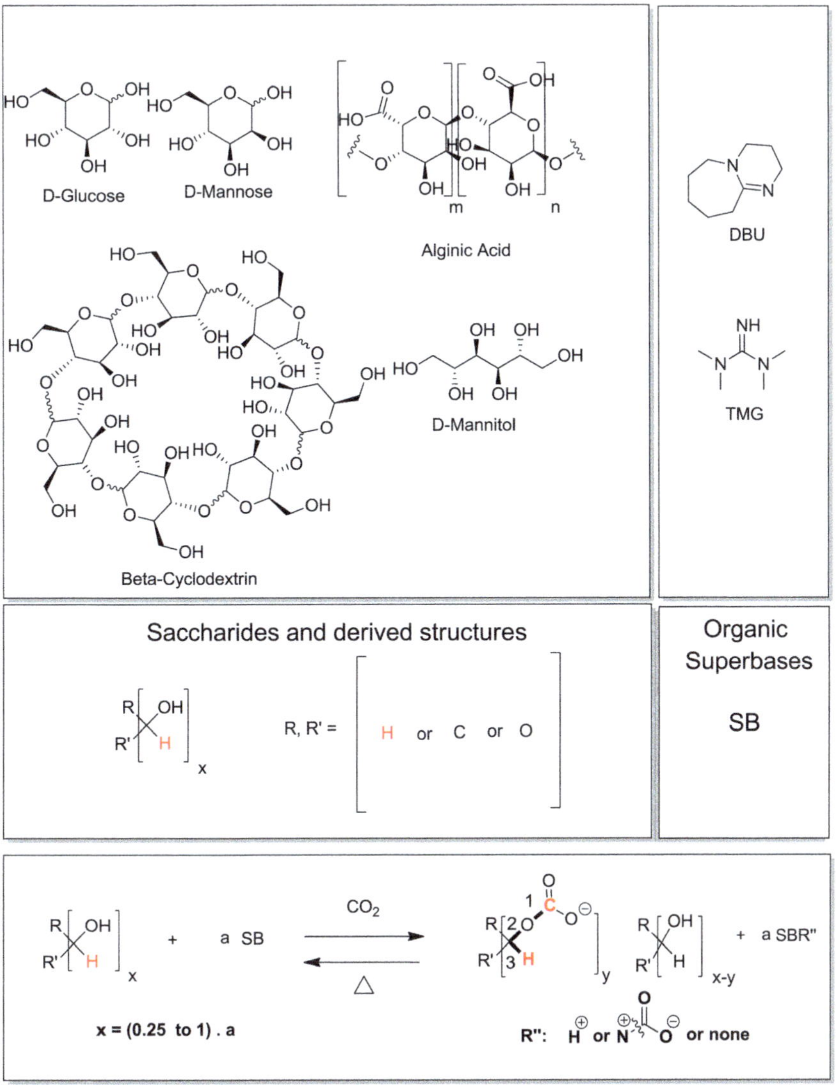

Fig. 1 Saccharides and derived compounds used in combination with an organic superbase to capture CO_2 leading to organic carbonates and also a carbamate when TMG is employed as organic superbase. CO_2 is released after heating. According to the change of functionalization from alcohol to carbonate, modification of the chemical shift in ^{1}H-NMR of the protons at a distance of three bonds from carbon and carbonate are detected as well as correlation by HMBC NMR spectroscopic techniques.

that the identity of the product was obtained by ^{1}H-NMR (by displacement of the $(N(CH_3)_2)_2C{=}N\underline{H}$ peak from 5.35 ppm in TMG to 5.93 ppm in the product $(N(CH_3)_2)_2C{=}N^+\underline{H}CO_2^-$), TOCSY (see ESI†; spectra of the product reveal a correlation $((C\underline{H}_3)_2N)_2C{=}N\underline{H}CO_2$ indication that the peak with a chemical shift of 5.93 ppm belongs to $TMGCO_2$), and ^{13}C-NMR spectra (the presence of an irregular peak at 165.8 ppm which can indicate the presence of two different quaternary

Table 1 Tetramethylguanidine (TMG) based systems in the capture of CO_2 and the effect of the presence of D-mannose

System	wt% CO_2	[OH to CO_3^-][b] (=NH of TMG to NCO_2^-)[c]	FTIR NCO_2^- bands (cm^{-1})	^{1}H-NMR H bond peak [TMG–$C\underline{H}_3$ peak] (δ, area)	^{1}H-NMR– CO_3^- peaks (δ, area)	^{13}C-NMR TMG quat. carbon [TMG–$\underline{C}H_3$ peak]
TMG	—	[—] (—)	—	5.35, 0.94 [2.62, 12]	—	166.20 [39.04]
TMG w/CO_2	29.5[a]	[—] (0.775 : 1[a])	3282 3107 1667	5.93, 1.09 [2.64, 12]	—	165.82 [39.04]
Man : TMG 1 : 1 eq.	—	[—] (—)	—	5.16, 9.38 [2.62, 60]	—	166.23 [39.07]
Man : TMG w/CO_2 1 : 1 eq.	22.8[a]	[2.32 : 5[d]] (1.58 : 5[a])	3129 1685	6.06, 10.75 [2.70, 60]	(5.25, 0.15) (4.92, 0.05) (4.84, 0.25) (4.48, 0.05) (4.36, 0.05) (4.29, 0.18) (4.20, 0.01) (4.12, 0.02) (3.94, 0.18) (3.86, 0.08) (3.79, 0.2) (3.51, 1.36)	164.71 [39.14]

[a] Determined by weighed mass. [b] Number of OH groups from D-mannose converted to carbonates. [c] Number of =NH groups converted to carbamates. [d] Determined by ^{1}H-NMR and HMBC. NMR spectra obtained in d_6-DMSO.

carbons: $((CH_3)_2N)_2\underline{C}{=}NH\underline{C}O_2$, one from TMG and the other from the carbamate functionality). Moreover the presence of carbamate functionality is confirmed by the presence of characteristic bands at 3107, 3282 and 1667 cm^{-1} detected in the FTIR spectra (bands at a higher frequency give indication of a N–H stretch while the band of lower frequency is related with C=O stretch, both from carbamate functionality[5,20]). Differently from the work of Pereira *et al.*,[18] in our experimental work TMG was dried prior to its use in CO_2 capture.

In the case of the systems based on TMG or mannose/TMG the weight percentages of CO_2 uptake are very high, nevertheless, in addition to the reactivity of the alcohol groups from D-mannose in the presence of CO_2, TMG also reacts, leading to the respective carbamate. In order to increase the average number of -OH groups of saccharides converted into carbonates and simplify the system, avoiding multiple products after reaction with CO_2, we decided to use DBU as the organic superbase instead of TMG. Initially, only dry DBU was tested in CO_2 capture. After reaction, a CO_2 uptake of 2.4% was obtained with 13% of DBU protonated (by ^{1}H-NMR – slight shift of quaternary (N)$\underline{C}$(C)(=N) carbon, 159.66 ppm compared with the initial DBU signal, 159.54 ppm). Moreover, any H/C correlation from a possible carbamate or even from bicarbonate was not detected by HMBC. The only hypothesis that explains this residual CO_2 uptake and low protonation of DBU is the formation of carbonate $[CO_3]^{2-}$ by the reaction of residual water present in DBU with CO_2. From this assay we conclude that DBU doesn't react directly with CO_2 to originate the respective carbamate and it seems a good candidate to replace TMG in the activation of the alcohol groups of the saccharides and further reaction with CO_2, leading to carbonates. In the sequence of this reaction the mixture D-mannose : DBU (ratio of 1 in equivalents) was used to capture CO_2. Such a mixture is very viscous precluding stirring during the reaction. As result the wt% of CO_2 uptake was approximately 0. In order to improve the efficiency of the system and attain maximum functionalization of the saccharide we tested different ratios D-mannose : DBU (0.25, 0.5, 0.625 – besides 1). Prior to reaction with CO_2, the mixture is liquid and afterwards the product is solid as shown in Fig. 2.

The possibility to have a solid instead of a viscous liquid facilitates the transport of captured CO_2. ^{1}H-NMR spectra were used to access the average number of alcohol groups of D-mannose converted into carbonates. The protons adjacent to a carbonate functionality (Fig. 1 – at a distance of three bonds from carbon of carbonate) have a different chemical shift when compared with equivalent protons when alcohol functionality is present. Considering this fact and comparing the ^{1}H-NMR spectra of the systems under discussion with the spectra of D-mannose : DBU prior to its reaction with CO_2 it is possible to highlight the differences (Fig. 3) and account for areas of protons with modified chemical shift with the aid of HMBC spectra (H/C correlation – Fig. 1 shows the carbon of carbonate and the protons at a distance of three bonds) estimate the average number of alcohol groups converted into carbonates.

Considering the results presented in Table 2, it is possible to conclude that decreasing the proportion of saccharide respective to DBU leads to a much more effective CO_2 capture (13.9 wt% of CO_2 uptake in the case of the ratio Man : DBU = 0.625 in equivalents). In any case it is not possible to establish a direct relationship between the ratio Man : DBU and the wt% of CO_2 uptake nor for a relationship between the Man : DBU ratio and the number of alcohol groups converted into carbonates or an exact concordance between the two methodologies of calculation of the number of alcohol groups converted into carbonates (Table 2 – by mass or ^{1}H-NMR). In order to explain these results it is important to consider that magnetic bar based stirring is not the ideal method to promote the reaction and also obtain reproducible results. Moreover traces of water in DBU and/or even in D-mannose may be present, and partial precipitation of the product of reaction from d_6-DMSO solution may occur during storage, delay and acquisition of the NMR spectra. Additionally, when weighing the reactor before and after reaction with CO_2 we detect differences in the first or second decimal place

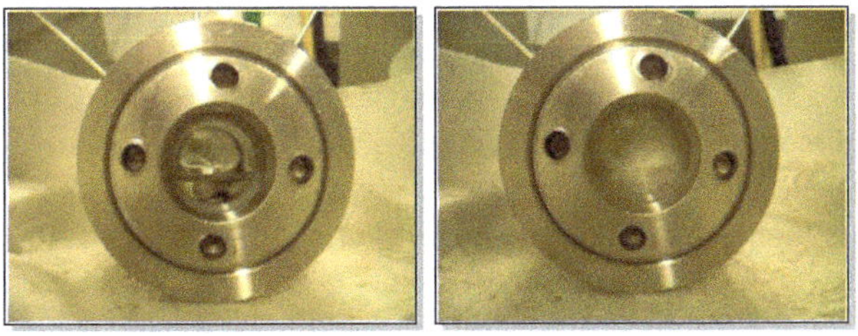

Fig. 2 System D-mannose : DBU (0.625 : 1 eq.). From left to right: before and after reaction with CO_2.

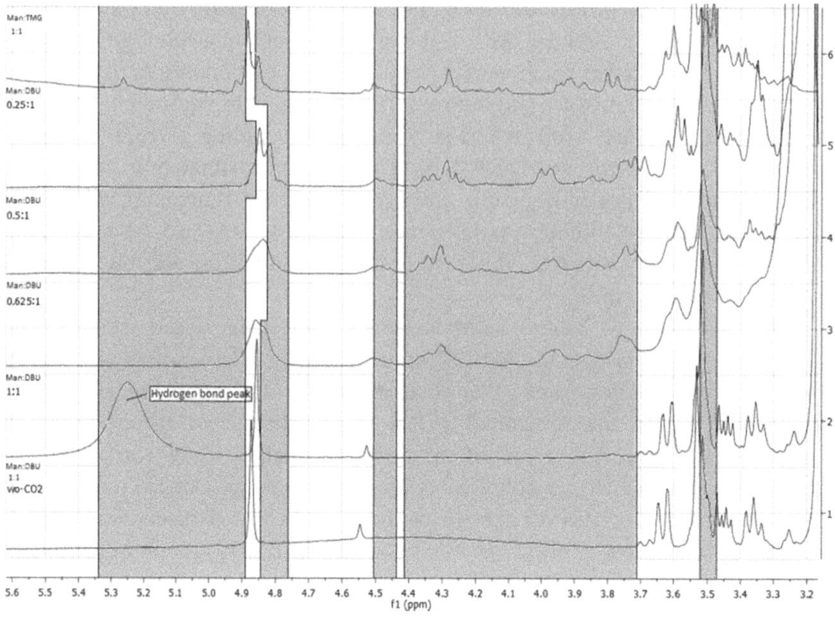

Fig. 3 ^{1}H-NMR spectra-systems based on mannose and an organic superbase (DBU or TMG). Highlighted in grey are the new peaks obtained after reaction with CO_2. From top to bottom: Man : TMG (1 : 1), Man : DBU (0.25 : 1), Man : DBU (0.5 : 1), Man : DBU (0.625 : 1), Man : DBU (1 : 1) and Man : DBU (1 : 1) prior to reaction with CO_2.

when the total mass of the reactor with the mixtures is around 1200 g. Nevertheless it is interesting to note by ^{1}H-NMR that the average number of alcohol groups converted into carbonates is ~3 out of 5 in all the systems studied except in the Man : DBU (ratio 1 in equivalents) system.

These results represent an improvement in the functionalization of D-mannose when compared with a similar system using TMG as the organic superbase (2.32 out of 5 – Table 1). Additionally, FTIR spectra confirm the presence of carbonates, as bands at 1585, ~1400 and 1271 cm^{-1} appear in the studied systems, similar to in the work of Heldebrant *et al.*[21] While in DBU a C=N stretch band centred at 1610 cm^{-1} is evident,[22] in our systems, DBU + D-mannose, the two bands at ~1650 cm^{-1} and ~1610 cm^{-1}, are replaced by a single band at ~1650 cm^{-1} after reaction with CO_2 (Table 2), which, according to Galezowski *et al.*[23] is indication that the DBU is protonated. Moreover, the value of the chemical shift of the quaternary carbon (C)(N)C(=N) is also indicative of the degree of protonation of DBU with the highest value (161.07 ppm) corresponding to the system Man : DBU = 0.625. Factors such as the proportion of D-mannose and the percentage of incorporation of CO_2 increment the chemical shift of this carbon. The results obtained with D-mannose and DBU encouraged us to study D-glucose based systems as this saccharide is highly abundant. Again, saccharide : DBU ratios (in equivalents) of 1, 0.625, 0.5 and 0.25 were tested (Table 3 and Fig. 4). Different from the D-mannose case, a D-glucose based system in addition to the magnetic bar based stirring, was also tested magnetically coupled stirring in a much bigger reactor, as

Table 2 Mannose series, effect of the mannose : DBU ratio on the performance of CO_2 capture

System	wt % CO_2	–OH to CO_3^{-a} mass [NMR]	FTIR CO_3^- [DBU] bands (cm^{-1})	^{1}H-NMR H adjacent to CO_3^- (δ, area)[b]	^{13}C-NMR DBU quat. carbon peak[b]	Man : SB eq. by ^{1}H-NMR
DBU	—	—	—	—	159.54	—
DBU w/CO_2	2.4	[—] — [—]	[1610] 1372.18 1090.23 863.19 [1649.24] [1608.43]	—	159.66	—
Man : DBU 1 : 1 eq.	—	— [—]	— [1648.89] [1611.13]	—	159.78	1
Man : DBU w/ CO_2 1 : 1 eq.	0	0 [0.21]	1585.77 1398.57 1271.80 [1651] [1615.08]	(3.74–3.89, 0.26)	160.12 160.05	1.008
Man : DBU w/ CO_2 0.625 : 1 eq.	13.9	4.37 [3.3]	1585.47 1386.10 1271.69 [1645.60]	(4.83, 0.40) (4.51, 0.19) (4.34, 0.46) (4.19, 0.08) (4.07, 0.05) (3.97, 0.35) (3.87, 0.24) (3.75, 0.69) (3.52, 2.71)	161.07	0.814
Man : DBU w/ CO_2 0.5 : 1 eq.	6.6	2.56 [3.06]	1585.79 1399.22 1271.89 [1642.71]	(4.84, 0.45) (4.46, 0.06) (4.36, 0.23) (4.19, 0.15) (3.98, 0.29) (3.86, 0.29) (3.74, 0.64) (3.52, 1.17)	160.69	0.492
Man : DBU w/ CO_2 0.25 : 1 eq.	7.9	5.75 [2.61]	1585.33 1399.66 1271.66 [1645.45]	(4.82, 0.29) (4.50, 0.01) (4.30, 0.25) (4.17, 0.03) (3.99, 0.19) (3.86, 0.09) (3.74, 0.66) (3.51, 2.01)	160.24	0.281

[a] Number of OH groups of D-mannose converted into carbonates. [b] NMR spectra obtained in d_6-DMSO. All the reactions were promoted using a magnetic bar.

a clear influence of stirring on the outcome of reaction of carbonation was detected.[2,13]

With this test, we could obtain the maximum thermodynamically permitted number of alcohol groups converted into carbonates for a glucose : DBU 1 : 1

equivalent ratio. In a previous report, we had indication that multi-anionic charged organic molecules, with the charges in close vicinity, obtained from functionalization with CO_2 are not very stable.[6]

In this context this test could be an indication of the limits of glucose based systems to accommodate density of charge. Comparing identical systems (Glu : DBU of 1 : 1 ratio) differing in the type of stirring, it was possible to obtain a much higher number of alcohol groups converted into carbonates with magnetically coupled rather than with magnetic bar based stirring (2.47/5 *vs.* 1.08/5 – Table 3) as expected. The magnetically coupled stirred system is a 250 mL reactor that for obvious reasons could not be weighed before and after reaction with CO_2. Considering this limitation, the value of 11.5 wt% of CO_2 uptake was estimated based on ^{1}H-NMR analysis. This was the maximum percentage obtained in this series highlighting the importance of the type of stirring.

Curiously, a clear trend for all the magnetic bar stirred systems was observed, with the lower Glu : DBU ratios resulting in a higher number of alcohol groups converted into carbonates (Table 3) with a maximum of 3.28 groups converted (estimated by ^{1}H-NMR) using the Glu : DBU ratio of 0.25 : 1 equivalents.

Two different arguments could explain the results: first, lower ratios of glucose in the system lead to lower viscosities, leading to more effective mixing between gaseous CO_2 and the liquid phase until an extended functionalization is attained; differently, from a thermodynamic point of view, following the Le Châtelier principle, the increment of the reagents will shift the equilibrium towards the formation of more product. In this case an excess of DBU respective to the number of equivalents of alcohol groups in glucose if decreasing the glucose : DBU ratio (eq.). Another relevant fact is that the values of wt% uptake determined by different methods (weighing and NMR) are very similar, indicating that the methodologies used are reliable.

The presence of carbonate based functionalities was confirmed by FTIR, with bands centred at ~1590, ~1400 and ~1270 cm^{-1} (similar to in the case of D-mannose). Moreover a shift of band from ~1610 to ~1650 cm^{-1} was also detected in Glu : DBU systems after reaction with CO_2, indicating that DBU is protonated in the obtained products. It is also important to note that the degree of protonation of DBU, influenced by the extension of CO_2 uptake and ratio of D-glucose to DBU, is correlated with the chemical shift of the quaternary carbon (C)(N)$\underline{C}$=N of DBU, which is inversely correlated with the frequency of the band ~1650 cm^{-1} detected by FTIR ($R^2 = 0.91$). These results lead us to an empirical equation that relates the Glu : DBU ratio, the number of OH groups converted into carbonates and the number of alcohol groups that remained unchanged with a calculated factor that is well-correlated with the frequency of the band at ~1650 cm^{-1} in the FTIR spectra ($R^2 = 0.98$).

The empirical equation is:

$$\begin{aligned}\text{Factor} = {} & (\text{ratio Glu : DBU in eq.})^{0.24} \times \text{nr. of } -\text{CO}_3 \text{ per glucose}^{0.7} \\ & \times \text{nr. of } -\text{OH per glucose}^{0.81}\end{aligned} \quad (1)$$

and the correlation equation that relates the FTIR frequency (cm^{-1}) with the determined factor in eqn (1):

$$\text{Frequency} = -4.9939 \times \text{factor} + 1661.9. \quad (2)$$

Table 3 Glucose based systems using DBU as the organic superbase in a CO_2 atmosphere. The effect of the ratio of glucose : DBU (in number of equivalents) and type of stirring[a]

System	wt% CO_2	OH to CO_3^{-c} mass [NMR]	FTIR CO_3^- [DBU] bands (cm^{-1})	^{1}H-NMR H adjacent to CO_3^- peaks (δ, area)	^{13}C-NMR DBU quat. carbon peak	Glu : SB eq. by ^{1}H-NMR
DBU	—	— [—]	— [1610]	—	159.54	—
DBU w/CO_2	2.4	— [—]	1372.18 1090.23 863.19 [1649.24] [1608.43]	—	159.66	—
Glu : DBU 1 : 1 eq.	—	— [—]	—	—	159.97	0.71
Glu : DBU w/CO_2 1 : 1 eq.	2.3	0.48 [1.08]	1589.12 1394 1274.55 [1645.99]	(5.01, 0.01) (4.86, 0.08) (3.99, 0.14) (3.90, 0.11) (3.82, 0.1) (3.77, 0.11) (3.54, 0.1)	160.38	0.5
Glu : DBU w/CO_2 1 : 1 eq. w/magnetic coupled stirred[b]	11.5[b]	— [2.47]	1585.97 1399.59 1271.89 [1642.34]	(5.04, 0.05) (4.87, 0.17) (4.30, 0.18) (3.96, 0.44) (3.86, 0.49) (3.55, 0.91)	161.31	0.75
Glu : DBU w/CO_2 0.625 : 1 eq.	6.1	1.94 [1.73]	1588.79 1378 1274.63 [1644.51]	(4.99, 0.02) (4.87, 0.1) (4.29, 0.19) (3.99, 0.28) (3.82, 0.24) (3.74, 0.29) (3.55, 0.25)	160.67	0.372
Glu : DBU w/CO_2 0.5 : 1 eq.	7.9	3.04 [2.85]	1588.84 1394 1274.59 [1645.17]	(5.03, 0.08) (4.86, 0.15) (4.29, 0.33) (3.99, 0.47) (3.86, 0.76) (3.75, 0.53) (3.55, 0.25)	160.81	0.375
Glu : DBU w/CO_2 0.25 : 1 eq.	4.9	3.59 [3.28]	1588.14 1365.92 1274.04 [1649.46] [1611.60]	(4.97, 0.02) (4.85, 0.13) (4.27, 0.23) (4.00, 0.24) (3.91, 0.09) (3.85, 0.08) (3.77, 0.42) (3.55, 0.24)	160.02	0.191

[a] In all entries, except when stated "magnetically coupled system", stirring was promoted using a magnetic bar. [b] The value of wt% CO_2 incorporation was estimated based on the number of equivalents of alcohol groups of glucose converted into carbonates. [c] Number of alcohol groups of D-glucose converted to carbonates.

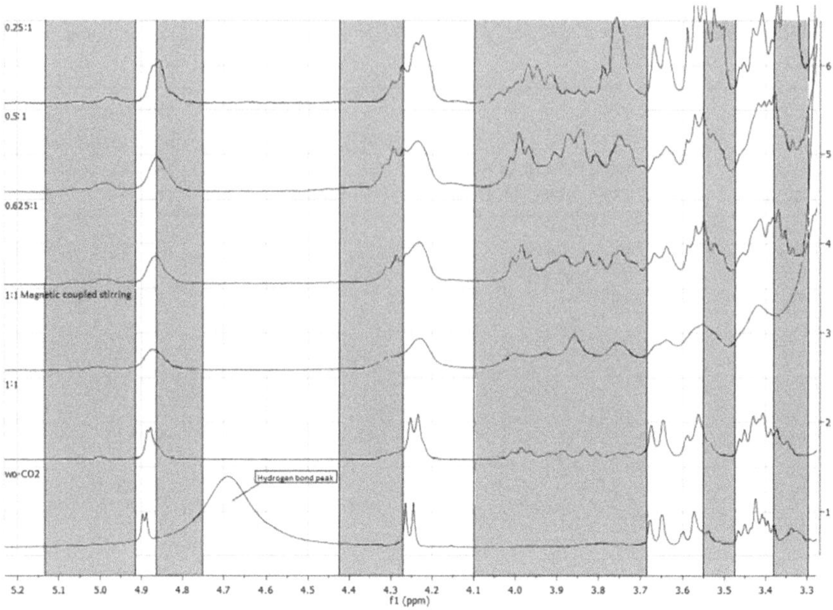

Fig. 4 ^{1}H-NMR spectra of the glucose : DBU systems. Highlighted are the new peaks obtained after reaction with CO_2. From top to bottom: Glu : DBU (0.25 : 1), Glu : DBU (0.5 : 1), Glu : DBU (0.625 : 1), Glu : DBU (1 : 1) with magnetically coupled stirrer, Glu : DBU (1 : 1) and Glu : DBU (1 : 1) prior to reaction with CO_2.

With both equations it is possible, by performing a FTIR spectra of the product of CO_2 capture and providing the Glu : DBU ratio (eq.), to access directly the average number of carbonate groups per molecule of functionalized glucose.

With D-glucose based systems, similarly to D-mannose, the estimation of the number of OH groups converted into carbonates was performed by highlighting the differences between Glu : DBU systems before and after reaction of CO_2 (Fig. 4) in ^{1}H-NMR spectra, with the aid of HMBC spectra (ESI† – long times of acquisition were required and low intensity correlations were detected in most cases, and in specific cases no correlation was detected). Other saccharides and related structures were tested for CO_2 capture such as β-cyclodextrin, alginic acid and mannitol with three, two and six alcohol groups per unit or molecule respectively. Again due to high viscosities it was necessary to use ratios of saccharide : DBU in equivalents lower than 1 in most cases (Table 4).

Relatively low values of wt% of CO_2 uptake were attained with these saccharides. The main reasons behind these results are the considerable excess of superbase used (in the case of β-cyclodextrin and mannitol), inefficient stirring at a relative early stage of reaction (all the reactions were promoted by the use of a magnetic bar) and the relatively low number of alcohol groups able to be functionalized (in the case of β-cyclodextrin and alginic acid). Again, determination by NMR of the number of alcohol groups converted into carbonates was based on the area of protons adjacent to carbonate functionalities that were unveiled with the aid of HMBC spectra. In this context, it is important to highlight the relative resemblance between the values determined by NMR and mass. In all the cases

Table 4 Other saccharides combined with DBU to capture CO_2

System	wt% CO_2	OH to CO_3^{-a} mass [NMR]	FTIR CO_3^- [DBU] bands (cm^{-1})	^{1}H-NMR H adjacent to carbonate (δ, área)	^{13}C-NMR DBU quat. carbon peak
DBU	—	— [—]	— [1610]	—	159.54
DBU w/CO_2	2.4	— [—]	1372.18 1090.23 863.19 [1649.24] [1608.43]	—	159.66
β-CD : DBU w/CO_2 0.5 : 1 eq.	1.8	3.08 : 21 [3.93 : 21]	1586.87 1384.56 1270.36 [1652.19] [1612.66]	(4.72, 1.69) (4.01, 0.85) (3.57, 1.98)	160.15
Alginic acid : DBU w/CO_2 1 : 1 eq.	4.73	0.68 : 2 [—][b]	1586.62 1384.62 1270.18 [1651.65]	—[b]	—[b]
Mannitol : DBU w/ CO_2 0.5 : 1 eq.	4.73	2.16 : 6 [1.95 : 6]	1585.87 1385.77 1272.17 [1646.05]	(3.37, 2.21) (3.58, 1.04)	160.53

[a] Number of OH groups of D-mannose converted into carbonates. [b] Solubility constraints didn't permit to obtain reliable NMR spectra. Magnetic bar stirring was used to promote reaction with CO_2 in all the cases.

Table 5 DSC thermal study of the saccharide based systems tested in this report after reaction with CO_2

Systems	T_d[e] (°C)	T_g[e] (°C)
(TMG)	56.52, 108.88[b]	—
(Man : TMG) 1 : 1 (eq.)	103.15[b]	—
(DBU)	118.01[b]	—
(Man : DBU) 1 : 1 (eq.)	128.01[b,d]	−44.39
(Man : DBU) 0.625 : 1 (eq.)	T_d > 107.10[b]	−50.16
(Man : DBU) 0.5 : 1 (eq.)	122.1[b]	—
(Man : DBU) 0.25 : 1 (eq.)	127.57[b]	−66.68
(Glu : DBU) 1 : 1 (eq.)	102.56, 122.25, 147.37[b]	−51.49
(Glu : DBU) 1 : 1 (eq.) magnetically coupled stirr	110.87, 145.48[b]	−56.70
(Glu : DBU) 0.625 : 1 (eq.)	116.71, 146.26[b]	—
(Glu : DBU) 0.5 : 1 (eq.)	119.21, 146.68[b]	—
(Glu : DBU) 0.25 : 1 (eq.)	108.22[b], 134[c]	−74.78
(β-CD : DBU) 0.5 : 1 (eq.)	127.78[b]	−68.38
(Alginic acid : DBU) 1 : 1 (eq.)	125.08[b]	−75.10
(Mannitol : DBU) 0.5 : 1 (eq.)	>112.70[a]	−66.16

[a] Onset. [b] Horizontal peak. [c] Curve point. [d] Exothermic peak. [e] T_d: decomposition temperature, T_g: glass transition temperature.

tested, carbonate characteristic bands in the FTIR spectra (Table 4) were detected similarly as in D-mannose and D-glucose series. In the same spectra it is possible to verify that in the cases where the functionalization reaction is more extended (alginic acid and mannitol) a single band centred a ~1650 cm^{-1} is obtained. In the case of β-cyclodextrin two bands arise at ~1652 and 1613 cm^{-1} indication that DBU is partially protonated[23] and that the reaction was less extended when compared with other saccharides. The chemical shift of the quaternary carbon of DBU (C)(N)C=N indicates that DBU is more protonated in the mannitol : DBU system with a ratio of 0.5 (eq.).

Additionally, a DSC thermal study of the products of reaction with CO_2 was performed (Table 5). The experiment consisted of lowering the temperature of the sample to −90 °C and increasing the temperature afterwards to 150 °C. T_g (glass transition temperature), mp (melting point) and T_d (decomposition temperature) measurements were performed, and finally a second cycle was performed to confirm CO_2 release by non-reproducibility of the second cycle respective to the first. In the case of the TMG based systems, two endothermic peaks were detected in the first cycle, one at 56.52 °C and the other at 108.88 °C. In the study by Pereira *et al.*,[18] two TGA losses were detected for the same system after CO_2 capture. According to the authors, the release at a lower temperature corresponds to carbamate, and the other is from bicarbonate. We agree with this interpretation however, considering also the T_d result of the DBU system (118.01 °C), that corresponds to CO_2 release from $[CO_3]^{2-}$, we hypothesise that instead of bicarbonate the reason behind the second peak in the TMG system is CO_2 release from carbonate.

Considering the mannose : DBU series, it is possible to verify that when the number of alcohol groups converted into carbonates is higher (Table 2 – by NMR) the value of T_d decreases (Table 5), indicating that a higher density of charge in a small molecule like D-mannose leads to higher destabilization of the product and consequently a lower temperature is required for CO_2 release.

It is also important to note that at this point it is premature to discard the influence of dilution of the functionalized saccharide respective to DBU in increasing the value of T_d.

Considering T_g analysis, it is possible to observe that for all the cases that were detected, the value of T_g increases with the increment of the Man : DBU (eq.) ratio, indicating that a possible extended hydrogen bond network may lead to an increased value of T_g. Regarding the D-glucose series, a similar approach can be performed, and differently from the D-mannose series, it is possible to verify in the great majority of ratios a peak at >145 °C that should correspond to the melt/decomposition of glucose[24] after release of CO_2 from carbonate functionalized saccharide. In order to explain the values obtained at lower temperatures along the series we have to consider both the density of charge and the dilution of the functionalized saccharide, with the latest leading to an increased value of T_d, and the former leading to a lower value of decomposition temperature, for the ratios, 1 : 1 (magnetically coupled), 0.625 : 1 and 0.5 : 1. The trend is clear because there's only one value of T_d associated to CO_2 release from carbonates. Differently for the ratios 1 : 1 and 0.25 : 1 two peaks associated to CO_2 release from the different carbonates present in the mixture could be detected. An average of both peaks could lead to a value that is in agreement with the hypothesis of the influence of the effect of density of charge and dilution on the value of T_d.

Concerning the T_g values, and similar to the D-mannose series, an extended hydrogen bond network provided by higher Glu : DBU ratios leads to higher values of T_g. Other saccharides lead to similar values of T_d associated to CO_2 release.

Conclusions

Effective systems for CO_2 capture based on a cheap organic superbase (TMG or DBU) and highly abundant natural saccharides were designed and tested. The main objectives of this work were to obtain maximal percentages of CO_2 uptake and the number of alcohol groups of saccharides converted into carbonates. With TMG systems, very high percentages of CO_2 uptake were obtained, nevertheless when D-mannose is present, competition between carbonate and carbamate products was detected. In order to attain maximal conversion of alcohol groups into carbonates and to avoid competition with the superbase to react with CO_2, DBU was tested as an alternative organic superbase. Moreover in the case of D-mannose and D-glucose different ratios of saccharide : DBU were tested in order to find an optimal balance between the average number of alcohol groups functionalized into carbonates and the total wt% of CO_2 uptake. With D-mannose : DBU (ratio eq. = 0.625) an optimal wt% of CO_2 uptake of 13.9% was obtained corresponding to 3.3/5 alcohol groups being converted into carbonates. In the case of glucose : DBU (ratio eq. = 0.5) a maximum wt% of CO_2 uptake of 7.9%, corresponding to 3.04/5 alcohol groups converted into carbonates, was obtained with a magnetic bar based stirring method. In order to test the thermodynamic limit of the reaction and obtain maximal CO_2 uptake, a Glu : DBU system (ratio eq. = 1) was stirred in a magnetically coupled system. In that case, 2.47/5 alcohol groups were converted into carbonates, corresponding to 11.5 wt% of CO_2 uptake. Factors, such as the stirring efficiency in promoting solubilisation of CO_2 gas into the liquid phase, saccharide : organic superbase ratio and type of superbase were unveiled as fundamental in the outcome of the reaction. Other saccharide based structures were also tested with interesting indications, especially mannitol, with six alcohol groups per molecule, where it's expected that the obtained results could improve significantly if magnetically coupled stirring is used instead of a magnetic bar. Thermal DSC analysis indicated that the stability of the obtained products are dependent on the dilution of the saccharide and the density of charge.

Acknowledgements

This work was supported by Fundação para a Ciência e Tecnologia (PEst-C/EQB/LA0006/2013, PTDC/CTM-NAN/120658/2010 and a postdoctoral fellowship GVSMC – SFRH/BPD/72095/2010). The authors acknowledge Prof. Madalena Dionisio and Dr Teresa Casimiro for providing lab facilities and fruitful discussions. The NMR spectrometers are part of The National NMR Facility, supported by Fundação para a Ciência e Tecnologia (RECI/BBB-BQB/0230/2012).

Notes and references

1 M. Mikkelsen, M. Jørgensen and F. C. Krebs, *Energy Environ. Sci.*, 2010, **3**, 43.

2 D. J. Heldebrant, C. R. Yonker, P. G. Jessop and L. Phan, *Energy Environ. Sci.*, 2008, **1**, 487.

3 M. S. Shannon and J. E. Bara, *Sep. Sci. Technol.*, 2012, **47**, 178.

4 C. Wang, H. Luo, X. Luo, H. Li and S. Dai, *Green Chem.*, 2010, **12**, 2019.

5 E. D. Bates, R. D. Mayton, I. Ntai and J. H. Davies Jr, *J. Am. Chem. Soc.*, 2002, **124**, 926.

6 B. E. Gurkan, J. C. de la Fuente, E. M. Mindrup, L. E. Ficke, B. F. Goodrich, E. A. Price, W. F. Schneider and J. F. Brennecke, *J. Am. Chem. Soc.*, 2010, **132**, 2116.

7 P. G. Jessop, D. J. Heldebrant, X. Li, C. A. Eckert and C. L. Liotta, *Nature*, 2005, **436**, 1102.

8 T. Yamada, P. J. Lukac, M. George and R. G. Weiss, *Chem. Mater.*, 2007, **19**, 967.

9 G. V. S. M. Carrera, M. Nunes da Ponte and L. C. Branco, *Tetrahedron*, 2012, **68**, 7408.

10 T. Yamada, P. J. Lukac, T. Yu and R. G. Weiss, *Chem. Mater.*, 2007, **19**, 4761.

11 T. Yu, T. Yamada, G. C. Gaviola and R. G. Weiss, *Chem. Mater.*, 2008, **20**, 5337.

12 G. V. S. M. Carrera, N. Jordão, M. M. Santos, M. Nunes da Ponte and L. C. Branco, *RSC Adv.*, 2015, **5**, 35564–35571.

13 G. V. S. M. Carrera, N. Jordão, L. C. Branco and M. Nunes da Ponte, *J. Supercrit. Fluids*, DOI: 10.1016/j.supflu.2015.02.015.

14 D. J. Heldebrant, C. R. Yonker, P. G. Jessop and L. Phan, *Chem.–Eur. J.*, 2009, **15**, 7619.

15 V. M. Blasucci, R. Hart, P. Pollet, C. L. Liotta and C. A. Eckert, *Fluid Phase Equilib.*, 2010, **294**, 1.

16 Q. Zhang, N. S. Oztekin, J. Barrault, K. de Oliveira Vigier and F. Jérôme, *ChemSusChem*, 2013, **6**, 593.

17 S. Abraham and R. G. Weiss, *Photochem. Photobiol. Sci.*, 2012, **11**, 1642.

18 F. S. Pereira, E. R. de Azevedo, E. F. da Silva, T. J. Bonagamba, D. L. S. Agostíni, A. Magalhães, A. E. Job and E. R. P. González, *Tetrahedron*, 2008, **64**, 10097.

19 M. C. Ozturk, O. Y. Orhan and E. Alper, *Int. J. Greenhouse Gas Control*, 2014, **26**, 76.

20 R. M. Silverstein and F. X. Webster, *Spectrometric Identification of Organic Compounds*, John Wiley & Sons, Inc, New York, 6th edn, 1997.

21 D. J. Heldebrant, P. K. Koech, M. T. C. Ang, C. Liang, J. E. Rainbolt, C. R. Yonker and P. G. Jessop, *Green Chem.*, 2010, **12**, 713.

22 Web page: http://sdbs.db.aist.go.jp/sdbs/cgi-bin/cre_index.cgi last check 23/04/2015.

23 W. Galezowski, A. Jarczewski, M. Stanczyk, B. Brzezinsky, F. Bartl and G. Zundel, *J. Chem. Soc., Faraday Trans.*, 1997, **93**, 2515.

24 http://www.chemicalbook.com last check 24/04/2015.

Faraday Discussions

PAPER

An enriched electroactive homoacetogenic biocathode for the microbial electrosynthesis of acetate through carbon dioxide reduction

Gunda Mohanakrishna, Jai Sankar Seelam, Karolien Vanbroekhoven and Deepak Pant*

Received 27th April 2015, Accepted 29th June 2015
DOI: 10.1039/c5fd00041f

In the direction of generating value added chemicals from carbon dioxide (CO_2) reduction through microbial electrosynthesis (MES), considering the crucial impact of the electrode material for the biofilm development and electron delivery, an attempt was made in this study to evaluate the efficiency of two different materials as biocathodes and their respective output in terms of electrosynthesis. The electrode material is a key component in the MES process. Several electrodes such as platinum, graphite foil, dimentionally stable anode (DSA) and graphite rod, and VITO-CoRE™ derived electrodes were tested for their suitability for ideal electrode combination in a three electrode cell setup. Bicarbonates (the dissolved form of CO_2) was reduced to acetate by a selectively developed biocathode under a mild applied cathodic potential of −400 mV (*vs.* SHE) in 500 mL of single chamber MES cells operating for more than four months. Among the two electrode combinations evaluated, VITO-CoRE™-PL (VC-IS, plastic inert support) as the cathode and VITO-CoRE™-SS (VC-SS, stainless steel metal support) as the counter electrode showed higher production (4127 mg L^{-1}) with a volumetric production rate of 0.569 kg per m^3 per d than the graphite rod (1523 mg L^{-1}) with a volumetric production rate of 0.206 kg per m^3 per d. Contrary to the production efficiencies, the coulombic efficiency was higher with the second electrode combination (40.43%) than the first electrode combination (29.91%). Carbon conversion efficiency to acetate was higher for VC-IS (90.6%) than the graphite rod (82.0%).

1. Introduction

Carbon dioxide (CO_2) reduction for the generation of renewable energy and value added chemicals is one of the emerging areas of present research developments. Microbial electrosynthesis (MES) is the recent application of bioelectrochemical systems (BES), that involves cathodic reduction reactions to generate biofuels and

Separation & Conversion Technologies, VITO – Flemish Institute for Technological Research, Boeretang 200, 2400 Mol, Belgium. E-mail: deepak.pant@vito.be; pantonline@gmail.com

speciality chemicals.[1–5] The bacteria that metabolizes on the conductive electrode surface acts as the catalyst for the chemical conversions in BES and are known as electroactive biocatalysts. Other forms of BES are microbial fuel cells (MFCs) and microbial desalination cells (MDC) where the biocatalyst functions on the anode[6–9] and in microbial electrolysis cells (MEC), the biocatalyst functions on the cathode.[10–12] On the whole, the BES processes are considered as a biotechnology with electrodes in which solid state electrodes can provide a stable and sustainable solution.[13] The anode serves as a terminal electron acceptor for the discharge of excess/surplus microbial reducing equivalents and the cathode provides reducing equivalents required for biochemical processes. The applied electrode potential tunes the strength of the electron acceptor or donor. It results in an electric current which determines the biocatalytic performance of electroactive biofilms.[14–18]

CO_2 plays a key role in the eventual climate change due to its accumulation in the atmosphere. The control of CO_2 emissions into the atmosphere requires novel ideas that can assure future generations with sustainable growth options. Considerable research is being carried out to convert CO_2 to various commercial products by thermochemical and electrochemical technologies, which need high energy input but also generate a wide variety of products.[19] MES is a novel biocathode-driven production technology for the reduction of CO_2 to chemicals and biofuels. It is characterised as a promising way for sustainable development and new carbon-consuming technologies.[3,20] The specificity of the product formation from CO_2 is largely influenced by the bacteria and its metabolic functions and electrical potential that drives the process.[2,12] Acetate is the basic intermediary product compound that is formed by the photosynthesis process from CO_2 conversion and is the most important step to generate available carbon forms for biological domains. Furthermore, acetate can be used as the substrate for various biological building blocks of life. Homoacetogenic bacteria are a specialized group of bacteria that are able to produce acetate from two molecules of CO_2 through the Wood–Ljungdahl pathway (WLP).[21,22] Acetogens are facultative autotrophs that can grow by the consumption of hydrogen (H_2) or carbon monoxide (CO) which is usually coupled to the reduction of CO_2. Also, a large variety of organic substrates, including hexoses, pentoses, alcohols, methyl groups and formic acid can be used as the substrates.[21,23] On the whole, acetogens are present in 23 different bacterial genera. Most acetogens are found in phylum Firmicutes, which are Gram-positive bacteria with low GC content. The genera *Clostridium* contains acetogenic as well as non-acetogenic species, whereas other genera such as *Acetobacterium* or *Sporomusa* contain only acetogens. Most known acetogens belong to the *Clostridium* and *Acetobacterium* genera.[21,22] Among the large number of acetogens, three organisms have been studied quite extensively (*Moorella thermoacetica*, *Acetobacterium woodii* and *Clostridium ljungdahlii*). These bacteria were primarily studied for biocathode driven CO_2 fixation. Another advantage of acetogens for use in the MES process is their metabolic flexibility. Acetogens are ubiquitously distributed in alkaline, high-salt and hot environments, in deep subsurface samples, the termite hindgut and in human intestines. They are also found in terrestrial soil, in freshwater and marine sediments.[22] This metabolic flexibility can be an advantage for their application in MES which is a hybrid system with biological electrochemical functions.

Proof of concept studies of the MES process reported by Nevin *et al.*,[17,24] showed the production of acetate, butyrate, oxy-buterate and formate by the bioelectrochemical production by *Sporomusa ovata*, *Clostridium ljungdahlii*, *Clostridium aceticum*, and *Moorella thermoacetica* at an applied potential of −400 mV (*vs.* SHE). Later Nie *et al.*,[25] achieved higher acetate production rates by *Sporomusa ovata*. Besides pure cultures, few researchers were able to produce acetate with mixed and adapted mixed cultures at various applied potentials ranging from the −400 mV to −950 mV.[18,20,26–29] Due to several obvious advantages of the process such as low energy input and CO_2 fixation, it is gaining much global attention. Along with whole cell biocatalysts, the use of specific enzymes in bioelectrocatalysis for CO_2 sequestration is also finding possible application for sustainable energy and chemicals generation.[3,30]

In the present study, an attempt was made to generate a sustainable homoacetogenic biocathode that reduces CO_2 to acetate under mild applied potential conditions where the hydrogen production can be completely limited. A detailed evaluation was also performed for the selection of electrode combination for the potential window analysis where no hydrogen can be produced and to avoid interference of the fermentation process. Furthermore, a detailed discussion is made on the benchmarking of MES process with respect to the type of bacterial catalyst, coulombic efficiency and production rate.

2. Experimental

2.1 Homoacetogenic electroactive bacteria

Homoacetogenic bacteria that have specific metabolic competence to produce acetate from the inorganic carbon source (CO_2) were enriched from granular activated sludge. The granular activated sludge was collected from an anaerobic digester that treats effluents of the potato processing industry (Cargill Sas van, Ghent). It contains 6% of total solid (TS) and 75% of volatile solid (VS) content along with <1% of inorganic carbon. A four stage selective enrichment methodology was applied to enrich the electroactive homoacetogenic bacteria for biocathode development. During the first stage the sludge collected from anaerobic digester was subjected to heat treatment (90 °C for 1 hour) to eliminate the methanogenic bacteria and to retain the acetogenic bacterial spores.[31] In the second stage, the heat treated sludge was subjected to heterotrophic growth using glucose as a substrate in nutrient broth medium at pH 6.0 under 35 °C to activate the metabolism of whole consortia. Heterotrophically grown active cultures of acetogenic bacteria were shifted to autotrophic conditions by providing a mixture of CO_2 and H_2 as a substrate, where only homoacetogenic bacteria were grown and they were enriched by repeated culturing on a CO_2 and H_2 mixture for four cycles. The homoacetogenic activity was confirmed through the acetate production and inhibition of methane. The resultant consortia was used as inoculum for electroactive biofilm formation.

2.2 Bioelectrochemical systems

2.2.1 Design. A MES reactor was designed with a three electrode setup (single chamber) and fabricated with a glass bottle (total/working, 0.62/0.5 L) which included suitable ports for the provision of electrodes and N_2 gas flushing (Fig. 1).

Initially, 6 different materials were screened to use as the anode/counter electrode to select the most suitable one. As the BES is single chambered, if the counter electrode evolves oxygen, the biofilm on the working electrode cannot sustain its anaerobic properties to carry out the reduction reaction. Henceforth, different electrode materials, *viz.*, platinum, dimensionally stable anode (DSA made of titanium coated with iridium mixed metal oxide, Magneto BV Netherlands), graphite foil, graphite rod, VITO-CoRE™ derived electrodes with black layer on both sides with stainless steel (VC-SS) and inert plastic support (VC-IS) were chosen against a VC-SS counter electrode in both abiotic and biotic (enriched bacteria in suspension) conditions under −400 mV of applied cathodic potential.

The performance of the electrodes was evaluated in terms of voltage reached and whether the electrode showed a lower than water electrolysis potential (1.23 *vs.* SHE, anode potential 0.80 V *vs.* SHE). Two MES reactors (MES1 and MES2) of similar configuration, having a VC-SS electrode[34] as the anode (selected from the screening experiment), were used in the current experiment. Two different cathode materials, *viz.*, VC-IS in MES1 and a graphite rod in MES2, were

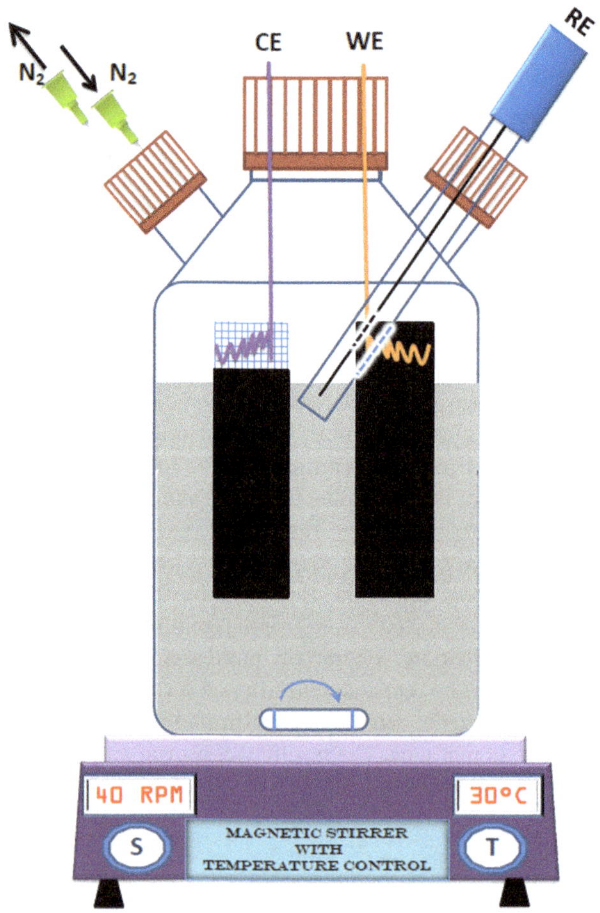

Fig. 1 Schematic diagram of the single chamber microbial electrosynthesis cell (CE, counter electrode; WE, working electrode and RE, reference electrode).

evaluated for their effective homoacetogenic activity as well as electron uptake for the acetate synthesis. The total surface area (37.5 cm^2) and the active surface area (the area in contact with the electrolyte, 30 cm^2) of all the electrodes (anodes and cathodes) was kept constant, except for the DSA (6 × 2.5 cm) and platinum (surface area, 2.37 cm^2) due to the unavailability of sizes. A fine stainless steel wire, weaved through the electrode and extended through the airtight passage of the reactor cap, was used as a current collector (Fig. 1). No separate current collector was used in case of platinum wire.

2.2.2 **Operation.** Both MES reactors were operated in batch mode at 30 °C in a temperature controlled oil bath. MES1 was operated for 127 days, while the MES2 was operated for only 47 days due to loosing biofilm on the cathode and instability of the output. The phosphate buffer media which was used during the enrichment of homoacetogenic bacteria was also used as media in MES operation. Phosphate buffer media containing NH_4Cl of 200 mg L^{-1}, $MgCl_2 \cdot 6H_2O$ of 200 mg L^{-1}, yeast extract of 10 mg L^{-1} along with the trace elements solution (per litre, nitrilotriacetic acid, 1.5 g; $MgSO_4 \cdot 7H_2O$, 3.0 g; $MnSO_4 \cdot H_2O$, 0.5 g; NaCl, 1.0 g; $FeSO_4 \cdot 7H_2O$, 0.1 g; $CoSO_4 \cdot 7H_2O$, 0.18 g; $CaCl_2 \cdot 2H_2O$, 0.1 g; $ZnSO_4 \cdot 7H_2O$, 0.18 g; $CuSO_4 \cdot 5H_2O$, 0.01 g; $KAl(SO_4)_2 \cdot 12H_2O$, 0.02 g; H_3BO_3, 0.01 g; $Na_2MoO_4 \cdot 2H_2O$, 0.01 g; $NiCl_2 \cdot 6H_2O$, 0.03 g; $Na_2SeO_3 \cdot 5H_2O$, 0.30 mg) and vitamin solution (per litre, biotin, 2 mg; pantothenic acid, 5 mg; B-12, 0.1 mg; *p*-aminobenzoic acid, 0.5 mg; thioctic acid (alpha lipoic), 5 mg; nicotinic acid, 5 mg; thiamine, 5 mg; riboflavin, 5 mg; pyridoxine HCl, 10 mg; folic acid, 2 mg). Sodium bicarbonate was considered as the substrate for BES at a concentration of 3.44 g L^{-1} which is equivalent to 2.5 g L^{-1} of bicarbonate (HCO_3^-). Bromoethanesulfonic acid (BESA) was also added to the medium at a concentration of 0.5 g L^{-1} to inhibit the possible methanogenic activity.[31–33] The final pH of the media was maintained at 7.0. During the start-up of the reactor, 10% of the enriched inoculum (50 mL) was added to the MES along with 450 mL of media. Both, MES1 and MES2 were continuously poised at −600 mV *vs.* Ag/AgCl (−400 mV *vs.* SHE) through a chronoamperometry (CA) technique using a potentiostat (BioLogic-VMP3 model, France). All the assays were performed *in situ* by considering the cathode as the working electrode and the anode as the counter electrode against an Ag/AgCl (3.0 M KCl) reference electrode. All the potentials mentioned further in the manuscript are *vs.* SHE unless otherwise stated.

Consumption of bicarbonates in the electrolyte (less than 0.5 g HCO_3^- per L) was considered as time for feed change. To avoid sudden changes in the electrolyte concentration, only 60% of feed was replaced with fresh feed using a siphon flow method where the planktonic cells were settled first not to remove any whole cell biocatalyst. In the case of MES1, after start-up, the feed was changed on the 21st, 28th, 35th, 43rd, 50th, 60th, and 70th day of operation. Later, a constant time interval of 7 days was maintained until the 126th day of operation. In the case of MES2, feed replacement was done during the 21st, 28th, 34th, 40th and 47th day of operation. Both the reactors were continuously flushed with N_2 gas to maintain the anaerobic environment.

2.3 Analysis

Liquid samples from MES1 and MES2 were collected on every fourth day. During enrichment of the homoacetogenic bacteria, along with liquid samples, gaseous samples were also collected. The liquid samples were analysed for organic acids (formic acid, acetic acid, propionic acid and butyric acid), ethanol and pH, whereas the gaseous samples were analysed for H_2, CO_2, CH_4 and O_2. Prior to analysis through HPLC, the samples were filtered through 0.45 μm Acrodisc syringe filters. HPLC analyses was performed using an Agilent HPLC 1200 series equipped with a RID detector (Agilent 1260) set at a wavelength of 215 nm. The column used was an Agilent Hi-Plex column 8u (3000 mm × 7.7 mm) operated at 60 °C equipped with a guard column of the same material. Phosphoric acid (0.05% in isocratic gradient) was used as the eluent at a flow rate of 1 mL min^{-1}. The injection volume of the samples and standards was 20 μL. EZchrom software provided by Agilent was used for HPLC data analysis.

Gaseous samples from the enrichment cultures and MES reactors were analysed using a Trace GC Ultra (Thermo Scientific) connected with 2 different thermal conductivity detectors. A carbosphere column (2m-1/800SS) was used for hydrogen, whereas for CO_2 and methane, both a Hayesep N column (1m-1/800SS) and a Molsive column A (1m-1/800SS) were used at a constant temperature of 60 °C. pH was analysed with a calibrated pH meter (WTW Multi 340i). The liquid samples from MES1 and MES2 reactors were also analyzed for total inorganic carbon (TIC) for quantitative analysis of bicarbonates through a TOC analyzer (Multi N/C 3100 of Analytik Jena) with an auto-sampler (APG 49 of Analytik Jena) according to the methodology developed in ISO 8245.

3. Results and discussion

3.1 Screening of electrodes for the anode

MES is a process that proceeds with the reduction of protons *via* electrons that are generated from the counter electrode or anode of the cell. As the present study was carried out in single chambered configuration, it was important to avoid the abiotic water electrolysis reaction. By maintaining cells with mild electrochemical activity, oxygen and hydrogen production can be controlled. These conditions help to maintain the anaerobic system which is a prerequisite for biocatalyst activity. However, the net counter electrode (anode) potential sustained in the cell also influences the cathode potential. Theoretically, efficient electrolysis process can continue at a potential of 1.23 V (*vs.* SHE).[35,36] However, it is also found to be influenced by the type of electrode material and the surface area considered for the electrode. Maintaining lower potentials on MES systems also warrants direct electron transfer rather than H_2 mediated electron transfer for acetate production.

To identify an efficient and suitable anode for MES, six different materials were screened under similar conditions of MES operation under both biotic and abiotic conditions (Fig. 2). Among the six different electrodes used as the counter electrode, at −400 mV of working electrode potential, three electrodes *viz.*, platinum, DSA and graphite rod, reached a potential higher than 0.80 V (*vs.* SHE), indicating the possibility of oxygen evolution. Platinum showed the highest net counter electrode potential of 5.05 V under abiotic conditions and 5.01 V under biotic conditions followed by DSA (abiotic, 2.01 V; biotic 1.87 V) and graphite rod (abiotic, 1.57 V; biotic 1.46 V). Even though graphite foil showed a lower anode potential of 0.78 V under abiotic conditions, a small rate of oxygen evolution was observed by the gas analysis from the reactor head space. In the case of biotic

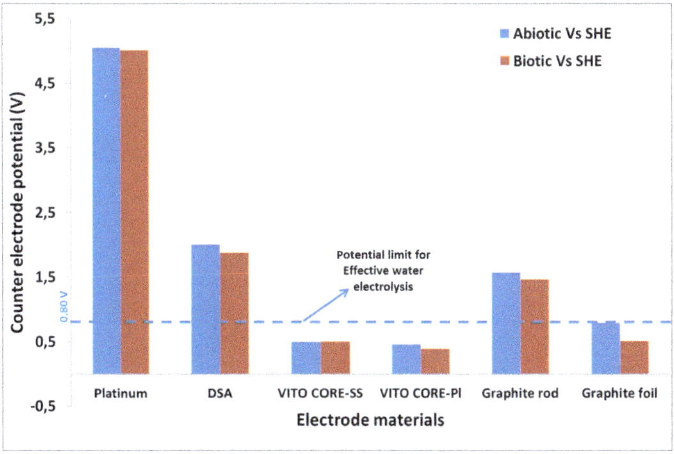

Fig. 2 Counter electrode potentials observed with various electrode materials against a VITO-CoRE™ derived electrode as the working electrode at an applied potential of −400 mV. (Surface area of all the electrodes (except platinum) was equal. It is considered as 39 cm^2 and effective/reactive surface area is 30 cm^2). In the case of platinum, as the economic and practical applicability of the platinum is expensive, a wire surface area consisting of 2.37 cm^2 (effective/reactive surface area) was used.

conditions, 0.51 V of counter electrode potential was observed with no visible oxygen evolution. The two types of VITO-CoRE™ electrodes (VC-SS and VC-IS) evidenced mild counter electrode potential (VC-SS, abiotic, 0.50 V and biotic 0.49 V; VC-IS, abiotic, 0.45 V and biotic 0.39 V) and no oxygen evolution was observed under both abiotic and biotic conditions. The mild electrochemical activity and similar performance under both biotic and abiotic conditions was indicated to select the VC-SS as the suitable counter electrode (anode) for further experimentation in MES towards acetate production.

3.2 Bioelectrosynthesis of acetate

Once the screening was done, two MES reactors were started up with the selected VC-SS (MES1) as the anode and VC-IS (MES2) and a graphite rod as the cathodes (working electrodes). During the start-up stage (first phase, 50 days) of MES1, electroactive biofilm formation from the enriched homoacetogenic bacteria and subsequent stabilization of biofilm was observed at −400 mV with bicarbonate as the carbon source and electron acceptor for the production of acetate. As the inoculum was enriched with CO_2 and H_2 for acetate production, initially MES1 showed an acetate concentration of 101.42 mg L^{-1} which later decreased to 67.12 mg L^{-1} on the 5th day with negligible current generation (Fig. 3). From the 8th day of operation, reduction of bicarbonate was confirmed with a reduction current as well as metabolite analysis in the electrolyte. By the end of the 20th day, the acetate concentration reached 755.67 mg L^{-1} (Fig. 3a) with a reduction current density of −104 mA m^{-2} (Fig. 3b). During the second batch the acetate production was also found to increase gradually and reached a concentration of 1231.43 mg acetate per L (30th day) with a reduction current density of −133 mA m^{-2}. In the first phase, a similar improvement in acetate production was observed with every feeding event and registered a maximum at the 35th day (4th cycle,

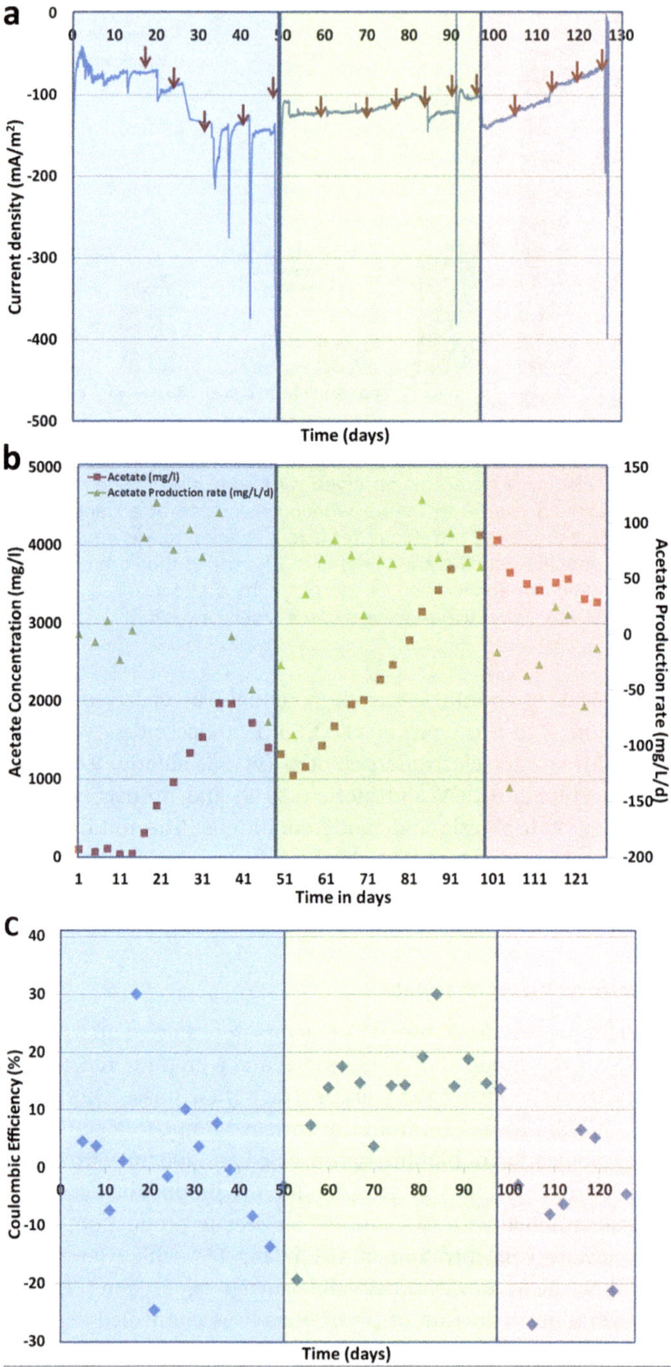

Fig. 3 The performance of the MES1 reactor during 127 days in batch mode operation with several feeding events (arrow mark depicts the feeding event). (a) Reduction current pattern during chronoamperometry (CA) used for the bicarbonate to acetate production; (b) acetate concentration and acetate production rate pattern during the 127 days of operation; (c) coulombic efficiency during the total operation time.

1972.86 mg L^{-1}; −165 mA m^{-2}). Furthermore, this phase continued for another two batch cycles (until 50 days) to identify the optimum retention time for the 60% of feed replacement. During this period, acetate production was found to fluctuate between 1959.36 mg L^{-1} (38th day, −146 mA m^{-2}) and 1320.02 mg L^{-1} (50th day, −128 mA m^{-2}). After continuing for another two batch cycling events, it was identified that seven days was the optimum retention time for one batch of 60% feeding, and the same retention time was respected to the end. From the 51st day to the 98th day of operation MES1 exhibited a stable performance where the acetate production gradually increased from 1320.02 mg L^{-1} to a maximum concentration of 4127.01 mg L^{-1} (98th day, −142 mA m^{-2}) with an average reduction current density of −117 mA m^{-2}. Later both the acetate concentration and reduction current density were found to decrease until the 127th day of operation and this period can be considered as the destabilization phase. Even after four feeding events, a decrease in acetate concentration was observed rather than an improvement. By the 126th day of operation, acetate concentration was found to have decreased to 3270.26 mg L^{-1}. Similarly, the reduction current density was also found to decrease (126th day −67 mA m^{-2}).

The acetate production rate, the amount of acetate produced during a day (mg per L per d), was analysed to identify the specific bicarbonate reduction rate to acetate. Acetate production rates were also correlated with bicarbonate reduction rates during start-up phase, stable operation and the destabilization phase. During the start-up phase the average acetate production rate was found to be 28.16 mg per L per d, which later increased to 59.94 mg per L per d (Fig. 3b). During the destabilization phase, the average acetate production rate registered a negative value (−33.61 mg per L per d). The observed negative rate is due to possible acetate oxidation at the anode of the single chamber design. A visible biofilm was observed on the anode electrode after 90 days of operation, supporting this phenomenon. When the oxidation of bio-electrochemically produced acetate at the cathode is considered, the actual production rates of the start-up and stabilization phases can be higher than visualized. As the media included was with 0.5 g L^{-1} concentration of BESA, the conversion of acetate to methane is completely inhibited.[31] Along with acetate, a negligible concentration of ethanol was also observed during the start-up phase (10.05 mg L^{-1}, 35th day) and the final phase (8.62 mg L^{-1}, 112th day). This could be due to further bioelectrochemical reduction of acetate to ethanol. Besides ethanol production, the consumed acetate can be attributed to the biomass production of electroactive biofilm.[12] pH is also an important factor that influences the activity of the electroactive biofilm. Homoacetogenic bacteria favours mild acidic to neutral pHs. The initial pH of the present MES1 was maintained at pH 7.0. It was found to increase to alkaline pH with the time of operation. During the stable phase of operation, the catholyte pH was registered between 7.8 and 8.4, whereas in the destabilization phase, it was found to be between 8.6 and 9.3. A high alkaline pH, that is unfavourable for homoacetogenic bacteria prevailing in the cathodic chamber, is also one of the reasons for biofilm loss.

In the case of MES2, initially, a negligible reduction current was registered for 10 days (Fig. 4a). The acetate transferred along with the enriched inoculum culture was (103.32 mg L^{-1}) observed degrading from day 5 (84.22 mg L^{-1}) and it reached a minimum concentration on the 11th day (33 mg L^{-1}) (Fig. 4b). Later, an increasing trend of acetate production associated with the reduction current was

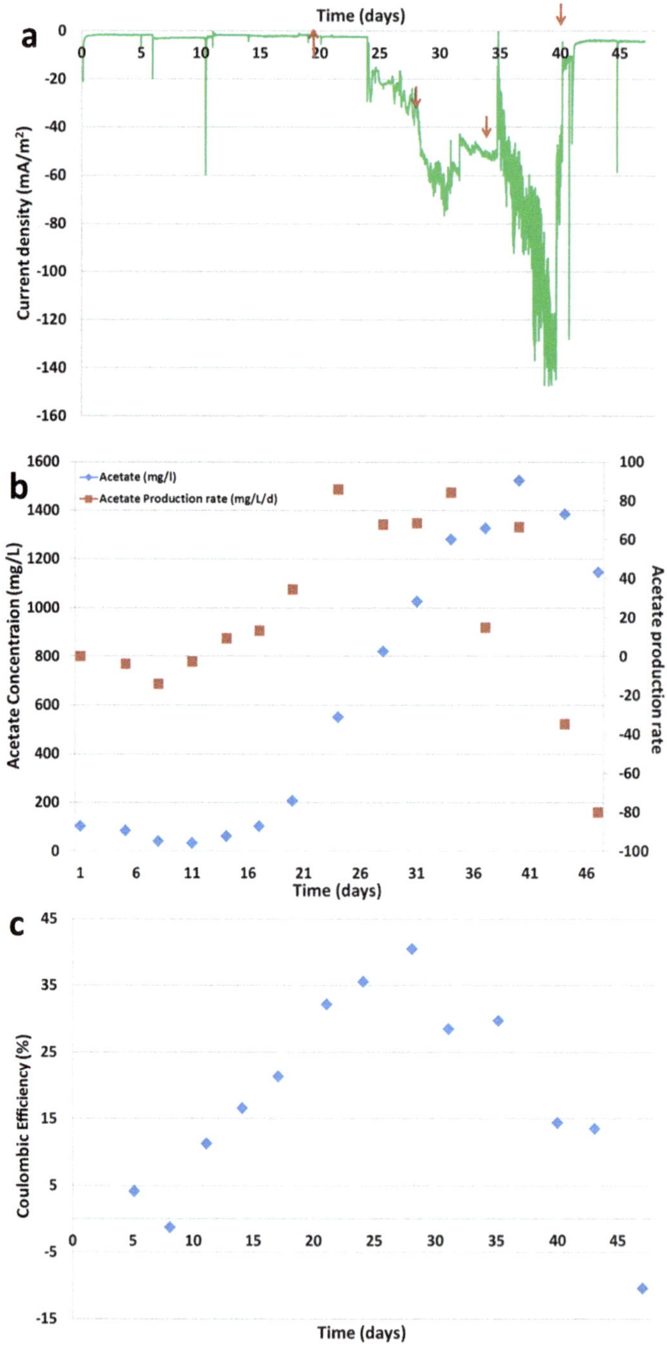

Fig. 4 The performance of MES2 reactor during 47 days in batch mode operation with several feeding events (arrow mark depicts the feeding event). (a) Reduction current pattern during chronoamperometry (CA) used for the bicarbonate to acetate production; (b) acetate concentration and acetate production rate pattern during 127 days of operation; (c) coulombic efficiency during the total operation time.

observed. This depicts the electroactive biofilm formation and concomitant conversion of bicarbonate to acetate. By the end of the first feeding cycle, the acetate production reached 206.49 mg L^{-1} with an average production rate of 5.15 mg per L per d. It gradually increased with every feeding event and reached 821.25 mg L^{-1} (75.62 mg per L per d) and 1279.42 mg L^{-1} (76.33 mg per L per d) respectively by the end of the second and third feeding cycles. The maximum concentration of acetate was observed on the 40th day (4th cycle, 1523.2 mg L^{-1}; 40.67 mg per L per d) with a current density of −53 mA m^{-2}. Among the 47 days of operation, the reduction current density was significantly increased from the 23rd day of operation and registered a maximum on the 39th day (−147 mA m^{-2}).

During this operation, no ethanol production was observed. Later, visible detachment of biofilm was observed. It is also correlated well with the drop in reduction current and acetate concentration in the system. At the end of this cycle, the acetate concentration was decreased to 1145.25 mg L^{-1} with an acetate production rate of −54.20 mg per L per d. Even after biofilm loss, the system continued operation with regular feeding cycles for 5 consecutive cycles but no further acetate production and no reduction current were observed confirming there was no biofilm formation. As MES2 was operated for a lesser period than MES1 and the loss of biofilm occurred immediately after reaching the maximum acetate production and reduction current, no specific phases such as stable and final phases were observed. Similar to MES1, the catholyte pH of MES1 was also found to increase to alkaline conditions. By the 30th day of operation (3rd feeding cycle), the pH had increased to 7.8 and later it was found to increase at a rapid rate and reached 8.7 by the 44th day of operation (5th cycle).

3.3 Coulombic efficiency

In MES, coulombic efficiency (CE%) is a relative expression that signifies the ratio between the total charge (coulombs) consumed in the reduction process and the actual charge contributed in the formation/production of desired product. It can be calculated with the formula CE (%) = C_P/C_T, where, C_P is the product of b (number of electrons consumed for the product, for acetate production it is 8), n (number of moles of product) and F (Faraday's constant, 96 485 C mol^{-1}). C_T is the total coulombs consumed that can be derived from integrating charge with the time.[20] Since ethanol was detected in negligible concentrations and the electron demand for acetate to ethanol conversion is lower ($4e^-$) than acetate, considering it for CE calculations do not show a significant difference on total CE. In the case of MES1, during the stable phase of operation, a maximum CE was registered as 29.91% and a minimum as 3.65% with an average value of 23.52%. Whereas during the start-up phase and the destabilization phase, the maximum CE was registered as 18.38% and 6.54% respectively. Biofilm formation phase demands a high amount of energy for biomass formation and related metabolic activities. In the start-up phase, the electrons accepted by the cathode might be involved more in metabolic functions to establish the biofilm on the cathode surface. The acetate that produced at biocathode also participates in the production of various molecules which decreases the observed acetate production in the electrolyte. So, the resultant CE is recorded lower. A well-established electroactive biofilm requires less energy for the metabolic functions that reflected in the higher CE than the start-up phase. The lowest CE values recorded in the destabilization phase might be attributed to two events. Firstly, in the destabilization phase, the biofilm observed on the anode electrode is involved in the oxidation of acetate produced due to a cathodic reduction reaction. Secondly, due to the unfavourable conditions that prevail in the system, the cathode electrode cannot efficiently participate in acetate production. In the case of MES2, CE increased gradually with the time of operation and reached a maximum of 40.42% on the 28th day (second cycle). Later a gradual drop was observed. During the second and third cycles of operations, it was limited to 29.11% and 13.93% respectively. As the biofilm had detached and no acetate production was observed in the fifth cycle of operation, a negative value of CE was registered −10.43% (47th day). On the whole, MES2 registered an average CE value of 17.71%. The CE was found to be higher for the MES2 with a graphite rod as the cathode compared to MES1 with VC-IS. However, the formed biofilm was not sustained for long term operation.

3.4 Carbon conversion efficiency

Carbon conversion efficiency is considered as one of the crucial parameters that shows effective conversion of substrate to product. The efficiency of the bioelectrochemical reduction of bicarbonates (electron acceptors) to acetate can be found by calculating the ratio between the carbon equivalents present in the initial concentration of the substrate and final product in each operating cycle (Table 1). It should be noted that the carbon conversion efficiency calculation is considered not only of those products observed in the electrolyte but also in the biomass produced for the biofilm and suspension. In the case MES1, during the first cycle operation, it was registered as 47.08%. Gradually a steady increase in carbon conversion efficiency was observed with every feeding event and a

Table 1 The consolidated representation of calculated parameters from the basic results obtained in the present study

		MES1			
Parameter	Units	Start-up	Stable phase	Final phase[c]	MES2
Maximum acetate concentration	mg L^{-1}	1972.86	4127.01	3568.41	1523.20
Production rate (maximum/average)	mg per L per d	117.96/26.97	120.90/56.94	24.37/—	85.96/20.68
Maximum current density	mA m^{-2}	−165	−142	−67	−147
Coulombic efficiency (CE)	%	29.97	29.91	6.54	40.43
Carbon conversion efficiency[b]	%	90.10	90.61	19.90	81.97
Volumetric production rate[a]	kg per m^3 per d	0.269	0.569	—	0.206
Surface based production rate	g per m^2 per d	4.497	9.490	—	3.447

[a] Calculated from the average performance during each phase. [b] Calculated based on the feeding event. [c] The values of average production rate, volumetric production rate and surface based production rates were negative. So the respective values are not presented in the table.

maximum of 90.10% was registered on the 35th day (start-up phase). During the stable phase of operation it was registered as 90.61% on the 84th day and with an average value of 62.38%. Due to the destabilization in the final phase, the carbon conversion efficiency was registered very low (19.90%). The trend observed for carbon conversion efficiency is found to be similar to CE. The carbon converted to acetate is involved in the biomass production. If we consider the participation of carbon in biomass production, the carbon conversion efficiency may be higher than others reported. In the case of MES2, a different pattern of carbon conversion efficiency is observed. In the first cycle only 8.25% of carbon was converted to acetate. Then a sharp increase in carbon conversion to acetate was observed in the second cycle and registered as 81.97%. It is the maximum efficiency for MES2. During the third and fourth feeding cycles, the efficiency is limited to 61.09% and 32.50% respectively. Later due to destabilization, no significant efficiency of carbon was converted to acetate. Compared to MES2, MES1 showed a higher bioelectrocatalytic conversion of carbon to acetate and the conversion process was sustained for a longer period of time. On the whole, the registered average values in the stable phase of MES1 are encouraging for the early stage research on the bio-electrochemical reduction process of CO_2 to organic molecule production (Table 1). The difference between MES1 and MES2 is in the electrode material used for the cathode (VITO CORE-PL in MES1 and a graphite rod in MES2). The better performance of MES1 could be attributed to the high BET surface area of the VITO CORE compared to graphite.[34] This also led to better biofilm growth in MES1 which could be operated for 127 days while MES2 saw a loss of biofilm after 44 days and had to be stopped.

3.5 Discussion

The present study was performed under mild operation conditions at 30 °C with a continuous applied potential of −400 mV. The thermodynamic potential required for the conversion of HCO_3^- to acetate is found to be −280 mV (SHE)[2] and the present study used still higher potential than the thermodynamic requirement. The potential losses related to the electrode and microbial interactions; microbial electron transfer and electrolyte conductivity determined the required potential for the MES. The design of the present study was based on characteristics that would help to understand enriched biocathode for acetate production at a lower cathodic potential and to maintain the biofilm for long term. It also helps for the upscaling aspects of the MES process. Compared to additional supplementation of H_2 in the reactor, direct feeding of electrons to acetogens through electrodes is likely to be practically more feasible. The lower energy input and renewable bacterial biocatalyst employment are advantageous aspects. The standard cathodic potential for bicarbonate/acetate redox couple is −280 mV.[2] A biofilm of *S. ovata* that was grown on H_2, when introduced into a cathodic chamber containing bicarbonate-based medium with no organic compounds except a vitamin mixture, resulted in acetate production by directly reducing the bicarbonate. Under an applied potential of −400 mV (*vs.* SHE) and in the presence of a well-developed *S. ovata* biofilm, CO_2 or bicarbonate acted as the sole electron acceptor and was reduced to acetate.[17,24] Along with acetate, 2-oxobutyrate was also produced in smaller concentrations. Current consumption by *S. ovata* was found to be constant and correlated well with acetate production with a coulombic efficiency of more than 86%. Production of both compounds was stopped when current supply was stopped. Furthermore, various acetogenic microorganisms such as *Sporomusa silvacetica*, *Sporomusa sphaeroides*, *Clostridium ljungdahlii*, *Clostridium aceticum* and *Moorella thermoacetica* as pure cultures showed their more than 88% efficiency in converting CO_2 to acetate.[24] Under the same applied potential of 0.4 V and with *S. ovata* as the biocatalyst, Nie *et al.*, could achieve a higher acetate production rate of 1.13 mM per d.[25] In the present study, MES1 showed 2.04 mM per d of maximum production rate (average, 0.97 mM per d) during the stable operating phase.

The efficiency of the total carbon equivalents supplied to total carbon resulted in the product and can be called carbon conversion efficiency. MES1 exhibited a maximum carbon conversion efficiency of 90.61% in the stable phase followed by the start-up phase. In the final phase, it was limited to 19.90%, which might be due to the biofilm degeneration phase. In the case of MES2, 81.97% of carbon conversion efficiency was registered. Carbon conversion efficiency determines the carbon footprint of the system and practical applicability of the MES process on upscaling. The use of bicarbonates, instead of CO_2 is the major advantage to achieve higher conversion efficiencies. For direct CO_2 application, it is important to choose a system that solubilizes the CO_2 at a higher rate or developing an electrolyte that favours both CO_2 solubility and biocathodic functions. Various reactor designs such as a continuous stirred tank reactor, hallow fiber membrane reactor, bubble column, and airlift reactors used for the syngas fermentation can be applied in MES design.[37–40] As the biocatalyst of MES is electrode dwelling and the bioelectrochemical conversion of CO_2 to acetate occurs in the electrode vicinity designing hybrid reactors with BES and a conventional reactor design can fetch these advantages, that in turn improves the bioelectrocatalytic reduction of CO_2 to products. Biological components such as using CO_2 solubilizing enzymes also could be considered for a higher carbon conversion efficiency. A carbonic anhydrase enzyme system to solubilize the CO_2 and further integrate the electrolyte with the acetogenic biocathode could be an interesting option for the improved productivities.[3,41] In the present study, the higher carbon conversion efficiency can be majorly attributed to the use of bicarbonates in the system. It also suggests that an electroactive biocathode can effectively reduce the dissolved CO_2 to acetate. Evaluation of various approaches to improve CO_2 solubility is also interesting to explore.

The electrodes used as the cathode, their conductive properties and the surface coatings that are involved in the electron transfer mechanism also influence the production efficiency. A detailed study was done using various types of electrodes as cathodes that enhance the electrode and microbe-electron transfer rate for CO_2 reduction using *S. ovata*.[13] Electron transfer rate enhancing strategies were validated by modifying the cathode surface properties which in turn resulted in improved acetate production from CO_2 reduction. Biofilms of *S. ovata* on carbon cloth functionalized with positive-charged modifications such as chitosan, cyanuric chloride, 3-aminopropyltriethoxysilane, metals and (gold and palladium) nickel nanoparticles resulted in 3–7 folds improved acetate production than plain carbon cloth. However, other positively charged surfaces tested with melamine or ammonia gas did not stimulate acetate electrosynthesis.[13] The chitosan coated carbon cloth showed a maximum surface based production rate (13.8 g per m^2 per day), which was higher than the VITO electrodes (9.49 g per m^2 per day). The VITO

CoRE™ electrode manufactured with the activated carbon functioned efficiently for CO_2 reduction and the coulombic efficiency can be compared to other studies that employed similar mixed cultures and other economic electrode materials but at cathodic potentials higher than −400 mV (Fig. 5a).

The volume of the bioelectrochemical system is also an important parameter that determines the efficiency and the losses associated with the input energy and process efficiency. In the case of MFCs, it was clearly observed that the reactor volume is inversely proportional to the electron transfer efficiency.[9,42] In MES, the maximum volume of the studies were limited to the 250 mL of cathodic volume in dual chambered configuration. For both MES1 and MES2, the electrolyte volume was considered as 500 mL with a single chambered configuration. The maximum acetate production achieved with the mixed culture in a dual chambered configuration at a poised potential of −590 mV, is 10.5 g L^{-1} [20] followed by Su *et al.*,[27] with 4.7 g L^{-1} at a cathodic potential of −900 mV (Fig. 5b). The present study achieved 4.13 g L^{-1} acetate with −400 mV which is almost equal to the

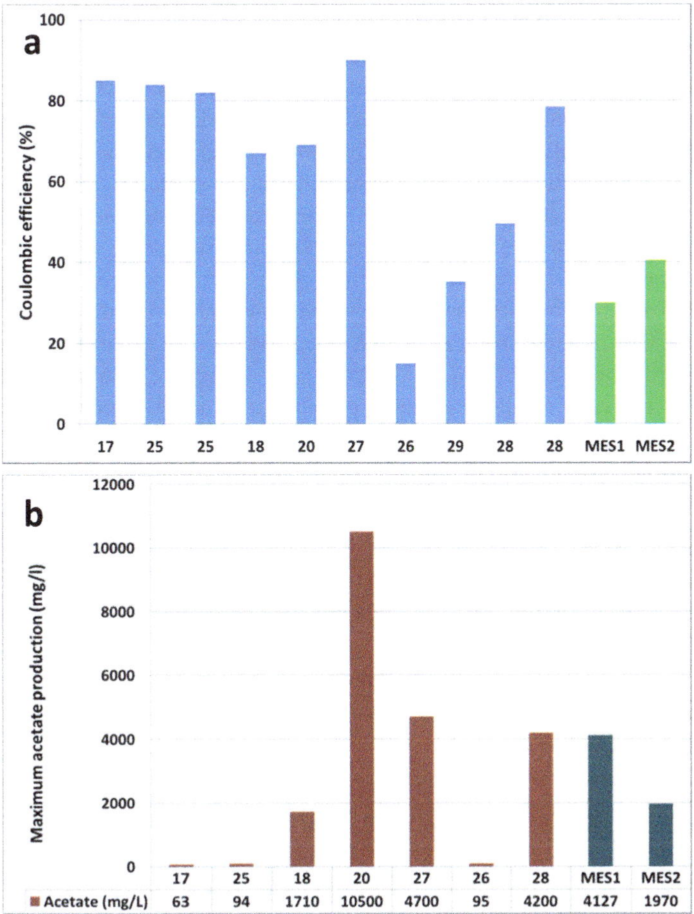

Fig. 5 Comparative evaluation of the results with the state of the art microbial electrosynthesis of acetate from CO_2 (*x*-axis represents studies with the references and their respective production efficiencies).

concentration achieved by Jourdin *et al.*,[28] at −850 mV. In the MES mechanism, the reaction is also influenced by the electrode surface area and/or the biofilm covered on the electrode to the bulk liquid (electrolyte), so the projected surface area and surface based production rate (g per m^2 per d) explains MES functionality. Jourdin *et al.*,[28] achieved a highest surface based production rate of 37 g per m^2 per d with nanoweb reticulated vitreous carbon at −850 mV followed by Jiang *et al.*,[26] (19 g per m^2 per d at −800 mV) and Su *et al.*,[27] (10 g per m^2 per d at −700 mV) with carbon felt. The VITO CoRE™ electrode registered 9.49 g per m^2 per d at −400 mV, where H_2 mediated electron transfer was completely limited and only direct electron transfer is possible (Table 1). Maintaining higher cathodic potentials can also lead to methane production.[26] MFCs were proven to sustain the lower potentials for long term operation and the present study showed interesting results for acetate from CO_2 and can warrant the direct integration of stable MFCs for acetate production through the biocathodic reduction reaction.

4. Conclusions

Microbial electrosynthesis driven by a biocathode offers a wide range of chemicals from inorganic carbon (CO_2). The selection of bacteria and operating conditions influences the product and production efficiency. Formation of an electroactive biofilm, stability and reproducibility of such biofilm delivers a future application of this process. This is the first study that proved single chambered operation for bicarbonate reduction with enriched bacteria and was successfully operated for more than four months. VITO-CoRE™ derived electrodes were found to be effective for CO_2 to acetate production. The biofilm became fragile at higher concentrations of acetate and identifying optimum concentrations of the biofilm and design of the required operational conditions to improve the stability of the MES is necessary. On integration with other MES systems where the produced acetate can be used for the synthesis of higher organic compounds such as ethanol, butyrate, ethylene, *etc.*, will provide sustainable competition between global warming and energy needs.

Acknowledgements

G. Mohanakrishna gratefully acknowledges the Marie-Curie Intra-European Fellowship (IEF) supported project BIO-ELECTRO-ETHYLENE (Grant No. 626959) from the European Commission.

References

1 D. R. Lovley, *Environ. Microbiol. Rep.*, 2011, **3**, 27–35.
2 K. Rabaey and R. A. Rozendal, *Nat. Rev. Microbiol.*, 2010, **8**, 706–716.
3 S. Srikanth, M. Maesen, X. Dominguez-Benetton, K. Vanbroekhoven and D. Pant, *Bioresour. Technol.*, 2014, **165**, 350–354.
4 R. Ganigué, S. Puig, P. Batlle-Vilanova, M. D. Balaguer and J. Colprim, *Chem. Commun.*, 2015, **51**, 3235–3238.
5 S. A. Patil, S. Gildemyn, D. Pant, K. Zengler, B. E. Logan and K. Rabaey, *Biotechnol. Adv.*, 2015, **33**, 736–744.

6 A. ElMekawy, H. M. Hegab and D. Pant, *Energy Environ. Sci.*, 2014, **7**, 3921–3933.
7 D. Pant, A. Singh, G. Van Bogaert, S. I. Olsen, P. S. Nigam, L. Diels and K. Vanbroekhoven, *RSC Adv.*, 2012, **2**, 1248–1263.
8 G. Mohanakrishna, S. Krishna Mohan and S. Venkata Mohan, *Appl. Energy*, 2012, **95**, 31–37.
9 G. Mohanakrishna, S. Venkata Mohan and P. N. Sarma, *J. Hazard. Mater.*, 2010, **177**, 487–494.
10 S. Venkata Mohan, G. Velvizhi, K. V. Krishna and M. L. Babu, *Bioresour. Technol.*, 2014, **165**, 355–364.
11 B. E. Logan, *Appl. Microbiol. Biotechnol.*, 2010, **85**, 1665–1671.
12 M. Sharma, N. Aryal, P. M. Sarma, K. Vanbroekhoven, B. Lal, X. Dominguez Benetton and D. Pant, *Chem. Commun.*, 2013, **49**, 6495–6497.
13 T. Zhang, H. Nie, T. S. Bain, H. Lu, M. Cui, O. L. Snoeyenbos-West, A. E. Franks, K. P. Nevin, T. P. Russell and D. R. Lovley, *Energy Environ. Sci.*, 2013, **6**, 217–224.
14 L. T. Angenent and M. Rosenbaum, *Biofuels*, 2013, **4**, 131–134.
15 D. Pant, S. Bajracharya, S. Srikanth, G. Mohanakrishna, H. de Wever, K. Vanbroekhoven and L. Diels, *Qatar Foundation Annual Research Conference 2014 (ARC'14)*, Qatar, 18–19th November 2014.
16 A. ElMekawy, H. M. Hegab, K. Vanbroekhoven and D. Pant, *Renewable Sustainable Energy Rev.*, 2014, **39**, 617–627.
17 K. P. Nevin, T. L. Woodard, A. E. Franks, Z. M. Summers and D. R. Lovley, *mBio*, 2010, **1**(2), DOI: 10.1128/mBio.00103-10.
18 C. W. Marshall, D. E. Ross, E. B. Fichot, R. S. Norman and H. D. May, *Appl. Environ. Microbiol.*, 2012, **78**, 8412–8420.
19 M. Aresta, A. Dibenedetto and A. Angelini, *Chem. Rev.*, 2013, **114**, 1709–1742.
20 C. W. Marshall, D. E. Ross, E. B. Fichot, R. S. Norman and H. D. May, *Environ. Sci. Technol.*, 2013, **47**, 6023–6029.
21 K. Schuchmann and V. Müller, *Nat. Rev. Microbiol.*, 2014, **12**, 809–821.
22 H. L. Drake, A. S. Gößner and S. L. Daniel, *Ann. N. Y. Acad. Sci.*, 2008, **1125**, 100–128.
23 M. R. Wilkins and H. K. Atiyeh, *Curr. Opin. Biotechnol.*, 2011, **22**, 326–330.
24 K. P. Nevin, S. A. Hensley, A. E. Franks, Z. M. Summers, J. Ou, T. L. Woodard, O. L. Snoeyenbos-West and D. R. Lovley, *Appl. Environ. Microbiol.*, 2011, **77**, 2882–2886.
25 H. Nie, T. Zhang, M. Cui, H. Lu, D. R. Lovley and T. P. Russell, *Phys. Chem. Chem. Phys.*, 2013, **15**, 14290–14294.
26 Y. Jiang, M. Su, Y. Zhang, G. Zhan, Y. Tao and D. Li, *Int. J. Hydrogen Energy*, 2013, **38**, 3497–3502.
27 M. Su, Y. Jiang and D. Li, *J. Microbiol. Biotechnol.*, 2013, **23**, 1140–1146.
28 L. Jourdin, S. Freguia, B. C. Donose, J. Chen, G. G. Wallace, J. Keller and V. Flexer, *J. Mater. Chem. A*, 2014, **2**, 13093–13102.
29 Z. Zaybak, J. M. Pisciotta, J. C. Tokash and B. E. Logan, *J. Biotechnol.*, 2013, **168**, 478–485.
30 P. K. Addo, R. L. Arechederra, A. Waheed, J. D. Shoemaker, W. S. Sly and S. D. Minteer, *Electrochem. Solid-State Lett.*, 2011, **14**, E9–E13.
31 E. J. Bouwer and P. L. McCarty, *Appl. Environ. Microbiol.*, 1983, **45**, 1408–1410.

32 S. Venkata Mohan, G. Mohanakrishna, S. V. Raghuvulu and P. N. Sarma, *Int. J. Hydrogen Energy*, 2007, **32**, 3284–3292.
33 S. Venkata Mohan, G. Mohanakrishna and P. N. Sarma, *Int. J. Hydrogen Energy*, 2008, **33**, 2156–2166.
34 Y. Alvarez-Gallego, X. Dominguez-Benetton, D. Pant, L. Diels, K. Vanbroekhoven, I. Genné and P. Vermeiren, *Electrochim. Acta*, 2012, **82**, 415–426.
35 D. G. Nocera, *Acc. Chem. Res.*, 2012, **45**, 767–776.
36 J. P. Torella, C. J. Gagliardi, J. S. Chen, D. K. Bediako, B. Colón, J. C. Way and D. G. Nocera, *Proc. Natl. Acad. Sci. U. S. A.*, 2015, **112**, 2337–2342.
37 M. D. Bredwell, P. Srivastava and R. M. Worden, *Biotechnol. Prog.*, 1999, **15**, 834–844.
38 S. D. Rajagopalan, R. P. Datar and R. S. Lewis, *Biomass Bioenergy*, 2002, **23**, 487–493.
39 R. P. Datar, R. M. Shenkman, B. G. Cateni, R. L. Huhnke and R. S. Lewis, *Biotechnol. Bioeng.*, 2004, **86**, 587–594.
40 P. C. Munasinghe and S. K. Khanal, *Bioresour. Technol.*, 2012, **122**, 130–136.
41 X. Dominguez-Benetton, S. Srikanth, Y. Satyawali, K. Vanbroekhoven and D. Pant, *Journal of Microbiology Biochemical Technology*, 2013, S6-007.
42 S. Srikanth, T. Pavani, P. N. Sarma and S. Venkata Mohan, *Int. J. Hydrogen Energy*, 2011, **36**, 2271–2280.

Faraday Discussions

DISCUSSIONS

Capture agents, conversion mechanisms, biotransformations and biomimetics: general discussion

Eryk Remiezowicz, Jeroen Spooren, Elizabeth Bay, Alexander Cowan, Ian Ingram, Pedro Abrantes, Manuel Nunes da Ponte, Michael North, Jonathan Albo, Peter Styring, Michael Priestnall, Katie Lamb, Michele Aresta, Elsje Alessandra Quadrelli, Richard Heyn, André Bardow, William Webb, Renata Silva, Carlos Alonso-Moreno, Csaba Janaky, Geoffrey Maitland, Seetharaman Vaidyanathan, Gonçalo V. S. M. Carrera, Daniel Reed, Karolien Vanbroekhoven, Cafer Tayyar Yavuz, Deepak Pant and Nathan Hollingsworth

DOI: 10.1039/C5FD90082D

Michael North presented some work related to the papers by Michael Priestnall, Christopher Hardacre and Cafer Tayyar Yavuz: I would like to share some recent work carried out jointly by my group and that of James Clark on the use of mesoporous carbons as CO_2 capture agents. These materials are able to capture 65% more CO_2 than activated carbon whilst retaining the rates of adsorption and desorption. The mesoporous materials also show much better selectivity for physisorption of CO_2 over N_2 than activated carbon (up to 20 : 1 compared to 5 : 1 for activated carbon) and also selectively adsorb CO_2 over methane with about 2.8 : 1 selectivity.

Geoffrey Maitland asked: You showed some very interesting results concerning the adsorption of CO_2 in the nanopores of a variety of carbonaceous materials, particularly the selectivity towards CO_2 over N_2, relevant to flue gas separation. Can you attribute the observed differences to variations in the hydrophobicity of potential adsorption sites or are there other factors at play in determining the loading and selectivity? Have you for instance considered the effects of confinement of the thermodynamics of adsorption in these systems?

Michael North responded: the hydrophobicity increases as the carbonisation temperature increases, but the CO_2 adsorption goes through a maximum, so we do not believe that hydrophobicity is the explanation.

Peter Styring remarked: Michael North is talking about assisted physisorption. Really it is self-assembly. The polyamine on its own has little selectivity as we have shown in an Energy and Engineering Science paper with Aschenbrenner as the lead author.[1] If you put it on a support (that has oxygenated groups in it) you get better activity. When you look at the support you are looking at self-assembly as well. That is why monoethanol amine is a better sorbent than amines in isolation: the hydroxyl motif aids the sorption process.

1 O. Aschenbrenner and P. Styring, *Energy Environ. Sci.*, 2010, **3**, 1106–1113.

Eryk Remiezowicz opened the discussion of the paper by Michael Priestnall: As I see, the necessary step is digestion with NaOH. NaOH is a great capture agent, how do you justify its further processing step? How do you balance your process with carbon capture by NaOH itself?

Michael Priestnall replied: NaOH is an excellent CO_2 capture agent – agreed – but more CO_2 is generated in its manufacture than in its carbonation to sodium carbonate, so it makes no life cycle sense to use NaOH to reduce net CO_2 emissions. In this work we are using a minimum amount of NaOH as an agent to split magnesium silicate into magnesium hydroxide and sodium silicate and have calculated the CO_2 captured on the basis of the carbonation of the magnesium hydroxide only. Industrially, sodium silicate solution is converted to silica powder and sodium carbonate by reaction with CO_2; or the sodium silicate solution can be converted electrochemically into silica and sodium hydroxide. Energy is required to recover NaOH. The basis of CO_2 balance in the paper does not include any benefit for capture of CO_2 by NaOH or sodium silicate, nor does it include any CO_2-cost for the recovery of NaOH, it simply assumes the NaOH is consumed. If we include the additional silica production part of the process and NaOH recovery, then the overall net reaction is: $Mg_2SiO + 4CO_2 + 2H_2O = 2Mg(HCO_3)_2 + SiO_2$.

Daniel Reed said: From a process point of view, how long does the process take from start to finish (feedstock to product)?

Michael Priestnall answered: The process has only been investigated in lab-scale experiments. We have used reaction times of 1–6 h and temperatures ~150–250 °C at ambient pressure. At 250 °C the conversion of olivine to brucite is mostly completed within 1 h. Higher temperatures and elevated pressure or smaller particle size would reduce reaction time further.

Wet carbonation of brucite (to scrub CO_2 emissions) is a separate process. This is a continuous real-time process.

Daniel Reed asked: Do you envisage this to be a batch process for real world applications?

Michael Priestnall replied: The process could be operated as batch or continuous. Most industrial processes are cheaper when configured as continuous processes. For example the solid-state reaction could be achieved continuously by feeding a pelletised premix of olivine powder and NaOH through a rotary oven. The separate CO_2 scrubbing process was tested experimentally with a single batch charge of brucite, but industrially this would certainly be a continuous process.

Daniel Reed enquired: Are you concerned about how much CO_2 you can capture or are you just happy to be using some CO_2 at all?

Michael Priestnall responded: For the scrubbing process to have commercial viability it is essential to have a very high utilisation (carbonation) of the brucite – uniquely we have shown close to 100% carbonation of the brucite at ambient temperature and pressure.

Again, to be industrially relevant, it is important to be able to remove >90% of the CO_2 from an exhaust gas and also to be able to do this from a wide range of gas compositions. In the work reported we achieved close to 100% removal of CO_2 from a $5\%CO_2/N_2$ gas mixture.

Elsje Alessandra Quadrelli noted: The kinetics of the process (and the connected necessity to grind and pre-condition the minerals) have been mentioned as a potential hurdle that can worsen the energy consumption and hence the feasibility of the inorganic silicate minerals route to large scale CO_2 capture processes. In your study, how does the size of the particle come into play in the overall process proposed?

Michael Priestnall answered: We have not yet investigated the optimisation of particle size *versus* reaction times and temperatures and life cycle energy. Our work has used virgin quarried and comminuted olivine rock; however we could also have used already finely ground magnesium silicate wastes from nickel, platinum or asbestos mining. Such mineral wastes are available in mega-ton quantities in many sites around the world.

Geoffrey Maitland questioned: Did you include the significant particle grinding and mixing energies in the overall energy balance for your process? What effect does this have on your conclusions regarding the cradle-to-grave energy requirements and near carbon-neutrality of the CO_2 storage process?

Michael Priestnall replied: In a separate (unpublished) life-cycle analysis for Alberta CCEMC, we estimated the full process and product life cycle from quarrying and comminution of virgin rock in China, on-site processing to silica powder and magnesium hydroxide, shipping 10 000 km and industrial wet-scrubbing to production of magnesium carbonate powder. Energy use in grinding and milling is a moderate factor, but by far the most significant contributors to the life cycle calculation are the amount of NaOH used and the extent to which the amorphous precipitated silica by-product can offset conventional production of APS. The result of that LCA calculation was a net 1 ton of CO_2 reduced per ton of brucite converted to solid carbonate powder or a net 2 tons CO_2 reduced per ton of brucite converted to dissolved magnesium bicarbonate.

Peter Styring asked: The process to get the carbonate from the CO_2 is exothermic. The reactions you're looking at include the generation of heat. Is there any attempt to recover the heat back into the process and does that have any savings on the whole life cycle?

Michael Priestnall answered: In the wet-scrubbing process we describe that the carbonation of brucite takes place at ambient temperature ~10 °C. There is minimal prospect of recovering useful heat energy at this temperature. However, in other work we have demonstrated the feasibility of recovering a proportion of the

~100 kJ mol^{-1} carbonation energy directly as electricity by configuring the carbonation reaction as redox half-cell reactions in an alkaline electrolyte fuel cell.[1]

1 M. A. Priestnall, Patent Appl. WO2011073621, 2009.

Richard Heyn communicated: My impression is that the investigated process is carbon neutral or carbon negative when the CO_2 is captured only as aqueous $Mg(HCO_3)_2$. Storage of CO_2 in an aqueous solution does not strike me as particularly effective in the long-term. Could you expound on the eventual application or further processing or disposal of the aqueous $Mg(HCO_3)_2$ solutions, and the impact of those downstream steps on the overall carbon balance?

Michael Priestnall communicated in reply: Rau and others report the stability of ocean dissolved magnesium bicarbonate as being of the order of 1000 years.[1,2] Large-scale storage of CO_2 as ocean alkalinity (bicarbonate), *via e.g.* ocean liming or accelerated weathering is a relevant related area or research. The process we report has been considered for use on ships to scrub acid gas emissions directly into seawater without raising ocean pH. Dissolved magnesium bicarbonate will endothermically decompose into solid magnesium carbonate phases and dissolved CO_2 as water temperature is increased or ambient pressure reduced or above saturation. Keener and others have devised cyclic process using brucite-bicarbonate-carbonate to separate pure CO_2 from exhaust gases.[3,4] Half as much CO_2 is captured as solid carbonate compared to bicarbonate. However solid carbonate products can then offset life cycle CO_2 from other production process routes.

1 Greg H. Rau, *Energy Procedia*, 2009, **1**, 823–828.
2 Greg H. Rau *et al.*, *PNAS*, 2013, **110**, 10095–10100.
3 L. Cheng, T. Li, T. C. Keener, J. Y. Lee, *Int. J. Greenhouse Gas Control*, 2013, **17**, 240–249.
4 H. K. Bharadwaja, J-Y. Lee, X. Li, Z. Liu, T. C. Keener, *J. Hazard. Mater.*, 2013, **250–251**, 292–297.

Jeroen Spooren asked: Related to the life cycle is cost analysis in order to obtain an idea on the feasibility of the (overall) process – what are the prices of the produced products and what is the overall cost of the process?

Michael Priestnall responded: The values of products are hugely dependent upon volume, purity, morphology and market application. At moderate (<1 Mt CO_2 per year) scales the process might be expected to be optimised to maximise product material value as advanced industrial minerals. At a larger scale of CO_2 sequestration, where product volumes may be much larger than available markets for industrial minerals, the process may be optimised for minimum input and process costs, maximum CO_2 capture and carbonated products configured for high-volume, low-price construction materials, agricultural and geo-materials or even zero-value storage/land-fill.

For process scales <1Mt CO_2 per year:

• Process inputs are olivine or serpentine powder at a delivered price of ~18 euros per ton or production cost of ~1.5 per euros ton. The NaOH purchase price is ~230 euros per ton.

• The sodium silicate products are valued at ~260 euros per ton, which can, optionally, be further processed into precipitated silica valued at ~600 euros per ton and recovered sodium hydroxide valued at ~230 euros per ton. The magnesium hydroxide product (internal value of ~260 euros per ton) is used to capture CO_2 (nominal sequestration value of ~35 euros per ton) to produce magnesium carbonate valued at 210 euros per ton.

• On the basis of input costs and output values, net profitability for the process has be estimated at ~800 euros per ton CO_2 sequestered.

Jeroen Spooren queried: The sodium silicate which is produced contains impurities as well? How does the quality of this impure sodium silicate compare with the raw sodium silicate material that is nowadays used in the process industry? Does it comply with the industrial requirements for this material?

Michael Priestnall answered: The Norwegian olivine used in this work contains ~6% iron oxides. We have not compared the purity of the sodium silicate digestion by-product with commercial sodium silicate. However, amorphous silica powder with low-Fe has been precipitated from the processed sodium silicate and supplied to tyre manufacturers for comparison against commercial silica powder.

Katie Lamb communicated: I have a question about the magnesium silicate resources quoted in this paper. I was wondering if you have considered the availability of dunite in terms of a global scale? Is this a mineral that is abundant across the globe, and therefore could potentially be utilised on a global scale, or is this potentially a niche method that would only be beneficial in certain areas, in terms of economics and other factors?

Michael Priestnall communicated in reply: Olivine (dunite) and serpentine is available around the globe in gigaton quantities. Lackner and others have published resource maps for these minerals.[1–3] Oman is reported to have 70 cubic-kilometres of accessible olivine deposits. We have used commercial olivine quarried in Norway and serpentine quarried in Austria; also serpentine from a 5 bn ton mine in China.

1 F. Goff, K. Lackner, *Environ. Geosci.*, 1998, **5**, 89–101.
2 K. Lackner *et al.*, *Energy*, 1995, **20**, 1153–1170.
3 A. Sanna *et al.*, *Chem. Soc. Rev.*, 2014, **43**, 8049–8080.

Michael Priestnall concluded: To comment on small-scale, high-value *versus* large-scale, low-value CDU products in the context of our paper (DOI: 10.1039/c5fd00047e): one needs to configure a CDU process very specifically to the individual customer/market application and need – if we are applying our mineral carbonation process to a GWe-scale power station, we would configure it to operate as cheaply as possible assuming the carbonate product is most likely to go to landfill. If it is being applied to remediate chrysolite asbestos mining waste we would use CO_2 from the air and maximise the yield of amorphous silica from silicate digestion.

Manuel Nunes da Ponte opened a general discussion of the paper by Christopher Hardacre†: I thought the process you describe in your paper is very

† Christoper Hardacre's paper was presented by Nathan Hollingsworth, University College London, United Kingdom.

interesting, but I was somewhat puzzled with the choice of materials for the cathode: silver, gold and platinum. I would have thought that maybe Cu or Ni would be the most obvious choices? And a second question: is it not surprising that the chemisorbed carbon dioxide is much more easily reduced (at a much lower potential) than the one that is only physisorbed in the ionic liquid?

Nathan Hollingsworth responded: Thank you for your questions. Ag, Au and Pt electrodes were chosen because they are inert within the ionic liquid/water/acetonitrile system. The authors agree that the use of Cu and Ni are a more obvious choice for CO_2 reduction, however, within the electrolyte Cu was found to be unstable. Ni is a good suggestion and a material we should endeavour to test. Physisorbed CO_2 is well reported to be difficult to reduce and we have postulated this reaction mechanism based on the interaction of the anion with the CO_2 DFT studies are being undertaken to explore the mechanism of the low energy pathway the findings of which will hopefully provide an answer to this important question.

Csaba Janaky enquired: Could you tell us the approximate current density you achieve during the stationary reduction experiment?

Nathan Hollingsworth answered: The approximate current density used for the Au electrode was in the range of 1.8–4 mA cm^{-2}.

Csaba Janaky continued: Could you comment on the stability of your system? Do you get back your ions unchanged after the CO_2-reduction?

Nathan Hollingsworth replied: We have seen no evidence for any decomposition *via* ^{1}H-NMR of the ionic liquid after electrolysis. It is therefore suggested the ionic liquid is returned unchanged after electrolysis.

Geoffrey Maitland asked: Concerning the electrodes, is the physical form (rod *vs* foil), microstructure or the structure at the nanoscale important for their performance, apart from surface area effects?

Nathan Hollingsworth responded: Aside from surface area effects, the different physical electrode form will allow for differing sites and surfaces. Electrodes composed of nanoparticles are particularly interesting in that an electrode could be produced with an abundance of a particular active site or surface. To this end I believe the physical form and surface termination is a key variable to be altered when optimising a catalytic process.

Richard Heyn commented: Reduction of both chemically bound CO_2 and physically bound CO_2 is observed, with the chemically bound CO_2 reduced at lower applied potentials. Do you have any thoughts or can you share any data about how this particular pathway can be improved upon?

Nathan Hollingsworth answered: I would attempt to improve this pathway by tailoring the anion. This would be to lower the reduction potential for the CO_2 anion complex. One way to would be to use *in silico* methods to screen a wide range of anions with chemical binding sites and thus direct the experimental effort.

Peter Styring said: It might be tricky in your system, but there is a problem with units: some people use % mass or % mol capture *etc.* Here we have mg g^{-1}. To try and get consistency, in terms of CO_2 capture, what is the efficiency in your systems?

Nathan Hollingsworth answered: In terms of conversion efficiency the reported Faradaic efficiencies are a good guide. The highest reported is using a Ag electrode at −0.7 V giving an efficiency of 95%.[1] With regards to capture efficiency, work needs to be done looking at capture efficiency when dissolved in MeCN.

1 N. Hollingsworth, S. F. R. Taylor, M. T. Galante, J. Jacquemin, C. Longo, K. B. Holt, N. H. de Leeuw and C. Hardacre, Reduction of Carbon Dioxide to Formate at Low Overpotential using a Superbase Ionic Liquid., *Angew. Chem. Int. Ed.*, 2015, **54**, 14164–14168.

Peter Styring remarked: The highest efficiency that we have found is 0.7 g g^{-1} (70%) but to caveat that, MEA costs 14 000 euros per ton. If we were to scale ours it would be 20 000 000 euros per ton so you've really got to look at the economics. We have seen some cheaper reagents which make economic sense. Often you have to make a compromise. If you were to consider your capture agents, do you have a value for their cost?

Nathan Hollingsworth replied: Traditionally ionic liquids outperform MEA for CO_2 capture (mole basis), however the high cost of ionic liquid based capture agents have been a barrier to market. The cost of the IL has been considered and we have purposely chosen a cheaper IL. The reagent used to make the ionic liquid reported is [P66614]Cl, which is an industrially produced fire retardant. Life cycle analysis of this process is currently being undertaken by Prof Adisa Azapagic's group in Manchester, once completed we should have a much better idea of cost. Realistically the IL will cost more than the MEA therefore the stability and regeneration of the system (both of which can be improved by the use of an IL) is key to competing with conventional technology.

Peter Styring enquired: What is the lifetime of these materials? Looking at MEA, you're looking at anything from 28 days to 3 months before you need to completely replace them due to evaporative loss and degradation: so something that is more expensive may be better in the long run.

Nathan Hollingsworth responded: As an IL [P66614][124Triz] has a negligible vapour pressure. The decomposition temperature is also >250 °C. Recyclability studies of the pure IL system have shown no decrease in CO_2 capture efficiency after 25 cycles.[1] CO_2 is removed from these systems at 80 °C compared to 120 °C for MEA and assuming the degradation is negligible the lifetime of this ionic liquid system should be longer than MEA. Given the higher CO_2 uptake and the ability to convert the CO_2 *in situ* you are correct in saying a potentially more expensive initial out lay may make more economic sense in the long run.

1 Wang *et al.*, *Angew. Chem. Int. Ed.*, 2011, **50**, 4918–4922.

Michael Priestnall commented: To comment further on capture efficiency, (DOI: 10.1039/c5fd00047e): the capture efficiency of $Mg(OH)_2$ we have shown in our paper is approximately 160% – *i.e.* we capture 1.6 g of CO_2 per 1 g of capture agent – looking at the costs, to ship a ton from Norway to UK is about 18 euros, and that material contains 50wt% $Mg(OH)_2$ – you're not reusing it but the storage time is permanent, so overall this makes for a very good comparison against reuseable capture agents.

Peter Styring remarked: One thing you have to take into account is that whilst ionic liquids can be more efficient, you need to look at the process conditions as well. While they have vanishing vapour pressure at room temperature and pressure they enter the vapour phase above 100 °C and decompose above 200 °C. We have carried out molecular modelling studies looking at the binding energy/bond distortion as you bring the ionic liquid molecules together with CO_2. The results allow us to predict which make good materials.

Michael North questioned: An issue for CO_2 capture from power stations is water – the paper by Cafer Tayyar Yavuz (DOI: 10.1039/c5fd00099h) looked at pure CO_2 and wet CO_2. For your ionic liquid based system, what effect will water have on the CO_2 capture? Will it form the bicarbonate instead of the ionic liquid CO_2 adduct?

Nathan Hollingsworth answered: Thank you for your question. Water has been deliberately added to this system to provide a source of H^+ and appears to still capture CO_2. A recent study[1] has reported the differences in CO_2 capture with wet and dry CO_2 streams. It should be noted that certain ionic liquids have increased CO_2 capture properties when hydrated CO_2 streams are employed. Good recyclability of [P66614][124Triz] for CO_2 capture is also reported under wet conditions. With regards to the formation of bicarbonate, it is likely that with a large excess of water bicarbonate would form. This may compete with the physically absorbed CO_2, however the main component of CO_2 capture in [P66614][124Triz] is through chemical absorption and this should be unaffected.

1 S. F. R. Taylor *et al.*, CO_2 Capture in Wet and Dry Superbase Ionic Liquids, *J. Solution Chem.*, 2015, **44**, 511–527.

Manuel Nunes da Ponte commented: In favour of ionic liquids – the paper presented by Nathan Hollingsworth (DOI: 10.1039/c5fd00099h) looks at capture and conversion *in situ*, while the currently available amine-based capture technology requires a desorption step afterwards, in order to make the carbon dioxide available. A direct comparison of the two processes is therefore unfair to the ionic liquid-based one. Of course, we have to think how to separate the electroreduction products from the ionic liquid solvent, but the same would be required for any CDU process that utilises carbon dioxide captured by amines.

Jonathan Albo communicated: Among the broad diversity of ILs, those based on acetate anions typically present a higher degree of CO_2 solubility (physical and chemical absorption) across a wide range of temperatures and applied pressures.[1]

Have you considered the application of these Ac-based ILs as a reaction media in your system?

1 J. Albo, T. Yoshioka, T. Tsuru, *Sep. Purif. Technol.*, 2014, **122**, 440–448.

Nathan Hollingsworth communicated in reply: We have not considered using acetate based anions in this system. Firstly because CO_2 reduction in acetate based ionic liquids is already reported,[1,2] and secondly because the CO_2 absorption properties are inferior to the super basic anion systems.

1 L. E. Barrosse-Antle and R. G. Compton, Reduction of carbon dioxide in 1-butyl-3-methylimidazolium acetate, *Chem. Commun.*, 2009, 3744–3746.
2 J. D. Watkins and A. B. Bocarsly, Direct Reduction of Carbon Dioxide to Formate in High-Gas-Capacity Ionic Liquids at Post-Transition-Metal Electrodes, *ChemSusChem*, 2014, 7, 284–290.

Jonathan Albo further communicated: ILs containing metals in their anion (also known as magnetic ionic liquids), have shown interesting properties as a reaction media for various catalytic reactions, as well as for electrochemical devices (*i.e.* flow battery systems, anti-glare rear, information displays, *etc.*).[1] Would you expect an improved CO_2 electroreduction performance by applying this new family of ILs?

1 E. Santos, J. Albo, A. Irabien, *RSC Adv.*, 2014, **4**, 40008–40018.

Nathan Hollingsworth communicated in reply: Including a catalytically active component into the ionic liquid (whether it be magnetic or non-magnetic) is in the future plan and should produce some very interesting results. We believe that to have the highest impact, the anion would have to directly interact with the CO_2 and in these cases it should result in further improvements of the electroreduction performance.

Geoffrey Maitland addressed Michael Priestnall and Nathan Hollingsworth: You have added nitrogen because you were trying to simulate flue gas to some extent. What about the tolerance of your systems or otherwise to other impurities such as NO_x, SO_x, H_2S *etc.*? The cost of a viable process will be very sensitive to the degree of purification of CO_2 from a flue gas stream that is required, so tolerance of other components would be advantageous.

Michael Priestnall responded: Another reason to add N_2 is so that the change in concentration of CO_2 can easily be measured. We haven't investigated the effects of other impurities but magnesia is used commercially to scrub SO_x, NO_x and H_2S. We anticipate that the system should be tolerant to and highly effective at scrubbing common acid gas impurities.

Nathan Hollingsworth continued: Indeed, tolerance of flue gas impurities is very important in the commercialisation of any capture conversion process. To this end, the effect of SO_x species on the capture process is currently being studied; initial results suggest that these IL systems are tolerant to SO_x. The effect of NO_x and H_2S at present has not been assessed, however we are looking into metal sulphide electrodes, where some H_2S may be advantageous.

Ian Ingram continued the discussion of the paper by Michael Priestnall: Your method for CO_2 capture using magnesium-based mineralisation seems to be very effective and obviously lends itself to application on a very large scale. However, on the RSC's sustainable periodic table,[1] magnesium is colour-coded yellow, signifying that supplies might be running short in the next 200 years or so. How does this potential scarcity affect the large scale application of your process going forward? Are the magnesium silicates going to become prohibitively expensive in the future?

1 A. J. Hunt *et al.*, *Green Chem.*, 2015, **17**, 1949–1950.

Michael Priestnall replied: Mg is the 8th most abundant element, comprising >2wt% of the Earth. Mg is mostly in the form of Mg-silicate rocks. There are no commercial processes for the production of Mg from magnesium silicate rocks and Mg-silicates are not included RSC's Mg resource base. The alkaline digestion process we describe enables the use of magnesium silicate as a commercial source of Mg metal for the first time. Serpentine and olivine are available in high purity and low cost from existing quarrying operations with accessible resources in the order of tens of billions of tons. The current global market for serpentine and olivine is very small.

Peter Styring said: Michael, you raised an interesting point about impurities and mentioned Carbon8 Systems. They take air pollution control residues and the toxic impurities are entrapped by accelerated carbonation using CO_2 to produce building materials. These not only use CO_2 but also reduce toxic landfill. In terms of public perceptions, these materials have gone through stringent building regulations to show they are fit for purpose in the building industry while having the environmental advantages discussed. Do you have any comment on this?

Michael Priestnall answered: A number of commercial mineral carbonation operations exist around the world. Currently, all of these are driven financially by either the value of the waste remediated (*e.g.* through avoiding landfill charges) or by the value of the end products (*e.g.* metals recovered, minerals or construction materials produced), rather than by the value of the CO_2 sequestered. In cases where pure CO_2 is utilised, this currently has to be paid for (at ~£100 per ton) as an input chemical and significant cost to the process. There is therefore an economic incentive to utilise and sequester the minimum amount of CO_2. This disincentive to utilisation of CO_2 probably exists in most CDU schemes including EOR. For it to become economically attractive to sequester CO_2 there has to be a sufficiently high CO_2 price.

Jeroen Spooren stated: Another example of CO_2 utilisation in construction materials is in the Carbstone project/product (Belgium) where we store CO_2 and produce high grade construction materials from otherwise landfilled steel slag materials. We store about 150 kg of CO_2 in 1 ton of product and the required process energy emits an equivalent of 80 kg of CO_2 per ton of product. This process will not capture large amounts of CO_2 but can be a piece of the puzzle to contribute to a wider CDU spectrum. For more information see http://www.carbstoneinnovation.be/en/.

Alexander Cowan continued the discussion of the paper by Christopher Hardacre: Thank you for your presentation. It is a most interesting study. Could you comment on the oxidation side of your cell? Do you think the ionic liquid has a role to play in oxidation reactions as well or would you envisage that the cathode and anode reactions will be carried out in two different electrolytes?

Nathan Hollingsworth replied: In the presented manuscript (DOI: 10.1039/c5fd00099h) the same electrolyte was employed in both the cathode and anode compartments. With regard to the oxidation reactions taking place at the counter electrode, the authors believe that the [P66614][124Triz] is stable to oxidation under these conditions. This is due to no [P66614][124Triz] oxidation products being detected by ^{1}H-NMR of the electrolyte after electrolysis. However, O_2 is detected by GC after electrolysis; one plausible explanation is the O_2 is a result of anodic processes involving H_2O.

Alexander Cowan remarked: I was particularly interested in the potential role of the IL in mediating water oxidation. I understand that there is some really interesting work from the group of Doug MacFarlane[1] where certain ionic liquid/catalyst combinations promote water oxidation. Could this be of interest for your system?

1 MacFarlane *et. al.*, *Energy Environ. Sci.*, 2012, 5, 9496.

Nathan Hollingsworth responded: The paper by MacFarlane *et al.* is very interesting; he reports the oxidation of water to hydrogen peroxide, which is amine stabilised. The hydrogen peroxide can then form oxygen and water *via* a disproportionation reaction in the presence of MnO_2 with current densities greatly increased by using the ionic butyl ammonium bisulphate. In our manuscript we have only been concerned with CO_2 reduction products, however, you raise a very pertinent question with regard to counter electrode oxidation reactions and suggest a very interesting alternative use of the ionic liquid system. Linking MacFarlane's study to this manuscript it is not possible to say if the oxygen we detect after electrolysis is from the decomposition of hydrogen peroxide or oxidation of water to O_2. Further tests would need to be carried out to confirm this, such as performing a potassium permanganate or diphenylamine test which would detect and quantify any H_2O_2 formed. In the same paper MacFarlane reported, through DFT calculations, a greater affinity for hydrogen bonding of H_2O_2 with amine over water, providing twice as much stabilisation for 4 amines over 4 water molecules. It was further reported that the solvation of H_2O_2 by amine is critical and sufficient enough to lower E_0. With the system presented, the ionic liquid [P66614][124Triz] is used without the addition of amine, as such a similar amine stabilised H_2O_2 complex would not be possible. However, it is plausible that the nitrogen groups on the $[124Triz]^-$ anion may provide hydrogen bonding sites and thus solvation to H_2O_2. As such, it is an area we should definitely look into and we thank you for this suggestion.

Geoffrey Maitland asked: You speculate on a mechanism that involves reduction of both chemically bound CO_2 *via* the $[124Triz]^-$ anion and of physically adsorbed CO_2, the latter being capable of being blocked by IL binding to the electrode. By 'easier CO_2 access to the electrode' on page 6 (DOI: 10.1039/c5fd00099h), are you referring to enhanced transport to the electrode or to proximity for electron transfer? Have you investigated the use of physico-chemical techniques for probing CO_2-IL-electrode interactions and investigating in more detail the underlying mechanisms resulting in the observed variations between electrode materials?

Nathan Hollingsworth replied: Here we are speculating a pseudo enhanced transport to the electrode through a reduced blocking effect from the IL that in turn may also lead to a closer proximity for the electron transfer. At present we have no evidence for this other than the difference in conversion, nor have we undertaken any further studies to yield mechanistic insight. Modelling is being carried out by the groups of Professor de Leeuw and Professor Catlow to model the reduction processes. More insight maybe gleaned by performing spectro electro chemistry, such as *in situ* IR studies which we will hopefully look into in the future.

William Webb communicated: I notice you talk about CO_2 adopting a bent configuration during the course of the reaction, namely with the 1,2,4-triz. Does this bent configuration allow for utilisation of CO_2 in other reactions. In much of the literature the formation of the bent conformation of CO_2 is essential in allowing reactions to be undertaken, could your system therefore be employed for reactions such as the addition of CO_2 to epoxides?

Nathan Hollingsworth communicated in reply: Thank you for your suggestion. Indeed this system is essentially a low potential way of generating the reactive $[CO_2]^-$ anion. Without the presence of H_2O the $[CO_2]^-$ radical anion could be reacted with alcohols to make carbonates or added to epoxides.

Geoffrey Maitland enquired: How generic is this chemical binding of CO_2 to ionic liquid anions other than $[124Triz]^-$? What is the potential for optimization of the anion structure for reducing CO_2 electrochemically and is there scope for tuning the composition of the resulting syngas?

Nathan Hollingsworth answered: The chemical binding is generic in that if an anion binding site is present it is possible the paper by Sheng Dai[1] reports various anions that allow this chemical binding. The tuning of compositions of resulting products by tuning the anion is a study that should definitely be looked into. Within a recent reported paper[2] we have shown that tuning the cation may also be a fruitful approach. For example, when the $[N4444]^+$ cation is used in place of $[P66614]^+$ with the $[124Triz]^-$ anion formaldehyde can be produced, which has not been seen previously.

1 C. Wang *et al.*, Tuning the Basicity of Ionic Liquids for Equimolar CO_2 Capture, *Angew. Chem., Int. Ed.*, 2011, **50**(21), 4918–4922.
2 N. Hollingsworth, S. F. R. Taylor, M. T. Galante, J. Jacquemin, C. Longo, K. B. Holt, N. H. de Leeuw and C. Hardacre, Reduction of Carbon Dioxide to Formate at Low Overpotential using a Superbase Ionic Liquid, *Angew. Chem. Int. Ed.*, 2015, **54**, 14164–14168.

Peter Styring stated: The chloride ion performs best in the molecular modelling studies using ionic liquids as sorbents. However, when you put it into a process it is not so good, especially in a humid environment because of the corrosion. Klaus Lackner uses chloride but he uses a humidity swing which uses water. This could lead to corrosion. You might you need to screen out the chloride ions because of the practicalities.

Nathan Hollingsworth remarked: Our system does not employ chloride and so this should not be an issue here. In addition our system out performs chloride based ionic liquids for capture. Furthermore, a chloride anion would not have the CO_2 binding site that is key to this study.

Renata Silva opened the discussion of the paper by Cafer Tayyar Yavuz‡ by communicating: Nowadays one of the main problems of valuability of the CDU is the high cost of the process. I think the idea of expanding the CDU concept is very important but this is not sufficient. What do you think remains to be done so that CDU can become real instead of just ideas to reduce problems caused by CO_2 emissions?

Cafer Tayyar Yavuz communicated in reply: In CDU efforts for CCS purposes, we tried to commercialize (and make a profit) graphene that is made directly from CO_2. We made graphene as cheap as 10 per kg and yet we couldn't find a single buyer who may be eager to scale up. The point we learned the hard way is that no matter how exciting your research-led product development is, the industry may turn a blind eye, just like that. It seems like graphene is not worth anything, at least not yet. Please see our company page: www.GrapheneTechnologies.com for more information.

Carlos Alonso-Moreno communicated: What is the reason of choosing just 800 and 25000 molecular weight polyethylenimines for the amine impregnation process? Is it just because they are commercially available compounds?

Cafer Tayyar Yavuz communicated in response: There are two reasons why we chose PEI 800 and PEI 25000 for this study. First, as you point out, is their commercial availability and second, this is a case study to show the potential of porous polymers for developing amine impregnated sorbents for carbon capture. Furthermore, we found that high molecular weight polymeric amines are more stable than low molecular weight amines (such as ethylenediamine and triethylenetetraamine) when considering repeated use of heat treatments for desorption, a leading cause for degradation of amines over long time.

Carlos Alonso-Moreno communicated: How have you achieved the calculation of the amount of PEI to impregnate the COP-19? Why was the solution of PEI and COP-19 only stirring for 24 h at room temperature? What is the maximum loading capacity of PEI for COP-13?

‡ The paper by Cafer Tayyar Yavuz was not presented at the meeting. Subsequently, any questions regarding this paper were submitted after the meeting.

Cafer Tayyar Yavuz communicated in reply: The amount of PEIs loaded on COP-19 was calculated from thermogravimetric analysis under a nitrogen atmosphere (Fig. 1a and b in DOI: 10.1039/c5fd00099h). The calculated PEI-800 in COP-97A, COP-97B, and COP-97C is 24.1%, 36.4%, and 46.2% respectively while PEI-25000 in COP-98A, COP-98B, and COP-98C is 22.4%, 32.3%, and 44.8% respectively. PEI-800 shows a slightly higher loading efficiency over PEI-25000 because of the smaller steric size of the low molecular weight aliphatic amines.

We focused on which loading provides greater CO_2 capture (at low partial pressure, *i.e.* 0.15 bar) rather than increasing PEI loading since a higher loading generally resulted in the blocking of the CO_2-philic sites. COP-97B (PEI 800 loading of 36.4 %wt.) showed CO_2 capture of 75 mg g^{-1} while COP-97C (PEI 800 loading of 46.2 %wt.) with a higher PEI loading actually demonstrated a decline in CO_2 capture capacity (Table 2 in DOI: 10.1039/c5fd00099h). In the case of PEI 25000 loading, the best CO_2 capture capacity was observed with COP-98A (PEI 25000 loading of 22.4 %wt.) which was 70.4 mg g^{-1}.

Carlos Alonso-Moreno communicated: As a conclusion, you proposed amine impregnated COPs as a commercially viable CO_2 sorbent. What can make COPs a commercially viable composition in comparison with other very cheap adsorbents such as silica, mesoporous silica or zeolites?

Cafer Tayyar Yavuz communicated in response: Sorbents to be used for amine impregnation should have larger pores, such as a mesoporous structure; therefore, zeolites are not suitable as amine-impregnated sorbents. The most popular mesoporous materials, mesoporous silicas (SBA or MCM), are not cheap since they need surfactant templation. They are as expensive as metal–organic frameworks. Conventional silica gels lack porosity and most of the amine is anchored on the surface of it resulting in a low CO_2 capture capacity. There have been many attempts made to synthesize hollow silica using polymeric surfactants that show exceptional CO_2 uptakes, however, the cost of surfactant also restricts their potential. For sorbents based on amine impregnation, the crucial component is the support material since CO_2 capture through amines is well documented.

Geoffrey Maitland communicated: How do you manipulate pore size and surface/connectivity of COPs, and how well can this be designed and controlled?

Cafer Tayyar Yavuz communicated in reply: The COPs are covalently linked amorphous networks and form their pores during the synthesis. It is, therefore, very difficult to make it more porous, unless one installs sacrificial porogens to be cleaved later (something we are working on at the moment). Pore filling is easier and that is what we pursued here. The idea is to make larger pores (mesoporous or bigger) and then shrink them with amine impregnation to achieve micropores with an optimum strength basic surface. Macroporosity in the order of microns is not desired either since you want the amines to be trapped inside. In general, COPs can be tuned by varying starting materials and the non-associating chemicals. Temperature and pressure also helps greatly since the hot solvents push away the walls of the growing polymer.

Geoffrey Maitland communicated: You say that CO_2-philicity is more important than surface area for optimised capture. Surely both are necessary for a successful system? If bicarbonate in water is much better than carbamate when dry, due to binding through one amine group rather than two, why is the loading difference not much closer to a factor of 2?

Cafer Tayyar Yavuz communicated in response: We agree with you on the contribution of both CO_2-philicity and surface area to the optimized capture. Of course, without surface area one cannot get high uptake, because of the unavailability of the contact surface, and the unspecific binding in a high surface area material won't help achieve high CO_2 capture since it will trap everything else as well. Our assessment is that the CO_2-philicity is more important in the sense that the return will be bigger compared to just increasing surface area. Bicarbonate formation is clearly better than carbamate, and Nature prefers that way as well. In solid systems, where we cannot easily claim a solution behavior (locally maybe), both mechanisms will take place and we suggest that the more we can shift to bicarbonate the better. Since none of our systems are perfect, an ideal, twice capacity – unfortunately – will never be reality. But you are absolutely right that we should do everything we can to make it more capacity per mass of sorbent material.

Geoffrey Maitland communicated: COP-19 has a lower CO_2 uptake but will have a lower cost – may not this material be the basis of a more viable CO_2 capture process?

Cafer Tayyar Yavuz communicated in reply: This is a perfect question, of which the answer will eliminate much clutter in the sorbent design. There are recent reports emerging on the compromises between cost–capacity interplay. I strongly suggest (and we are attempting to answer this on our limited capacity as well) that a team of collaborators assess the leading solid sorbents to be checked on all the process parameters. I'd be happy to participate. In the case of COP-19, my fear is the well-known competition of CO_2 with H_2O in physisorptive solids. COP-19 is not purely physisorptive as it has basic nitrogens, but it is not extremely selective towards CO_2 either. Suspicions don't count, obviously, and we will be reporting in the near future the applicability of our structures. In fact, we discovered a similar structure with cheaper production costs and now in talks with a major Korean company for large-scale applications.

Elsje Alessandra Quadrelli opened a general discussion of the papers by Michele Aresta, Gonçalo V. S. M. Carrera and Deepak Pant§ by commenting: Acidification of the oceans is a major issue. The potential solutions being talked about are indeed being explored: for example there are large projects around Offshore Membrane Enclosures for Growing Algae in the context of CO_2 capture.

The stability of the strain has been mentioned often as a recurrent bottleneck in the industrialization of your type of approach. Do you have a comment in general and more pointedly related to your specific strains?

§ Deepak Pant's paper was presented by Karolien Vanbroekhoven, VITO – Flemish Institute for Technological Research, Belgium.

Karolien Vanbroekhoven answered: Specifically the work I was talking about actually did focus on bacteria, not on algae! That means, in my case, the energy comes as electricity or chemical energy directly to combine with CO_2 to produce chemicals; the algae work is using direct sunlight as energy and I can imagine that increasing the strains robustness and stability is a big challenge, the same for all bioprocesses! Making a specific suggestions for algael strain selection is not my piece of cake.

Michele Aresta remarked: The use of biomass as an indirect way for CO_2 utilization is one of the most investigated topics today. I wish to make two preliminary considerations:

1) Nature is not efficient nor selective, it does not need to be. It uses a perennial and abundant source of energy: the sun.

2) Plants produce chemicals for themselves not for us. What we want is not what they need.

There have been many attempts in the last forty years to change by genetically engineering the selectivity of the most abundant enzyme: RuBisCO, which is implied in the conversion of CO_2, such as for making a C6 sugar from a C5. RuBisCO catalyzes the oxidation and carboxylation reactions and the selectivity is 50%. If one could be able to improve the carboxylation to 55% most probably the anthropogenic CO_2 would be fixed by Nature. Unfortunately, the natural system produced in a million years cannot yet be easily modified in monocellular organisms. So far, no exploitable results have been obtained.

Biomass can be categorized into two main categories: terrestrial (*i.e.*: trees, which mainly produce lignocellulosic materials and oily-seed plants, which give oils), and aquatic (micro- and macro-algae which produce proteins, carbohydrates, lipids, fine chemicals, *etc.*). The microbial biomass, which can be found in soils and aquatic environments, is a separate case that is now under investigation.

So far, lignocellulosic materials have been gasified, destroying many complex structures for making a C1 molecule (CO) with a high entropy change (increase). CO was then converted into structured C_n molecules, with large entropy reduction. Such a conversion consumes energy, reducing the net use of carbon fixed in biomass. Today, there is a different approach: namely, the use of a cascade of technologies for recovering useful molecules first (that may have a direct use or may be used as monomers for polymers), with gasification of the residual inert portion. Cellulose is used for recovering monomers (C6 and C5) that are a source of new platform molecules or can be converted into ethanol biotechnologically. In the case of algae, their fractionation allows to separate proteins (for animal feeding), carbohydrates (only cellulose, without lignin, used for ethanol production or other applications), lipids and fine chemicals. Lipids can be used for making biodiesel (saturated long chains) or for making monomers for polymers. From a C18 monounsaturated long chain such as that contained in oleic acid (contained in olive oil and other seeds), one can produce by oxidative splitting of the double bond, two interesting molecules: a) azelaic acid, a C9 dicarboxylic acid used for making polymers, and b) a C9 monocarboxylic acid, pelargonic acid, that has herbicide properties. Mono- and di- poly-unsaturated fatty acids can be used, *via* epoxidation, for making new polyurethanes. The use of algae only for the production of biodiesel cannot be economically acceptable nor environmentally viable due to their growing, separation, drying and conversion costs. It must be noted that so far microalgae have been mainly used as feed (in fisheries or for livestock) or for producing fine chemicals. In such cases the product will have a price that will pay the producer for all costs faced: no consideration is made about CO_2 emissions. If we wish to produce fuels from microalgae with the target of reducing the CO_2 emission, then the boundary conditions are very strict: the entire process cannot emit more CO_2 than is fixed. Therefore, all steps (growing, collection, drying and conversion) must be performed minimizing the CO_2 emission. This means, for example, that it is impossible to add all N, P and micronutrients the microalgae need for growing as the commercial products bear on themselves a CO_2 emission load and it has been found such to overcome the amount of fixed CO_2. Therefore, process waters or municipal waters must be used as source of N, P and micronutrients: this poses the problem of contamination to which microalgae are very sensitive. The use of biomass is a very interesting option for CO_2 reduction (leave Nature to convert CO_2 and we convert biomass) but it requires a new approach that is mainly based on the exploitation of the concept of biorefinery applied to biomass. Additionally, a new catalysis must be developed as biomass is quite different form hydrocarbons used so far as raw materials. The future is very challenging under many different aspects as we shall move away from fossil carbon.

Seetharaman Vaidyanathan continued: I am in agreement with Michele Aresta's comment. I wish to add that in Nature, algae predominantly encounter low levels of CO_2 (in relative terms) and have evolved to sequester this. However, they have inbuilt mechanisms to cope with excess excursion of nutrients that they are able to deal with to differing degrees. Simply taking the natural process and trying to implement it as a plausible CDU solution is bound to fail. The specific capability of algae to fix CO_2 and convert to organic carbon of value (in a broad sense) has to be systematically examined, even in a scientific sense. Provision of N, P, nutrients and water is key and the uptake of CO_2 along with these key nutrients for proper conversion to products of value needs further investigation for the perceived potential to materialise.

Peter Styring addressed Michele Aresta and Karolien Vanbroekhoven: Michele has a good point. Nature is very good at providing the energy it needs to survive: it produces a soup of chemicals. From a processing aspect, the energy costs of extracting the products is where it falls down. The mixture is too complex to identify and extract products, even using HPLC. Bearing in mind they produce chemicals for their own needs and not necessarily what we need. Are there any considerations on modifying the genomes rather than using the whole cells to obtain what we need?

Michele Aresta replied: Genetic engineering of monocellular organisms is an ongoing research: it may work with microalgae and less with superior plants (at the moment). It is noteworthy that genetically modified organisms and their derivatives may not find acceptance in all countries. With microlage one can use physical stress for pushing the production of lipids, if one wants to target high lipid production. Starvation in N (growing algae in a low amount of N) decreases the production of proteins and increases the production of carbohydrates and lipids, used as an energy reserve by the microorganism. This approach is giving interesting results in controlled conditions and, if it is applied to large ponds/PBR, can result in a way to address the enhanced production of large volumes of targeted products.

Karolien Vanbroekhoven continued: Indeed, that is exactly what we are considering! The final concentrations need to be high enough, but many fermentation based processes do run on a commercial scale, where titers or concentrations are in the range of 10–300 g l^{-1}, depending on the chemical produced. The toolbox of genome engineering has increased tremendously and opens up a lot of new opportunities to play with Nature and get out what we want.

Pedro Abrantes added: I'd like to explain some figures about trees. A forest (average 2000 trees per hectare) absorbs about 1.8 tons of C per year; 6.5 tons of CO_2. That means 3.6 kg CO_2 per tree per year. With the agriculture under tunnels and micro-algae, we hope to consume 300–700 tons of CO_2 per year with CO_2 levels from 1200 to 1500 ppm.

The great difference between the micro-algae and trees – if you cut a tree that's it, no more CO_2 consumed. You have to wait 10 years for a new tree. We need to work on micro-algae and agriculture. It needs to be a goal to keep humans on earth.

Michele Aresta responded: Terrestrial (fast growing trees, 2 years cut) and aquatic biomass have different and complementary roles, as they afford different raw materials. Trees will produce lignocellulosic materials, oily-seed plants will produce oil and cellulose, microalgae will give proteins, lipids and carbohydrates. Aquatic biomass is much more productive in terms of mass/surface used and may not require land, that often is a motif of argument about food–energy use. Moreover, aquatic biomass has not been exploited for energy so far, but only for food or fine chemicals. The exploitation for energy completely changes the boundary conditions for production: it is not a matter of how much the product will cost but how much CO_2 is avoided. Aquatic biomass can be grown in marginally salty waters, with the correct offshore technology. In my group we are now applying to these micro-/macro-algae the biorefinery concept, that means their fractionation into proteins, carbohydrates, lipids and other chemicals for their complete use. This approach may bring about the correct assessment of the potential of aquatic biomass: not only biofuels, but also fine and bulky chemicals. Such an integrated approach may make economic the use of aquatic biomass on a large scale for the production of diesel (from saturated lipids) together with raw materials for polymers (monounsaturated fatty acids) or functional materials for polymers, proteins and so on. The use of microalgae for making biodiesel alone is not competitive with fossil fuels (and maybe will never be). A key point with aquatic biomass is not to supply all micronutrients, N and P microalgae need for their growth: if we have to supply all such chemicals the environmental benefit will be negative, as chemicals bring with them large emissions, and economic criteria will never be matched. Therefore, the use of process waters (we have used with success water from biogas digesters) or municipal water is essential, supposed that they do not bring contaminants that may kill the pond.

If we are successful in the exploitation of microalgae, then large volumes of C can be recycled at an industrial level.

Elsje Alessandra Quadrelli enquired: I was under the impression that the operation of a bioreactor for an algae producing plant is very expensive (open and closed sites). Even more so for plant operation, and the only viable business model is linked with waste remediation. Are there any other business models? Are you at liberty of discussing yours?

Pedro Abrantes answered: You are right, it's expensive. We're trying to build a new process and find solutions to reach our equilibrium target. There is a way. For the moment, I can't explain our position. We need to wait a few months for the first results. I hope that we'll be able to present a paper next year.

Elsje Alessandra Quadrelli opened the discussion of the paper by Michele Aresta: The direct photocarboxylation is a very interesting reaction. Can you comment on your choice of the photocalyst chosen (zinc sulphides, possibly modified by Ru)? Are you driven by band gap modification toward visible?

Michele Aresta replied: I can confirm that the choice of the semiconductors is not made only on the basis of their band gap but also considering their electrochemical potential. Two approaches can be used with the selected material that must be able to activate the C–H bond and produce CO_2 anion radicals: either push the light absorption towards visible or enhance the ability of semiconductors to use the UV fraction of solar light. In the case of ZnS A and B we have used the second approach. In fact their activity is quite high around 375–390 nm, near UV on the boundary with the visible region. In this case we have improved their operational performance by decorating with Ru, which does not affect the light adsorption. In other cases, as with TiO_2, we have decorated the photomaterial with an organic or inorganic motif that changes the band gap.

Richard Heyn asked: This is a very nice reaction. Is the reaction selective for carboxylation at the carbon atom of the allyl radical alpha to the oxygen atom? If so, what is the reason for the selectivity?

Michele Aresta replied: The selectivity and yield depend on the solvent: in chlorinated solvents we see two carboxylic acids, in methanol only the 2-CO_2H. This can be explained in terms of solvent stabilization of reactive forms of the substrate. We are studying further the solvent effect, including solventless reactions.

Richard Heyn continued: Related to this, it seems that the 2-COOH-2,3-DHF (product B in Fig. 12) could arise from a 1,3 proton transfer after carboxylation at the allyl radical carbon atom alpha to the oxygen. The product distribution suggests that this transfer is more likely with the protic solvent MeOH, which is not too surprising. Do you have any data that might support this mechanism?

Michele Aresta responded: It is correct that the solvent plays a key role in the carboxylation, but also in the product distribution of isomers derived from the photoisomerization of the substrate in absence of CO_2. The 1,3 proton shift in 5-membered rings is not easy and can be assisted by solvents. We are investigating further the reaction mechanism in order to collect useful information that may be applied to other systems under study. A reaction that we observe is the dimerization of the isomerized 2,3-DHF: this influences the yield of carboxylic species as it subtracts the substrate. This is the beginning of a story that I am sure will be quite long.

Michael North questioned: You have managed to add one acid group to the furan ring, but have you any thoughts about how to introduce a second so that you make monomers for polymer synthesis?

Michele Aresta answered: In my short presentation I have shown phtalic acid as an example of a chemical that would be interesting to make *via* carboxylation of benzene, more than by oxidation of another organic substrate with carbon loss. Obviously, mono- and di-carboxylic acids have different fields of application. You correctly point at the use of the latter as co-monomers for polymers. The former, if they are long chain acids, may have application as surfactants. The great advantage of photocarboxylation is that the free acid is produced, instead of salts, as it happens in chemical catalysis. At the moment, the introduction of a single $-CO_2H$ moiety is not an easy task, the introduction of two of them in specific positions is a problem of superior difficulty, if we want to target high selectivity and conversion. Using high energy radiation one can produce mixtures in which several species are found, including dicarboxylic derivatives, with serious separation difficulties. It is not impossible, but more difficult to perform selectively.

Gonçalo V. S. M. Carrera said: In a different perspective from a photocatalyst based carboxylation reaction mechanism, photochemical reactions using the proper wavelength can eventually afford carboxylated products, as in excited states of olefins there's the option of singlet/triplet states, leading to biradical and zwiterionic intermediate, respectively, leading to a possible reaction with CO_2. Have you tried these options? A suggestion to make this type of synthesis more sustainable could be the replacement of methanol or chloroform by ethanol. Have you considered this option?

Michele Aresta replied: We have expressly used solvents with different polarities as they can stabilize different states of the excited/isomerized substrate. As I have discussed, methanol and $CHCl_3$ influence the reaction to a different extent, methanol being more selective towards the 2-carboxylic acid. This is in agreement with your comment. We have tried several other solvents (even water) or solventless reactions. There are three key issues for the solvents and these are:

i) the solubility of CO_2 in the solvent (we have developed a photoreactor for working under pressure, which is under continuous improvement trying to find the best geometry that allows the full use of the light);

ii) the fact that CO_2 can be converted into other species upon reaction with the solvent (in water, CO_2 undergoes multiple equilibria affording species such as HCO_3^- and CO_3^{2-}, the concentration of which depends on the pH);

iii) the reaction of the solvent with radiation (the solvents used were not reactive; ethanol gives some problems).

Pedro Abrantes enquired: You are talking about PV cells which deliver about 12% of solar energy. Do you follow the Israeli work about a PV converter that produced a 4m^2 solar converter which reaches 25% solar energy? They are trying to reduce it to 1m^2 and try to reach a 42% efficiency converter. Is this the solution for the next years to obtain a significant solution for our projects?

Michele Aresta responded: Converting solar energy into PV is a different approach. There are two ways to use solar energy: either direct photochemical conversions of CO_2, about which I have been speaking, or prior conversion into electricity and use of the latter for producing hydrogen or CO_2 electro-/photo-electro-catalytic conversion. The use of PV, that today can be produced with 20% efficiency and in perspective may reach 42%, is perfect for H_2 production and its use in methanol and other oxygenates such as DME, methane or high HC synthesis from CO_2.Today the limitation is the cost of PV-H_2. Electrocatalytic conversion of CO_2 has its own role. Photocatalysis has a different potential. These are ways of using solar energy for CO_2 conversion: I believe that if we can exploit their potential, and integrate such applications of solar with biomass conversion, we shall be able to reach an interesting level of carbon recycling, mimicking photosynthetic processes.

Jonathan Albo communicated: A previous report showed that Bi_2WO_6 hollow microspheres possess high activity for CO_2 conversion to methanol, without the aid of any co-catalyst, under visible light irradiation.[1] In addition, the material seems to present a considerably high capacity for CO_2 adsorption and charge separation. I wonder if its use could be interesting for the photocarboxylation of 2,3-dihydrofuran using CO_2.

1 H. F. Cheng, B. B. Huang, Y. Y. Liu, Z. Y. Wang, X. Y. Qin, X. Y. Zhang and Y. Dai, *Chem. Commun.*, 2012, **48**, 9729.

Michele Aresta communicated in reply: According to our experience catalysts which are able to reduce CO_2 to other C1 or even C_n molecules are not so good for the formation of C–C bonds and fix CO_2 onto organic substrates such as a carboxylic group. So Ru@ZnS is a catalyst of carboxylation while other p-type semiconductors based on Ni or Cu are poor as carboxylating agents and good for CO_2 reduction to methanol or even methane. I'll try the catalyst you have mentioned and will let you know.

Elsje Alessandra Quadrelli addressed Michele Aresta, Karolien Vanbroekhoven and Gonçalo V. S. M. Carrera :One criticism we hear often towards bio- and photo-based technologies revolves around the purported (limited) efficiency of the processes involved. Yet, if benchmarking against Nature, these artificial processes are often as efficient and selective (if not more so) than natural ones. What are the efficiencies and selectivities in your systems and do you consider them serious bottlenecks to the deployment of your technology?

Gonçalo V. S. M. Carrera answered: In the specific instance of the work presented (DOI: 10.1039/c5fd00044k), initially limitations regarding selectivities were issues to consider, as mentioned in the paper, afterwards it was possible to optimize the systems and solve selectivity constraints even at the cost of losing optimal capacity to capture CO_2. It was a matter of compromise. After optimization we consider that CO_2 capture capacity is competitive with benchmark systems, there is even the possibility to explore other alternatives. Nature presents a specific time scale and complexity that is certainly different from the necessities of the society where we live, in this context task-specific systems similar to those presented at this conference are highly sought and optimization and additional research will give new perspectives.

Karolien Vanbroekhoven responded: Specifically towards photobiobased systems, indeed efficiencies like for algae are only 3%; however, for biocatalytic processes where other chemicals are provided as energy carriers (instead of light) the efficiencies are higher. A typical product yield of 80% has been reported for polyhydroxybutyric acid using H_2 as the energy carrier. That's not too bad! Therefore, I think efficiencies and selectivities (typically known for biocatalytic systems to be higher) are not the biggest concern, it is the speed of the reaction.

Michele Aresta continued: There are not many examples of photocarboxylation in the literature: Those I have discussed are the first examples. We see that by mastering the photocatalytic system we can improve the yields and selectivities. We do not think we have reached the best performance, we are just at the start of research. The thermodynamics of the reaction poses a limit to the equilibrium concentration. We are now working to understand if we are already at the thermodynamic limit, and if not where it is and how to reach it.

André Bardow noted: The question has been asked whether we should benchmark biomimetic and photocatalytic processes against Nature and I want to comment on this question. In my view, we should not use Nature as a benchmark. Nature has a very different business model from ours and Nature is thus following very different (and multiple) objective functions. However, I agree that we should benchmark the proposed processes; I even regard such a benchmark as mandatory if we are aiming for a practical impact. I would suggest to benchmark against using the best photovoltaic (PV) cells available and further using the PV-generated electricity in the catalytic conversion. If the novel route has not even the theoretical potential to ever outperform an existing PV-based system, we cannot expect to make a practical impact and should regard and assess the work as fundamental research.

Geoffrey Maitland opened the discussion of the paper by Gonçalo V. S. M. Carrera: You appear to have reached the situation where you can achieve less than optimal CO_2 uptakes and are limited by the high viscosity, and hence poorer mixing, of systems at higher saccharide : superbase ratios. As the viscosity of the liquid phase decreases and more efficient mixers are used, mass transfer rates seem to increase and close to stoichiometric uptakes are observed. Would you please comment on the extent to which more effective mixing systems will be needed to give optimal conversions and whether mass transfer is likely to be a limiting factor for large-scale implementation of this approach?

Gonçalo V. S. M. Carrera replied: At an initial stage we found that tetramethylguanidine, as superbasebase, in combination with mannose lead to high yields and low viscosity before reaction with CO_2, nevertheless the resultant mixture is complex and not usefull for our purposes. In order to circumvent this problem we tested another base, DBU, in combination with carbohydrates. At the pressures tested, DBU alone doesn't react with CO_2, and at that point we verified that it was a good candidate for a straightforward reaction. Of course we found that the combination DBU : saccharide can be problematic, leading to high viscosities and incomplete reactions with CO_2. Nevertheless with an optimization effort in terms of ratios and more efficient stirring we could attain relatively satisfactory results. We think that we can improve our results by two valid paths: find a more effective base, that similarly to tetramethylguaninide, could lead to acceptable viscosities, and similarly to DBU lead to less complex mixtures. The other modification could be the type of mixing with CO_2. An alternative that we are working on is bubbling CO_2 directly to our mixtures instead of admission of this component from above. Considering the presented results and alternatives to improve, we consider this work valid for an eventual application at large scale.

Pedro Abrantes questioned: At this initial level of your work, how many years do you need to have a flagship technology?

Gonçalo V. S. M. Carrera answered: The work is at an initial stage of development, based on the same concept it can be optimized for improved performances in CO_2 capture and reduction of energetic and material costs. Of course this system is relevant and can be a good platform to build a viable alternative to commercial systems for CO_2 capture already in the market, however it is difficult to anticipate the time required to obtain a top technology as this depends on many factors.

Seetharaman Vaidyanathan opened a general discussion of the papers by Deepak Pant by communicating: Have you measured the residual bicarbonate in the media and monitored how it varies over time? This would give an idea of the rate of carbon uptake by the organism and help in attempting to improve this process.

Deepak Pant communicated in reply: We have measured the bicarbonate concentration in the media with time. We have also observed that the bicarbonate stripping is due to the continuous N_2 supply. In order to quantify the bicarbonate stripping, we have operated blank reactors with and without bacterial inoculation in the MES reactors. The average bicarbonate rate was calculated by deducting the stripping rate of bicarbonates in the blank reactors from the apparent bicarbonate reduction rate. We have observed that the bicarbonate uptake rate of 203 ± 42 mg HCO_3^- L^{-1} during the stable phase of operation. We completely agree that improving the bicarbonate uptake rate by the biocathode/bacteria will improve the process. Presently we are working on similar objectives.

Seetharaman Vaidyanathan communicated: How does your MES approach compare with other chemical and biochemical approaches for the reduction of CO_2 to acetate?

Deepak Pant communicated in response: The MES research is at its infancy, the process itself being discovered only 5 years ago. A lot of development is needed towards improving the higher concentrations and production rates of the process. At present, the research is going in several directions from all corners of the world. Compared to autotrophic fixation of CO_2, MES has showed about 10 times less acetate production rate (compared to Straub *et al.*[1]). In the case of a chemical approach, it is much lower. At the same time, compared to chemical processes, the MES approach requires very less energy input, where the MES is found as advantageous. Also by using the CO_2 directly as a feedstock it does not get into the food *vs* fuel debate that the terrestrial plants or algae have to face.

1 Straub *et al.*, *J. Biotechnol.*, 2014, **178**, 67–72.

Seetharaman Vaidyanathan communicated: What are the challenges you perceive in the large scale implementation of such MES systems?

Deepak Pant communicated in reply: This is a very important question where we are focused on our research. As microbial electrosynthesis (MES) is a combination of microbiological and electrochemical principles involving many disciplines, many challenges are associated:

1) Biocatalyst: the selection of a suitable biocatalyst for the electrochemical cathodic reduction reaction.

2) Electron transfer mechanism: in the case of microbial fuel cells (MFCs) the electron transfer mechanism was understood, and it works through direct and mediated electron transfer mechanisms (DET and MET). In the case of the biocathode, the electron transfer mechanism and the role of cell wall embedded cell organelles such as pili and cytochromes in reduction process still need to be explored. This is a crucial phenomenon to understand microbial electrosynthesis.

3) CO_2 solubility rate: the CO_2 solubility rate is very low. As this is the substrate for the process, higher solubility rates helps for the improved production rates. Presently we are focused on our state of the art technology of gas diffusion electrode (GDE) development which is expected to improve the availability of CO_2 to the biocathode.

4) Selectivity of the product: physiological conditions and several other factors control product selectivity. A more selective product improves the conversion efficiency and purity. It also results in decreasing energy demand on down stream processing.

5) Effective and stable biofilm: a stable biofilm that can work under varied physiological and electrochemical conditions is important for the MES process. The biofilm growth rates and energy dissipation towards biofilm production and maintenance is also important.

6) Integrating the system with cheap/economic energy generating system such as wind and solar energy systems rather than grid, *etc.* is one of the sustainable options for microbial electrosynthesis. On the other hand, integration of MFC to utilize the energy generation from the biomass oxidation/wastewater treatment at the anode is also a possible way to decrease the energy demand for the MES process.

7) Apart from the above stated challenges there are several other challenges such as energy efficiency, carbon yield, bio-electro-kinetics, electrode materials that support biofilm formation, reactor engineering, *etc.*, associated with the MES process.

Seetharaman Vaidyanathan communicated: Have you characterised the microbial consortia in your MES?

Deepak Pant communicated in reply: We have sent the samples of the biocathode to an external agency for microbial diversity analysis. Very recently we have received data for the analysis. The biofilm sample contains 6 different clones (*Chelativorans spp*, *Erysipelothrix spp*, *Natronincola peptidivorans*, *Mesorhizobium sp.*, *Delftia acidovorans* and *Clostridium felsenium*) along with several other unidentified bacteria.

Faraday Discussions

PAPER

Perspectives and visions on CO_2 capture and utilisation

M. North*[a] and P. Styring*[b]

Received 21st September 2015, Accepted 21st September 2015
DOI: 10.1039/c5fd90077h

This article summarises and contextualises the debates which occurred during the Carbon Dioxide Utilisation Faraday Discussion meeting. The utilisation of carbon dioxide is discussed in terms of both conversion to fuel, with a potential impact on atmospheric carbon dioxide levels, and conversion to chemicals with a potential impact on sustainability.

Introduction

Every year, BP publish a statistical review of world energy[1] and the 2015 review indicated that conventional reserves of oil, gas and coal will be consumed in 52, 54 and 110 years respectively, if we keep consuming them at the present rate. Of course, in reality this will not happen, as fossil fuel usage is influenced by two opposing factors. The ever-increasing human population and quality of life means that fossil fuels are being consumed at an increasing rate, but eventually demand will outstrip supply resulting in substantial price increases and a resulting fall in demand. It should also be noted that fossil fuels not only provide the energy needed to power mankind's current standard of living, they also provide around 90% of all commercially available organic chemicals. About 5–10% of crude oil is used for chemicals production, but that provides about 50% of the profit of an oil refinery. Thus, during the 21st century, mankind will need to find alternative, renewable and sustainable sources of both energy and chemicals and the reason for this Faraday Discussion was to explore the ways in which carbon dioxide could (partially) fulfil this role.

Carbon dioxide utilisation comprises of a broad portfolio of technologies and this was reflected in the diverse range of subjects covered at this meeting. While the majority of the delegates were from the United Kingdom, as to be expected for a conference on home soil, it was amazing to see that delegates from 23 different countries in total were present, emphasising the global interest in this subject.

[a]*Green Chemistry Centre of Excellence, Department of Chemistry, The University of York, York, YO10 5DD, UK. E-mail: michael.north@york.ac.uk*

[b]*UK Centre for Carbon Dioxide Utilisation, Department of Chemical & Biological Engineering, The University of Sheffield, Sir Robert Hadfield Building, Sheffield S1 3JD, UK. E-mail: p.styring@sheffield.ac.uk*

The very fact that there was such a diversity of subject areas covered is not surprising in the global context. At the 2015 International Conference on Carbon Dioxide Utilisation (ICCDU) in Singapore, Professor Paul Sharrett from the A-Star research facility likened CDU to not only a jigsaw with many pieces, but to several jigsaws that had been put in a bag and shaken up!

So where do we need to look for solutions to the puzzle? This Faraday Discussion provided many interesting suggestions and comments. Professor Boagerts was clear when she said that we need to invest in many technologies in order to cover all options. As Professor Aresta stated "we need to investigate the forest in order to find the right tree".

The discussion highlighted the need to take into account the catalyst in any process we consider. Not only its activity and selectivity, but also its preparation and post-reaction processing and separation. Life Cycle Analysis (LCA) and Techno Economic Analysis (TEA) studies will play a major role in catalyst selection. This will include risk analysis of metal availability, however, as Dr. Dowson stated, "please don't take away our catalyst metals any faster than is needed". In order for this to be achievable we must consider catalyst metal recovery and recycling. Public acceptance was also highlighted as an increasing consideration, whilst remembering there will always be environmental intransigence in any survey or focus group conducted: some people will never accept a new technology and this will have to be anticipated. We must emphasise the positives of CDU, however we need to be transparent and not hide any problems that might surface, technically, environmentally or economically.

One of the biggest economic (and indeed environmental) hurdles will be the need to capture and concentrate carbon dioxide prior to reaction. In any LCA the environmental impact will need to be considered, as depending on current capture-release technology, there is a 20–30% energy penalty on including the technology in any power generation system. Likewise, there will be energy and emissions penalties on industrial or even direct air capture. Dr. Priestnall showed that sorbents for mineralisation processes can be sourced at low cost, even if they need to be transported into the processing plant. Of course full environmental impact studies are still needed on these processes. While challenging, this offers considerable scope for research innovation.

The use of carbon dioxide

As previously stated, carbon dioxide utilisation is a broad subject and many approaches are needed to achieve our key aims of a sustainable supply of carbon for the chemical industry and to contribute to lowering global carbon emissions. If carbon dioxide is to be used as a starting material for the production of fuels and/or chemicals, then the availability of waste carbon dioxide should be considered. Table 1 presents 2008 data on the large scale sources, quantity and purity of carbon dioxide.[2] In addition to these sources, atmospheric carbon dioxide levels have now risen to 400 ppm,[3] which corresponds to about 3 teratonnes of carbon dioxide which is freely available to anyone who wants it anywhere on the planet. Globally, urea is the organic chemical that is produced on the largest scale, with an annual production of *ca.* 150 Mtonnes, so it is apparent from Table 1 that there is ample waste carbon dioxide available to satisfy all the potential needs of the chemicals industry. It is also apparent from Table 1

Table 1 Major sources of waste carbon dioxide[2]

Source	2008 global CO_2 emissions (10^6 t CO_2 per annum)	CO_2 purity (volume%)
Coal	14 200	12–15
Natural gas	6320	3–5
Refineries	850	3–13
Cement production	2000	14–33
Ethylene production	260	12
Iron and steel production	1000	15
Natural gas production	50	5–70
Ammonia production	150	100

however, that converting carbon dioxide into chemicals cannot make any significant impact on anthropogenic carbon dioxide emissions or on atmospheric carbon dioxide levels due to the major difference in scale between energy production (>20 Gtonnes of carbon dioxide per annum) and chemicals production (a few 100 Mtonnes per annum).

In contrast, conversion of carbon dioxide into fuel could substantially cut anthropogenic carbon dioxide emissions, as it would replace all of the emissions from the combustion of coal, oil and natural gas. However, the challenge in this area is to find a financially viable, non-fossil fuel source of hydrogen which can be used to reduce carbon dioxide to fuels such as methanol, dimethyl ether, methane and other hydrocarbons *via* Fischer–Tropsch chemistry. In this case, the carbon dioxide is playing the role of a convenient hydrogen carrier. The fact that many processes have been proposed that assume a supply of renewable-hydrogen (r-H_2) has led to the term "magic hydrogen". r-H_2 needs renewable energy to carry out production and many competitor technologies have also stated that they need r-H_2 and renewable energy: demand will exceed supply considerably. Therefore, accurate and transparent supply chain analyses will need to be carried out to consider viability. It may seem obvious, but the cost of the product must exceed that of the reactants for the process to make economic sense.

Table 1 also shows that most sources of waste carbon dioxide do not provide pure carbon dioxide. Thus, in most cases the carbon dioxide will require separation from other gases before it can be used. For carbon dioxide obtained from post-combustion fossil fuel power stations, the challenge is to separate carbon dioxide from mostly nitrogen at temperatures of 50–60 °C. In contrast, for carbon dioxide obtained from natural gas production, the carbon dioxide needs to be separated from mostly methane at ambient temperatures. It may be considered that nitrogen is inert to most processes, so it should be possible to use a mixture of carbon dioxide and nitrogen without separation. While in principle the nitrogen can be regarded as acting as a dilutant, many of the reactants discussed at this Faraday Discussion used nitrogen as an inert atmosphere. Therefore, a range of carbon dioxide purities need to be investigated that are relevant to the individual reaction and process being considered. Of course using dilute flows may affect the kinetics and increase reactor sizes and costs, however there needs to be consideration of the balance between process costs. Thus there is a conflict between enhancing the life cycle analysis, which will favour the use of unpurified waste carbon dioxide, and techno-economic costs, which will tend to favour purifying the carbon dioxide prior to use. When the impact on public perception of the process is also included, the result is a complex three-way analysis which led to a particularly lively discussion session during this Faraday Discussion.

From the data in Table 1 it would seem sensible to utilise the purest forms of carbon dioxide first, and in particular, those resulting from ammonia production. However, political priorities are to minimise emissions from fossil fuel power stations, so research grants tend to focus on utilising the less pure forms of carbon dioxide. Thus, in recent years the North group have demonstrated the use of raw flue gas[4] or even atmospheric carbon dioxide[5] in cyclic carbonate synthesis, whilst the Williams group and Bayer Material Science (now Covestro) have shown that carbon dioxide from power station flue gas (after a carbon capture process) can be used for polycarbonate–polyol synthesis.[6]

One of the biggest challenges in utilising waste carbon dioxide in power station flue gas is getting access to the gas. It cannot be bottled due to its water content and cannot be readily simulated in the laboratory as the temperatures obtained in a commercial boiler would not be reached on a small scale. A solution to this problem would be the establishment of an international centre for flue gas utilisation studies associated with a carbon capture and storage pipeline (Fig. 1). This would allow researchers access to at least three grades of waste carbon dioxide and allow the effect of impurities in the flue gas on their catalysts and processes to be determined. It is the view of the authors that any government (national or European) funding for a carbon capture and storage scheme should include support for a centre for flue gas utilisation studies to facilitate the development of processes which would allow as much of the carbon dioxide as possible to be utilised rather than dumped.

CCS and enhanced oil recovery

On the first day of the Faraday Discussion meeting there was a major debate around the role that carbon dioxide utilisation could play in providing future

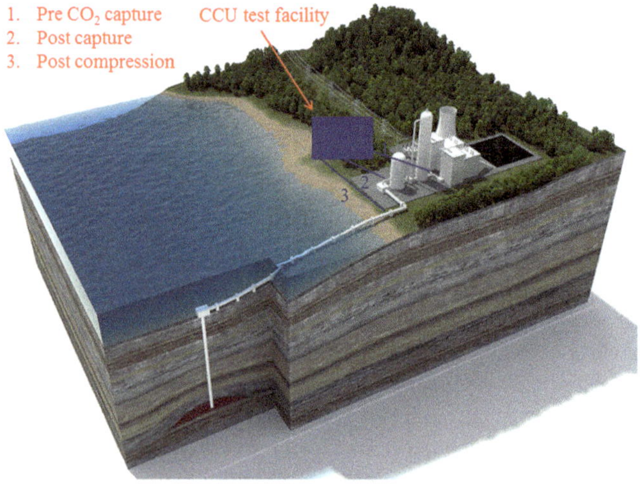

Fig. 1 A centre for flue gas utilisation studies (modified after ref. 7).

energy needs, with a particular focus on enhanced oil recovery and carbon capture and storage. It should be remembered that Carbon Dioxide Utilisation (CDU) and Carbon Capture and Storage (CCS) are not enemies. They are in fact part of a family of technologies. While they are not twins, and never will be, they are siblings. Of course there will be sibling rivalries, but that is to be expected. Both need to be nurtured and while both will potentially have different futures we must allow both to thrive. However, we must be careful that any technology shows a benefit in terms of economics and the environment. Technologies that result in net increased emissions will not give overall benefits and should not be considered. Carbon dioxide-enhanced oil recovery, in which carbon dioxide is forced into partially depleted oil fields to drive out more oil, is one process where the overall emissions far out-weigh the amount of CO_2 sequestered.[8] Depending on the molar ratio of the amount of carbon dioxide pumped in to the amount of oil extracted, there is likely to be a pyramid scheme as subsequent combustion of the oil derived fuels will generate more carbon dioxide than was sequestered in the process of extracting the oil. Indeed, Stephen Bygrave from UNSW in Australia has said that "CCS doesn't work if you're digging up more fossil oil".[9] Fabrizio Sibilla has said previously[10] that CDU is no different to synthetic biology or bio processing, "...you have proteomics, genomics, epi-genomics. In the end, only one thing wins; economics". A life cycle analysis of enhanced oil recovery confirmed that it would result in a net increase in carbon dioxide emissions unless the oil recovered could be used to displace an energy producer with even greater carbon dioxide emissions per mole of carbon combusted (*i.e.* a coal fuelled power station).[11] CCS, in particular the CC part, is needed in numerous technologies including CDU. There will need to be a price for carbon dioxide which is set at least at a European if not a global level.

Scheme 1 shows the chemical equations and thermodynamic data for combustion of coal, petrol and gas. The amount of energy obtained per mole of carbon dioxide emitted increases as the hydrogen/carbon ratio in the fuel increases so that combustion of natural gas produces more than twice as much energy as combustion of coal per mole of carbon dioxide emitted.

Coal:

$$\underset{(s)}{C} + \underset{(g)}{O_2} \longrightarrow \underset{(g)}{CO_2}$$

$\Delta H_r = -394\ \text{kJ mol}^{-1}$

Petrol:

$$\underset{(l)}{2C_8H_{18}} + \underset{(g)}{25O_2} \longrightarrow \underset{(g)}{16CO_2} + \underset{(g)}{18H_2O}$$

$\Delta H_r = -10{,}160\ \text{kJ mol}^{-1}$
($-635\ \text{kJ mol}^{-1}$ per CO_2 emitted)

Gas:

$$\underset{(g)}{CH_4} + \underset{(g)}{2O_2} \longrightarrow \underset{(g)}{CO_2} + \underset{(g)}{2H_2O}$$

$\Delta H_r = -803\ \text{kJ mol}^{-1}$

Scheme 1 Fossil fuel combustion data.

Since the amount of carbon dioxide emitted by fossil fuel power stations is orders of magnitude greater than the amount that the chemicals industry could utilise, some other means of preventing the vast majority of this carbon dioxide from entering the atmosphere will have to be found, and at present the only solution seems to be some form of carbon capture and sequestration. This could involve carbon dioxide mineralisation or underground/undersea storage of the carbon dioxide.

This led to a debate at this Faraday Discussion during which the hypothetical question 'which would you prefer: enhanced oil recovery or some other unconventional fossil fuel source such as fracking or tar sands?' was asked. The easy answer is none of them, as they will all result in increased atmospheric carbon dioxide levels as they are being implemented well ahead of any full-scale carbon capture and storage scheme. However, if faced with the reality that at least one of these technologies will be needed to bridge the gap whilst renewable energy schemes are implemented, then the data in Scheme 1 indicates that the least bad option is fracking for gas, followed by enhanced oil recovery with use of tar sands as the absolute worst option due to the large scale additional pollution that it causes and the low hydrogen/carbon ratio of tar.

CO_2 utilisation and sustainable chemistry

Of the many possible transformations of carbon dioxide, only its reduction to carbon monoxide does not involve its reaction with another chemical (two of the papers from this Faraday Discussion deal with the use of cold plasma to induce this transformation: DOI: 10.1039/c5fd00053j and 10.1039/c5fd00045a). Therefore, for carbon dioxide utilisation to make sense from a sustainability perspective, all of the other chemicals involved as reactants or catalysts must be sustainably sourced. Scheme 2 illustrates this for the industrially important production of ethylene carbonate. In this synthesis, ethylene carbonate is prepared by the reaction of carbon dioxide with ethylene oxide in the presence of a suitable catalyst. However, ethylene oxide does not occur naturally and is industrially produced by the oxidation of ethene by oxygen over a silver catalyst. At present, ethene is produced by the cracking of crude oil, a process which is both highly energy intensive and which relies on a non-sustainable feedstock. Fortunately, there is an alternative route to ethene; the dehydration of ethanol which has already been commercialised in Brazil. Ethanol is available by a number of sustainable routes involving the fermentation of sugar or waste biomass. Hence,

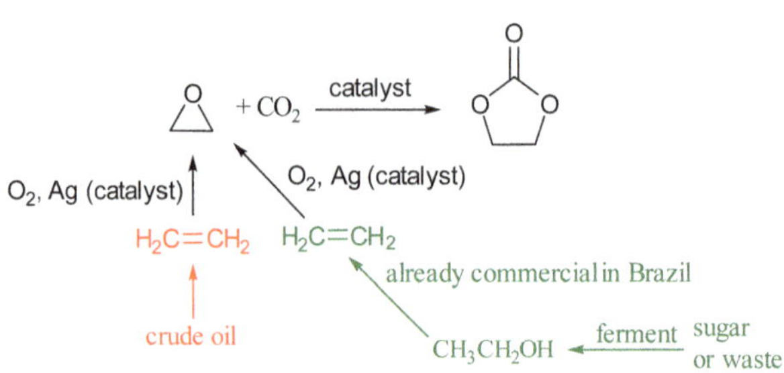

Scheme 2 Sustainable and non-sustainable synthesis of ethylene carbonate.

as currently operated, the production of ethylene carbonate is not sustainable, but it does have the potential to be sustainable and it therefore makes sense to further improve the latter steps in its synthesis.

The last two steps in the synthesis of ethylene carbonate are both catalytic, and whilst this is consistent with the principles of green chemistry, the nature of the catalyst should also be considered. Most commercially used catalysts are based around a metal and just like fossil fuels, supplies of these metals are being consumed. Elemental sustainability is a major topic in its own right,[12] but Fig. 2 provides a simple pictorial representation of the sustainability of the chemical elements. The oxidation of ethene to ethylene oxide uses a silver catalyst and silver is one of the elements whose supply is under serious threat over the next 100 years. Whether this is important or not very much depends on the scale of the process being considered. A catalyst, by definition, is only required in sub-stoichiometric amounts and therefore for the production of a pharmaceutical for example, only a few kilograms may be needed each year so even catalysts based on metals such as rhodium or iridium are likely to be viable. However, for the bulk chemicals industry where Mtonnes of product are being made, many tonnes of catalyst will be required. Most applications of carbon dioxide fall into the latter class, so it is important to develop catalysts based on the most abundant metals: sodium, potassium, calcium, aluminium, titanium or iron; or to develop metal-free catalysts. Substantial progress has been made in this area,[13] some of which are included in this Faraday Discussion. Notably, the North group have developed the catalyst shown in Fig. 3, which contains only the most sustainable elements and allows the synthesis of cyclic carbonates to be carried out at room temperature and pressure.[14] It has also been shown that the aluminium needed for the catalyst can be sourced from waste aluminium in the form of a chocolate bar wrapper.

Critics of carbon dioxide utilisation often point to thermodynamics as a reason why it will not work. It is certainly true that as the most highly oxidised form of

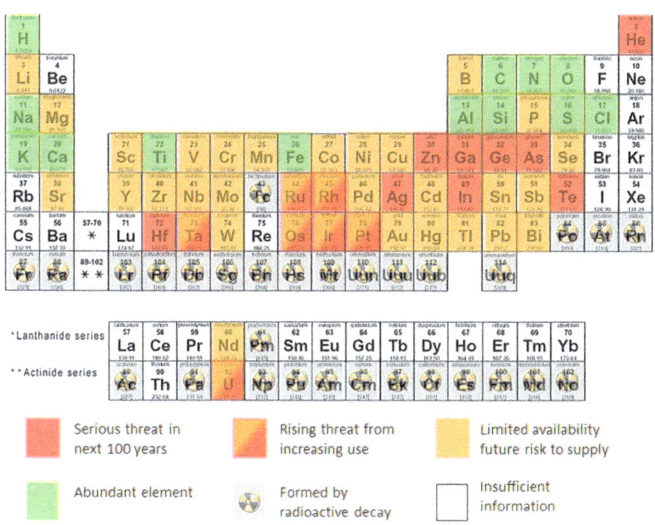

Fig. 2 A periodic table colour coded according to elemental sustainability.[12]

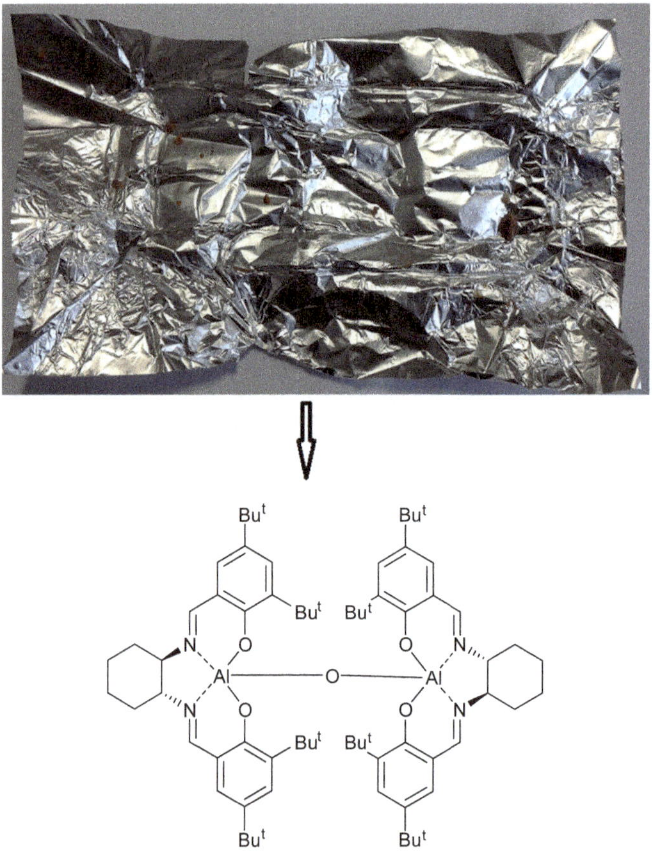

Fig. 3 Using recycled aluminium chocolate wrappers in the synthesis of a CDU catalyst.

carbon, carbon dioxide has a large negative heat of formation (-394 kJ mol^{-1}). However, by choosing appropriate co-reactants and products, there are reactions of carbon dioxide which are thermodynamically feasible. Four of these, all of which have been commercialised, are shown in Scheme 3.[15]

The highly exothermic synthesis of urea from carbon dioxide and ammonia is the largest scale chemical process operated by mankind and has been a commercial process for almost 100 years. The synthesis of salicylic acid from carbon dioxide and phenol is the first step in the industrial synthesis of aspirin and is an even older process having been first commercialised in the 19th century. The synthesis of ethylene carbonate (and other cyclic carbonates) from epoxides and carbon dioxide has been a commercial process since the 1950's and is currently undergoing a significant expansion due to the use of cyclic carbonates as electrolytes in lithium ion batteries. Most recently, the production of methanol from carbon dioxide and hydrogen has been commercialised by Carbon Recycling International in Iceland where the availability of cheap geothermal electricity allows the production of hydrogen by electrolysis of sea water to be commercially viable.

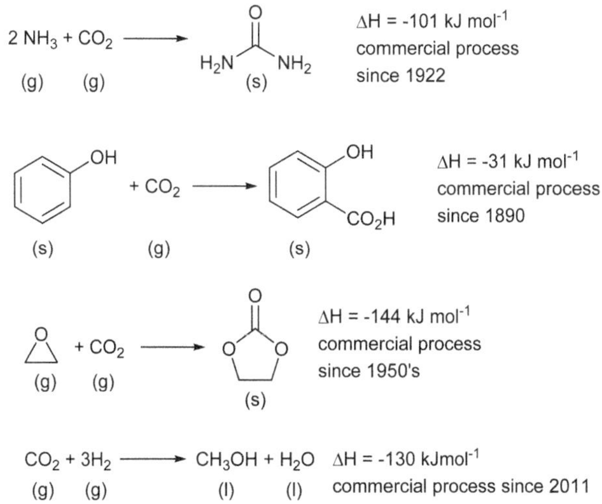

Scheme 3 Exothermic, commercial applications of carbon dioxide.

Many other reactions of carbon dioxide are exothermic including the synthesis of polycarbonates from carbon dioxide and epoxides, and the synthesis of acrylic acid from ethene and carbon dioxide ($\Delta H = -43$ kJ mol^{-1}). Thus, in many cases the challenge in carbon dioxide utilisation is not one of thermodynamics, but rather one of kinetics, requiring the development of a suitable catalyst to lower the activation energy of the desired transformation.

One of the most promising routes for carbon dioxide utilisation is through polymer synthesis. This is because polymers are prepared on a large scale and have the potential to sequester the carbon dioxide for a significant period of time. Two classes of carbon dioxide incorporating polymers are close to commercialisation: aliphatic polycarbonates and polyurethanes derived from polycarbonate–polyols and this Faraday Discussion contains papers on both (DOI: 10.1039/c5fd00073d and 10.1039/c5fd00067j). Aliphatic-polycarbonates are prepared by the copolymerisation of epoxides and carbon dioxide, and are being developed as replacements for aromatic polycarbonates (which themselves can incorporate carbon dioxide as a phosgene replacement[16]). At least three catalyst systems are being commercialised: those of Coates, Lee and Williams, and the work of Darensbourg in this area is also notable. The challenges are to prepare colourless polymers with glass transition temperatures as high as those of aromatic polycarbonates under mild reaction conditions, using sustainably sourced epoxides and catalysts.

Polyurethanes are manufactured on an 8 Mtonne per annum scale from a polyol and bis-isocyanide. In the traditional Bayer process, the polyol is a polyether derived from epoxides such as propylene oxide (Scheme 4). However, Bayer Material Science (now Covestro) are currently at demonstration plant scale with an alternative process (their DREAM process) in which the polyol is a polyether–polycarbonate–polyol in which some of the propylene oxide is replaced by carbon dioxide (Scheme 5). One paper in this Faraday Discussion which received a lot of interest was a life cycle analysis of this new process showing all the benefits of

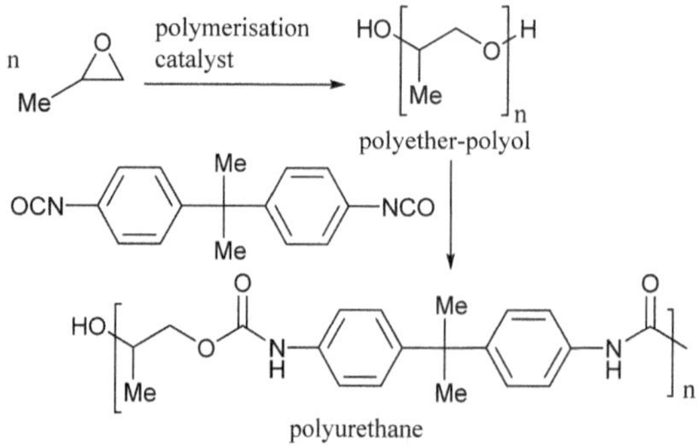

Scheme 4 Traditional Bayer route to polyurethanes.

carbon dioxide incorporation (DOI: 10.1039/c5fd00067j). Notably, the life cycle benefits come not just from the carbon dioxide incorporated into the polymer, but from the carbon dioxide avoided by not having to prepare as much propylene oxide. For every kg of carbon dioxide incorporated into the polyurethane, up to 3 kg of carbon dioxide emissions can be avoided.

This is not the only route to partly bioderived polyurethanes. In America, Ford have teamed up with Heinz to prepare polyurethanes from soybean oil derived polyols (Scheme 6). Whilst this may not seem to have anything to do with carbon dioxide utilisation, every carbon atom in soybean oil is derived from carbon dioxide through a natural biosynthesis pathway, thus illustrating that chemists working in this area should not ignore the ability of nature to assist in converting carbon dioxide into useful chemicals with sunlight as the ultimate green and sustainable energy source.

A future vision

Visions are interesting. One hundred years ago a vision would be considered a sign of madness! Many roadmaps are published with far-reaching visions. A typical one is the "2050 Vision", 35 years in the future. Jeroem van der Veer, the former CEO of Shell has stated that when planning scenarios in industry there will be at least a paradigm shift in any given technology in a 10 year period.[17] It makes sense that any 2050 vision will need to work with at least three paradigm shifts. How can such an accurate vision be realised? To show this, look back

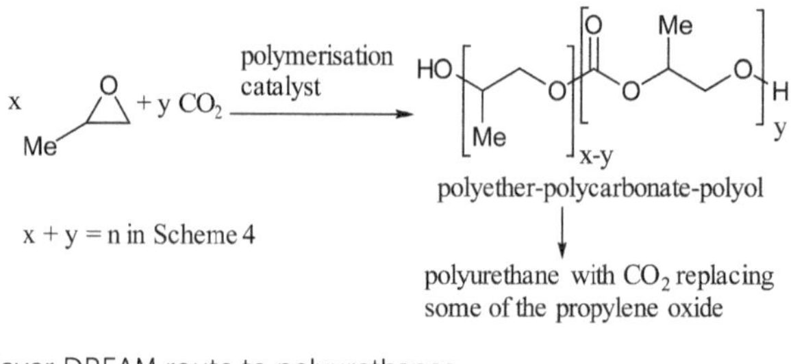

Scheme 5 Bayer DREAM route to polyurethanes.

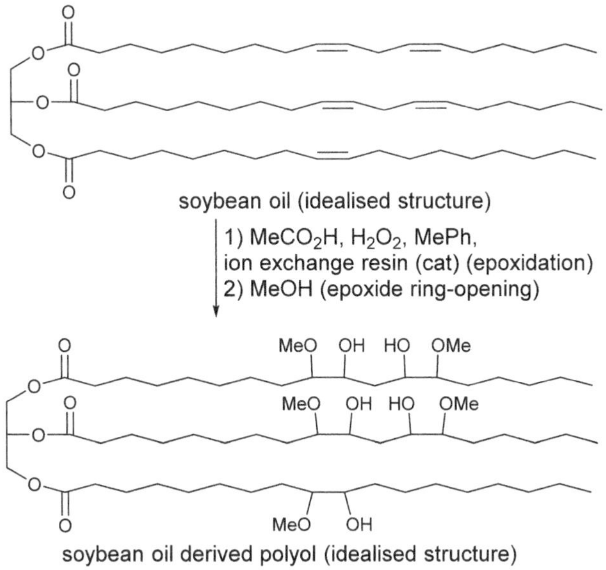

Scheme 6 Soybean oil derived polyols.

35 years to 1980 and consider whether you could have predicted then the technologies we have now. I doubt that many of us would have envisaged that we could be carrying the 1980s equivalent of a super-computer with full communications facilities including video conferencing, global data sharing and high definition photography and videography around in our pockets. Oh yes, it acts as a mobile telephone as well! In fact who even had a cell phone in 1980? The transport sector has also seen huge changes. Concord has come and gone, we now have hybrid and full electric cars. We even have hydrogen cars. However, one thing has not changed and is unlikely to in the near future – aviation and long-haul road transportation will still need high energy density hydrocarbon fuels to operate; Jet A-1 and diesel in particular. This is why fuels are a major target within the CDU community. Indeed, the RAC Foundation has identified that the UK could soon suffer diesel shortages if action is not taken now.[18]

In order to progress CDU over the forthcoming years we need to take emerging technologies from the discovery phase in the laboratory to the commercial arena. This means crossing the so-called valley of death where many technologies fail to progress the technology readiness levels (TRLs).[19] This requires a policy change by governments, however this needs to be evidence based including LCAs, TEAs and real data input from sub-pilot systems. Boundary conditions that are enforced must be transparent and consistent between different technologies. The European Commission now consider this to be the Valley of Hope and have recognised CDU/CCU as being a key future technology. A Horizon Prize in CO_2 re-use has also been established and is to be announced at the European SET Conference in Luxembourg in September 2015. This initiative shows that CDU is reaching a new maturity where commercial activities are emerging and innovation is being rewarded. The multifaceted approach of this meeting demonstrates not only that the pieces of the jigsaw are beginning to come together, but that we have complementary teams working to assemble the bigger picture. Bayer Material Science (now Covestro) and Sunfire in Germany,[20] and Novomer in the USA,[21] have demonstrated that CDU is possible in a commercial context. The German Ministry for Research and Education and the United States DOE have been instrumental in this by providing financial support to a wide base of technologies at an early stage in their development. In the UK, the Engineering and Physical Sciences Research Council (EPSRC) are beginning to fund more projects at the fundamental research stage, while Innovate UK is supporting the development of pilot plants and demonstrators. The same is true worldwide, but it is perhaps through consolidated support for European member states by the European Commission and Parliament, particularly under the Horizon 2020 programme, that CDU will make the greatest impact. Time will tell. If the enthusiasm and engagement of the participation of this discussion are anything to go by then the future is certainly looking bright.

Conclusions

Carbon dioxide utilisation is a key enabling technology in the transition from an unsustainable economy based on combustion of fossil fuels to a sustainable, circular economy based on the principles of sustainable development. The synthesis of liquid and gaseous fuels by hydrogenation of carbon dioxide will enable the phasing out of fossil fuel based power generation and its replacement by energy obtained from sustainably sourced hydrogen with carbon dioxide as a convenient hydrogen carrier. Other transformations of carbon dioxide can provide alternative feedstocks (or alternative routes to existing feedstocks) for the chemicals industry. These cover basic chemicals such as methanol and formic acid, more specialised chemicals such as linear and cyclic carbonates and polymers such as polycarbonates and polyurethanes. The papers presented in this Faraday Discussion showcase many of the leading developments in this area.

As with many new technologies, there are many barriers to overcome. Funding will always be an early limiting factor, particularly in the transition to high TRLs. However, the legislative framework in which CDU will need to operate needs to be addressed so that it has the mandate for success. There needs to be a carbon price established that gives carbon dioxide a value as a commodity so that it is not treated as a waste. Subsidies, if any, need to be on a level playing field against other carbon dioxide mitigation or avoidance technologies. When considering what to do with the carbon dioxide that cannot be used by the chemicals industry, care must be taken to assess the life cycle of the technology. Mineralisation is attractive as depending on the mineral produced and scale of operation, this may provide a useful construction material or just a way to sequester carbon dioxide for a long period of time. Conventional carbon capture and storage will also be essential to sequester the carbon dioxide that is beyond the capacity of the chemicals industry to utilise. However, enhanced oil recovery (and other non-conventional fossil fuels) cannot be part of efforts to reduce atmospheric carbon dioxide levels or develop a sustainable economy. Rather, they will make the situation worse by providing even more fossil fuel for mankind to burn. This will delay the transition to a sustainable economy both by providing additional years of fossil fuels and diverting limited research and development funding away from sustainable processes. When the inevitable transition to a sustainable economy is made it will therefore be starting from a worse situation and the transition will be that much harder. Thus enhanced oil recovery should be seen as a dangerous distraction from the critical process of developing a sustainable global economy based on renewable materials such as waste biomass and carbon dioxide.

In his opening paper of this Faraday Discussion, Martyn Poliakoff presents twelve principles of carbon dioxide utilisation (DOI: 10.1039/c5fd90078f). We would entirely agree with these and point out that if some of them are ignored (renewable energy in particular), the results could be very COSTLY:

Catalysis is crucial.
Origin of the CO_2?
Sustainability is essential.
Thermodynamics cannot be beaten.
Large energy input (as opposed to reasonable).
Your enthusiasm is not enough.

But, as the quality of the discussions at this meeting testify, CDU research is on the right path, avoiding COSTLY pitfalls. The opportunities that carbon dioxide utilisation could provide for a sustainable chemical industry are great and the future of CDU looks promising.

Acknowledgements

We thank the Engineering and Physical Sciences Research Council for funding the CO_2Chem Grand Challenges Network (EP/H035702/1 & EP/K007947/1) who were central to the development of this Faraday Discussion through the Scientific Committee: without which this conference may never have taken place.

Notes and references

1 http://bp.com/statisticalreview, accessed 12 September 2015.
2 J. Wilcox, *Carbon Capture*, Springer, New York, 2012.
3 http://esrl.noaa.gov/gmd/ccgg/trends/weekly.html, accessed 12 September 2015.
4 I. S. Metcalfe, M. North, R. Pasquale and A. Thursfield, *Energy Environ. Sci.*, 2010, **3**, 212–215; M. North, B. Wang and C. Young, *Energy Environ. Sci.*, 2011, **4**, 4163–4170.
5 C. Beattie, M. North, P. Villuendas and C. Young, *J. Org. Chem.*, 2013, **78**, 419–426.
6 A. M. Chapman, C. Keyworth, M. R. Kember, A. J. J. Lennox and C. K. Williams, *ACS Catal.*, 2015, **5**, 1581–1588; N. von der Assen and A. Bardow, *Green Chem.*, 2014, **16**, 3272–3280.
7 http://www.fastcompany.com/1704105/problem-carbon-capture-co2-doesnt-always-stay-captured, accessed 18 September 2015.
8 K. Armstrong and P. Styring, Assessing the potential of utilization and storage strategies for post-combustion CO_2 emissions reduction, *Frontiers in Energy Research*, 2015, **3**, 8, DOI: 10.3389/fenrg.2015.00008.
9 S. Bygrave, *The Conversation*, 15 October 2014.
10 F. Sibilla, *CO_2 as a Feedstock for Chemicals and Fuels*, Essen, Germany, 2 December 2014.

11 P. Jaramillo, W. M. Griffin and S. T. McCoy, *Environ. Sci. Technol.*, 2009, **43**, 8027–8032.

12 *Element Recovery and Sustainability: RSC Green Chemistry No. 22*, ed. A. J. Hunt, Royal Society of Chemistry, Cambridge, 2013.

13 J. W. Comerford, I. D. V. Ingram, M. North and X. Wu, *Green Chem.*, 2015, **17**, 1966–1987.

14 M. North, *ARKIVOC*, 2012, 610–628.

15 M. North, *What is CO_2? Thermodynamics, Basic Reactions and Physical Chemistry in Carbon Dioxide Utilisation: Closing the Carbon Cycle*, ed. P. Styring, A. Quadrelli and K. Armstrong, Elsevier, Oxford, 2015.

16 S. Fukuoka, M. Kawamura, K. Komiya, M. Tojo, H. Hachiya, K. Hasegawa, M. Aminaka, H. Okamoto, I. Fukawa and S. Konno, *Green Chem.*, 2003, **5**, 497–507.

17 J. van der Veer, *TU/e Energy Days*, Technical University of Eindhoven, The Netherland, June 2014.

18 http://www.bbc.co.uk/news/business-34262990, accessed 16 September 2015.

19 https://www.nasa.gov/directorates/heo/scan/engineering/technology/txt_accordion1.htm, accessed 16 September 2015.

20 http://www.sunfire.de, accessed 16 September 2015.

21 http://www.novomer.com, accessed 18 September 2015.

Poster titles

A Dual-electrolyte Microfluidic Reactor for Electro-Reduction of CO_2, **X. Lu, D. Y. C Leung, H. Wang, J. Xuan,** *Heriot-Watt University, UK*

Electrochemical conversion of carbon dioxide to methanol at Cu_2O/ZnO gas diffusion electrodes, **J. Albo, P. Castaño, A. Irabien,** *University of Basque Country(UPV/EHU), Spain*

Atmospheric CO_2 fixation by a new type of complexes of inexpensive first row metals and bis(amine)-bis(phenolate) ligands: Mechanism and conversion into value derivatives, **L. Rodríguez-Silva, R. Carballido, M. Maneiro,** *Universidade de Santiago de Compostela, Spain*

The role of CO_2 utilisation in climate change mitigation, **N. MacDowell, P. S. Fennell, N. Shah, G. C. Maitland,** *Imperial College, UK*

Cyclic Carbonate Monomers from 1,3-diols and low pressure CO_2: a novel, mild and safe phosgene-free route, **G. Gregory, G. Kociok-Köhn, A. Buchard,** *University of Bath, UK*

Probing the electrode/electrolyte interfacial structure of CO_2 electrochemical reduction to CH_3OH catalyzed by the Pt-pyridine system, **X.-J. Chen, S. Yu, D.-Y. Wu, Z.-Q. Tian,** *Xiamen University, China*

Guanidine catalysts in carbon dioxide utilisation, **R. Nicholls, B. N. Nguyen,** *University of Leeds, UK*

Design of experiments as a valuable tool for rational catalyst development in methanation of syngas containing hydrocarbons, **F. Krebs, J. H. Baik, K. Simeonov, R. Palkovits,** *RWTH Aachen University, Germany*

Process conditions study via design of experiments for dry reforming of methane, **N. Pegios, J. H. Baik, J.-H. Park, K. Simeonov, R. Palkovits,** *RWTH Aachen University, Germany*

Structure-Dependent Electroreduction of CO_2 at Au-Pd Core-Shell Nanostructures, **J. Humphrey, D. Plana, V. Celorrio, S. Sadasivan, R. Tooze, P. Rodríguez, D. Fermín,** *University of Bristol, UK*

Methyl esters as efficient and green water suppressors in the synthesis of dimethyl carbonate from CO_2, **A. Marciniak, C. Mota,** *Universidade Federal do Rio de Janeiro, Brazil*

Production of Methanol and dimethyl ether from CO_2 hydrogenation over bifunctional catalysts, **R. Jorge de Silva, R. S. Monteiro, C. J. A. Mota,** *Universidade Federal do Rio de Janeiro, Brazil*

Processing of methane to synthesis gas over the fiberglass catalysts, **A. Mironenko, A. Kazieva, Z. Mansurov, J. Kudyarova, K. Dossumov,** *Institute of Combustion Problems, Kazakhstan*

Carbon dioxide conversion of methane over the polyoxide catalysts, **K. Dossumov, Z. Mansurov, A. Mironenko, G. Yergazyieva, D. Churina, A. Kazieva,** *Institute of Combustion Problems, Kazakhstan*

Carbon dioxide reforming of methane, **K. Dossumov, G. Yergazieva, D. Kh. Churina, U. Suynbayev,** *Institute of Combustion Problems, Kazakhstan*

The direct carbonation of glycerol with CO_2 using heterogeneous catalysts, **L. Ozorio, C. Mota,** *Universidade Federal do Rio de Janeiro, Brazil*

High pressure CO_2 reduction: *in operando* heterogeneous catalysis characterisation, **H. Reymond, A. Lauper, P. Rudolf von Rohr,** *ETH Zürich, Switzerland*

Polymers Derived from Biomass and Incorporating Carbon Dioxide as Replacements for Petroleum Based Plastics, **I. Ingram, J. Beament, E. M. O'Brien, C. R. McElroy, J. Dodson, T. Farmer, M. North, J. Clark,** *University of York, UK*

Cyclic Carbonates from Diols and CO_2, **J. Comerford, M. North, R. Heyn,** *University of York, UK*

Conversion of CO_2 via the utilisation of atmospheric pressure plasmas, **A. Foote, M. North, T. Gans,** *University of York, UK*

Ionic liquids as catalysts for CO_2 transformation in dimethyl carbonate, **M. Vieira, F. Bernard, I. Grillo, M. Seferin, S. Einloft,** *Pontifical Catholic University of Rio Grande do Sul, Brazil*

Electrochemically driven carbon dioxide mineralisation, **K. Lamb, M. North, P. Aguiar, A. Parkin,** *University of York, UK*

Hydrothermal Conversion of CO_2 into Higher Hydrocarbons and Oxygenates, **L. Quintana Gomez, M. Connolly, J. McGregor,** *The University of Sheffield, UK*

Fresh Water Generation by Oxidising Methane with Carbon Dioxide to Combat the Effects of Global Warming, Potential Uses of the Graphite By-product, **A. Coogan,** *Retired*

Effect of $NaNO_3$ Promoter on Desorption of CO_2 of MgO Absorbent, **S.-I. Jo, K.-Y. Kim, J.-S. Kwak, Y.-I. An, Y.-U. Kwon,** *Sungkyunkwan University, Republic of Korea*

Effect of Alkaline-earth Metal Carbonate Salts on CO_2 absorption on $NaNO_3$ –MgO Absorbent, **K.-Y. Kim, S.-I. Jo, J.-S. Kwak, Y.-I. Ahn, Y. U. Kwon,** *Sungkyunkwan University, Republic of Korea*

Photoelectrochemical Conversion of CO_2 to Useful Fuels: Opportunities and Challenges, **C. Janáky, D. Hursán, A. Kormányos, K. Rajeshwar,** *MTA SZTE, Szeged, Hungary*

Waste CO_2 and biomass-based chemical synthesis for the green economy, **J. R. Dodson, J. B. Barbero, C. Topi, C. Mota, T. Farmer, M. North, J. Clark,** *University of York, UK*

Using Low Cost Solid Sorbents for CO_2 Capture and Separation using a Pressure Swing Process, **D. Reed, G. Dowson, P. Styring,** *The University of Sheffield, UK*

Modelling of CO_2 reduction by Pulsed Corona Discharge, **M. Moss, K. Yanallah, R. Elder, R. Allen,** *The University of Sheffield, UK*

A hybrid QM/MM embedded cluster approach to modelling the reactivity of carbon dioxide over metal oxide surfaces, **M. Farrow, C. A. Downing, G. Dutta, D. Chaopradith, T. W. Keal, P. Sherwood, A. A. Sokol, C. R. A. Catlow,** *University College London, UK*

Study of the influence of the ionic liquid cation on the electrochemical reduction of CO_2, **R. Michez, J. Christophe, T. Doneux, C. Buess-Herman,** *Université Libre de Bruxelles, Belgium*

Multivariate Models for Ionic Liquids Catalytic Systems in CO_2 Cycloaddition to Propylene Oxide, **I. Grillo, M. O. Vieira, F. L. Bernard, S. Einloft, M. Seferin,** *Pontifical Catholic University of Rio Grande do Sul, Brazil*

Carbon dioxide utilisation in oxidative carboxylation of 1-decene, **R. Alsaiari, E. Nowicka, P. J. Miedziak, S. A. Kondrat, J. K. Edwards, G. J. Hutchings,** *Cardiff University, UK*

Synthesis of cyclic carbonates catalysed by aluminium heteroscorpionate complexes, **J. A. Castro-Osma, J. Martínez, C. Alonso-Moreno, A. Lara-Sánchez, M. North, A. Otero,** *Universidad of Castilla-La Mancha, Spain*

Understanding the Potential of CO_2 Utilization, **K. Armstrong, P. Styring, A. V. Zaragoza, I. A. Grant Wilson,** *The University of Sheffield, UK*

Improved CO_2 conversion using microchanneled cathode supports for SOECs, **S. Xu, X. Shao, D. Dong, G. Parkinson, L. Hucker, J. Marin, T. Pham, C.-Z. Li,** *Curtin University, Australia*

Carbon Dioxide Utilisation for Methanol as Chemical Feedstock: Life Cycle Analysis, **A. V. Zaragoza, P. Styring, K. Armstrong,** *The University of Sheffield, UK*

Challenges for Solar Fuel Production via Potocatalysis, **E. Bay, O. Ola, J.-W. Bos, M. Maroto-Valer,** *Heriot Watt University, UK*

Photoautotrophic carbon uptake by microalgae for biotransformations, **S. Vaidyanathan, R. Kapoore, K. Okurowska, D. Pandey,** *The University of Sheffield, UK*

CO_2 hydrogenation on Pd/ZnO catalysts, **H. Bahruji, M. Bowker, N. Dimitratos, P. Wells, E. Gibson, C. Brookes, W. Jones,** *Cardiff University, UK*

Effects of particle size and gas composition on discharge phenomena in a packed bed plasma for CO_2 reduction, **T. Butterworth, R. Elder, R. Allen,** *The University of Sheffield, UK*

The Skinner Prize for the best poster was awarded to Miss Georgina Gregory of the Centre for Sustainable Chemical Technologies, University of Bath, UK, for her poster on a novel, mild and safe phosgene-free route for carbonate monomers.

List of participants

Mr Emad Abdel Hadi, *SABIC Corporate Research and Innovation Center KAUST, Saudi Arabia*
Mr Pedro Abrantes, *CO2track, France*
Dr Jonathan Albo, *Universidad de Cantabria, Spain*
Dr Carlos Alonso-Moreno, *University of Castilla-La Mancha, Spain*
Mrs Raiedhah Alsaiari, *Cardiff University, United Kingdom*
Miss Meital Amoyal, *Ben-Gurion University, Israel*
Professor Michele Aresta, *NUS and CIRCC, Italy*
Mrs Katy Armstrong, *University of Sheffield, United Kingdom*
Dr Hasliza Bahruji, *Cardiff Catalysis Institute, United Kingdom*
Professor Dr André Bardow, *RWTH Aachen University Institute of Technical Thermodynamics, Germany*
Ms Elizabeth Bay, *Heriot Watt University, United Kingdom*
Professor Dr Annemie Bogaerts, *University of Antwerp, Department of Chemistry, Belgium*
Dr Antoine Buchard, *University of Bath, United Kingdom*
Mr Thomas Butterworth, *University of Sheffield, United Kingdom*
Professor Eleanor Campbell, *University of Edinburgh, United Kingdom*
Professor Carmine Capacchione, *Dipartimento di Chimica e Biologia-Università di Salerno, Italy*
Dr Gonçalo Carrera, *FCT/UNL, Portugal*
Dr José A. Castro-Osma, *University of Castilla-La Mancha, Spain*
Professor Gabriele Centi, *Univ. Messina, ERIC and INSTM/CASPE, Italy*
Miss Xue-Jiao Chen, *College of Chemistry and Chemical Engineering, Xiamen University, China*
Dr Dina Churina, *Center of Physical-chemical methods of Investigations ana Analysis, Kazakhstan*
Dr Peter Clark, *Knowlege Transfer Network, United Kingdom*
Dr James Comerford, *Green Chemistry Centre of Excellence, University of York, United Kingdom*
Mr Anthony Coogan, *Retired, Ireland*
Dr Marie Cote, *Royal Society of Chemistry, United Kingdom*
Dr Alexander Cowan, *University of Liverpool, United Kingdom*
Mr Yakov Dinburg, *ben gurion university, Israel*
Dr Jennifer Dodson, *University of York, United Kingdom*
Professor Kusman Dossumov, *Institute of Combustion Problems, Kazakhstan*
Dr George Dowson, *University of Sheffield, United Kingdom*
Mr Xian Du, *University of Oxford, United Kingdom*
Professor Sandra Einloft, *PUCRS, Brazil*
Dr Matthew Farrow, *University College London, United Kingdom*
Mr Alexander Foote, *Department of Physics, United Kingdom*
Professor Eric Fraga, *University College London (UCL), United Kingdom*

Miss Georgina Gregory, *Centre for Sustainable Chemical Technologies, University of Bath, United Kingdom*
Mr Derek Harris, *Retired, United Kingdom*
Miss Amy Hazlehurst, *Royal Society of Chemistry, United Kingdom*
Professor Dr Moti Herskowitz, *Ben-Gurion University of the Negev, Israel*
Dr Richard Heyn, *SINTEF Materials and Chemistry, Norway*
Dr Nathan Hollingsworth, *University College London, United Kingdom*
Miss Jo Humphrey, *University of Bristol, United Kingdom*
Dr Ian Ingram, *University of York, United Kingdom*
Professor Csaba Janaky, *MTA TKI, Hungary*
Ms Seung-Ik Jo, *Sungkyunkwan University, South Korea*
Dr Christopher Jones, *University of Sheffield, United Kingdom*
Ms Kangyeong Kim, *Sungkyunkwan University, South Korea*
Professor Christopher Kozak, *Memorial University of Newfoundland, Canada*
Mr Florian Krebs, *Chair of Heterogeneous Catalysis and Chemical Technology, ITMC, RWTH Aachen University, Germany*
Ms Katie Lamb, *University of York, United Kingdom*
Miss Sarah Latham, *Royal Society of Chemistry, United Kingdom*
Dr Niall MacDowell, *Imperial College London, United Kingdom*
Miss Mary Macleod, *Royal Society of Chemistry, United Kingdom*
Professor Geoffrey Maitland, *Imperial College London, United Kingdom*
Dr Troy Manning, *University of Liverpool, United Kingdom*
Miss Aryane Marciniak, *UFRJ-Universidade Federal do Rio de Janeiro, Brazil*
Dr Niklas Meine, *Covestro Deutschland AG, Germany*
Mrs Nora Meiri, *Ben-Gurion University of the Negev, Israel*
Dr Rebecca Melen, *Cardiff University, United Kingdom*
Professor Dr Vera Meynen, *University of Antwerp, Belgium*
Mr Roman Michez, *Université Libre de Bruxelles, Belgium*
Mr Koen Michiels, *VITO NV, Belgium*
Dr Alessandro Morbi, *CTG- Italcementi Group, Italy*
Mr Matthew Moss, *University of Sheffield, United Kingdom*
Dr Alexander Navarrete, *University of Valladolid, Department of Chemical Engineering and Environmental Technology, High Pressure Processes Group, Spain*
Dr Piero Negro, *Italcementi S.p.A., Italy*
Miss Rachel Nicholls, *University of Leeds, United Kingdom*
Professor Al Nielson, *Massey University, New Zealand*
Professor Michael North, *Green Chemistry Centre of Excellence, University of York, United Kingdom*
Professor Manuel Nunes da Ponte, *FCT/UNL-LAQV-REQUIMTE, Portugal*
Dr Rhodri Owen, *University of Sheffield, United Kingdom*
Mr Leonardo Ozorio, *Universidade Federal do Rio de Janeiro, Brazil*
Professor Gordon Parkinson, *Curtin University, Australia*
Mr Nikolaos Pegios, *ITMC RWTH Aachen, Germany*
Professor Martyn Poliakoff, *The University of Nottingham/The Royal Society, United Kingdom*
Mr Michael Priestnall, *Innovate UK, United Kingdom*
Dr Elsje Alessandra Quadrelli, *CNRS-CPE Lyon, France*
Ms Laura Quintana Gomez, *The University of Sheffield, United Kingdom*
Mr Adriano Randi, *Loughborough University, United Kingdom*
Mr Daniel Reed, *The University of Sheffield, United Kingdom*
Dr Eryk Remiezowicz, *ACP Polska, Poland*
Miss Helena Reymond, *ETH Zurich, Switzerland*
Dr Laura Rodríguez-Silva, *Universidade de Santiago de Compostela, Spain*

Mr Andrew Sadler, *The University of Sheffield, United Kingdom*
Professor Dr Marcus Seferin, *PUCRS, Brazil*
Miss Renata Silva, *UFRJ-Universidade Federal do Rio de Janeiro, Brazil*
Mr Steven Spall, *University of Sheffield, United Kingdom*
Dr Jeroen Spooren, *VITO nv, Belgium*
Ms Claire Springett, *Royal Society of Chemistry, United Kingdom*
Professor Peter Styring, *The University of Sheffield, United Kingdom*
Mr Volodymyr Tabas, *Loughborough University, United Kingdom*
Dr Rebecca Taylor, *Queen's University Belfast, United Kingdom*
Dr Seetharaman Vaidyanathan, *University of Sheffield, United Kingdom*
Professor Dr Gerard van Rooij, *DIFFER, Netherlands*
Dr Koarolien Vanbroekhoven, *VITO, Belgium*
Ms Ana Villa, *University of Sheffield, United Kingdom*
Mr Viswanath Vinu, *SAUDI BASIC INDUSTRIAL CORP. (SABIC), Saudi Arabia*
Dr Huizhi Wang, *Heriot-Watt University, United Kingdom*
Mr William Webb, *University of Southampton, United Kingdom*
Professor Charlotte Williams, *imperial college london, United Kingdom*
Dr Jin Xuan, *Heriot-Watt University, United Kingdom*
Dr Tsung-Hsien Yu, *Green Technology Research Institute, CPC Corporation, Taiwan, Chinese Taipei*